Information Security and Cryptography

Information Security – protecting information in potentially hostile environments – is a crucial factor in the growth of information-based processes in industry, business, and administration. Cryptography is a key technology for achieving information security in communications, computer systems, electronic commerce, and in the emerging information society.

Springer's Information Security & Cryptography (IS&C) book series covers all relevant topics, ranging from theory to advanced applications. The intended audience includes students, researchers and practitioners.

More information about this series at https://link.springer.com/bookseries/4752

Laurent Chuat · Markus Legner · David Basin ·
David Hausheer · Samuel Hitz · Peter Müller ·
Adrian Perrig

The Complete Guide
to SCION

From Design Principles to Formal Verification

 Springer

Laurent Chuat
Network Security Group
ETH Zürich
Zürich, Switzerland

Markus Legner
Network Security Group
ETH Zürich
Zürich, Switzerland

David Basin
Information Security Group
ETH Zürich
Zürich, Switzerland

David Hausheer
Otto von Guericke University Magdeburg
Magdeburg, Germany

Samuel Hitz
Anapaya Systems
Zürich, Switzerland

Peter Müller
Programming Methodology Group
ETH Zürich
Zürich, Switzerland

Adrian Perrig
Network Security Group
ETH Zürich
Zürich, Switzerland

ISSN 1619-7100   ISSN 2197-845X   (electronic)
Information Security and Cryptography
ISBN 978-3-031-05290-3   ISBN 978-3-031-05288-0   (eBook)
https://doi.org/10.1007/978-3-031-05288-0

This Springer imprint is published by the registered company Springer Nature Switzerland AG
The registered company address is: Gewerbestrasse 11, 6330 Cham, Switzerland

To our families, friends,
and loved ones

Contents

Foreword by Joël Mesot   xi

Foreword by Fritz Steinmann   xiii

Preface   xv

How to Read This Book   xvii

Acknowledgments   xix

1 Introduction   **1**
   1.1 Today's Internet 2
   1.2 Goals for a Secure Internet Architecture 9

I SCION Core Components   **15**

2 Overview   **17**
   2.1 Infrastructure Components 20
   2.2 Authentication 21
   2.3 Control Plane 23
   2.4 Data Plane 28
   2.5 ISD and AS Numbering 31

3 Authentication   **35**
   3.1 The Control-Plane PKI (CP-PKI) 36
   3.2 DRKey: Dynamically Recreatable Keys 52
   3.3 SCION Packet Authenticator Option 61

4 Control Plane   **65**
   4.1 Path-Segment Construction Beacons (PCBs) 66
   4.2 Path Exploration (Beaconing) 69
   4.3 Path-Segment Registration 71
   4.4 PCB and Path-Segment Selection 73
   4.5 Path Lookup 80
   4.6 Service Discovery 87
   4.7 SCION Control Message Protocol (SCMP) 89

5   Data Plane   **93**
   5.1   Inter- and Intra-domain Forwarding 94
   5.2   Packet Format 95
   5.3   Path Authorization 96
   5.4   The SCION Path Type 101
   5.5   Path Construction (Segment Combinations) 104
   5.6   Packet Initialization and Forwarding 115
   5.7   Path Revocation 120
   5.8   Data-Plane Extensions 124

II   Analysis of the Core Components   **127**

6   Functional Properties and Scalability   **129**
   6.1   Dependency Analysis 130
   6.2   SCION Path Policy 135
   6.3   Scalability Analysis 148
   6.4   Beaconing Overhead and Path Quality 150

7   Security Analysis   **157**
   7.1   Security Goals and Properties 158
   7.2   Threat Model 161
   7.3   Overview 162
   7.4   Control-Plane Security 165
   7.5   Path Authorization 170
   7.6   Data-Plane Security 172
   7.7   Source Authentication 174
   7.8   Absence of Kill Switches 176
   7.9   Other Security Properties 179
   7.10 Summary 181

III   Achieving Global Availability Guarantees   **183**

8   Extensions for the Control Plane   **185**
   8.1   Hidden Paths 185
   8.2   Time Synchronization 190
   8.3   Path Metadata in PCBs 197

9   Monitoring and Filtering   **203**
   9.1   Replay Suppression 204
   9.2   High-Speed Traffic Filtering with LightningFilter 207
   9.3   Probabilistic Traffic Monitoring with LOFT 217

10 Extensions for the Data Plane   **227**
   10.1 Source Authentication and Path Validation with EPIC 228
   10.2 Bandwidth Reservations with COLIBRI 237

11 Availability Guarantees   **267**
   11.1 Availability Goals and Threat Landscape 268
   11.2 Overview 270
   11.3 Defense Systems 271
   11.4 Traffic Prioritization 278
   11.5 Protected DRKey Bootstrapping 283
   11.6 Protection of Control-Plane Services 288
   11.7 AS Certification 294
   11.8 Security Discussion 297

IV SCION in the Real World   **301**

12 Host Structure   **303**
   12.1 Host Components 303
   12.2 Future Approaches 307

13 Deployment and Operation   **317**
   13.1 Global Deployment 319
   13.2 End-Host Deployment and Bootstrapping 327
   13.3 The SCION–IP Gateway (SIG) 332
   13.4 SIG Coordination Systems 336
   13.5 SCION as a Secure Backbone AS (SBAS) 345
   13.6 Example: Life of a SCION Data Packet 354

14 SCIONLAB Research Testbed   **361**
   14.1 Architecture 362
   14.2 Research Projects 366
   14.3 Comparison to Related Systems 368

15 Use Cases and Applications   **371**
   15.1 Use Cases 372
   15.2 Applications 382
   15.3 Case Study: Secure Swiss Finance Network (SSFN) 385
   15.4 Case Study: SCI-ED, a SCION-Based Research Network . . 389

16 Green Networking with SCION   **393**
   16.1 Direct Power Savings with SCION 394
   16.2 SCION Enables Green Inter-domain Routing 399
   16.3 Incentives for ISPs to Use Renewable Energy Resources . . 404

17 Cryptography   **407**
   17.1 How Cryptography Is Used in SCION 408
   17.2 Cryptographic Primitives 409
   17.3 Local Cryptographic Primitives 410
   17.4 Global Cryptographic Primitives 412
   17.5 Post-Quantum Cryptography 415

V   Additional Security Systems   **417**

18 F-PKI: A Flexible End-Entity Public-Key Infrastructure   **419**
   18.1 Trust Model 421
   18.2 Overview of F-PKI 423
   18.3 Policies 424
   18.4 Verifiable Data Structures 426
   18.5 Selection of Map Servers 428
   18.6 Proof Delivery 428
   18.7 Certificate Validation 430

19 RHINE: Secure and Reliable Internet Naming Service   **431**
   19.1 Background 433
   19.2 Why a Fresh Start? 437
   19.3 Overview of RHINE 440
   19.4 Authentication 444
   19.5 Data Model 452
   19.6 Secure Name Resolution 455
   19.7 Deployment 457

20 PILA: Pervasive Internet-Wide Low-Latency Authentication   **461**
   20.1 Trust-Amplification Model 463
   20.2 Overview of PILA 464
   20.3 ASes as Opportunistically Trusted Entities 464
   20.4 Authentication Based on End-Host Addresses 465
   20.5 Certificate Service 466
   20.6 NAT Devices 467
   20.7 Session Resumption 467
   20.8 Downgrade Prevention 468

VI Formal Verification   **471**

21 Motivation for Formal Verification   **473**
   21.1 Local and Global Properties 474
   21.2 Quantitative Properties 475
   21.3 Adversarial Environments 475

21.4 Design-Level and Code-Level Verification 476

22 Design-Level Verification **477**
22.1 Overview of Design-Level Verification 478
22.2 Background on Event Systems and Refinement 482
22.3 Example: Authentication Protocol 488
22.4 Verification of the SCION Data Plane 494
22.5 Quantitative Verification of the N-Tube Algorithm 510

23 Code-Level Verification **519**
23.1 Why Code-Level Verification? 520
23.2 Introduction to Program Verification 522
23.3 Verification of Go Programs 533
23.4 Verification of Protocol Implementations 547
23.5 Secure Information Flow 555

24 Current Status and Plans **563**
24.1 Completed Work 563
24.2 Ongoing Work 566
24.3 Future Plans and Open Challenges 567

VII Back Matter **573**

25 Related Work **575**
25.1 Future Internet Architectures 575
25.2 Deployment of New Internet Architectures 580
25.3 Inter-domain Multipath Routing Protocols 582

Bibliography **585**

Glossary **641**

Abbreviations **645**

Index **651**

Foreword by Joël Mesot

PRESIDENT ETH ZURICH

The pandemic impressively demonstrated to all of us how strong the interdependencies in the world economy have become with its globalized flows of goods and supply chains. The pandemic was a warning sign of the vulnerability of an increasingly connected world, in which exchange between people, companies, and organizations is only possible thanks to an extremely sophisticated IT infrastructure and digitalized processes. But what if some of these communication channels fail due to faulty manipulation, what if confidential data falls into the wrong hands, and what if the networks of critical infrastructures such as hospitals, energy companies, or airports are hacked?

Unfortunately, these are not scenarios from a science fiction movie, but already reality today. Cyberspace has become a playground for criminal activities; the targeted extortion of companies and organizations through ransomware, DDoS attacks, or data hijacking has become a billion-dollar business. The forms of threat from cyberspace pose challenges to authorities, organizations, corporations, and SMEs alike. Information security has become a priority for the C-level decision makers. Not only is it crucial for our economies, it is also paramount to our trust in democratic processes and institutions.

Today's Internet is undoubtedly a success story. The idea of a distributed packet-switched network, as it evolved from the early days of the ARPA network to the world wide web, has not only connected millions of people, but also created unimagined opportunities for innovation. But the Internet has vulnerabilities that are virtually inherent in its basic architecture. For example, Internet-based data traffic still uses the Border Gateway Protocol (BGP), which is error-prone and easy prey for attacks. The protocol was introduced in 1989 to better handle the enormous growth of the Internet. Its basic outline was sketched out by American computer engineers at a luncheon—on three napkins, which is why it was nicknamed the "Three-Napkins Protocol". The BGP was intended as a patch, a temporary solution to a problem. It still assigns paths to routers for traffic today, more than 30 years after its introduction. The BGP problem is emblematic of the current Internet's tendency to patch vulnerabilities with stopgap solutions rather than fixing them in principle.

This is where SCION comes into play, the new Internet architecture developed at ETH Zurich to which this book is dedicated. SCION stands for Scalability, Control and Isolation on Next-Generation Networks. It is the result of more than ten years of research work, which has evolved from conceptual rethinking to scientific verification to actual implementation of the new technology. The project has now reached a level of maturity so that it is already being used in various areas. The ETH Domain was able to contribute to the scale-up, with its four research institutes and two universities—ETH Zurich and EPFL—testing various applications over the SCION network since the end of 2019. The successful tests have led to the next step and the enablement of the SCION infrastructure by SWITCH for all Swiss universities. Among many other use cases that will rely on the SCION network in the future, the Secure Swiss Financial Network (SSFN) is worthwhile mentioning here. Initiated by the Swiss National Bank (SNB) and the SIX Group, the SCION-based communication network connects the players of the Swiss banking system. The ISPs Swisscom, Sunrise, and SWITCH are responsible for setting up and operating the SSFN. It is an impressive example of a public–private partnership that strengthens the competitiveness and innovative prowess of Swiss industry.

SCION technology offers both the sender and receiver of data as well as the ISPs full control over which parties are allowed to use the communication network and over which path the data is transported. This is made possible by protocols specially developed for SCION and the concept of isolation domains, which are synonymous with a protected network association between trusted partners. SCION relies on multipath communication, which brings important advantages: high availability of the network, fast failover in case of failures, increased end-to-end bandwidth, dynamic traffic optimization, and resistance to DDoS attacks. SCION addresses vulnerabilities of today's Internet while ensuring full connectivity to existing IT infrastructure. Investment costs for upgrading to the SCION network are correspondingly low.

Another strength of SCION technology turns out to be particularly relevant in times of global warming. We are all called upon to move the digital transformation in a sustainable direction. The huge increase in data traffic due to cloud computing and the integration of more and more devices into the global Internet of Things present us with major challenges. How do we manage to keep the CO_2 footprint of digitalization as small as possible? SCION could make a difference: ETH researchers estimate that the increased efficiency of the SCION network could save up to 2.8 GW of electricity globally compared to the energy consumption of today's networks.

As a university we have a responsibility towards society and want to contribute to the achievement of the Sustainable Development Goals of the UN. Security and sustainability are part of them. ETH Zurich continuously expanded the area of information security and has been cooperating with industry through the Zurich Information Security & Privacy Center (ZISC) for 20 years. Since 2019 and together with EPFL in Lausanne, ETH offers a Master's program in Cybersecurity. SCION fits perfectly into these strategic priorities. I hope that SCION will be able to capitalize on the momentum and convince more partners of the added value of the new Internet architecture. It would mean the continuation of a Swiss success story and bring the world closer to the goal of a securer and more sustainable data communication.

Foreword by Fritz Steinmann

IT Solution Architect SIX

Communication networks are a wonderful innovation. I grew up in a time when people around me started to hijack their parents' phone lines for hours, usually to call the nearest available electronic mailbox (or bulletin board systems, as they were called back then) using a 2400 bit/s modem, just to exchange news, chat with others, or download the newest games on our low-resolution text terminals.

Later I got involved with global networks and X.25 and I was amazed that I could establish a connection from Europe to Singapore within the blink of an eye, which was blazing fast—at least in the terms of those days.

The rapid progress of innovation has continued until today, as you probably know if you read this book. Some of the innovations, however, seem to arise and get deployed in a haphazard way, introducing considerable chance into what eventually succeeds as the deployed solution for a given problem. In one such seemingly random turn of events, it happened that BGP emerged as the ubiquitous Internet routing protocol, which we all have been using for over two decades whether we want to or not.

I vividly remember my first encounter with a radically new approach to how packets get directed from source to destination. It was in 2017, and our financial infrastructure company got involved with the Zurich Information Security & Privacy Center (ZISC) of ETH Zurich. I was encouraged to meet with a research team at ETH who seemingly had developed "a network technology that you need to look at," as I was told by our Chief Security Officer. My thoughts when looking at their website were more like "ah, yet another academic approach with no means to bridge the gap between now and the proposed future."

So I organized this meeting with a couple of engineers from our company and the Network Security group at ETH, headed by Prof. Adrian Perrig. He introduced us to the concepts of SCION, told us about implementations and deployments they already completed, outlined transition paths from today's Internet structure to a SCIONized Internet, and described plans and ideas that would become possible if SCION succeeded.

The design was electrifying and I became immediately energetic about what I heard. This was the first time in 25 years that somebody not just had developed a radically new approach to routing, but also had it implemented and was able to cover the gap between the existing and the new Internet. The meeting sparked new ideas as we spoke, and as I went back to my office I thought about where we could immediately make use of the new technology.

As a company we are always looking for ways to improve connectivity between financial market businesses—so it became clear rather quickly that we would do something in that respect. We immediately got in contact with the governing board of the Swiss National Bank, as we knew that they shared the vision of using technology not just for technology's sake, but to ever improve all layers of the financial infrastructure, which eventually leads to a more secure, reliable, and stable environment. The Swiss National Bank was the first company in Switzerland to actively test and later productively use SCION within their own network. Together and with other parties we embarked on a journey to create the first commercially used multi-party SCION isolation domain for the Swiss finance market.

At the beginning of the project, I used the picture of a raft trip, where the involved parties would neither know in detail what type of raft they were building nor where exactly they were headed to or perhaps where they may be washed ashore. Over time we succeeded on the trip without anybody going overboard or having to master rough waters. I might even say that the raft was converted to a solid boat, and the ambition is to eventually create a fleet of large ships to sail the SCION journey.

The 21st century Internet deserves a better foundation for packet transport than we have right now. It is time for a change, and I firmly believe that with SCION we have the right ingredients to replace the current routing infrastructure. Let's embark on that journey and share this excitement together!

Preface

DAVID BASIN, LAURENT CHUAT, MARKUS LEGNER, ADRIAN PERRIG

The first SCION book, published in 2017, summarized the results of eight years of work by over a dozen researchers. Now, five years later, we have completely rewritten the book. It includes new authors and new parts and chapters describing the substantial changes and improvements to SCION that were achieved since. Why all the changes?

The goal of the SCION project remains unchanged: To take a clean-slate approach to designing a better, more secure, and more performant Internet. What has changed is that, over the past years, SCION has made the transition from a largely academic project focused on this design and building research prototypes, to a robust Internet architecture with rapidly growing international deployment and adoption. This is the result of the architecture and its open-source foundation maturing. It is also the result of the maturing SCION ecosystem where companies, in particular Anapaya Systems, have built highly scalable industrial-strength routers and edge devices based on the SCION architecture.

More concretely, since the first SCION book, all the systems described have been fully implemented: all within the research testbed SCIONLAB, but even more importantly, most are now also used in production, deployed by Internet service providers (ISPs) and used by end customers for production traffic. The SCION Internet became operational in 2017 for selected use cases, first within Switzerland deployed by selected Swiss ISPs, and later internationally. As always with new technologies, important and necessary feedback occurs as experience with the technology is gained. For SCION, this included, for example, improvements in how isolation domains and autonomous systems are numbered, improvements to the PKI used for the control plane, and numerous improvements and generalizations to the control and data plane that further enhance operation, deployability, security, and efficiency. Discussions with different real-world customers of SCION also led to developing new services that were possible within the SCION architecture but difficult to realize in the conventional Internet. For example, high-speed firewalls that leverage SCION's ability to achieve highly efficient source authentication, or the use of SCION's multipath features to achieve data transfers with unprecedented transmission rates. We provide an account in this book of both the updated architecture and all the resulting extensions.

A second major addition is the new part on verification. We believe that every critical architecture and system should come with rigorous evidence that it works as intended. Namely that it functions correctly and securely. By evidence, we do not mean merely some combination of code review, testing, and hand-waving. Rather, the arguments should have a rigorous mathematical foundation and be supported by formal proofs. Moreover, given how delicate correctness arguments are, these proofs should ideally be machine checked. In other words, we would like our new Internet architecture to not only build in security from the ground up, but to have provable security guarantees as well. This is certainly not the case for today's Internet, which has neither security-by-design nor the existence of formal proofs. The work for SCION on this is therefore particularly exciting! Some parts are still work-in-progress and we describe what has been accomplished to date and the underlying verification technology that we have developed, in part, for this project.

A third addition is the expanded part on SCION in the real world. This part reports on our experience working with ISPs to deploy SCION infrastructure on a global scale. It also reports on our work with end customers from within government and private industry, to deploy edge devices that enable them to harvest the many benefits that SCION offers. We believe that this part of the book will be particularly exciting for many readers as it shows how SCION has really come of age and that end customers are already reaping its many benefits.

This book, and the SCION project, would not have been possible without the generous support of numerous individuals and institutions. We would like to express our sincere gratitude to all who contributed; for a detailed overview, see Acknowledgments.

How to Read This Book

This book is a complete guide to the SCION Internet architecture. It includes detailed descriptions of SCION's core components, protocols, and extensions, as well as several chapters dedicated to formal verification. Readers who want to gain in-depth knowledge about the SCION architecture should read the entire book. We recommend reading the chapters and sections listed in Table 1 to readers who want to obtain a basic understanding of SCION.

The back matter of the book contains a glossary (page 641), a list of common abbreviations (page 645), and an index (page 651). Throughout the book, we indicate terms with an entry in the glossary as follows:

glossary term*

A gray bar in the margin indicates the presence of an example:

This is an example.

Additional resources (research papers, videos, presentations, source code, links, answers to frequently asked questions (FAQ), and reasons to use SCION) are available on our web page:

https://www.scion-architecture.net

We also encourage interested readers to sign up to the SCION newsletter (through the above website) and the SCION mailing list for questions and support regarding the development and deployment of SCION.

Table 1: We recommend reading the following chapters to obtain a basic understanding of SCION.

Chapter		What to Read
Chapter 1	Introduction	The entire chapter
Chapter 2	Overview	The entire chapter
Chapter 3	Authentication	All sections except for §3.1.4
Chapter 4	Control Plane	The entire chapter
Chapter 5	Data Plane	§5.1 through §5.5; §5.8
Chapter 6	Functional Properties and Scalability	§6.1
Chapter 7	Security Analysis	§7.1
Chapter 8	Extensions for the Control Plane	The entire chapter
Chapter 9	Montoring and Filtering	§9.2
Chapter 10	Extensions for the Data Plane	§10.1; §10.2 through §10.2.3
Chapter 11	Availability Guarantees	§11.1 through §11.3; §11.7 and §11.8
Chapter 12	Host Structure	The entire chapter
Chapter 13	Deployment and Operation	§13.1 through §13.3; §13.5
Chapter 15	Use Cases and Applications	The entire chapter
Chapter 16	Green Networking with SCION	The entire chapter
Chapter 18	F-PKI	§18.2
Chapter 19	RHINE	§19.1 through §19.3
Chapter 21	Motivation for Formal Verification	The entire chapter

Acknowledgments

This book is the result of a vast collaborative effort over several years. Most chapters are based on research papers. At the beginning of each chapter, people who had a major role in the elaboration of the chapter are listed.

Special thanks go to the following individuals from ETH for providing feedback that helped improve the book (in alphabetical order):

Sofia Giampietro
Christelle Gloor
Jean-Pierre Smith
Fabio Streun
Piet De Vaere

We are especially indebted to Corine de Kater and Nicola Rustignoli, who tirelessly worked on the book during the final six months of the project, greatly enhancing the quality of the book.

We are also very grateful for the support and suggestions provided by the following individuals (in alphabetical order):

Patrick Bamert
Jeffrey Barnes
William Boye
René Merz
Bong Seob Shin
Pawel Szalachowski

We would like to express our sincere gratitude to all who contributed.

1 Introduction

Laurent Chuat, Markus Legner, Adrian Perrig*

Chapter Contents

1.1   Today's Internet   2
   1.1.1   The Internet Protocol (IP)   2
   1.1.2   The Border Gateway Protocol (BGP)   3
   1.1.3   RPKI and BGPsec   5
   1.1.4   Lack of Authentication   6
   1.1.5   Attacks   7
1.2   Goals for a Secure Internet Architecture   9
   1.2.1   Availability in the Presence of Adversaries   9
   1.2.2   Transparency and Control   9
   1.2.3   Efficiency and Scalability   11
   1.2.4   Extensibility and Algorithm Agility   12
   1.2.5   Deployability   12
   1.2.6   Formal Verification   13

The Internet has been successful beyond even the most optimistic expectations. It permeates and is intertwined with many aspects of our society. This success has created a fundamental dependency on communication, as many of the processes underpinning modern society would grind to a halt should the Internet become unavailable. Unfortunately, the security of today's Internet is far from commensurate with its importance as critical infrastructure.

Although we cannot precisely determine what impact a global Internet outage would have on our society, anecdotal evidence indicates that even minor connectivity outages can have severe consequences. To make matters worse, the Internet has not primarily been designed for high availability in the presence of malicious actors, and recent proposals to improve Internet security and availability have been constrained by the design of the current architecture. A new Internet architecture should offer availability and security by default, and SCION was designed from the outset to meet this objective.

---

*This chapter reuses content from Barrera, Chuat, Perrig, Reischuk, and Szalachowski [49].

© The Author(s), under exclusive license to Springer Nature Switzerland AG 2022
L. Chuat et al., *The Complete Guide to SCION*, Information Security
and Cryptography, https://doi.org/10.1007/978-3-031-05288-0_1

Furthermore, experience with developing large-scale distributed systems over several decades has shown that informal reasoning and testing are insufficient to ensure the secure operation of such a system in an adversarial environment. This is why the development of SCION was accompanied by the VerifiedSCION project, which aspires to formally prove key properties, from protocol specification to the code running on SCION routers.

1.1 Today's Internet

Witnessing the rapid advancement of Internet-based technologies, it might seem that the Internet itself is relentlessly evolving. In reality, only parts of the protocol stack benefit from constant innovation. The physical and application layers have adapted to new needs and trends, the transport layer has become more efficient and secure,[1] but the network layer has remained relatively static over the past few decades. This situation is sometimes referred to as the "Internet Hourglass" meaning that a handful of protocols form a thin— and seemingly irreplaceable—waist in the protocol stack, while both ends of the stack continue to grow.

Two protocols effectively define today's Internet architecture: the Internet Protocol (IP) [158, 417] and the Border Gateway Protocol (BGP) [432]. Nobody could have predicted how impressively these protocols would stand the test of time, as they have remained virtually unchanged since the standardization of IPv6 [157] and BGP-4 [294] in the 1990s. However, as the Internet continued to expand and needed to accommodate new uses, numerous issues came to light. Since a comprehensive treatment of the Internet's problems would require an entire book on its own, we only present an overview of salient issues that demonstrate the need for a new architecture.

1.1.1 The Internet Protocol (IP)

IP is one of the fundamental protocols of the Internet, as it enables the forwarding of packets between end hosts. Its first major version, IPv4, was specified in 1981 [417] and its (non-backward compatible) successor, IPv6, was introduced in 1998 [157]. IP enables the transmission of packets between a source and a destination along a single path that is opaque from the end host's perspective. To forward packets, end hosts (as well as routers) do not need a complete path, but only a table to determine the next hop solely based on the destination address. Neither senders nor receivers can typically influence the path that their packets take. This approach is simple, but it also comes with many drawbacks:

---

[1] In fact, even the evolution and development of transport protocols is severely impeded by *ossification* mainly caused by the deployment of middleboxes in the network [399, 416]; but protocols such as TLS and QUIC have been successfully deployed by building on top of TCP and UDP, rather than trying to replace them.

- **Lack of transparency and control:** Being able to select and verify the path that packets take is desirable in many situations. End hosts might want to avoid packets being routed through adversarial or untrusted networks, or they might want to choose the most suitable path with regard to a specific metric (e.g., latency or bandwidth). Unfortunately, IP does not offer such an option. Although systems that enable loose and strict source routing have been proposed, these extensions are not commonly supported in today's networks. It is also not possible to simultaneously use multiple distinct paths towards the same destination—even though multipath communication offers numerous benefits, which we will encounter throughout the book.

- **Stateful routers:** IP routers maintain forwarding tables to determine the next hop of a received packet. This basic requirement has undesirable consequences. Performing a forwarding-table lookup for every packet is a time-consuming operation. Therefore, high-performance networking equipment typically relies on ternary content-addressable memory (TCAM) hardware, which is expensive and energy-intensive. Moreover, the constantly growing size of forwarding tables, partially due to the slow but steady deployment of IPv6, poses a problem for routers, as the storage capacity of TCAM hardware is limited. Routers that keep state for network information can also suffer from denial-of-service (DoS) attacks exhausting the router's state [460].

1.1.2 The Border Gateway Protocol (BGP)

BGP is the routing protocol that provides connectivity between independently operated networks or **autonomous systems (ASes)**⋆ such as Internet service providers (ISPs).[2] Each AS advertises its reachability information as a list of **IP prefixes**⋆ through a BGP update message. Such BGP updates accumulate the sequence of ASes through which they have passed, and they contain a list of attributes characterizing the advertised routes.

BGP enables ISPs to perform traffic engineering and select routes based on policies that reflect their business relationships[3] through an intricate decision process that is used to select the best route to a destination [91]. Unfortunately, BGP comes with a number of shortcomings:

- **Outages:** Since the **control plane**⋆ and the **data plane**⋆ are not clearly separated in today's Internet, forwarding may suddenly fail during route changes. By attacking routing, an adversary can thus interfere with

---

[2]The definition of words marked with a star can be found in the glossary starting on page 641.
[3]The two most common business relationships between ASes are *customer–provider* relationships (the customer AS pays the provider AS to forward traffic) and the *peering* relationship where two ASes agree to forward traffic between their own customers and the peer and its customers without payment.

packet forwarding. Furthermore, when BGP update messages are sent, the network may require up to tens of minutes to converge to a stable state [313], which can lead to intermittent outages. As an indicator of these problems, a study has shown, for example, that a sudden degradation in user-perceived quality of voice-over-IP (VoIP) calls is highly correlated with BGP updates [307].

- **Lack of fault isolation:** BGP is a globally distributed protocol, running among all BGP speakers in the entire Internet. BGP update messages are thus disseminated globally. Due to the lack of any routing hierarchy or isolation between different areas, a single faulty BGP speaker can affect routing in the entire world, as occurred in the AS 7007 incident, which disrupted global connectivity due to a single faulty router [369].

- **Poor scalability:** The amount of work required to be performed by BGP is proportional to the number of destinations. Moreover, path changes are disseminated profusely and sometimes throughout the entire Internet. This reduces scalability and prevents BGPsec (a proposal for a secured version of BGP that we discuss in Section 1.1.3) from frequently disseminating freshly signed routing updates.

- **Convergence:** ASes must have a consistent view of the network topology and agree on the set of paths to use for packet forwarding. Otherwise, a situation could arise where an AS A configures AS B as the next hop for a particular destination, while B uses A as a next hop for the same destination. In this case, a packet would be sent back and forth between the two ASes, which constitutes a forwarding loop.

  Unfortunately, convergence to a consistent and stable state depends on the policies of individual ASes. It has been shown that for certain situations, BGP will never converge to a stable state [221] and other topologies cause *BGP wedgies*, where BGP converges but non-deterministically [220]. In general, even if BGP converges after the topology changes, this process can require several minutes [447] and users may experience outages during this process. In addition, BGP convergence constitutes an attack vector for malicious actors and makes verifying security and availability properties highly challenging.

- **Single path:** At the end of the BGP decision process used to determine how to reach a given destination, a single path is selected. Although some multipath protocols allow simultaneous use of multiple network interfaces, BGP does not provide path control to end hosts and does not allow use of multiple AS-level paths. This can even lead to outages when BGP selects a legitimate but inefficient route through a link that is too small to satisfy the demand (bottleneck routing). In such a situation, end hosts have no choice but to wait until ASes in the Internet manually modify policies such that a more appropriate path is chosen.

- **Lack of security:** BGP has no built-in security mechanisms and does not provide any tools for ASes to authenticate the information they receive through BGP update messages. This opens up a multitude of attack opportunities (some of which we describe in §1.1.5) and has only been addressed in recent years through RPKI and BGPsec, which have problems of their own, as we discuss in the next section.

1.1.3 RPKI and BGPsec

The issues arising from the lack of security mechanisms in core Internet protocols were recognized early. Researchers already started working on security improvements to BGP in the late 1990s and early 2000s. The fundamental idea of the later standardized RPKI [328] and BGPsec [329], a security-enhanced version of BGP, was to cryptographically certify ownership of IP address ranges and AS numbers, and to authenticate BGP messages. While change was very slow initially (RPKI and BGPsec were only standardized in 2012 and 2017, respectively), there has been a substantial increase in RPKI deployment in recent years, which rekindled the hope that the Internet can actually be secured in the near future. Unfortunately, this hope may be premature, as we will discuss in the following sections, and a more radical change to today's Internet architecture will be necessary to fundamentally resolve its security issues.

1.1.3.1 RPKI and Route Origin Authorizations

RPKI provides certificates to ASes, and certificates for the IP addresses they own, which are called route origin authorizations (ROAs). The attestation process follows the delegation of IP addresses from the Internet Corporation for Assigned Names and Numbers (ICANN) and regional registries down to individual ASes. When an AS announces that it owns a particular IP prefix through BGP, other ASes can check if it has a valid ROA; if not, the recipient of this announcement can conclude that it is fraudulent and reject it. Unfortunately, ROAs only prevent the simplest form of BGP hijacks, see §1.1.5.

1.1.3.2 Problems with BGPsec in Partial Deployment

BGPsec was designed to prevent hijacking attacks by cryptographically protecting paths in BGP messages. However, BGPsec was only standardized in 2017, and it will likely take many years until it reaches global deployment. A detailed analysis of the protocol has shown that it provides very little security benefits unless all ASes consistently use and enforce BGPsec [347]. In a partial deployment (i.e., if not all ASes use it), BGPsec can cause instabilities and is prone to downgrade attacks (in which an attacker causes other ASes to accept standard BGP messages even if a BGPsec-secured path exists).

1.1.3.3 Problems with BGPsec in Full Deployment

Even if all ASes in the world were to deploy BGPsec, many issues remain. Attackers would still be able to create wormholes and cause forwarding loops [336]. Even more worrying is the fact that RPKI and BGPsec may introduce circular dependencies, where communication depends on cryptographic keys and certificates, which in turn require existing communication paths to be exchanged [137].

RPKI and BGPsec also cause issues for network sovereignty [441]. As very few organizations are at the root of the RPKI hierarchy, these organizations have the power to create or revoke certificates. Depending on the jurisdiction, local courts of some countries may gain the power to shut down parts of the Internet, which makes some ISPs reluctant to deploy RPKI.

Finally, BGPsec further exacerbates BGP's scalability issues. To provide global connectivity, every one of the currently about 75,000 ASes in the world needs to know how to reach every other AS. This requires a large number of BGP update messages, the processing of which requires many more resources in BGPsec due to the additional cryptographic operations. Furthermore, prefix aggregation, which is used to combine multiple IP prefixes to reduce the number of routes and announcements, no longer works in BGPsec as the digital signatures are not aggregated. This is particularly cumbersome as the increasing fragmentation of the IP address space and the trend towards announcing ever smaller IP address ranges have caused a strong growth of the number of paths that Internet routers need to store and exchange. As we show in §6.3, SCION can reduce the routing overhead in the Internet by several orders of magnitude.

1.1.4 Lack of Authentication

The necessity of authenticating digital data is becoming increasingly prevalent, as adversaries exploit the absence of authentication to inject malicious information to attack the network. Unfortunately, not only BGP but virtually all original protocols used in the Internet lack authentication features.

Infrastructures to provide authentication have been added in an ad hoc manner: RPKI provides the roots of trust for the authentication of BGPsec messages; TLS [433] allows browsers to authenticate web servers; and DNSSEC [33] provides authentication for DNS. Nevertheless, the current situation is still unsatisfactory in many regards. For example, all these protocols are sensitive to the compromise of a single entity. DNSSEC and RPKI rely on a single or very small number of roots of trust, while TLS is based on an oligopolistic trust model in which any one of hundreds of authorities can issue a certificate for any domain name. The Internet Control Message Protocol (ICMP) [135, 417] does not even have an authenticated counterpart, thus allowing the injection of fake ICMP packets. The Internet also lacks a general infrastructure to enable two end hosts to establish a shared secret key for end-to-end encrypted and authenticated communication; the simplest mechanism today is to rely on trust-on-first-use (TOFU) approaches [538], which opportunistically send the public key unauthenticated to the other communicating party.

1.1.5 Attacks

There exist several attacks against which the current Internet architecture offers little to no protection. Among these attacks are prefix hijacking, spoofing, denial of service, DNS hijacking, and composed attacks (which use a combination of vulnerabilities, or use a vulnerability in one protocol to compromise another protocol).

1.1.5.1 Prefix Hijacking

Due to a lack of authentication and fault isolation in BGP, numerous Internet outages are caused by a malicious or erroneous announcement of IP address space, a problem called prefix hijacking. A famous case of prefix hijacking happened in February 2008 when Pakistan's internal censorship attempt resulted in a global outage of YouTube that took close to two hours to resolve [434]. This was not the first nor the last such event.

Prefix hijacking can also be used for interception: An adversary who wants to eavesdrop on traffic towards a destination can hijack its prefix to receive its packets but also engineer BGP updates such that the packets finally do reach the intended destination. Renesys (now Dyn) documented such cases of prefix redirection, where the adversary managed to re-direct traffic to take a detour across another continent [142]. This problem is exacerbated by the fact that defining BGP routing policies is often a complicated, manual, and thus error-prone process.

Unfortunately, BGP hijacks are still possible when RPKI is deployed and are only resolved in a full deployment of BGPsec: With RPKI, a malicious AS trying to hijack a particular IP prefix can still send a BGP update message claiming that it is directly connected to its legitimate owner (for which there exists a valid ROA). Recipients of such an announcement would accept it, since the legitimate owner of the addresses is noted as the last AS in the BGP message. They would then start sending traffic intended for those IP addresses to the attacker, who can inspect, reroute, or drop it.

In settings where route origin validation (ROV) is deployed, Morillo et al. [373] recently point out several new attacks: hidden hijack (where a non-deploying entity will not propagate an announcement upstream if the next-hop address is the same, so upstream the hijack remains invisible, creating an inconsistency between the observable BGP path and the actual traffic flow (data-plane path)), non-routed prefix hijack (where non-routed prefixes are hijacked), and super-prefix hijack of non-routed prefixes (circumventing ROV's defense for non-routed prefix hijacks).

1.1.5.2 Spoofing and DDoS Attacks

ICMP can be employed to send error or diagnostic messages (used by tools such as *ping* or *traceroute*). Because ICMP packets are not authenticated, the source address can easily be spoofed, which can lead to distributed denial-of-service (DDoS) attacks [306], or be used to disconnect two BGP routers from each other [214]. Since regular IP packets are not authenticated either, they suffer from the same problem, i.e., the source IP address can be spoofed.

DDoS attacks have been widely used to prevent access to servers or network resources. For example, a large-scale attack against Estonia made much of the country's critical infrastructure inaccessible during one week in April 2007 [231], and an attack with a previously unobserved amount of attack traffic—exceeding 1 Tbps—on Dyn's DNS infrastructure rendered numerous websites unavailable in 2016 [301, 409].

1.1.5.3 Deceiving Certification Authorities with BGP

Certification authorities (CAs) are responsible for issuing digital certificates to domain owners so they can protect their website against man-in-the-middle attacks with HTTPS. Ironically, CAs themselves are vulnerable to attacks, as they rely on exchanging unencrypted and unauthenticated packets during domain validation. If attackers can redirect these packets, for example if they have access to an AS's border router and are thus able to launch a prefix-hijacking attack, they can obtain fraudulent certificates [75]. One of the countermeasures proposed by Birge-Lee et al. is to perform domain validation from multiple vantage points [76]. Unfortunately, this does not fully solve the problem, as it only protects CAs from attackers who control a limited number of border routers.

1.1.5.4 DNS Hijacking and Cross-layer Attacks

We are witnessing a renewed wave of DNS security concerns that threaten the integrity of the Internet's phone book. Over the past few years, researchers have discovered a flurry of subtle vulnerabilities that can poison DNS caches or hijack name servers. For instance, the SAD DNS attack exploits unforeseen side channels to infer the randomness in DNS transactions and corrupt resolver caches [351]. Zheng et al. identify flaws in the real-world DNS dynamism and craft a cache poisoning attack against non-validating forwarders [568]. The Zaw attack targets popular DNS hosting services and demonstrates the feasibility of taking over domains through their stale NS records [16]. An alarming fact is that DNS hijacking does not stay at a conceptual level but has been widely observed in the wild [244], as evidenced by large-scale hijacking campaigns such as Sea Turtle [7].

Because of its essential functionality, DNS also plays a central role in broader contexts of Internet security. A recent measurement research shows that DNS vulnerabilities can be leveraged to subvert a range of Internet systems and services including but not limited to PKIs, time synchronization, email, VPN, and web applications [148]. A secure DNS is thus indispensable to a secure Internet. Unfortunately, as will be analyzed later (§19.2), DNSSEC as the standardized solution has not achieved widespread deployment, in part also because it introduces numerous additional limitations.

1.2 Goals for a Secure Internet Architecture

In this section, we present high-level goals that an inter-domain point-to-point communication architecture should accomplish. We illustrate why these goals are important, why they are challenging, and hint at how they can be achieved.

1.2.1 Availability in the Presence of Adversaries

Our overarching goal is the design of a communication infrastructure that remains highly available even in the presence of distributed adversaries: As long as an attacker-free path between end hosts exists, that path should be discovered and provide some guaranteed amount of bandwidth between these end hosts.

Availability in the presence of adversaries is an exceedingly challenging property to achieve. An on-path adversary may drop, delay, or alter packets that it should forward, or inject additional packets into the network (e.g., to cause congestion). An off-path adversary could hijack traffic and then perform on-path attacks. Traffic hijacking can take various forms: For instance, an adversary could announce a desirable path to a destination by using forged paths or attractive network metrics. Conversely, an adversary could render paths not traversing its network less desirable (e.g., by inducing congestion). An adversary controlling a botnet could also perform DDoS attacks, congesting selected network links, for example employing a Coremelt [495] or Crossfire [279] attack. Finally, an adversary could interfere with the discovery of legitimate paths (e.g., by flooding the control plane with bogus paths).

Part III is wholly devoted to describing several defense mechanisms—including better and more efficient packet authentication in the network and at end hosts, probabilistic traffic-monitoring techniques, and bandwidth-reservation architectures—and how their interplay provides strong guarantees for network availability.

1.2.2 Transparency and Control

We seek to achieve greater transparency and control for the forwarding paths of network packets, and the trust roots used for authentication. When the network offers path transparency, end hosts can predict (and verify) the forwarding path taken by packets.

Taking transparency of network paths as a first property, we aim to additionally achieve **path control**[*], a stronger property that (1) enables ASes to control the incoming **path segments**[*] through which they are reachable and (2) allows senders to then create and select end-to-end paths. This seemingly benign requirement has several repercussions—beneficial but also fragile if implemented incorrectly. The beneficial aspects of path control for senders and receivers include the following:

- **Separation of control plane and data plane:** To enable path control, the control plane (which determines networking paths) must be separated from the data plane (which forwards packets according to the determined paths). The separation ensures that forwarding cannot retroactively be influenced by control-plane operations, e.g., routing changes. The separation contributes to enhanced availability.

- **Multipath communication:** Path control lets any sender select multiple paths to carry packets towards the destination. Multipath communication is a powerful mechanism to enhance availability [24] and enables fast failover, where end hosts can immediately switch to a backup path in the event that a link on their currently used path fails.

- **Geofencing:** Applications that transmit sensitive data can benefit from path control by ensuring that packets only traverse certain trusted ASes and avoid others. This is important for certain industries like the financial or healthcare sectors, where regulatory requirements demand that certain data not leave a particular jurisdiction.

- **Defending against network attacks:** If the packet's path is carried in its header (which is one way to achieve path control), then the destination can reverse the path to return its response to the sender, mitigating reflection attacks. Path control also enables circumvention of malicious network entities or congested network areas, providing a powerful mechanism against DoS and DDoS attacks.

The fragile aspects that need to be handled with care are the following:

- **Respecting the forwarding policies of ISPs:** If senders have complete path control, they may violate ISPs' forwarding policies. We thus need to ensure that ISPs offer a set of policy-compliant paths which senders can choose from.

- **Preventing malicious path creation:** A malicious sender could exploit path control for attacks, for example by forming malicious forwarding paths such as loops that consume increased network resources.

- **Scalability of path control:** Source routing allows a sender of a packet to specify the route the packet takes through the network. Consequently, source routing does not scale to inter-domain networks, as a source would need to know the full network topology to determine paths. With SCION, sources do not compute an end-to-end path based on a topology, but select among a set of paths provided by the control plane. We call this path control. To make path control scale, we thus rely on source-selected paths and packet-carried forwarding state instead of full-fledged source routing.

- **Permitting traffic engineering:** Fine-grained path control would inhibit ISPs from operating and performing traffic engineering. We thus seek to provide end-host path control at the granularity of links between ASes, allowing ISPs to fully control internal paths. ISPs can further perform traffic engineering based on per-path bandwidth allocations, which can be encoded in the forwarding information, or dynamically adjusted in the network.

- **Network stability:** Past research has shown that uncoordinated path selection by end hosts can lead to persistent *oscillations*, i.e., an alternating grow-and-shrink pattern of traffic volumes on links [188, 467]. This is often raised as a key concern and represents one of the biggest obstacles to deployment of path-aware networks [153]. It is thus imperative that a future Internet architecture take this into account and develop mechanisms preventing these instabilities [452].

1.2.3 Efficiency and Scalability

Aside from the lack of availability and transparency, today's Internet also suffers from a number of stability deficiencies. For instance, BGP encounters stability issues in cases of network fluctuations, where routing protocol convergence can can take several minutes [447]. A 2006 earthquake in Taiwan that severed several undersea communication cables caused Internet outages throughout Asia for several days [61]. Moreover, forwarding tables have reached the limits of their scalability due to IP prefix de-aggregation (i.e., announcement of more specific prefixes) and multihoming [256]. Unfortunately, extending the memory size of forwarding tables is challenging as the underlying TCAM hardware is expensive and power-hungry, consuming on the order of a third of the total power consumption of a router [81].

Security and high availability come at a cost, usually resulting in lower efficiency and potentially diminished scalability. High performance and scalability, however, are essential to make a system such as an Internet architecture viable. We therefore explicitly seek high efficiency as a goal, so that packet forwarding is at least as efficient (in terms of latency and throughput) as current IP forwarding, in the most common cases. Moreover, we seek improved scalability compared to the current Internet, in particular with respect to BGP and the size of forwarding tables.

An approach to achieving efficiency and scalability is to avoid storing forwarding state on routers wherever possible. We thus aim to encode state into packet headers and to protect that state cryptographically, enabling simpler router architectures compared to today's IP routers. We observe that modern block ciphers such as AES can be computed faster than performing memory lookups. For example, on current PC platforms, performing one AES encryption requires on the order of 30 cycles while fetching a byte from main memory requires around 200 cycles [13]. Moreover, a modern block cipher can be implemented in hardware with on the order of ten thousand gates [15], which is sufficiently small to replicate it profusely, which in turn enables high parallelism—the high complexity of a high-speed memory system prevents such replication at the same scale. Besides higher efficiency, avoiding state on routers also prevents state-exhaustion attacks [460] and state inconsistencies across routers.

This shows that our goal of efficiency and scalability is in line with the design rationale of providing path control to end hosts. A selected path is communicated to the network by packet-carried forwarding information, which in turn removes the need for inter-domain forwarding tables at border routers and makes forwarding simpler and more efficient.

Furthermore, end-host path selection is in line with the *end-to-end principle*, which states that a network functionality should be implemented by the entity that has the required information, and is thus in the best position to correctly implement the functionality [449]. Since the end host has the most information about its internal state, functions such as bit-error recovery, duplicate suppression, or delivery acknowledgments are most efficiently handled by the end host itself. Similarly, the end host has the knowledge of preferred or undesirable network paths and thus should be involved in path selection.

1.2.4 Extensibility and Algorithm Agility

To future-proof SCION, its core architecture and codebase are designed to be extensible, such that additional functionality can be easily built and deployed. SCION clients and routers should (without overhead or expensive protocol negotiations) discover the minimum common feature set supported by all intermediate nodes.

Algorithm agility allows a protocol to easily migrate from one algorithm to another. It is especially important in the context of cryptographic algorithms, which only become weaker over time. Since it is not possible to predict advances in cryptanalysis techniques, every future-proof protocol that employs cryptographic algorithms should provide a mechanism for algorithm agility.

1.2.5 Deployability

Incentives for deployment are important to overcome the resistance to upgrading today's Internet. A multitude of features is necessary to offer the initial impulse: high availability even under control-plane and data-plane attacks (e.g., built-in DDoS defenses), path transparency and control, trust-root transparency and control, application-specific path optimization, robustness to configuration errors, fast recovery from failures, increased network capacity, high forwarding efficiency, multipath forwarding, and so on.

If early adopters cannot obtain sufficient benefits from migrating to a new network architecture, even initial deployment is unlikely to be successful. So ideally, the first deploying ISP should already gain a competitive advantage through the ability to sell a service that is desirable even for the initial customers.

Migration to the new architecture should require only minimal added complexity. Deployment should be possible by reutilizing the internal infrastructure of an ISP, and only require installation or upgrade of a few border routers. Moreover, configuration of the new architecture should be similar to that of the existing architecture, such as in the configuration of BGP policies, minimizing the amount of additional personnel training.

Economic and business incentives are also of critical importance. ISPs should be able to define new business models and sell new services. Users should derive a business advantage from the new architecture, for example by obtaining properties similar to a leased line at a lower cost. Migration cost should be minimal, requiring only the deployment of low-cost routers. Finally, a new architecture should not disrupt current Internet business models, but maintain the current Internet topology and business relationships (e.g., support peering).

1.2.6 Formal Verification

Modern distributed systems reach a scale that eludes people's mental capacities for considering all possible states and interactions, thus necessitating automated protocol verification techniques. Such formal verification achieves a high level of assurance for protocol correctness. Vulnerabilities in widely-used protocols such as BGPsec, TLS, and 5G authentication [159, 171, 336] illustrate that established engineering approaches—largely based on reviews and testing—are insufficient to ensure the security of advanced distributed protocols. Moreover, even if a protocol is secure, its implementation might introduce vulnerabilities. It is, therefore, imperative that the security and availability of a new Internet architecture be ensured by formal proofs.

Part I

SCION Core Components

2 Overview

LAURENT CHUAT, MARKUS LEGNER, ADRIAN PERRIG*

Chapter Contents

2.1 **Infrastructure Components** **20**
2.2 **Authentication** **21**
   2.2.1 The Control-Plane PKI (CP-PKI) 21
2.3 **Control Plane** **23**
   2.3.1 Path Exploration and Registration 24
   2.3.2 Path Lookup 26
   2.3.3 PCB and Path-Segment Selection 27
   2.3.4 Link Failures 27
2.4 **Data Plane** **28**
   2.4.1 Path Construction via Segment Combination 28
   2.4.2 Path Authorization 30
   2.4.3 Forwarding 31
   2.4.4 Intra-AS Communication 31
2.5 **ISD and AS Numbering** **31**
   2.5.1 ISD Numbers 31
   2.5.2 AS Numbers 32
   2.5.3 Assignment of ISD and AS Numbers 32

To briefly reiterate the vision we presented in the first chapter, our main goal is to design a network architecture that offers highly available and efficient point-to-point packet delivery—even in the presence of actively malicious entities. SCION introduces the concept of **isolation domains (ISDs)***ISD constitutes a logical grouping of autonomous systems (ASes), as depicted in Figure 2.1. Each ISD is administered by a few distinguished ASes that we refer to as **core ASes***, and usually also contains multiple non-core ASes. An AS can join an ISD by connecting to another AS in the ISD.

There are three types of links in SCION: (1) core links, (2) parent–child links, and (3) peering links. A core link can only exist between two core ASes.

---

*This chapter reuses content from Barrera, Chuat, Perrig, Reischuk, and Szalachowski [50], Chuat, Perrig, Reischuk, and Trammell [115].

© The Author(s), under exclusive license to Springer Nature Switzerland AG 2022
L. Chuat et al., *The Complete Guide to SCION*, Information Security
and Cryptography, https://doi.org/10.1007/978-3-031-05288-0_2

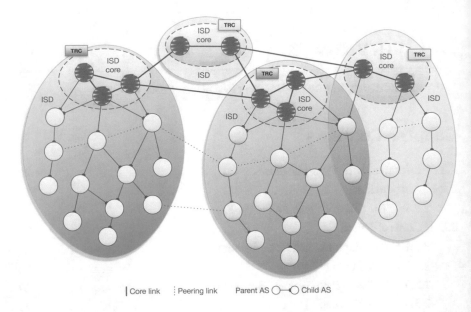

| Core link   Peering link   Parent AS ○——◖ Child AS

Figure 2.1: ASes grouped into four ISDs. Core ASes are connected via core
           links. Non-core ASes are connected via parent–child links or peer-
           ing links. An AS can belong to multiple ISDs. The directionality
           of links is generally given from top (parent) to bottom (child), but
           to avoid ambiguity we denote a child AS with a half-circle at the
           end of the line representing the link to its parent.

On the other hand, parent–child and peering links require that at least one
of the two connected ASes be a non-core AS. ASes with a parent–child link
typically either belong to the same entity or have a provider–customer relationship (where the customer pays the provider for traffic). Peering links exist
between ASes with a (standard or paid) peering relationship. Core links can exist for various underlying business relationships, including provider–customer
and peering relationships.

Each ISD groups ASes that span an area with a uniform trust environment
or a common jurisdiction. A possible model is for ISDs to be formed along
national boundaries or federations of nations, as entities within a legal jurisdiction can enforce contracts. ISDs can also overlap, so an AS may be part of
several ISDs.

ISDs serve several related purposes:

- They allow SCION to support trust heterogeneity, as each ISD can independently define its roots of trust (see §3.1);

- They provide transparency for trust relationships;

- They isolate the routing process within an ISD from external influences such as attacks and misconfigurations; and

- They improve the scalability of the routing protocol by separating it into a process *within* and one *between* ISDs.

Although ISDs may seem to lead to Internet fragmentation at first glance, they neither impede connectivity nor facilitate censorship (see §7.9); instead, ISDs provide openness and transparency, as we hope to elucidate in this book.

SCION operates on two routing levels: intra-ISD and inter-ISD. Both levels use **path-segment construction beacons (PCBs)**[*] to explore network paths. The creation of PCBs is initiated by core ASes. Each PCB is then disseminated as a policy-constrained multipath flood either within an ISD (to explore intra-ISD paths) or among core ASes (to explore core paths, potentially across different ISDs). We refer to this process as *beaconing*. PCBs accumulate cryptographically protected AS-level path information as they traverse the network, including efficiently encoded and protected forwarding information, in the form of **hop fields (HFs)**[*], and metadata about properties of links and traversed ASes. The HFs are subsequently used by end hosts to create end-to-end forwarding paths for data packets, which traverse a sequence of ASes. Packets thus contain AS-level path information, which avoid the need for border routers to maintain inter-domain forwarding tables. We refer to this concept as **packet-carried forwarding state (PCFS)**[*].

A few steps are required to obtain a forwarding (i.e., end-to-end) path. During the *path exploration* (or beaconing) phase, ASes discover paths to other ASes. *Path registration* then allows ASes to transform a few selected PCBs into path segments, and register these path segments with a path infrastructure (making them available for other ASes). The *path resolution* process allows end hosts to create an end-to-end forwarding path to a destination; it consists of (1) a *path lookup* step, where the end host obtains path segments, and (2) a *path combination* step, where a forwarding path is created from the path segments. We briefly present these steps in the remainder of this chapter and will describe them in more detail in the sections referred to in Figure 2.2.

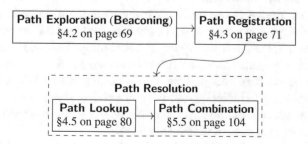

Figure 2.2: Process leading to the creation of a forwarding path.

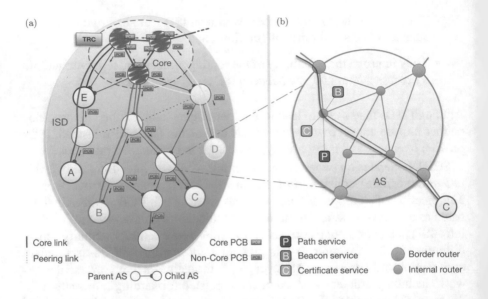

Figure 2.3: (a) ISD with path-segment construction beacons (PCBs) that are propagated from the ISD core to child ASes, and path segments for ASes *A, B, C, D*, and *E*. (b) Magnified view of a SCION AS with its routers and services.

2.1 Infrastructure Components

Figure 2.3b shows the main infrastructure components in SCION: The **beacon service**⋆ discovers path information, the **path service**⋆ disseminates path information, and the **certificate service**⋆ assists with validating path information. Depending on the size and type of the AS, there can be multiple servers for each of these services to load-balance requests and provide failover capability, or they can be combined into one or more *control services*. In addition, *border routers* provide the connectivity between ASes, while *internal routers* forward packets inside the AS.

The beacon service is responsible for the beaconing process, that is, generating, receiving, and propagating PCBs (see Figure 2.3a) to construct path segments. SCION utilizes two types of beaconing: intra-ISD beaconing (to construct path segments from core ASes to non-core ASes within an ISD) and core beaconing (to construct path segments among core ASes within an ISD and across ISDs). Periodically, the beacon service generates a set of PCBs, which it forwards to its child ASes (in case of intra-ISD beaconing) or neighboring core ASes (in case of core beaconing).

Core beaconing in SCION is similar to BGP's route-advertising process, although in SCION the process is periodic and PCBs are flooded over policy-compliant paths to discover multiple paths between any pair of core ASes. All

PCBs contain signatures of on-path ASes. In contrast to BGPsec, where expensive asymmetric cryptography must be performed at routers, SCION relies on dedicated services to generate and verify signatures. SCION's beacon service can be configured to implement all BGP route selection policies, as well as additional properties (e.g., control of upstream ASes) that BGP cannot express; this is discussed in detail in §6.2.

The path service stores mappings from AS identifiers to sets of announced path segments. The path service is organized as a hierarchical caching system similar to that of DNS. Through the beacon service, ASes select the set of path segments through which they want to be reached, and they upload them to the path service in the ISD core.

The certificate service keeps cached copies of certificates and manages keys and certificates for securing inter-AS communication. The certificate service is queried by the beacon service when validating the authenticity of PCBs (i.e., when the beacon service lacks a certificate).

Border routers connect different ASes supporting SCION. The main task of border routers is to forward packets to the next border router or the destination host within the AS. Since SCION can operate using any intra-AS routing protocol (e.g., IS-IS, OSPF, SDN) and communication fabric (e.g., IP, MPLS), the internal routers do not need to be changed to support SCION.

2.2 Authentication

SCION's control plane relies on a public-key infrastructure we call the **control-plane PKI (CP-PKI)***, in which each ISD defines its own roots of trust and policies in a file called **trust root configuration (TRC)***. A TRC is a signed collection of certificates, which also contains ISD-specific policies, for example, specifying how many signatures an updated TRC should contain to be valid. Figure 2.4 illustrates the TRC content and update process.

Each SCION AS must hold a private key (to sign PCBs) and a certificate attesting that it is the rightful owner of the corresponding public key. One of the main roles of the TRC is thus enabling the verification of **AS certificates*** and PCBs. However, other systems (such as DRKey) also rely on the CP-PKI, as we discuss in Chapter 3.

2.2.1 The Control-Plane PKI (CP-PKI)

Trust within an ISD is anchored in a TRC. Each TRC contains root certificates, which are used to sign CA certificates, which are in turn used to sign AS certificates. The TRC can be seen as a collection of root certificates, which also contains policies regarding its usage, validity, and future updates. TRCs are the main components of the CP-PKI. Initial TRCs constitute trust anchors; however, in contrast to other PKIs where any change to root certificates requires a

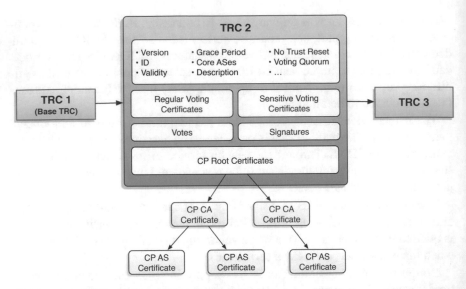

Figure 2.4: TRC chain and associated certificates. TRC 2 must be verified using the voting certificates in TRC 1. Control-plane (CP) root certificates are used to verify other CP certificates (which are in turn used to verify PCBs).

manual or out-of-band action (such as a software update), SCION includes a straightforward process to update TRCs.

2.2.1.1 Dissemination of TRC Updates

Information about a TRC update is disseminated via SCION's beaconing process. Each PCB contains the version number of the currently active TRC. If the TRC version number of a received PCB is higher than the locally stored TRC, a request is sent to the AS that sent the PCB to obtain the new TRC. The new TRC is verified on the basis of the current one, and is accepted if it contains at least the required quorum of correct signatures by trust roots defined in the current TRC. This simple dissemination mechanism has two major advantages: It is very efficient (as fresh PCBs rapidly reach all ASes), and it avoids circular dependencies with regard to the verification of PCBs and the beaconing process itself (as no server needs to be contacted over unknown paths in order to fetch the updated TRC).

2.2.1.2 TRC Update and Verification

In each ISD, an initial TRC called **base TRC**⋆ must first be signed (during a signing ceremony) and distributed throughout the ISD. With a base TRC as trust anchor, TRCs can be updated in a verifiable manner. We assume that all entities within an ISD can initially obtain an authentic TRC, for example, with an offline mechanism such as a USB flash drive provided by the ISP, or with an online mechanism that relies on a trust-on-first-use (TOFU) approach.

There are two kinds of TRC updates: regular and sensitive updates. A TRC update is sensitive if and only if critical sections of the TRC are affected (for example, if the set of core ASes is modified). For both regular and sensitive updates, a number of votes (in the form of signatures) must be cast to approve a TRC update. This number of votes is dictated by the **voting quorum*** (a parameter that must be defined within each TRC).

At most two TRCs per ISD can be considered active at the same time. Another TRC parameter called the **grace period*** indicates for how long the previous unexpired version of the TRC can still be considered active after a new TRC is disseminated. The grace period starts at the beginning of the validity period of the new TRC. An older TRC can only be considered active until either (1) the grace period has passed, or (2) yet a newer TRC is announced. AS certificates are validated by following the chain of trust up to an active TRC.

2.2.1.3 Revocation and Recovery from a Catastrophic Event

The TRC dissemination mechanism also enables rapid revocation of trust roots. When a trust root is compromised, the other trust roots can remove it from the TRC and disseminate a new TRC alongside a PCB with a new version number.

In case of a catastrophic event—such as several private root keys being disclosed due to a critical vulnerability in a cryptographic library—SCION is equipped with a recovery procedure called **trust reset***. The procedure consists in creating a new TRC with fresh trustworthy keys (and potentially new algorithms), and re-distributing this TRC out of band. A trust reset effectively establishes a new base TRC for the ISD. It is possible for ISDs to opt out and disable trust resets by setting a "no trust reset" Boolean to true in their TRC, with the effect that the entire ISD would have to be abandoned in the event of such a catastrophic compromise (this abandonment would also have to be announced out of band).

The partition of the SCION network into ISDs guarantees that no single entity can take down the entire network. Even if several entities formed a coalition to carry out an attack, the effects of that attack would be limited to one or a few ISDs. Moreover, all actions are publicly visible, which deters misbehavior.

2.3 Control Plane

The control plane is responsible for discovering path segments and making them available to end hosts, alongside the certificates needed to verify those path segments.

2.3.1 Path Exploration and Registration

Core beaconing enables core ASes to learn paths to other core ASes. Through intra-ISD beaconing, non-core ASes learn path segments leading to core ASes (and vice versa). Figure 2.3a shows path segments from ASes *A*, *B*, *C*, *D*, and *E* to the core. The beaconing process is asynchronous and each AS can independently set policies dictating which PCBs are sent in which time intervals, and to which neighbors. In particular, PCBs do not need to be propagated immediately upon arrival.[1] Figure 2.5 shows how PCBs originate from the core beacon service and are propagated to non-core child ASes. The non-core beacon service receives these PCBs and re-sends them to child ASes, which results in AS-level path segments. At every AS, metadata as well as information about the ingress and egress interfaces (i.e., link identifiers) of the AS is added to the PCB.

Paths at AS-level granularity are insufficient for diversity; ASes often have several connection points, and thus a disjoint path is possible despite the AS sequence being identical. For this reason, SCION encodes AS ingress and egress interfaces in the form of interface IDs as part of the path, exposing a finer level of path diversity. These interface IDs identify connections to neighboring ASes and can be chosen and encoded by each AS independently and without any need for coordination as the IDs only need to be unique within each AS. Figure 2.5 demonstrates this feature: AS *F* receives two different PCBs via two different links from a core AS. Moreover, AS *F* uses two different links to send two different PCBs to AS *G*, each containing the respective egress interfaces. AS *G* extends the two PCBs and forwards both of them over a single link to a child AS.

An AS typically receives several PCBs representing several path segments to various core ASes. Figure 2.3a shows two path segments for AS *D*, for example. There are three types of path segments:

- A path segment from a non-core AS to the core is an *up-segment*.

- A path segment from the core to a non-core AS is a *down-segment*.

- A path segment between core ASes is a *core-segment*.

However, path segments are typically bidirectional and thus support packet forwarding in both directions. In other words, up-segments and down-segments are invertible: An up-segment can be converted to a down-segment and vice versa. The path service in non-core ASes learns up-segments by extracting them from PCBs they obtain from the local beacon service. Path services in core ASes obtain down-segments when they are registered by beacon services in non-core ASes and in addition store core-segments to reach other core ASes.

---

[1] However, during bootstrapping and in case a PCB containing a previously unknown path is obtained, it should be forwarded immediately such that other ASes learn about it quickly.

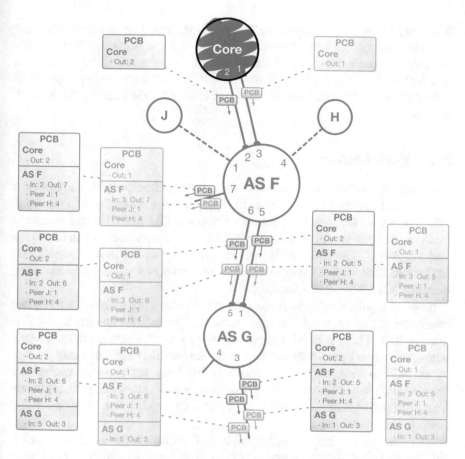

Figure 2.5: Intra-ISD PCB propagation from the ISD core down to child ASes. For the sake of illustration, the interfaces of each AS are numbered with integer values. In practice, each AS can choose any encoding for its interfaces; in fact, only the AS itself needs to understand its encoding.

The beacon service in an AS selects the down-segments through which the AS desires to be reached, and registers these path segments at the core path service. When links fail, segments expire, or better segments become available, the beacon service keeps updating the down-segments registered for their AS.

Peering links are of paramount importance for an efficient Internet and to implement the existing economic relationships between ISPs. Therefore, SCION has a built-in mechanism for peering links between two non-core ASes or between a core AS and a non-core AS. To reduce beaconing overhead and prevent possible forwarding loops, PCBs do not traverse peering links. Instead, peering links are announced along with a regular path in a PCB. Figure 2.5 shows how AS F includes its two peering links in the PCB. If the same peering link is announced in the path segments by both ASes adjacent to the peering link, then the peering link can be used to shortcut the end-to-end path (i.e., without going through the core). SCION also supports peering links that cross ISD boundaries, which also adheres to SCION's path transparency property: A source knows the exact set of ASes and ISDs traversed during the delivery of a packet.

2.3.2 Path Lookup

An end host (source) who wants to start communication with another host (destination), requires up to three path segments: An up-segment to reach the ISD core, a core-segment to reach the destination ISD, and a down-segment to reach the destination AS. The source host queries the path service in its AS for such segments. The path service has up-segments stored in its database and furthermore checks if it has appropriate core- and down-segments in its cache; in this case it returns them immediately.

If not, the path service in the source AS queries core path services (using locally stored up-segments) in the source ISD for core-segments to the destination ISD. Then, it combines up-segments with the newly retrieved core-segments to query core path services in the remote ISD to fetch remote down-segments. To improve overall efficiency the local path service caches the returned path segments and uses parallelism when requesting path segments from core path services. Finally, the local path service returns all path segments to the source host.

This recursive lookup strongly simplifies the process for end hosts (which only have to send a single query, similar to stub DNS resolvers). The caching strategy ensures that path lookups are fast for frequently used destinations (similar to caching in recursive DNS resolvers).

Example.   Consider a source host in ISD 1 sending a path lookup request to its local path service. If the destination AS is within ISD 1, the local path service queries the core path services in ISD 1 for down-segments. Each core path service responds returning down-segments to the local path service. If needed, the local path service also requests intra-ISD core-segments to connect up- and down-segments that do not start and end in the same core AS. The local path service then returns the set of up-, down-, and (potentially) core-segments to the source.

If the destination AS is in ISD 2, then the local path service first requests core-segments to ISD 2 from the core path services in ISD 1. Combining the returned core-segments with the existing up-segments, the local path service then requests down-segments from core path services in ISD 2. Finally, the set of up-, core- and down-segments is returned to the requesting source.

All path segments learned by the local path service are cached.

2.3.3 PCB and Path-Segment Selection

Among the received PCBs, ASes must choose a set of PCBs to propagate further, and a set of path segments to register. These PCBs and path segments are selected based on a path quality metric with the goal of identifying consistent, diverse, efficient, and policy-compliant paths. *Consistency* refers to the requirement that there exists at least one property along which the path is uniform, such as an AS capability (e.g., support for COLIBRI (§10.2)) or link property (e.g., high bandwidth). *Diversity* refers to the set of paths that are announced over time being as path-disjoint as possible to provide high-quality multipath options. *Efficiency* refers to the length, bandwidth, latency, utilization, and availability of a path, where more efficient paths are naturally preferred. *Policy compliance* refers to the requirement that the path adheres to the AS's routing policy. Based on past PCBs that were sent, the beacon service scores the current set of candidate path segments and sends the best segments in the next beaconing interval.

Core beaconing operates similarly to intra-ISD beaconing, except that core PCBs only traverse core ASes. The same path selection metrics apply, where an AS attempts to forward the set of most desirable paths to its neighbors. A difference, however, is that each AS forwards a number of PCBs *per source AS*. This process, which is similar to how BGP operates, is susceptible to scalability issues as the overhead grows linearly with the number of core ASes. However, as shown by the scalability analysis in §6.3, on a per-path basis SCION's diversity-based beaconing algorithm (§4.4.6) is around two orders of magnitude more efficient than BGP due to the following reasons: SCION beaconing does not need to iteratively converge, SCION makes AS-based announcements instead of BGP's prefix-based announcements (there are around 10 times more prefixes than ASes in today's Internet), and diversity-based beaconing is optimized to reduce the number of transmitted beacons.

2.3.4 Link Failures

Unlike in the current Internet, link failures are not automatically resolved by the network, but require active handling by end hosts. Since SCION forwarding paths are static, they break when one of the links fails. Link failures are handled by a two-pronged approach that typically masks link failures without any outage to the application and rapidly re-establishes fresh working paths:

- The SCION Control Message Protocol (SCMP) (the SCION equivalent of ICMP) is used for signaling connectivity problems. Instead of relying on application- or transport-layer timeouts, end hosts get immediate feedback from the network if a path stops working and can quickly switch to an alternative path.

- SCION end hosts are encouraged to use multipath communication by default, thus masking a link failure with another working path. As multipath communication can increase availability (even in environments with very limited path choices [24]), SCION beacon services attempt to create disjoint paths, SCION path services attempt to select and announce disjoint paths, and end hosts compose path segments to achieve maximum resilience to path failure. Consequently, most link failures in SCION remain unnoticed by the application [235], unlike the frequent (albeit mostly brief) outages in the current Internet [285, 307].

2.4 Data Plane

While the control plane is responsible for providing end-to-end paths, the data plane ensures packet forwarding on the selected path. SCION border routers forward packets to the next AS based on the AS-level path in the packet header (which is augmented with ingress and egress interface identifiers for each AS), without inspecting the destination address and also without consulting an inter-domain forwarding table. Only the border router at the destination AS needs to inspect the destination address to forward it to the appropriate local end host.

An interesting aspect of this forwarding is enabled by the split of locator (the path towards the destination AS) and identifier (the destination address) [185]: The identifier can have any format that the destination AS can interpret, since only the destination needs to consider that local identifier. In other words, an AS can select an arbitrary addressing format for its hosts, e.g., a 4-byte IPv4, 6-byte media access control (MAC) address, 16-byte IPv6, or any other up to 16-byte addressing scheme. A valuable consequence is that hosts with different address types can directly communicate, e.g., an IPv4 host can communicate with an IPv6 host over SCION.

In the next two sections, we describe how an end host combines path segments into an end-to-end forwarding path, and how border routers forward packets efficiently.

2.4.1 Path Construction via Segment Combination

Through the path lookup, the end host obtains path segments that must be combined into an end-to-end path. A valid SCION **forwarding path*** can be created by combining up to three path segments, in the following ways (all combinations are illustrated with sample intra-ISD paths depicted in Figure 2.6):

- **Immediate combination of path segments** (e.g., $B \rightarrow D$): The last AS on the up-segment (core AS Z_3) is also the first AS on the down-segment. In this case, the simple combination of an up-segment and a down-segment creates a valid forwarding path.

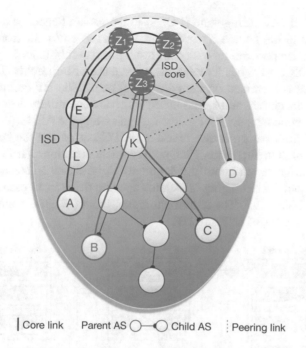

Core link   Parent AS ◯—◉ Child AS   ⋮ Peering link

Figure 2.6: ISD with path segments for ASes A, B, C, D, and E.

- **AS shortcut** (e.g., $B \to C$): The up-segment and down-segment intersect at a non-core AS (e.g., K). In this case, a shorter forwarding path can be created by removing the extraneous part of the path.

- **Peering shortcut** (e.g., $A \to B$): A peering link (e.g., $L \to K$) exists between the two segments, so a shortcut via the peering link is possible. As in the *AS shortcut* case, the extraneous path segment is cut off. The peering link could be traversing to a different ISD.

- **Combination with a core-segment** (e.g., $A \to D$): The last AS on the up-segment is different from the first AS on the down-segment. This case requires an additional core-segment (e.g., $Z_1 \to Z_2$) to connect the up- and down-segment. If the communication remains within the same ISD ($A \to D$), a local ISD core-segment is needed; otherwise (e.g., $A \to I$ in Figure 2.1), an inter-ISD core-segment is required.

- **On-path** (e.g., $A \to E$): The destination AS is part of the up-segment or the source AS is part of the down-segment; in this case, a single up- or down-segment, respectively, is sufficient to create a forwarding path.

Once a forwarding path is chosen, it is encoded in the SCION packet header, which makes inter-domain routing tables unnecessary for border routers: Both the egress and the ingress interface of each AS on the path are encoded as packet-carried forwarding state (PCFS) in the packet header. The destination can respond to the source by reversing the end-to-end path from the packet header, or it can perform its own path lookup and combination.

The SCION packet header contains a sequence of hop fields (HFs), one for each AS that is traversed on the end-to-end path. The HF contains interface numbers of the ingress and egress links, which are essentially descriptors of the links across which the packet is entering and exiting the AS. Figure 2.5 depicts how the HF information is assembled in the PCB as part of the beaconing process. In addition to hop fields, each path segment contains an **info field (INF)**⋆ with basic information about the segment. A host can create an end-to-end forwarding path by extracting info fields and hop fields from path segments, as depicted in Figure 2.7. The additional meta header (META) contains pointers to the currently active INF and HF.

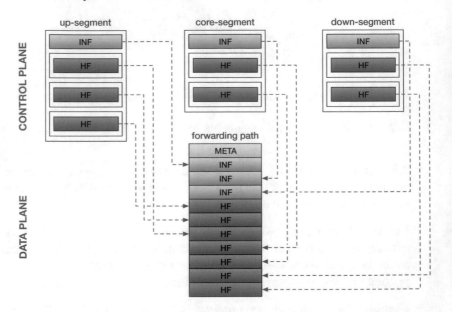

Figure 2.7: Example showing the construction of a forwarding path through the combination of three path segments.

2.4.2 Path Authorization

A crucial requirement for the data plane is that end hosts can only use paths that were constructed and authorized by ASes in the control plane. In particular, end hosts should not be able to craft HFs themselves, modify HFs in authorized path segments, or combine HFs of different path segments (path splicing). We refer to this property as **path authorization**⋆ [297, 327].

SCION achieves path authorization by creating message-authentication codes (MACs) during the beaconing process. Each AS calculates these MACs using a local secret key (that is only shared between SCION infrastructure elements within the AS) and chains them to the previous HFs. The MACs are then included in the forwarding path as part of the respective HFs.

2.4.3 Forwarding

Routers can efficiently forward packets in the SCION architecture. In particular, the absence of inter-domain routing tables and the absence of complex longest-IP-prefix matching performed by current routers enables the construction of more efficient routers.

During packet forwarding, a SCION border router at the ingress point of the AS first verifies that the packet entered through the correct ingress interface corresponding to the information in the HF, that the HF is not expired yet, and that the MAC in the HF is correct. If the packet has not yet reached the destination AS, the egress interface defines the egress SCION border router, in which case native intra-domain forwarding (e.g., IP or MPLS) is used to send the packet from the ingress SCION border router to the egress SCION border router. In the destination AS, the border router inspects the destination address and sends the packet to the corresponding host instead.

2.4.4 Intra-AS Communication

Communication within an AS is handled by existing intra-domain communication technologies and protocols such as IP with Software-Defined Networking (SDN), or Multiprotocol Label Switching (MPLS). Figure 2.3b shows one possible intra-domain path through the magnified AS.

2.5 ISD and AS Numbering

In this book, we label ISDs and ASes in one of two ways: either (1) using a short symbolic notation, where our convention is to assign numbers to ISDs and letters to ASes, or (2) using the numbering scheme and notation used in practice, where each ISD-AS is identified with a 64-bit number; the most significant 16 bits identify the ISD (represented in decimal), the least significant 48 bits identify the AS in a format similar to that of IPv6 addresses, and a hyphen is used to separate the two—e.g., `4-ff00:1:f`. In the remainder of this section, we elaborate on the latter (i.e., the concrete numbering).

2.5.1 ISD Numbers

ISDs are represented with decimal numbers, ranging from 0 to 65535. Table 2.1 shows the current allocation of ISD numbers. The ISD number 0 is reserved to mean "any ISD" and used notably during path lookup (see §4.5).

Table 2.1: Allocation and description of ISD number ranges.

ISD Number	Description
0	Wildcard ISD (stands for "any ISD").
1–15	Reserved for documentation and sample code [252].
16–63	Private use (can be used for testing, analogously to RFC6996 [370]).
64–4094	Public ISDs, should be assigned in ascending order, without gaps and without vanity numbers.[2]
4095–65535	Reserved.

2.5.2 AS Numbers

The SCION numbering scheme uses a superset of the existing BGP AS numbering scheme [264]. The default format for AS numbers is similar to IPv6. It uses a 16-bit colon-separated lower-case hexadecimal encoding with leading zeros omitted. However, the double colon (: :) zero-compression feature of IPv6 is not supported. As for ISD numbers, 0 represents the wildcard AS and stands for "any AS", which may be used during path lookup (see §4.5).

The range from 1 to $2^{32} - 1$ is dedicated to BGP AS numbers [264]. If a BGP AS supports SCION, it has the same AS number for both BGP and SCION. To facilitate the comparison with BGP AS numbers, any number in the BGP AS range is represented as a decimal. While it is legal to write a BGP AS number using the SCION syntax, programs should use the decimal representation for display. For example, if a program receives 0 : 1 : f, it should display it as 65551.

Currently, the 2 : 0 : 0/16 range is allocated to public SCION-only ASes (i.e., ASes that are not existing BGP ASes). AS numbers in that range should be assigned in ascending order, without gaps and without vanity numbers.[2]

2.5.3 Assignment of ISD and AS Numbers

Ideally, each AS number should be globally unique (partly to facilitate the comparison and transition from BGP), but the actual requirement is only that each AS number be unique within an ISD. Since an AS can be part of several ISDs, picking a globally unique AS number also facilitates joining new ISDs.

In principle, ISD numbers can be self-assigned: A number that is not yet used by any other ISD is picked by the constituents of the new ISD before the first TRC is signed and distributed, then other ISDs are free to accept or reject the new TRC. Our vision, however, is that an organization such as ICANN or

---

[2]A vanity number is a sequence of digits chosen specifically to be aesthetic or easily remembered.

a regional Internet registry (e.g., RIPE NCC) will take on the responsibility of assigning ISD and AS numbers. In Chapter 3, we will discuss in more detail how TRCs are signed and how trust in base TRCs is established.

3 Authentication

LAURENT CHUAT, SAMUEL HITZ, MARKUS LEGNER, ADRIAN PERRIG*

Chapter Contents

3.1   **The Control-Plane PKI (CP-PKI)** **36**
    3.1.1   Trust Model 37
    3.1.2   Overview of Roles, Certificates, and Keys 37
    3.1.3   Certificates 38
    3.1.4   Trust Root Configuration (TRC) Format 41
    3.1.5   TRC Signing Ceremony 44
    3.1.6   TRC Update 46
    3.1.7   Signing and Verifying Control-Plane Messages 48
    3.1.8   TRC Bootstrapping 49
    3.1.9   TRC Dissemination and Discovery 51
    3.1.10 Substitutes to Certificate Revocation 51
3.2   **DRKey: Dynamically Recreatable Keys** **52**
    3.2.1   Use Cases 54
    3.2.2   Overview 55
    3.2.3   Key Hierarchy 57
    3.2.4   Key Establishment 60
    3.2.5   Limitations of DRKey 61
3.3   **SCION Packet Authenticator Option** **61**
    3.3.1   Option Format 62
    3.3.2   Network Address Translation 62
    3.3.3   Preventing Replay Attacks 63

In contrast to the legacy Internet, SCION was designed and built with security as one of the primary objectives. Therefore, all control-plane messages are authenticated. The verification of those messages relies on a public-key infrastructure (PKI) called the control-plane PKI or CP-PKI. In short, the CP-PKI is a set of mechanisms, roles, and policies related to the management and

---

*This chapter reuses content from Chuat, Perrig, Reischuk, and Szalachowski [113, 114], Chuat, Perrig, Reischuk, and Trammell [115], Rothenberger, Roos, Legner, and Perrig [442].

© The Author(s), under exclusive license to Springer Nature Switzerland AG 2022   35
L. Chuat et al., *The Complete Guide to SCION*, Information Security
and Cryptography, https://doi.org/10.1007/978-3-031-05288-0_3

usage of certificates, which enable the verification of signatures, e.g., on path-segment construction beacons (PCBs)[1].

The CP-PKI has undergone an important redesign since the publication of the first SCION book [413]. In particular, SCION certificates now follow the X.509 standard, as opposed to the first version where certificates were in JSON format. Moreover, the CP-PKI has been simplified (notably by abandoning the concept of cross-signatures) and decoupled from name resolution and the end-entity PKI, which are discussed in Chapters 18 and 19.

This chapter also describes DRKey, a system for efficient derivation and distribution of symmetric keys, which relies on the CP-PKI. DRKey allows SCION border routers and other entities to efficiently derive symmetric keys on the fly from a local secret. These keys can then be used to achieve source authentication for data packets (without requiring a previous handshake) and prevent denial-of-service attacks.

3.1 The Control-Plane PKI (CP-PKI)

The control-plane PKI is the foundation of all authentication procedures in SCION. In comparison to today's authentication infrastructures such as RPKI [20], the CP-PKI offers the following benefits:

- **Efficient updating of trust roots:** Even after key loss, disclosure, or compromise, trust roots can be rapidly updated, without software updates.

- **Resilience to compromised entities and keys:** Compromised or malicious trust roots outside an ISD cannot affect operations that stay within that ISD. Moreover, each ISD can be configured to withstand the compromise of any single voting key.

- **Decentralized trust model:** Authentication relies on local trust roots, limiting the scope of authorities and offering local sovereignty [441].

- **Trust flexibility:** Each ISD can define its own trust policy. ASes must accept the trust policy of the ISD(s) in which they participate, but they can decide which ISDs they want to join, and they can participate in multiple ISDs.

- **Scalability:** The authentication infrastructure scales to the size of the Internet and is adapted to the heterogeneity of today's Internet constituents.

---

[1]The CP-PKI has already several professional industry-grade implementations, e.g., by SIX, the main financial infrastructure and service provider in Switzerland [477]. For more information on an industrial implementation of a SCION CP-PKI based on HashiCorp Vault, see `docs.anapaya.net/en/latest/user-guide/ca-service/`.

- **No circular dependencies:** One major challenge in designing a PKI is to avoid circular dependencies, where a communication path is necessary to establish authenticity, but authenticity must be verified to establish a communication path. The control-plane PKI is designed to avoid any such circular dependencies.

3.1.1 Trust Model

In a public-key infrastructure, trust is established through one or more trust anchors. A trust anchor is a certificate, public key, or set thereof that is axiomatically trusted. In other words, a trust anchor is a cryptographic object for which trust is assumed rather than derived. In SCION, trust anchors are special trust root configurations (TRCs) called **base TRCs***. Non-base TRCs are updates, which can be verified with the cryptographic material and policies contained in the previous TRC.

The two predominant trust models in today's Internet are monopolistic (single root of trust; e.g., in DNSSEC) and oligopolistic (multiple roots of trust; e.g., in the Web PKI). Typically, in both models, some or all certification authorities are omnipotent. If their key is compromised, then the security of the entire system collapses.

The SCION trust model is different from classic PKI models in mainly two ways. First, no entity is omnipotent. The capabilities of ISDs (authentication-wise) are limited to communication channels in which they are involved. Second, the trust roots of each ISD are co-located in a single file, the TRC, which is co-signed by multiple entities in a process we call voting.

If several important keys have been compromised, it may be necessary for an ISD to re-establish all trust roots. This is possible with a process we refer to as a **trust reset***. When the base TRC of an ISD is created, the ISD can decide to forbid trust resets (with a dedicated TRC field) to avoid any external re-establishment of trust roots, with the risk of losing control of the ISD in case of a major compromise.

3.1.2 Overview of Roles, Certificates, and Keys

All SCION ASes must hold at least one AS certificate and the corresponding private key to sign PCBs, but certain ASes have additional roles:

- **Core ASes**: In each ISD, the core ASes are listed in the TRC. Core ASes have links to other core ASes (in the same or in different ISDs), and they initiate beaconing.

- **Certification authorities (CAs)**: CAs are responsible for issuing AS certificates to other ASes and/or themselves. CAs are typically SCION ASes themselves, but can also be independent entities.

- **Voting ASes**: Only certain ASes within an ISD may sign TRC updates. We refer to the process of appending a signature to a new TRC as "casting a vote" and we designate ASes that hold a certificate whose private key can be used to sign a TRC update as "voting ASes." Votes must be listed in the new TRC before it is signed, to avoid any manipulation of the set of signatures.

Trust within an ISD is anchored in the trust root configuration (TRC). Each TRC contains one or more root certificates, which are used to verify CA certificates, which are in turn used to verify one or more AS certificates. In addition to those certificates, voting certificates are used to verify TRC updates.

As opposed to the previous version of the CP-PKI (described in the first SCION book [413]), TRCs contain self-signed certificates instead of plain public keys. Self-signed certificates have several advantages over plain public keys: They make the binding between name and public key explicit; that binding is signed to prove possession of the corresponding private key, and using certificates allows us to more easily make use of the X.509v3 standard and existing cryptographic software.

There are two types of TRC updates (and thus two types of voting certificates): regular and sensitive. A regular TRC update is a periodic re-issuance of the TRC where the entities and policies listed in the TRC remain unchanged, whereas a sensitive update is an update that modifies critical aspects of the TRC, such as the set of core ASes.

In addition to the two types of voting certificates, there are three types of CP certificates (forming a chain of trust, which is anchored in a TRC): root certificates, CA certificates, and AS certificates. Figure 3.1 illustrates at a high level the relationships between TRCs and the five types of certificates, while Table 3.1 provides a more formal overview of the different types of keys and certificates in the control-plane PKI. In §3.1.3 and §3.1.4, we present in more detail the contents of certificates and TRCs, respectively.

3.1.3 Certificates

All certificates in the control-plane PKI are in X.509v3 format [136]. Every certificate has a subject (the entity that owns the certificate) and an issuer (the entity that signed the certificate, usually a CA). The subject and the issuer can be the same entity.

3.1.3.1 Voting Certificates

Regular and sensitive voting certificates define which keys are allowed to cast votes, respectively, in a regular update and a sensitive update. In X.509 terms, both regular and sensitive certificates are self-signed certificates (i.e., issuer and subject are the same entity, and the key within the certificate can be used to verify its signature). Voting certificates can be also considered end-entity

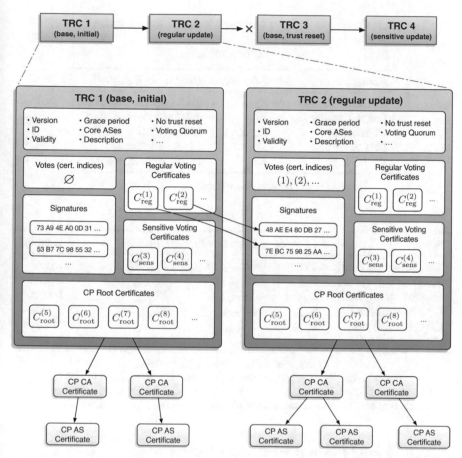

Figure 3.1: TRC update chain and the different types of associated certificates. Arrows show how signatures are verified; in other words, they indicate that a public key contained in a certificate or TRC can be used to verify the authenticity of another item.

certificates (in X.509 terms) because they cannot be used to verify signatures on other certificates; however, they are used to verify TRC signatures.

3.1.3.2 Control-Plane (CP) Root Certificates

Root certificates define which entities act as certification authorities (CAs) in an ISD. In X.509 terms, a root certificate is a self-signed CA certificate (i.e., issuer and subject are the same entity, and the key within the certificate can be used to verify its signature). To bootstrap trust for root certificates, they are embedded in TRCs.

Table 3.1: Overview of keys and certificates in the control-plane PKI.

(a) Key pairs.

Name	Notation[1]	Used to verify/sign
Sensitive voting key	K_{sens}	TRC updates (sensitive)
Regular voting key	K_{reg}	TRC updates (regular)
CP root key	K_{root}	CP CA certificates
CP CA key	K_{CA}	CP AS certificates
CP AS key	K_{AS}	PCBs, path segments

[1] $K_x = PK_x + SK_x$, where PK_x = public key, SK_x = private key

(b) Certificates.

Name	Notation	Signed with	Contains	Validity[2]
TRC (trust root conf.)	TRC	SK_{sens}, SK_{reg}[3]	$C_{root}, C_{sens}, C_{reg}$[3]	1 year
Sensitive voting cert.	C_{sens}	SK_{sens}	PK_{sens}	5 years
Regular voting cert.	C_{reg}	SK_{reg}	PK_{reg}	1 year
CP root certificate	C_{root}	SK_{root}	PK_{root}	1 year
CP CA certificate	C_{CA}	SK_{root}	PK_{CA}	11 days[4]
CP AS certificate	C_{AS}	SK_{CA}	PK_{AS}	3 days

[2] recommended maximum validity period
[3] multiple signatures and certificates of each type may be included in a TRC
[4] A validity of 11 days with 4 days overlap between two CA certificates is recommended to enable best possible operational procedures when performing a CA certificate rollover.

3.1.3.3 Control-Plane (CP) CA Certificates

CA certificates are used to verify AS certificates. In X.509 terms, a CA certificate is a self-issued CA certificate (i.e., issuer and subject are the same entity, but the key used to verify the certificate is in a different certificate, a root certificate in this case). CA certificates do not bundle the root certificate needed to verify them. In order to allow the CA certificate to be verified, a pool of root certificates must first be extracted from one or more active TRCs (we discuss this mechanism in more detail in §3.1.7).

3.1.3.4 Control-Plane (CP) AS Certificates

AS certificates are used to verify control-plane messages such as PCBs. In X.509 terms, an AS certificate is an end-entity certificate (i.e., the key inside the certificate cannot be used to verify the signature on another certificate). Typically, SCION CAs are also SCION ASes. This means that a CA may also create an AS certificate for itself. The steps required to create a new AS certificate are the following:

1. The AS creates a new key pair and a certificate signing request (CSR) using that key pair.

2. If the AS and CA are different entities, the AS sends the certificate signing request to the CA.

3. The CA uses its CA key and the CSR to create the new AS certificate.

4. If the AS and CA are different entities, the CA sends the AS certificate back to the AS.

3.1.4 Trust Root Configuration (TRC) Format

The trust root configuration (TRC) is a signed collection of X.509v3 certificates that also contains ISD-specific policies encoded in a Cryptographic Message Syntax (CMS) [245] envelope. This collection contains a set of CP root certificates that build the roots of the verification path for the AS certificates of an ISD. The other certificates in the TRC are solely used for signing the next TRC, a process we refer to as voting. The verification of a new TRC thus depends on the policies and voting certificates defined in the previous TRC.

We use several qualifiers to describe the state or nature of a TRC:

- **Valid:** The validity period of a TRC is defined in the TRC itself. A TRC is considered valid if the current time falls within its validity period.

- **Active:** An active TRC is a valid TRC that may still be used for verifying certificate signatures, i.e., either the latest or the previous TRC if it is still in its **grace period*** (as defined in the new TRC). No more than two TRCs can be considered active at the same time for any ISD.

- **Base:** Trust for a base TRC cannot be inferred by verifying a TRC update; base TRCs are trusted axiomatically, similarly to how root CA certificates are trusted by clients in the Web PKI.

- **Initial:** The very first TRC of an ISD is the initial TRC of that ISD. It is a special case of the base TRC, where the number of the ISD is chosen (cf. §3.1.4.2).

In the remainder of this section, we describe the different fields of a TRC.

3.1.4.1 Version (TRC Field)

This field describes the version of the TRC format specification. For now, this version is always v1. Note that the version field has different semantics compared to the JSON-format TRCs in the previous version of the control-plane PKI (where the version field was an incrementing counter). The JSON-style version is moved to the serial number field in the ID sequence below.

3.1.4.2 ID (TRC Field)

This field is a unique identifier of the TRC. It is a sequence of ISD number, serial number, and base number. The ISD number must be an integer between 1 and 65535 (inclusive). The serial and base number both must be positive integers. The base number indicates the starting point of the TRC update chain. A TRC where the serial number is equal to the base number is a base TRC. The initial TRC is a special case of a base TRC. It must have a serial number of 1 and a base number of 1. With every TRC update, the serial number must be incremented by one. This facilitates uniquely identifying the predecessor and successor TRC in a TRC update chain starting in the same base TRC. If a trust reset is necessary, a new base TRC is announced to start a new and clean TRC update chain. The base number should be the number following the serial number of the last TRC produced before the trust reset.

3.1.4.3 Validity (TRC Field)

The TRC validity period is the interval during which the TRC may be considered in the valid state. This interval sets the lower and upper bound for which a TRC can be *active*.

3.1.4.4 Grace Period (TRC Field)

The grace period indicates an interval during which the previous valid version of the TRC should still be considered active. The field encodes the grace period as a number of seconds (positive integer). The start of the grace period is equal to the beginning of the validity period of this TRC. The predecessor of this TRC, if any, should be considered active until either (1) the grace period has passed, (2) the predecessor's expiration time is reached, or (3) the successor TRC of this TRC has been announced.

The grace period of a base TRC must be zero. The grace period of a non-base TRC should be non-zero and long enough to provide sufficient overlap between the TRCs in order to facilitate interruption-free operations in the ISD. If the grace period is too short, some AS certificates might expire before the subject can fetch an updated version from its CA.

3.1.4.5 No Trust Reset (TRC Field)

The "no trust reset" Boolean indicates whether a trust reset is forbidden by the ISD. Within a TRC update chain, this value must not change. Thus, the base TRC decides on the value. This field is optional and defaults to false.

A trust reset represents a special use case where a new base TRC is created. It therefore differs from a TRC update (regular or sensitive), as signatures cannot be verified with the certificates contained in the predecessor TRC. Instead, a trust reset must be axiomatically trusted, similarly to how the initial TRC is trusted. On trust resets, the "no trust reset" value may change. However, once a trust reset is disallowed, it cannot be re-enabled. ISDs should set this value to false, unless they have assessed the risks and implications.

3.1.4.6 Voting Quorum (TRC Field)

The voting quorum indicates the number of necessary votes on a successor TRC for it to be verifiable. A voting quorum greater than one will prevent any single entity from creating a malicious TRC update.

3.1.4.7 Votes (TRC Field)

This field contains a sequence of indices of the voting certificates in the predecessor TRC. In a base TRC, this sequence is empty. Every entry in this sequence must be unique. If index i is part of the votes field, then the voting certificate at position i in the certificates sequence of the predecessor TRC casts a vote for this TRC. Further restrictions on the votes are discussed in §3.1.6 on page 46.

This sequence is included to prevent stripping voting signatures from the TRC. If this sequence were not included, a TRC that has more voting signatures than the voting quorum could be transformed into multiple verifiable TRCs with the same payload, but different voting signature sets. This would violate the uniqueness of a TRC, without the consent from a voting quorum.

3.1.4.8 Core ASes (TRC Field)

This field contains a sequence of AS numbers that are the core ASes in this ISD. To revoke or add the core status for a given AS, a TRC update is necessary. The entries in this sequence must be unique.

3.1.4.9 Description (TRC Field)

The description contains a UTF-8 encoded string that describes the ISD. This value should not be empty. It may contain information in multiple languages, but an English description of the ISD is mandatory.

3.1.4.10 Certificates (TRC Field)

Voting ASes and CAs are not listed directly in the TRC. Instead, this information is given by the list of voting and root certificates. The certificates field is a sequence of self-signed X.509 certificates. Each certificate must fall under one of the three following categories:

- sensitive voting certificate,
- regular voting certificate, or
- CP root certificate.

The certificates can be distinguished, based on their listing location within the TRC.

3.1.5 TRC Signing Ceremony

The TRC contains policy information about an ISD and acts as a distribution mechanism for the trust anchors of that ISD. It enables securing the control-plane interactions, and is thus an integral part of the SCION infrastructure.

In the signing ceremony, the voters of the ISD meet offline to sign an agreed-upon TRC. As part of the ceremony, the public keys of all voters are exchanged. There are two types of signing ceremonies: the ceremony to sign a base TRC and the ceremony to sign a non-base TRC.

The base TRC builds the anchor point for a TRC update chain. All voters need to take part in this ceremony. The very first TRC for an ISD number, the initial TRC, is a special case of the base TRC where the ISD number is chosen. Another base TRC can only be created in a trust reset, which is a recovery procedure that must be followed after a catastrophic event where the high-security keys of multiple voting members of the ISD have been simultaneously compromised. The likelihood of such a compromise is low if keys are adequately handled and stored.

A non-base TRC is the result of a TRC update. Only a quorum of voters needs to partake in a non-base TRC signing ceremony.

3.1.5.1 Ceremony Participants

A signing ceremony typically includes participants from various organizations that will fulfill different roles.

For example, a signing ceremony might include the following participants:

- **Ceremony Administrator:**
    - Jane Doe, Pear Inc.

- **Voting AS Representatives:**
    - Paul Smith, Lemon GmbH
    - John Anderson, Grape Inc.
    - Marie Williams, Limette SA

- **Witnesses:**
    - Marc Jones, Berry GmbH
    - Jessica Miller, Kiwi Inc.

The *ceremony administrator* is in charge of moderating the signing process and guides all participants through the steps they need to take. The ceremony administrator may also act as an intermediary between participants when they share information with each other. A *voting AS representative* is capable of creating voting signatures on the TRC. This means the voting representative is in possession of a device with the private keys of the respective certificates in the TRC. A *witness* is any person that participates in the ceremony as a passive entity. They observe the ceremony execution.

3.1.5.2 Ceremony Preparations

Prior to the ceremony, participants decide on the physical location of the ceremony, the devices that will be used during the ceremony and the policy of the ISD. Furthermore, the location should provide a monitor (or projector) that allows the ceremony administrator to screen cast. Each party brings their own device that is provisioned with the required material.

The voting entities must agree on the ISD policy, before the ceremony can be executed. Specifically, they must agree on the following parameters:

- validity of the TRC,
- grace period (except for base TRCs),
- voting quorum,
- core ASes,
- description, and
- list of CP root certificates.

When these values are agreed upon, a quorum of voters needs to execute the signing ceremony. The set of needed keys depends on whether a base TRC, or a TRC update is signed. For the base TRC, all voting entities need to be present with both their sensitive and regular voting keys. The ceremony process is structured in multiple rounds of data sharing. The ceremony administrator leads the interaction and gives instructions to each participant.

3.1.5.3 Phase 1: Certificate Exchange

All entities share the certificates they want to be part of the TRC with the ceremony administrator, who aggregates and bundles them. The bundle is then shared with all voters. The ceremony administrator displays the hash value of each bundled certificate on the monitor. Each voting representative verifies that the certificates they contributed have the same hash value as what is displayed on the monitor. Further, all voting representatives confirm that the bundled certificates on their machine all have matching hash values.

3.1.5.4 Phase 2: TRC Generation

The ceremony administrator generates the TRC based on the bundled certificates and the agreed-upon ISD policy. The result is displayed on the monitor along with a hash value. The TRC is distributed to all voting representatives, and they must verify that the hash value matches.

3.1.5.5 Phase 3: TRC Signing

Each voting representative attaches a signature created with each one of their new voting keys to the TRC. When signing a non-base TRC, the voting representatives further cast a vote with the voting key present in the last TRC. In other words, it may be necessary to cast votes with both old and new keys; the purpose of signing a TRC with keys contained in the previous TRC is to certify the update, whereas the purpose of signing a TRC that contains a newly introduced public key with the corresponding private key is to prove the possession of that private key.

3.1.5.6 Phase 4: TRC Validation

All voting representatives share the signed TRC with the ceremony administrator, who aggregates them in a single signed TRC file. The signed TRC is validated by inspecting its contents on the monitor and verifying the signatures based on the exchanged certificates in Phase 1. The ceremony administrator then shares the signed TRC with all participants. Each of them must then inspect it once more, and verify it based on the certificates exchanged in Phase 1. At this point, the ceremony is completed. All participants have the signed TRC, and can use it to distribute the trust anchors for their ISD.

3.1.6 TRC Update

TRC updates are split into two categories: sensitive and regular updates. The type of update is inferred from the information that changes in the updated TRC. Based on the category of the update, a different set of voters is necessary to create a verifiable TRC update.

The following rules must hold for both update categories:

- The ISD and base number in the ID field must not change.

- The serial number in the ID field must be incremented by one.

- The "no trust reset" field must not change.

- Votes can only be verified with sensitive voting certificates or regular voting certificates present in the predecessor TRC. This means that the votes sequence must only contain indices of sensitive voting certificates or regular voting certificates.

- The number of votes must be greater than or equal to the voting quorum of the predecessor TRC.

- Every sensitive voting certificate and regular voting certificate that is new in the TRC attaches a signature to the TRC. This ensures that the freshly included voting entity agrees with the contents of the TRC, and approves being part of it.

In the context of a TRC update, we identify a certificate as changing, if the certificate is part of the certificate sequence in the predecessor TRC, but no longer part of the certificate sequence in the successor TRC. Instead, there is a certificate of the same category and distinguished name in the certificates of the successor TRC. We identify a certificate as new, if there is no certificate of the same category and distinguished name in the certificates of the predecessor TRC.

3.1.6.1 Regular TRC Update

A TRC update qualifies as regular, if all of the following restrictions apply:

- The voting quorum does not change.

- The core ASes section does not change.

- The number of sensitive voting certificates, regular voting certificates, and CP root certificates and their distinguished names does not change.

- The set of sensitive voting certificates does not change.

- For every regular voting certificate that changes, the regular voting certificate in the predecessor TRC is part of the voters on the successor TRC.

- For every CP root certificate that changes, the CP root certificate in the predecessor TRC attaches a signature to the signed successor TRC. These signatures are added as part of the CMS envelope just as any other voting or proof-of-possession signature of voting ASes.

In order for a regular TRC update to be verifiable, all votes must be cast by a regular voting certificate.

3.1.6.2 Sensitive TRC Update

If a TRC update does not qualify as a regular update, it is considered a sensitive update. In order for sensitive updates to be verifiable, all votes must be cast by a sensitive voting certificate.

3.1.7 Signing and Verifying Control-Plane Messages

In SCION, most control-plane messages are signed. For example, each hop information in a path segment is signed by the respective AS. All **relying parties*** must be able to verify signatures with the help of the CP-PKI.

3.1.7.1 Signing Process

To sign a message, the signing entity chooses an AS certificate that authenticates their private key. With the private key, they sign the message and attach the following information as signature metadata:

- **ISD-AS number:** The ISD-AS number of the signing entity.

- **Subject key identifier:** The key identifier of the public key used to verify the message.

This is the minimum information a relying party requires to identify which certificate to use to verify the signed message. Additionally, the signer should include the following information:

- **Serial and base number of the latest TRC:** Including this information allows relying parties to discover TRC updates and trust resets.

- **Timestamp:** For many messages, the time at which it was signed is useful information to ensure freshness.

3.1.7.2 Verification Process

When the relying party receives a message that they want to verify, they first need to identify the certificate that authenticates the corresponding public key.

AS certificates are bundled together with the CA certificate that signed them into certificate chains. For efficiency, these certificate chains are distributed decoupled from the signed messages. A certificate chain is verified against a root certificate. However, the root certificate is not bundled in the chain, it is bundled with the TRC. This allows a TRC to be updated while extending the validity period of a root certificate without modifying the certificate chain.

To verify a control-plane message, the relying party first must build a collection of root certificates from the latest TRC from the ISD referenced in the signature metadata of the message. If the grace period introduced by the latest TRC is still on-going, the root certificates in the second to latest TRC are also included. If the signature metadata contains the serial and base number, the relying party must check that they have at least that TRC.

After the relying party has constructed the pool of root certificates, they must select a certificate chain used to verify the message. The AS certificate included in this certificate chain must have the following properties:

- The ISD-AS number in the subject of the AS certificate must match the ISD-AS number in the signature metadata.

- The subject key identifier of the AS certificate must match the subject key identifier in the signature metadata.

- The AS certificate is valid at verification time. Normally, this will be the current time. In special cases, e.g., auditing, the time can be set to the past to check if the message was verifiable at the given time.

The relying party then verifies the certificate chain (by executing the regular X.509 verification) and the control-plane messages, using the set of root certificates. In addition, the relying party checks that all subjects carry the same ISD number, that each certificate is of the correct type, and that the CA certificate validity period covers the AS certificate validity period.

If any cryptographic material is missing in the process, the relying party queries the originator of the message for the missing material. If it cannot be resolved, the verification process fails.

An implication of this is that path segments should be verifiable at time of use. We cannot simply rely on them being verified on insert, since TRC updates that change the root key can invalidate a certificate chain.

3.1.8 TRC Bootstrapping

Base TRCs are trust anchors and are thus axiomatically trusted. All ASes must be pre-loaded with at least the current base-version TRC of their own ISD. In this section, we discuss multiple options for distributing TRC base versions of other ISDs and adding them to the trust store of relying parties. The problem of announcing a trust reset is essentially equivalent to that of distributing the initial version of a TRC, and can thus be solved with the same mechanisms.

3.1.8.1 Manual Mode

In manual mode, operators are responsible for adding all TRCs manually to their trust store. They can receive TRCs through multiple channels (e.g., through an out-of-band mechanism, or by discovering them during beaconing).

This method has the advantage that it is technically simple (although it incurs operational overhead) and has been a de-facto standard for establishing trust roots in other public-key infrastructures, such as the Web PKI, for years.

3.1.8.2 Trust on Multiple Announcements (TOMA)

In the TOMA mode, initial TRCs and trust reset TRCs that are discovered during the beaconing process are put in quarantine for a specified amount of time. If, during this quarantine period, the TRCs have been received on a pre-defined number of distinct paths with exactly the same content, the TRC is trusted and added to the trust store. Alternatively, the quarantined TRC can be sent to an operator for review, who then accepts the TRC. This combines the TOMA and manual methods for easier operations, as it limits operator involvement.

Direct neighbors of ASes in a new ISD still need to manually accept the new TRC as part of the procedure to set up their inter-AS link. These direct neighbor ASes only propagate beacons originating from the new ISD if they have authenticated and approved the new entity. To be trusted by TOMA, a new ISD thus needs to be approved by multiple established core ASes. This poses a significant hurdle for the creation of ISDs and trust reset TRCs.

TOMA is only applicable for core ASes, since they are the only ones that are able to receive beacons with unknown TRCs. Non-core ASes use an authentic channel to the core, based on the preloaded ISD-local TRC, in order to fetch newly discovered TRCs.

3.1.8.3 TRC Attestation

The previously mentioned methods induce a certain amount of operational overhead. With the TRC attestation method, human involvement is reduced to a minimum. A numbering authority coordinates the attribution of identifiers to ISDs globally. This authority may delegate and allocate ISD ranges to regional authorities, which then assign specific identifiers to ISDs in their region. An attestation is issued for every base TRC and indicates to any relying party that the base TRC is considered the trust anchor for the respective ISD at the time of signing. Whenever an ISD registers itself with a regional numbering authority, it simultaneously obtains an attestation for bootstrapping purposes. Similarly, whenever an ISD must engage in a trust reset, a new attestation must be obtained.

Attestations are used by relying parties to validate initial or trust-reset TRCs of remote ISDs. They are stored in the trust store alongside TRCs. When a relying party encounters an unknown base TRC during a TRC request, it fetches the attestation from the same node. Therefore, even infrastructure nodes operating in manual, TOMA or any other mode should be able to provide attestations for every base TRC in their store. In fact, they should include attestation verification into their decision process for added security.

Numbering authorities must maintain an append-only log of TRCs and attestations in their ISD range(s) for audit purposes. This also allows all entities to quickly discover new base TRCs, either caused by a new ISD joining the network or by a trust reset. They discover new TRCs by periodically fetching the newest TRCs in their ISD range.

A numbering authority can only issue an attestation for a trust reset if the initial TRC does not prevent it with the "no trust reset" flag. This prevents authorities from triggering a kill switch on ISDs who decided to enable this feature. However, this also renders an ISD number unusable if a large amount of keys are lost or compromised.

3.1.9 TRC Dissemination and Discovery

Relying parties must have recent TRCs available. They should discover TRC updates within a short time frame. In this section, we describe two mechanisms for discovering these updates. Regardless of the employed discovery method, the following requirement must be satisfied:

Requirement.   Any entity that sends information that is secured through the CP-PKI (be it during beaconing or path lookup) *must* be able to provide all the necessary trust material to verify said information.

As it is always possible to communicate with the sender of a packet (either via path reversal or one-hop paths), this requirement avoids circular dependencies between authentication and packet forwarding, see §6.1 for further details.

3.1.9.1 Beaconing Process

The TRC version is announced in the beaconing process. Each AS announces what it considers to be the latest TRC. Furthermore, each AS includes the hash value of the TRC contents to facilitate the discovery of discrepancies. Therefore, relying parties that are part of the beaconing process discover TRC updates passively. That is, the beacon service in a core AS notices TRC updates for remote ISDs that are on the beaconing path. The beacon service in a non-core AS only notices TRC updates for the local ISD through the beaconing process.

The creation of a new TRC triggers the generation of new PCBs, as the propagation of PCBs will help other ASes rapidly discover the new TRC.

3.1.9.2 Path Lookup

In every path segment, all ASes reference the latest TRC of their ISD. Therefore, when resolving paths, every relying party will notice TRC updates, even remote ones. This mechanism only works, however, when there is an active communication between the relying party and the ISD in question.

3.1.10 Substitutes to Certificate Revocation

The CP-PKI does not explicitly support certificate revocation. Instead, we rely on the following mechanisms to effectively invalidate compromised keys in case of security breach:

- **Short-lived certificates:** We recommend that both CA and AS certificates be created with a validity period on the order of a few days (see
  Table 3.1), but they can be arbitrarily short-lived. Short-lived certificates
  constitute an attractive alternative to a revocation system for the following reasons:

  – Both short-lived certificates and revocation lists must be signed by
    a CA. Instead of periodically signing a new revocation list, the
    CA can simply re-issue all the non-revoked certificates. Although
    the overhead of signing multiple certificates is greater than that
    of signing a single revocation list, the overall complexity of the
    system is greatly reduced. Short-lived certificates are a simple and
    effective solution when the number of certificates that each CA
    must issue is manageable. This condition holds in the CP-PKI:
    Each CA is responsible for the issuance of a small number of AS
    certificates.

  – Even with a revocation system, a compromised key cannot be in-
    stantaneously revoked. Through their validity period, both short-
    lived certificates and revocation lists implicitly define an attack
    window (i.e., a period during which an attacker who managed to
    compromise a key could use it before it becomes invalid). In both
    cases, the CA must consider a tradeoff between efficiency and se-
    curity when picking this validity period.

- **TRC updates:** The remaining types of certificates (voting and root certificates) can effectively be revoked through a TRC update. The grace
  period dictates for how long certificates defined in the previous TRC remain valid. A tradeoff must be considered here as well: While picking
  a short grace period reduces the attack window, it can disrupt legitimate
  connections (if non-compromised keys that were used to authenticate
  paths become invalid).

- **Trust reset:** As a last line of defense, a trust reset can be triggered to
  re-establish all trust roots in an ISD (and thus revoke all certificates).
  The exact process depends on the trust bootstrapping method used by
  the relying parties (see §3.1.8). In any event, a trust reset would require
  out-of-band communication and extensive coordination among administrators. This would be an exceptional event that numerous safeguards
  are meant to prevent.

3.2 DRKey: Dynamically Recreatable Keys

In addition to the authentication of control-plane messages, many use cases
also exist for authentication in the data plane, such as source authentication,

providing authenticity of network control messages, or to efficiently filter out unauthenticated packets. However, while authentication based on digital signatures works well for the relatively low message rates in the control plane, they do not meet the performance requirements for the high message rate of the data plane: The authentication of data-plane traffic and control messages requires a highly efficient and ideally stateless system to achieve bandwidths of several Gbps on commodity hardware, and to avoid creating opportunities for DoS attacks.

A tight processing budget only permits symmetric cryptographic operations, such as the computation of a MAC or pseudorandom function (PRF). However, these in turn rely on a key-distribution system to establish pairwise shared keys. Unfortunately, performing a Diffie–Hellman key exchange with each host would take much effort for routers and recipients of connectionless protocols like DNS. Additionally, storing these keys would require much state—both aspects could be exploited to mount DoS attacks. For example, Schuchard et al. [460] show how exhausting router state can be used to mount DoS attacks to paralyze today's Internet.

In this section, we present the *dynamically-recreatable-key* (DRKey) system, which enables high-speed data-plane components, like border routers, to derive symmetric cryptographic keys from local secrets only. More precisely, infrastructure components of a specific AS can use a secret key—known only to privileged and trusted components in that AS—to derive an individual symmetric key for any other AS or end host, using an efficient PRF, without keeping per-AS or per-host state. Entities in remote ASes and untrusted hosts in the local AS obtain their keys by posing an explicit key request. The control logic for requesting keys is offloaded to the AS's control plane—more precisely, to the certificate service (see §2.1)—and is secured based on the CP-PKI. The DRKey derivation is organized hierarchically, so that the derivation of keys for hosts in remote ASes can be safely delegated to their respective local certificate service.

DRKey was initially proposed in the paper "Lightweight Source Authentication and Path Validation" [290] and described in the first SCION book with an additional key-distribution system [413]. DRKey was later extended by PISKES [442] (pragmatic Internet-scale key-establishment system), which added efficient mechanisms to delegate specific keys to services in an AS, and fixed recently discovered vulnerabilities of DRKey's key-exchange mechanisms [268]. In this book, we describe the system with several minor modifications compared to the PISKES paper.

We start with use cases enabled by DRKey (§3.2.1), then give an overview of the system (§3.2.2), and then explain it in further detail (Sections 3.2.3–3.2.5).

3.2.1 Use Cases

Efficient on-the-fly key establishment is useful in a variety of SCION contexts. We briefly discuss several of them here; the details will be provided in dedicated sections throughout the book. For example, due to the efficient key establishment on intermediate routers, SCION provides the first control message protocol (SCMP) that supports the authentication of network control messages (see §4.7 on page 89 for further details). As discussed above, the naïve approach of adding digital signatures to control messages would create a processing bottleneck at routers and thus a DoS attack vector, as an attacker would be able to send specific packets that trigger SCMP error messages. Thus, efficient symmetric cryptographic keys are necessary and constitute an important building block for the efficient and authentic propagation of network control messages.

Another use case is name resolution, where DRKey can provide source authentication of requests to prevent reflection and amplification attacks. In addition, authenticating the reply with the same key ensures that it cannot be modified or spoofed by on-path attackers.

In Part III, we describe several systems (including EPIC and COLIBRI) that provide strong availability guarantees for communication. To achieve these guarantees, they rely on highly efficient source authentication at every on-path AS, which is provided by DRKey.

DRKey can also be used for source and packet authentication at end systems to enable per-AS rate limiting and firewall rules based on otherwise spoofable header fields like source address and port. In §9.2, we describe the *Lightning-Filter* system that integrates these modules for highly efficient traffic filtering to defend against (D)DoS attacks. This system can also be used as a "first line of defense" of control-plane services, which might otherwise be susceptible to DoS attacks like signature flooding.

Finally, while DRKey should not be used directly for end-to-end encryption (see §3.2.5), they can be leveraged to authenticate an explicit key exchange to subsequently encrypt and authenticate end-to-end traffic (e.g., via TLS). This approach provides increased security for opportunistic encryption because an attacker can only mount a MITM attack if it controls either the source or destination AS—as opposed to anonymous Diffie–Hellman, where every on-path attacker can mount such an attack. Higher levels of security can be achieved by using more expensive asymmetric cryptography to ensure accountability of the ASes. This approach based on asymmetric cryptography, which we call Pervasive Internet-Wide Low-Latency Authentication (PILA), is described in Chapter 20.

Table 3.2: Notation used to describe DRKey.

$\parallel$	bitstring concatenation
A, B	ASes identified by ISD and AS number (IA)
H_A, H_B	end hosts identified by host address
CS_A, CS_B	certificate services located in a specific AS
SV_A^p	AS A's local secret value for protocol p
$SV_A^\star$	AS A's local secret value for the generic hierarchy
$K_{A \to B}^p$	first-level key between ASes A and B for protocol p
$K_{A \to B:H_B}^p$	second-level "to" key between AS A and end host H_B in AS B for protocol p
$K_{A:H_A \to B}^p$	second-level "from" key between end host H_A in AS A and AS B for protocol p
$K_{A:H_A \to B:H_B}^p$	third-level key between end host H_A in AS A and end host H_B in AS B for protocol p
$K_\bullet^{\star,p}$	symmetric key for non-standard protocol p derived (indirectly) from $SV^\star$
$h(\bullet)$	non-cryptographic hash operation
$MAC_K(\bullet)$	message authentication code using key K
$PRF_K(\bullet)$	pseudorandom function using key K

3.2.2 Overview

In this section, we will provide a simplified[2] overview of the system. We introduce notation on-the-fly, but a summary is also provided in Table 3.2.

While used for symmetric cryptography, the derivation and distribution of DRKey is inherently *asymmetric*: There is one entity (on the *fast* side) that can efficiently derive the key from local secrets, and another entity (on the *slow* side) that needs to fetch the key with an explicit request through the control plane. The decision of which side is *fast* and which is *slow* depends on the application: For example, if a border router must generate an error message because it was unable to forward a data packet, it is the fast side and can authenticate an SCMP message using a key efficiently derived from a local secret; the end host receiving this SCMP message is the slow side and needs to query the local certificate service to obtain the key required to verify the authenticity of the message (if the key is not already cached). SCMP will be discussed in further detail in §4.7. In client–server interactions, the client typically represents the slow side and the server, which needs to be able to process requests quickly, is the fast side.

Figure 3.2 shows a typical use case of DRKey, where a host H_A in AS A desires to communicate with a server S_B in AS B, and S_B wants to efficiently authenticate the network address of H_A. We denote the certificate services in ASes A and B CS_A and CS_B, respectively.

---

[2]Protocol-specific keys and additional hierarchy levels will be described later.

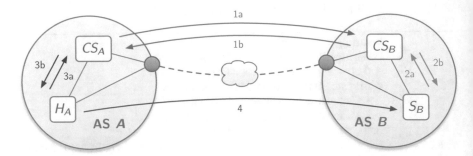

Figure 3.2: Basic topology and key-establishment procedure for communication between an end host H_A in AS A and a server S_B in AS B.
ASes A and B have deployed the certificate services CS_A and CS_B,
respectively.

Each AS randomly selects a local secret value, SV_A and SV_B, which is only
shared with trustworthy entities (in particular, the certificate service) in the
same AS; it is never shared outside the AS. Since end hosts are typically not
trustworthy, they would not obtain the local secret value; however, in some
cases a trusted end host could obtain it, as we are going to assume in this
overview for simplicity—later we describe how to delegate symmetric values
to less trusted services. The secret value will serve as the root of a symmetric-
key hierarchy, as illustrated in Figure 3.3, where keys of a level are derived
from a key of the preceding level.[3] In DRKey, the keys are derived using an
efficient PRF [284].

The key derivation used by CS_B in our example is $K_{B \to A} = \mathrm{PRF}_{SV_B}(A)$.
Thanks to the key-secrecy property of a secure PRF, $K_{B \to A}$ can be shared with
another entity without disclosing or leaking information about SV_B. The arrow
notation emphasizes the asymmetry described above: $K_{B \to A}$ is derived based
on B's secret value, meaning that B is the fast side and can efficiently derive
the key on the fly. A is the slow side, so A's certificate service needs to fetch
the key from B and store it locally.

To continue with our example depicted in Figure 3.2, CS_A will pre-fetch
keys $K_{* \to A}$ from certificate services in other ASes, including $K_{B \to A}$ from CS_B.
In Figure 3.2, this action is illustrated with the arrows labeled 1a and 1b. Also,
we assume that server S_B is trusted within AS B and can thus obtain the secret
value SV_B, indicated with the arrows labeled 2a and 2b. When H_A desires to
send authenticated packets to S_B, it contacts its local certificate service CS_A
and requests a host-to-host key $K_{B:S_B \to A:H_A}$, which CS_A can locally derive from
$K_{B \to A}$ (arrows 3a and 3b). H_A can now directly use this symmetric key for
authenticating to S_B (arrow 4).

---

[3]We emphasize that this key hierarchy is different from a PKI key hierarchy: In DRKey, higher-
level keys are *derived* from a lower-level key, whereas in a PKI, private keys sign public-key
certificates of the next entity in the certificate chain.

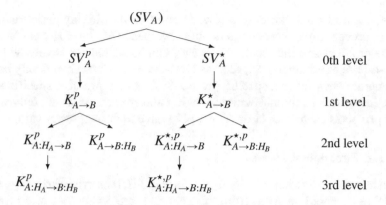

Figure 3.3: DRKey multi-level key hierarchy derived by AS A. The notation $K_1 \rightarrow K_2$ indicates that K_2 is derived from K_1 through a PRF.

The important property of DRKey is that S_B can rapidly derive $K_{B:S_B \rightarrow A:H_A}$, by using SV_B and performing two PRF operations. This illustrates an important design aspect of DRKey: The one-wayness of the key-derivation function allows a certificate service to delegate key derivation to specific entities. In contrast to other systems that rely on a dedicated server for key generation and distribution (such as Kerberos [387]), this delegation mechanism allows "fast" entities to directly derive a symmetric key without communication to the certificate service.

The design of DRKey is based on the following additional systems:

- The authentication of key-exchange messages relies on the CP-PKI.

- We assume that there exists an authentication mechanism for end hosts within an AS, which is needed for access control at the certificate service.

3.2.3 Key Hierarchy

The DRKey key-establishment framework uses a key hierarchy consisting of four levels, illustrated in Figure 3.3.

3.2.3.1 Zeroth Level (AS-Internal)

On the zeroth level of the hierarchy, each AS A randomly generates local *AS-specific secret values* SV_A^p for a set of standardized protocols p. These can optionally be generated from a single root secret value SV_A. There is a special secret value $SV_A^\star$, which is used for deriving DRKeys for non-standard protocols in a separate hierarchy. The secret values represent the per-AS basis of the key hierarchy and are renewed frequently (e.g., daily). The different secret values can have different validity periods depending on the requirements of the respective protocols; however, a lower bound on the validity period is necessary to prevent excessive communication overhead for global key exchanges.

The protocol-specific secret values SV_A^p can be shared selectively with individual services within the AS, such as DNS servers, which should only be able to generate keys for that specific protocol. As every protocol-specific secret value introduces additional communication and storage overhead, only widely used protocols such as DNS or NTP would have their own secret values.

3.2.3.2 First Level (AS-to-AS)

By using key derivation, an AS A can derive different symmetric keys using a PRF (e.g., based on AES) from the generic secret value $SV_A^\star$ or a protocol-specific secret value SV_A^p. These derived keys, which are shared between AS A and a second AS B, form the first level of the key hierarchy and are called *first-level* keys:

$$K_{A \to B}^x = \mathrm{PRF}_{SV_A^x}(B). \tag{3.1}$$

The input to the PRF is the combination of ISD and AS number of AS B.[4]

The general and protocol-specific first-level keys are periodically exchanged between certificate services of different ASes, see §3.2.4.1.

3.2.3.3 Second Level (Host-to-AS, AS-to-Host)

Using the symmetric keys of the first level of the hierarchy, *second-level keys* are derived between an AS on the fast side and a host on the slow side or vice versa. In the case of protocol-specific keys, the PRF can be applied directly to the host's address:

$$K_{A \to B:H_B}^p = \mathrm{PRF}_{K_{A \to B}^x}(0 \parallel H_B), \tag{3.2a}$$

$$K_{A:H_A \to B}^p = \mathrm{PRF}_{K_{A \to B}^x}(1 \parallel H_A). \tag{3.2b}$$

For generic first-level keys, an ASCII encoding of the protocol needs to be included as well:

$$K_{A \to B:H_B}^{\star,p} = \mathrm{PRF}_{K_{A \to B}^x}(0 \parallel H_B \parallel \text{``}p\text{''}), \tag{3.3a}$$

$$K_{A:H_A \to B}^{\star,p} = \mathrm{PRF}_{K_{A \to B}^x}(1 \parallel H_A \parallel \text{``}p\text{''}). \tag{3.3b}$$

Second-level "host on the slow side" keys $K_{A \to B:H_B}^x$ are used when the fast side is an infrastructure component in AS A and the slow side is a host H_B— e.g., for SCMP. Second-level "host on the fast side" keys $K_{A:H_A \to B}^x$ are not used directly for packet authentication; instead, they are provided to the host H_A such that it can locally derive third-level keys (see below) for authenticating packets it receives from other hosts.

---

[4]Using the ISD-AS tuple makes the derivation resilient against non-unique AS numbers.

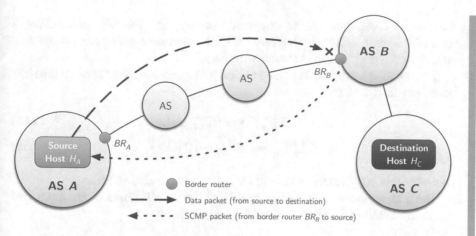

Figure 3.4: Border router sending SCMP error message to end host.

3.2.3.4 Third Level (Host-to-Host)

Host-to-host (third-level) keys are derived from second-level "host on the fast side" keys by applying the PRF to the slow host's address:

$$K^x_{A:H_A \to B:H_B} = \mathrm{PRF}_{K^x_{A:H_A \to B}}(H_B). \tag{3.4}$$

These keys are used to authenticate packets sent by host H_B on the slow side to host H_A on the fast side.

The purpose of the intermediate second-level "host on the fast side" keys is to make authentication of packets more efficient for the receiver: Instead of having to fetch the keys for each individual source *host*, the receiver only needs to fetch and store keys for each source *AS*. If H_A has already cached the second-level key for the corresponding protocol and source AS, it can locally derive this key and authenticate packets with no additional communication. This is particularly interesting for setting up a LightningFilter (see §9.2) without access to the AS-internal secret value.

3.2.3.5 Example of Key Derivation

Consider the situation depicted in Figure 3.4 where a host H_A in AS A sends a data packet to a host in AS C, but forwarding fails (e.g., due to an expired hop field) at the ingress interface of AS B's border router BR_B.

The router generates an SCMP message indicating the error. When the packet is created, BR_B derives a key for authenticating packets destined to the end host (source), based on the key shared with AS A, and calculates a MAC over the packet.

As the router needs to be able to process packets at very high rates, it uses a DRKey where it is on the fast side. As an infrastructure component it can use the second-level "host on the slow side" key for this. Furthermore, SCMP is one of the main protocols that rely on DRKey with its own protocol-specific secret value SV_B^{SCMP} that is shared with BR_B.

Thus, BR_B uses this secret value to derive the second-level "host on the slow side" key in two steps,

$$K_{B \to A}^{SCMP} = \text{PRF}_{SV_B^{SCMP}}(A), \tag{3.5a}$$

$$K_{B \to A:H_A}^{SCMP} = \text{PRF}_{K_{B \to A}^{SCMP}}(0 \parallel H_A), \tag{3.5b}$$

and uses this key to authenticate the SCMP message sent to H_A.

The authentication mechanism itself and the SCMP protocol are described in further detail in Sections 3.3 and 4.7, respectively.

3.2.4 Key Establishment

3.2.4.1 First-Level Key Establishment

Key exchange is offloaded to the certificate service deployed in each AS. The certificate service is not only responsible for first-level key establishment, it also derives second- and third-level keys and provides them to hosts within the same AS.

To exchange a first-level key, the certificate services of corresponding ASes perform the key-exchange protocol, which is initiated by the certificate service on the slow side. The protocol uses TLS to provide a secure channel, where the two participants are authenticated based on the CP-PKI.[5]

The initiating certificate server sends a request specifying a point in time at which the requested key is valid, *valTime*, and one or multiple protocols. The requested keys may not be valid at the time of request, either because they already expired or because they will become valid in the future. For example, pre-fetching future keys allows for seamless transition to the new keys. On the other hand, fetching a key that was valid in the past may be necessary to authenticate already received packets.

The responder—i.e., the certificate service of the AS on the fast side—replies with the requested keys and their respective validity periods (which include the specified *valTime*). If *valTime* is too far in the past or future such that the responder does not have any valid secret values for that time, an error message is sent instead. Once the requesting certificate service has received the keys, it can now respond to queries by entities within its AS requesting second- or third-level keys (see below).

Pre-fetching First-Level Keys. To be able to derive second- and third-level keys without additional delay, an AS can keep a history of frequently

---

[5]In the following description, we ignore the required TLS handshake messages and simply rely on the fact that TLS provides a secret and authentic channel.

contacted ASes and fetch first-level keys for those ASes proactively. For ASes that are not contacted regularly, on-demand key exchange is also possible. For example, if a certificate service is missing a first-level key that is required for the derivation of a second-level key, the certificate service initiates a key exchange.

Today's Internet consists of around 75,000 active ASes [59]; in a SCION Internet, this is likely going to increase. However, even ASes that contain a large number of end hosts are unlikely to communicate with every other AS. Assuming a validity period of one day for first-level keys, performing one key exchange per second is sufficient to exchange first-level keys with 86,400 other ASes. Furthermore, these requests can be easily distributed using multiple certificate servers, which ensures that the system remains scalable.

To avoid explicit revocation, the shared keys are *short-lived* and new keys are established frequently (e.g., daily). Subsequent key exchanges to establish a new first-level key can use the current key as a first line of defense to avoid signature-flooding attacks in the TLS handshake.

3.2.4.2 Second/Third-Level Key Establishment

End hosts request a second- or third-level key from their local certificate service specifying the precise type of key, the remote AS and host, and a *valTime*. The certificate service only replies with a key if the querying host is authorized to use the key. An authorized host must either be an endpoint of the communication that is authenticated using the second- or third-level key or authorized separately by the AS. Both the request and the reply must be sent over a secure channel—e.g., over TLS, similar to the first-level key exchange.

3.2.5 Limitations of DRKey

It is important to note that all second- and third-level keys are known to both the source and destination ASes in addition to the end hosts. These ASes can thus trivially encrypt and authenticate messages with these keys, impersonate end hosts, and modify messages in transit. Therefore, the use of DRKey must be limited to network-layer security protocols and applications where these limitations are acceptable; for example, for authenticating DNS requests to prevent reflection attacks, it is sufficient to be able to authenticate the source AS. An end host can only authenticate another end host under the assumption that both source and destination ASes are honest.

3.3 SCION Packet Authenticator Option

The SCION Packet Authenticator Option (SPAO) is an end-to-end option that can be added to the end-to-end options header, see §5.8. Its purpose and functionality are similar to the IP Authentication Header (AH) [287]. However, in contrast to the AH, the SPAO does not require explicit key exchange and key management as it can leverage the DRKey infrastructure.

3.3.1 Option Format

The option contains the following fields:

- *algorithm* defines the cryptographic algorithms to be used and what parts of the packet header and payload are protected.

- *keyID* specifies the key used for computing the *authenticator*; for DRKey, this includes the protocol identifier, type of key, and directionality (i.e., is the source or receiver the fast side). The AS numbers and host addresses required to uniquely identify the key are part of the address header (see §5.2).

- *timestamp* identifies the time the packet was sent. The timestamp serves two purposes: it (i) allows dropping old packets, and (ii) facilitates uniquely identifying packets to identify duplicates (see §3.3.3).

- *authenticator* contains a cryptographic authenticator of the packet, which may consist of a MAC, a hash, a signature, or a combination thereof, depending on the *algorithm*.

The *input* to the authentication algorithm depends on the keys used for computing the tag. In general, it includes all immutable header fields; when used with DRKey, the addresses are not included in the input as they are already used to derive the DRKey.[6] Optionally, the input to the authentication algorithm can additionally include mutable but predictable header fields and/or the packet payload.

Source authentication and packet integrity can also be separated and parallelized by calculating a cryptographic hash of the payload and storing it in the *authenticator* field together with a MAC computed over the packet header and the payload hash. This has the advantage that performing source authentication is highly efficient and the payload hash only needs to be computed for packets that have a valid MAC. This separation is used, for example, for the LightningFilter system, see §9.2.

3.3.2 Network Address Translation

As we discuss in §13.1.6, network address translation (NAT) is not necessary in a SCION Internet. However, if NAT *is* performed between the source host and the border router in its AS, the destination may see a different source address than the source host. To enable successful source-address verification in this

---

[6]When using a second-level DRKey, only one address is included in the key derivation; the other address must be included in the input to the authentication algorithm.

case, the local certificate service always provides the second- or third-level key based on the host's *public* end-host address to the host.[7]

In addition, it is important that the input to the MAC calculation does *not* include the source of the packet as this might be changed along the way. With this mechanism, the destination host uses the translated source end-host address to derive the symmetric key, which coincides with the key used by the source to create the authentication tag.

3.3.3 Preventing Replay Attacks

In order to rule out replay attacks, the destination can deploy a duplicate-detection system: It can use the *timestamp* to directly filter out outdated packets and use Bloom filters to keep track of recently received packets. Such a system is explained in further detail in §9.2.

---

[7]Note that carrier-grade NAT is neither necessary nor supported in SCION.

4 Control Plane

Laurent Chuat, Samuel Hitz, Markus Legner, Adrian Perrig, Seyedali Tabaeiaghdaei*

Chapter Contents

4.1   **Path-Segment Construction Beacons (PCBs)** **66**
   4.1.1   Components of a PCB   66
   4.1.2   Beacon Extensions   68
4.2   **Path Exploration (Beaconing)** **69**
   4.2.1   Initiating Beaconing   70
   4.2.2   Propagating PCBs   70
4.3   **Path-Segment Registration** **71**
   4.3.1   Intra-ISD Path-Segment Registration   71
   4.3.2   Core Path-Segment Registration   72
4.4   **PCB and Path-Segment Selection** **73**
   4.4.1   Beacon Store   73
   4.4.2   Selection Properties   74
   4.4.3   Selection Policy   75
   4.4.4   Filtering PCBs   76
   4.4.5   Selecting PCBs and Path Segments   76
   4.4.6   Diversity-Based Path Selection   76
4.5   **Path Lookup** **80**
   4.5.1   Requirements and Design Goals   80
   4.5.2   Lookup Process   81
   4.5.3   Alternative Lookup Strategies   86
4.6   **Service Discovery** **87**
   4.6.1   Service-Level Anycast   87
   4.6.2   Discovery Service   88
4.7   **SCION Control Message Protocol (SCMP)** **89**
   4.7.1   Goals and Design   89
   4.7.2   SCMP Message Format   89
   4.7.3   SCMP Authentication   90

---

*This chapter reuses content from Hitz, Perrig, Shirley, and Szalachowski [237], Krähenbühl, Tabaeiaghdaei, Gloor, Kwon, Perrig, Hausheer, and Roos [305].

© The Author(s), under exclusive license to Springer Nature Switzerland AG 2022   65
L. Chuat et al., *The Complete Guide to SCION*, Information Security
and Cryptography, https://doi.org/10.1007/978-3-031-05288-0_4

This chapter describes SCION's control plane, whose main purpose is to create and manage path segments, which can then be combined into forwarding paths to transmit packets in the data plane. We first discuss how path exploration is realized through beaconing and how path segments are registered. We then discuss path lookup and present the SCION Control Message Protocol (SCMP).

4.1 Path-Segment Construction Beacons (PCBs)

SCION uses path-segment construction beacons (PCBs) for path exploration and registration. PCBs are used for intra-ISD and inter-ISD path exploration, and contain topology and authentication information. They can include additional metadata that helps with path management and selection. Broadly speaking, a PCB represents a single path segment that can be used to construct end-to-end forwarding paths. Formally, a PCB is defined as follows:

$$PCB = \langle\, INF \parallel ASE_0 \parallel ASE_1 \parallel \dots \parallel ASE_n \,\rangle, \tag{4.1}$$

where INF is an info field and ASE_i is an AS entry that contains all information about a particular AS on the path segment represented by the PCB.

4.1.1 Components of a PCB

Each PCB is thus composed of one info field and several AS entries. Info fields and AS entries themselves are made of several components, which we describe in this section.

4.1.1.1 Info Field (INF)

The first component of every PCB is the info field (INF), which provides basic information about the PCB. Specifically, the info field contains the following elements:

$$INF = \langle\, Flags_{INF} \parallel SegID \parallel TS \,\rangle, \tag{4.2}$$

where $Flags_{INF}$ contains flags, most of which are only required in the data plane to describe the type and the direction of the constructed end-to-end path, $SegID$ is set to a random value and is required for the MAC-chaining mechanism, and TS is a timestamp that indicates when the PCB's propagation started.

4.1.1.2 AS Entry (ASE)

The complete information about an AS in a PCB is called an AS entry and consists of a signed component $ASE^{(\text{signed})}$, a signature Σ, and an unsigned component $ASE^{(\text{unsigned})}$:

$$ASE = \langle\, ASE^{(\text{signed})} \parallel \Sigma \parallel ASE^{(\text{unsigned})} \,\rangle. \tag{4.3}$$

The signed component has the following format:

$$ASE^{\text{(signed)}} = \langle\, Local \parallel Next \parallel HE \parallel PE_0 \parallel \ldots \parallel PE_m \parallel MTU \parallel Ext \,\rangle, \quad (4.4)$$

where *Local* is the ISD-AS number of the AS corresponding to this entry, and *Next* is the ISD-AS number of the AS to which the PCB is forwarded. Each AS entry also contains a single hop entry *HE*, a list of optional peer entries $PE_0 \ldots PE_m$, the size of the maximum transmission unit (MTU) within the AS's network, and optional beacon extensions *Ext*. The unsigned component is optional and consists of a variable number of unprotected extensions *ExtUnsigned*. Both signed and unsigned beacon extensions are discussed in §4.1.2.

Each AS entry is signed with a private key K_i that corresponds to the public key certified by the AS's certificate. The corresponding signature Σ includes the PCB's info field *INF*, the signed component of the current AS entry ASE_i, and the signed component and signatures of all previous AS entries in the PCB. The signature Σ_i of an AS entry ASE_i in a PCB is computed as follows:

$$\Sigma_i =$$
$$\text{Sign}_{K_i}\!\left(INF \parallel ASE_0^{\text{(signed)}} \parallel \Sigma_0 \parallel \ldots \parallel ASE_{i-1}^{\text{(signed)}} \parallel \Sigma_{i-1} \parallel ASE_i^{\text{(signed)}} \right). \quad (4.5)$$

The signature metadata minimally contains the ISD-AS number of the signing entity and the key identifier of the public key that should be used to verify the message. For more information on signing and verifying control-plane messages, see §3.1.7.

4.1.1.3 Hop Entry (HE)

A hop entry has the following format:

$$HE = \langle\, HF \parallel IngressMTU \,\rangle, \quad (4.6)$$

where *IngressMTU* specifies the MTU of the ingress interface and HF_H is a hop field that contains the authenticated information of the ingress and egress interfaces to be used in the data plane.[1] Hop entries only contain the ingress MTU because it is sufficient to specify one MTU per link.

To allow end hosts to explicitly select paths to reach other end hosts, hop fields are propagated with the corresponding topology information to the end hosts (see below).

---

[1] Here, ingress and egress refer to the beaconing direction; when describing the data plane in Chapter 5, we will make this more explicit to avoid confusion with the forwarding direction.

4.1.1.4 Peer Entry (PE)

Through a peer entry, an AS can announce that it has a peering link to another AS. Peer entries have a format similar to that of hop entries, but with additional fields:

$$PE = \langle\, HF \parallel Peer \parallel PeerInterface \parallel PeerMTU \,\rangle, \qquad (4.7)$$

where *HF* contains the information necessary to create a data-plane hop, *Peer* is the ISD-AS number of the peering AS, *PeerInterface* is the interface identifier of the peering link on the remote AS side, and *PeerMTU* specifies the MTU on the peering link.

4.1.1.5 Hop Field (HF)

Finally, the hop field (HF), which is contained in hop entries and peer entries, is used directly in the data plane for packet forwarding: It specifies the incoming and outgoing interfaces of the ASes on the forwarding path. To prevent forgery, this information is authenticated with a message authentication code (MAC). A hop field has the following format:

$$HF = \langle\, Flags_{HF} \parallel ExpTime \parallel ConsIngress \parallel ConsEgress \parallel HFAuth \,\rangle, \quad (4.8)$$

where $Flags_{HF}$ may be used to indicate processing options for the AS; *ExpTime* specifies for how long the hop field is valid;[2] *ConsIngress* and *ConsEgress* identify, respectively, the ingress and egress interfaces (in the direction of construction, i.e., in the direction of beaconing); and *HFAuth* is a MAC.

We show how hop fields are authenticated (i.e., how the MAC is computed) in §5.3.3. While the format of a SCION hop field is standardized, each AS can independently choose the algorithm used to calculate the MAC.

4.1.2 Beacon Extensions

In addition to basic routing information like hop entries and peer entries, PCBs can be used to communicate additional metadata.

4.1.2.1 Signed and Unsigned Extensions

There are two fields in PCBs that can contain optional metadata: The field *Ext* contains extensions that are included in the signed component of an AS entry and thus protected by the AS's signature; the field *ExtUnsigned* contains extensions that are not protected by the signature.

---

[2]The expiration time of a hop field is an offset relative to the PCB's absolute info field timestamp *TS*. The combination of the two thus allows an AS to compute the absolute expiration time of the hop field. Data-plane packets using the hop field after the expiration time can be dropped.

4.1.2.2 Detachable Extensions

Metadata that is important enough to be added to PCBs is typically important enough to be protected cryptographically. Unfortunately, signed extensions in the *Ext* field have a disadvantage: As they are an integral part of the input to every AS's signature, they need to be processed by every AS and end host that creates or verifies a signature. This is a problem if the size of these extensions is large as they must be transmitted and stored with the PCB at all times, creating a processing overhead. In addition, it is not possible to selectively communicate the data contained in these extensions to some entities but not others—we will see an example of this in the context of hidden paths and EPIC, see Sections 8.1 and 10.1.

This is why SCION also supports detachable extensions for PCBs. These detachable extensions add their data *DetachableExtData* to the unsigned extensions, *ExtUnsigned*, and additionally add a cryptographic hash of the extension data to a *DigestExt*, which is part of the signed extensions, *Ext*. A PCB with a detachable extension can be signed and verified without the *DetachableExtData*, only based on the *DigestExt*. Additionally, the *DetachableExtData* can be verified by checking the cryptographic hash which is contained in the PCB's signatures through the *DigestExt*.

Thus, detachable extensions combine the benefits of both signed and unsigned extensions: The extension data is protected through the signature of AS entries but the PCB remains verifiable even if the extension data is removed.

4.2 Path Exploration (Beaconing)

In this section, we describe how path exploration is realized, i.e., how beaconing is initiated and how PCBs are propagated by the beacon service.

Once per propagation period, each AS initiates the PCB generation process. PCBs are then immediately propagated. When a PCB is received by an AS, its beacon service registers the contained path segment at the path service, extends the PCB, and sends the PCB to the next AS (if there is one). The propagation period is a parameter specified by each AS.

There are two types of beaconing: intra-ISD beaconing (for creating up-segments and down-segments) and core beaconing (for creating core-segments). The main difference between the two is that the goal of core beaconing is to create paths that connect every pair of core ASes; therefore, core PCBs are flooded to all other core ASes through core links (which do not have a defined directionality). In the intra-ISD case, PCBs are only disseminated along parent–child links (forming a more limited distribution tree).

4.2.1 Initiating Beaconing

Each core AS, through its beacon service, periodically initiates the path exploration process by creating an initial PCB and propagating it. The PCB is either
sent to a child AS (in the case of intra-ISD beaconing) or to other core ASes
(in the case of core beaconing). The beacon service inserts (among other information) the initial AS entry ASE_0 in the PCB. In the intra-ISD case, the initial
PCB can optionally contain peer entries to non-core ASes. The hop entry HE
inside ASE_0 includes an initial hop field with the ingress interface identifier set
to • (which indicates an empty value):

$$HF_0 = \langle\, Flags_{HF} \parallel ExpTime \parallel \bullet \parallel ConsEgress \parallel HFAuth \,\rangle. \qquad (4.9)$$

The initial hop field denotes the extremity of a path segment and authenticates a forwarding decision for every packet that

- enters the AS through the interface *ConsEgress* and terminates in the
  AS;

- originates from the AS and exits through the interface *ConsEgress*; or

- switches to another path segment at this AS (using one of the possible
  path-segment combinations, as described in §5.5).

The beacon service then signs the PCB and sends it to a border router (which
corresponds to the *ConsEgress* identifier as specified in the hop field).

PCBs are disseminated within packets addressed to the beacon service using
the corresponding service address (see §4.6). Furthermore, the special one-
hop path is used to initiate the communication to a neighboring beacon service
(see §5.4.1). This is necessary because there may not be a full forwarding
path available for beaconing. Indeed, the discovery of such paths in turn relies
on beaconing. The purpose of one-hop paths is thus to break this circular
dependency.

During core beaconing, the neighboring AS that receives the PCB can be
in the same or in a different ISD. The ISD identifier included in the PCB's
signature metadata describes only the ISD of the PCB's originator.

4.2.2 Propagating PCBs

After beaconing is initiated, each PCB is propagated in the following way:
The ingress border router of the next AS in the beaconing path receives the
PCB, detects that the destination is a SCION service address, and sends it
to the AS's beacon service. The beacon service verifies the structure and all
signatures on the PCB. The PCB contains the version numbers of the TRC(s)[3]
and certificate(s) that must be used to verify the signatures. This enables the

---

[3]Even within a single ISD, there can be multiple valid TRCs at the same time, see §3.1.6.

beacon service to check whether it has the relevant TRC(s) and certificate(s); if not, they can be requested from the beacon service of the sending AS, and then forwarded to the local certificate service. If the PCB verification is successful, the beacon service adds the PCB to its local database.

Every propagation period (as configured by the AS), the beacon service selects the best combinations of PCBs and interfaces connecting to a next AS (i.e., a child AS or a core AS, depending on the type of beaconing) and continues path exploration by sending each selected PCB to the selected egress interface(s) associated with it. PCB and egress interface selection criteria are set according to local AS policies. The selection process is presented in §4.4. In the core beaconing, several core PCBs are selected for each originating core AS, to ensure connectivity. In some cases, for example when a PCB is received from a previously unknown AS, the PCB can be immediately extended and forwarded, to enable rapid connectivity establishment.

For every selected PCB and egress interface combination, the AS creates a new PCB by adding an AS entry to the selected PCB. The AS entry includes hop fields that authenticate the permission to send traffic between ingress and egress interfaces. It can also contain peer entries. The peer entry of a non-core AS represents a peering link to either a core or non-core AS. The new PCBs are then sent to the next AS's beacon service (through to the egress interface specified in the hop field) using a one-hop path and the appropriate service address.

As core beaconing is based on flooding without a defined direction, it is necessary to avoid loops during path creation. A core beacon service avoids loops at both the AS and ISD levels as follows:

- It discards PCBs that include an AS entry created by itself,

- It can be configured to discard PCBs that re-enter an already visited ISD.[4]

Finally, the beacon service adds the PCB to its local database.

4.3 Path-Segment Registration

After beaconing, path segments can be created from PCBs. These path segments must then be registered to make them available to other entities. We distinguish between two cases: intra-ISD and core path-segment registration.

4.3.1 Intra-ISD Path-Segment Registration

Beaconing provides regular ASes with paths to communicate with core ASes in their ISD. To make paths available to their own and remote end hosts, path

---

[4]There are legitimate reasons for crossing the same ISD multiple times: For example, if the ISD spans a large geographical area, a path transiting another ISD may constitute a shortcut.

segments must be registered. Every *registration period* (determined by each AS), the beacon service selects two sets of path segments:
- up-segments, which allow local infrastructure entities and end hosts to communicate with core ASes; and
- down-segments, which allow remote entities to reach the local AS.

An AS can set different selection policies for these two sets (see §4.4). More specifically, in every registration period, the beacon service performs the following operations:

1. From the cached PCBs, select PCBs that will be used as up-segments and PCBs that will be used as down-segments.

2. Optionally, the AS can decide to discard unsigned extensions and detachable extension data.

3. To every selected PCB, add a new AS entry where the *Next* field is empty and with a final hop field in which only the ingress interface is specified (since the path ends at the AS):

$$HF = \langle\, Flags_{HF} \parallel ExpTime \parallel ConsIngress \parallel \bullet \parallel HFAuth \,\rangle. \quad (4.10)$$

4. If the AS has peering links, for each peering link add to the AS entry a hop field with an empty egress interface (as in Equation (4.10)).

5. Sign every selected beacon and append the computed signature. Such modified PCBs are then called *path segments*.

6. Register the resulting up-segments with the local AS's path service, and the down-segments with the core path service in the AS that originated the PCB.

Path-segment registrations are sent as packets addressed to the path service.

4.3.2 Core Path-Segment Registration

The core beaconing process creates path segments from core AS to core AS. These path segments must be registered at the path service of local ASes so that local and remote end hosts can obtain and use them. In contrast to the intra-ISD registration procedure, there is no need to register core-segments with other ASes (as each core AS will receive PCBs originated by every other core AS).

In every registration period, the core beacon service performs the following operations:

1. Select the best PCBs towards each core AS observed so far.

2. To every selected PCB, add a new AS entry where the *Next* field is empty and with a hop field in which the egress interface is empty, as in Equation (4.10).

3. Sign the modified PCBs, which are then called *core-segments*.

4. Register the core-segments with the local AS's path service.

4.4 PCB and Path-Segment Selection

As an AS receives a series of intra-ISD or core PCBs, it must select the PCBs it will use to continue beaconing and to register path segments at path services. A non-core AS must select (1) a subset of PCBs to propagate downstream, (2) up-segments to register at the local AS's path service, and (3) down-segments to register at a core path service. A core AS must select (1) a subset of PCBs to propagate to neighbor core ASes, and (2) core-segments to register at the local AS's path service. Core ASes do not register core-segments at remote path services because all core ASes find a set of paths to all other core ASes.

The selection process can be based on path properties (e.g., length, disjointness across different paths) as well as PCB properties (e.g., age, remaining lifetime of sent instances). In this section, we describe the process by which an AS evaluates and selects PCBs. The beacon service of an AS maintains a data structure of received PCBs under consideration for downstream propagation and registration at path services. Each AS can specify how PCBs are evaluated or eliminated from consideration through a local policy.

4.4.1 Beacon Store

Each time the beacon service receives a PCB, it chooses whether or not the PCB will be stored as a candidate (i.e., under consideration for propagation and registration). To manage the set of candidate PCBs, the beacon service maintains a database of PCBs called the beacon store. The beacon store has a fixed capacity and supports the following operations:

- add: Add a new PCB to the beacon store after checking if it complies with the selection policy. If the beacon store is full, remove the least desirable PCB.

- remove: Remove a PCB from the beacon store.

- select: Select a number (specified as parameter) of PCBs.

Through the above operations, the beacon store is thus implicitly responsible for applying the AS's selection policy (described below). In addition to storing PCBs, the beacon store also stores metadata for each PCB, for example the expiration time of the most recently sent PCBs per egress interface.

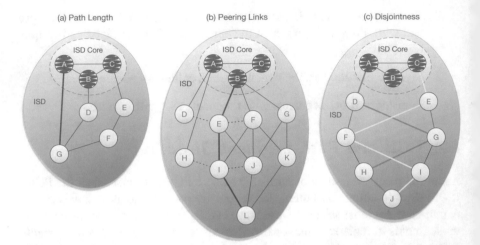

Figure 4.1: Example networks to illustrate beacon and path-segment selection
based on different path properties.

4.4.2 Selection Properties

As mentioned before, a selection of PCB(s) occurs at the beacon service when
adding PCBs received from a neighbor AS, when registering PCBs at the path
server, and when selecting which PCBs to forward on which egress interface.
At each of these selection steps, different policies with different selection properties can be used. Below is a non-exhaustive set of metrics that represent a
range of desirable properties in a path or received or propagated PCB:

- **Path length:** The first property we consider is path length. In this case,
  path length is defined as the number of hops from the originator AS to
  the local AS. This can give an indication of the path's latency (although
  there are many other factors affecting latency).

- **Peering ASes:** For intra-ISD beaconing, we also consider peering ASes,
  defined as the number of different peering ASes from all non-core ASes
  on the PCB. This metric is useful because a greater number of peering
  ASes on a PCB increases the likelihood of finding a shortcut using that
  path segment.

- **Disjointness:** Unlike other properties, the disjointness of candidate
  PCBs (illustrated in Figure 4.1c) is calculated relative to other PCBs
  and thus depends on PCBs that have been previously sent. We use the
  two following definitions of disjointness: Paths can be *vertex-disjoint*
  (i.e., they have no common upstream/core AS for the AS the beacon
  store is in) or *edge-disjoint* (i.e., they do not share any AS-to-AS link).
  Depending on the objective of the AS, both definitions can be used:

Vertex-disjointness allows path diversity in the event that an AS becomes unresponsive, and edge-disjointness provides resilience in case of link failure.

- **Last reception:** The last reception of a PCB is defined as the time that has elapsed since the PCB arrived at the AS's beacon store. This metric is important because a short elapsed time indicates that upstream ASes found the PCB desirable and fewer errors (e.g., a failing link) can have affected the segment since it was propagated. Because upstream ASes may propagate the same PCB multiple times, a beacon store may receive a PCB from its upstream AS that it has already received before. In this case, the beacon store simply updates the PCB's arrival time. On the other hand, new paths (never seen before) can also be desirable and should be propagated quickly to announce new paths.

- **Propagated paths' lifetime:** The beacon store keeps track of the propagated paths on each egress interface as well as the expiration time of the PCBs associated with the propagated paths. The expiration time of a propagated path is important: It allows the beacon service to renew paths that will expire soon and propagate new paths with a long remaining lifetime.

- **Feature support:** Beacon selection can be extended to support richer criteria, such as bandwidth reservations, consistent support for a certain SCION extension on a path, support for a specific cryptographic algorithm, geographic coordinates, latency information, or carbon footprint, for instance. SCION beaconing naturally supports routing on multiple optimality criteria [480], e.g., computing a path quality metric based on bandwidth, latency, and carbon footprint.

4.4.3 Selection Policy

Each AS has a selection policy, which governs the storage and selection of PCBs at the beacon service in the AS. In particular, a selection policy specifies the following:
- the maximum number of candidate PCBs to store;
- the number of up-segments to register at the local path service each registration period;
- the number of down-segments to register at a core path service (specified only by non-core ASes) each registration period;
- the number of PCBs to propagate (downstream or to core ASes) each propagation period;
- a list of blacklisted ASes that must not appear in any PCB sent downstream or registered;
- a set of minimum and maximum allowable values for properties; and

- a set of weights representing the relative importance of the previously mentioned properties in evaluating and selecting PCBs.

Beacon policies are local to the AS, and it might be in the commercial interest of the AS to keep them private.

4.4.4 Filtering PCBs

When the beacon service receives a PCB, the beacon store first checks the path against a series of filters defined by a selection policy. These filters check whether any ASes in the segment are blacklisted, and whether the path properties fall between the minimum and maximum allowable values specified in the selection policy. The latter type of filtering allows paths with certain undesirable properties, such as being longer than a threshold number of hops, to be ignored as a candidate PCB.

4.4.5 Selecting PCBs and Path Segments

The beacon store computes the overall quality of a PCB, for instance by computing a weighted sum, using the weights specified in the selection policy. Once it has computed the quality of all candidate PCBs, the beacon service selects the top-ranked PCBs. Time-based path properties, such as remaining lifetime, must be recomputed when the beacon store selects PCBs. Disjointness is based on previous operations and must also be computed when PCBs are selected (i.e., every propagation or registration period).

4.4.6 Diversity-Based Path Selection

In this section, we describe a path-exploration algorithm that at the same time ensures scalability and optimizes path diversity. This algorithm, which we call diversity-based path selection, is a distributed greedy algorithm maximizing the disjointness of paths, while reducing the overhead by inhibiting redundant path retransmissions. The rationale of the algorithm is to prefer PCBs with link-disjoint paths, PCBs with new paths, and PCBs with a long remaining lifetime. This is enabled by keeping track of recently sent PCBs on each egress interface. We choose link-disjointness instead of AS disjointness as a metric for diversity, since failures of entire ASes are unlikely events. The algorithm can be modified to additionally take into account other criteria such as those described in §4.4.2.

The diversity-based path selection (Algorithm 1) is triggered periodically by the beacon service to select extended paths with a path quality above a certain threshold. An extended path consists of an existing path combined with an egress interface.

The algorithm selects the best extended paths iteratively. At each iteration, it selects at most one extended path from each origin AS to each eligible neighbor AS (i.e., a child or a neighboring core AS, depending on the type of beaconing) among all possible extended paths. First, it constructs the set of all possible extended paths by appending all egress interfaces connecting to the neighbor AS to all paths extracted from all PCBs in the beacon store that are originating from the origin AS. Then, it calculates the score of each extended path as a function of its link-disjointness with regard to previously disseminated extended paths from the same origin AS towards the same neighbor AS, the age and the lifetime of its PCB in the beacon store, and the lifetime of the most recently sent extended PCB with the same extended path (if exists).

At the end of each iteration, if the best extended path's score is above the threshold, the beacon service extends the PCB associated with the path and propagates it to the neighboring AS on the selected interface. Finally, it updates the algorithm's data structures (i.e., the metadata associated with sent extended PCBs in the beacon store). The algorithm iterates until either the number of selected paths reaches a maximum threshold, or the best path's score in the last iteration is below the score threshold.

Link Diversity Score Calculation. To perform the *link diversity score* calculations, the algorithm stores a *Link History Table* per *[origin AS, neighbor AS]* pair. Each table is a one-to-one map from *link_ids* to their associated *counters* where the *link_id* is an identifier for every link between two ASes, and the *counter* counts the number of times the link is part of a valid extended path from the origin AS to the neighbor AS.

When a path originating from an origin AS is disseminated to a neighbor on an outgoing link, the counters associated with every link on the extended path (including the outgoing link itself) in the *Link History Table* of that (*origin AS, neighbor AS*) pair are incremented. If a link has not been visited on any previously disseminated path from the origin AS to the neighbor AS, a new entry for that link is created. With the help of this *Link History Table*, the algorithm calculates the *link diversity score* of an extended path from an origin AS to a neighbor AS by finding the geometric mean of counters associated with all links on the path. This number is then normalized to the maximum acceptable geometric mean to get a score between 0 and 1. The geometric mean of links' counters shows the path's degree of jointness with regard to previously sent extended paths as the counter of each link is equal to the number of paths having that link in common with the current path.

To keep the *link diversity scores* of previously sent extended paths constant over time, the algorithm stores their *link diversity score* when they were sent for the first time, and uses the stored value afterwards. The algorithm stores the *link diversity score* and the expiration time of the most recently sent extended PCBs associated with extended paths in the *Sent PCBs Metadata Table*. If a path is sent again, its corresponding expiration timer in *Sent PCBs Metadata Table* gets updated.

Algorithm 1: Path-diversity-based path selection

For f, see Equation (4.12b)

For g, see Equation (4.12a)

---

sent_PCBs_cnt ← 0;
while *sent_PCBs_cnt < max_PCBs_to_send* **do**
    max_score ← 0;
    best_extended_path ← ⊥;
    best_PCB ← ⊥;
    best_eg_iface ← ⊥;
    best_PCB_diversity_score ← 0;
    for *PCB ∈ received PCBs with origin o* **do**
        p ← extract_path(*PCB*);
        for *eg_iface ∈ egress interfaces to neighbor n* **do**
            p_{ext} ← [p, eg_iface];
            diversity_score ← calculate_diversity_score(p_{ext}, o, n);
            if p_{ext} ∈ *sent_PCBs_metadata*[*eg_iface*].*keys( )* **then**
                score ← diversity_score^g;
            else
                score ← diversity_score^f;
            if *score > score_threshold* **and**
              score > max_score **then**
                max_score ← score;
                best_extended_path ← p_{ext};
                best_PCB ← *PCB*;
                best_eg_iface ← eg_iface;
                best_PCB_diversity_score ← diversity_score;
    if *best_extended_path = ⊥* **then**
        break;
    else
        PCB$_{ext}$ ← extend_PCB(best_PCB, best_eg_iface);
        propagate(PCB$_{ext}$);
        sent_PCBs_metadata[best_eg_iface][best_extended_path] ←
          <PCB$_{ext}$.exp_time, best_PCB_diversity_score> ;
        for *link ∈ best_extended_path* **do**
            link_history_table[o][n][link] ←
            link_history_table[o][n][link] +1
        sent_PCBs_cnt ← sent_PCBs_cnt +1

PCB Age Score Calculation. Although the *link diversity score* shows the disjointness of paths, it is not sufficient for selecting high-quality paths, since the age and lifetime of the most recently received PCB (the one in the beacon store) and the lifetime of most recently sent extended PCB associated with an extended path is an important factor of path quality; if a previously sent extended PCB is about to expire soon, its associated extended path needs to be re-sent to preserve connectivity in downstream ASes; otherwise, re-sending paths wastes bandwidth. If a received PCB in the beacon store is about to expire soon, disseminating it is not useful either. Therefore, for each extended path, the algorithm also considers the age and the lifetime of the most recently received PCB and the lifetime of the most recently sent extended PCB associated with that extended path. The final score is calculated by an exponential function where its base is proportional to the *link diversity score* as shown in Equation (4.11).

If an extended path has been sent before, the exponent is proportional to a power of the ratio of the remaining lifetime of the most recently sent extended PCB to the remaining lifetime of the most recently received PCB associated with the extended path as shown in Equation (4.12a). If an extended path has not been propagated to a particular egress interface before, the exponent is proportional to the ratio of the most recently received PCB's age to its lifetime as shown in Equation (4.12b). The different functions are due to the following three objectives, which cannot be satisfied with a single function or two functions of the same form.

- **Preserve connectivity** by prioritizing previously sent extended paths over new ones from the same origin AS, when the previously sent extended PCB will soon expire.

- **Discover new paths** by prioritizing new extended paths for a given origin AS whenever the previously sent extended PCBs expiration time is far away.

- **Save bandwidth** by not re-sending very recently sent extended paths, by lowering their score.

The parameters α, β, and γ are set in the AS configuration such that the above three objectives are achieved. The values should be chosen depending on the topology and the configured lifetime of a PCB.

$$\text{score} = \begin{cases} (\text{diversity score})^g & \text{if extended path was previously sent} \\ (\text{diversity score})^f & \text{otherwise} \end{cases} \quad (4.11)$$

g and f are calculated using Equations (4.12a) and (4.12b) respectively:

$$g = \left(\beta \frac{\text{most recently sent extended PCB's remaining lifetime}}{\text{most recently received PCB's remaining lifetime}} \right)^\gamma \quad (4.12a)$$

$$f = \alpha \frac{\text{most recently received PCB's age}}{\text{most recently received PCB's total lifetime}} \quad (4.12b)$$

We evaluate the performance of the diversity-based path selection in §6.4.3, analyzing its optimality with respect to failure resilience, and measuring its overhead.

4.5 Path Lookup

Path lookup is a fundamental building block of SCION's path management architecture, as it enables end hosts to obtain path segments found during path exploration. They can then construct end-to-end paths from a set of possible path segments returned by the path lookup process.

4.5.1 Requirements and Design Goals

We considered the following requirements and design goals that led to the design of SCION's path-lookup mechanisms.

Low Latency. In the absence of a cached path at end hosts, a path lookup needs to be performed before a packet can be sent to a new destination. It is therefore performance-critical that a path lookup can be performed as fast as possible. Effective caching is critical for the performance and scalability of path lookup, as it can decrease the latency of path lookups. To minimize the number of path lookups, path services and end hosts should also cache paths to exploit the temporal locality of network destinations.

Scalability. Path lookup not only has to scale with respect to the number of users, but also to an increasing number of paths available in an ever-expanding network such as the Internet. Caching supports scalability given the typical Zipf-like distribution of destinations of requests [6].

Availability. If the path lookup infrastructure experiences outages, end hosts might be unable to look up new paths. The path lookup infrastructure should therefore be distributed and replicated to guarantee high availability even when single parts of the system fail or are under attack, e.g., during a DDoS attack. This goal is described as part of more general availability properties across SCION in Chapter 11.

Security. In terms of security, the following properties are critical for the path lookup infrastructure to function properly in the presence of an attacker. First, end hosts should be able to verify the authenticity of paths they receive from path lookup, i.e., that path segments were registered by the true destination and have not been altered since registration. This prevents an attacker from tricking an end host into using a fake path (similar to cache-poisoning attacks in DNS [371]). Second, a path should only be removed from the path lookup infrastructure with proper authorization (apart from expiration). Otherwise, an attacker could disconnect an AS from the Internet by repeatedly revoking all paths to that AS (§5.7). Third, not all paths should be public. While path services facilitate the retrieval of paths, it should be possible to distribute paths out

Table 4.1: Path services responsible for providing different types of path segments.

Segment Type	Responsible Path Service(s)
Up-segment	Local path service
Core-segment	Core path services in local ISD
Down-segment	Core path services in destination ISD

of band directly to potential senders. SCION supports non-registered (or hidden) paths (§8.1), which can serve as an important ingredient in DDoS attack defense.

4.5.2 Lookup Process

Our path-lookup strategy has evolved since the publication of the first SCION book [413]. In summary, two main changes were made to this process: First, path segments are now only registered with the core AS that originated the segment in question, whereas our previous lookup scheme required synchronization of path segments between all core path services in an ISD. Second, the lookup for destinations in a remote ISD is not recursive anymore; that is, the path service in the source AS contacts the core ASes of the destination ISD directly instead of relying on the core AS in the source ISD for this. However, we describe alternative strategies and the tradeoffs they entail in §4.5.3.

An application must resolve paths with a sequence of segment requests to the AS-local path service. Typically this is orchestrated by the SCION daemon.[5] The local path service answers directly, or forwards these requests to the responsible core path services. Then, the replies to these forwarded requests are cached.

The use of caching is essential to ensure that the path-lookup process is scalable and can be performed with low latency. However, caching can introduce consistency problems. If a cache delivers stale paths, then the performance of the path lookup and all upper layers are negatively impacted. Our lookup process was designed to strike a balance between simplicity and performance.

Table 4.1 shows which path services are responsible for providing the host with the different types of path segments. Once path segments are obtained, they are combined in the data plane (see §5.5).

The overall sequence of requests required for the SCION daemon to resolve a path is the following:

1. Request up-segments.

---

[5]The SCION daemon is software that runs on end hosts and handles control-plane messages (see §12.1.2).

2. Request core-segments starting at core ASes reachable with up-segments, to the core ASes in the destination ISD. If the destination ISD is the local ISD, this step requests segments to core ASes that are not directly reachable with an up-segment.

3. Request down-segments starting at core ASes in the destination ISD.

Wildcard addresses (of the form I-0 to designate any AS in ISD I, see §2.5) can be used by the SCION daemon to designate any core ASes in path-segment requests. These wildcard addresses are then "expanded" (i.e., translated into one or more actual addresses):

- **Up-segment requests:** The destination (i.e., the wildcard replacing the AS where the segment should end) is expanded by the local path service.

- **Core-segment requests:** The source is expanded by the local path service to all provider core ASes (i.e., cores for which an up-segment exists), and the destination is expanded by the core path service.

- **Down-segment requests:** The source is expanded by the local path service to all core ASes of the specified ISD.

In this path-lookup process, the resolver (i.e., the SCION daemon) can employ different strategies:

- **Breadth-first search:** Attempt to resolve all paths. The SCION daemon will request all up-segments, then request all core-segments, then request all down-segments.

  Using wildcard addresses, the requests for all three segment types become independent and can be queried concurrently. This is the approach that is currently implemented.

- **Depth-first search:** Attempt to resolve a single path with as few requests as possible. The SCION daemon will pick a local core AS; see what remote core ASes are reachable; pick one; see if the remote destination AS is reachable; if unreachable, track back and pick another remote core-AS; see if the remote destination AS is reachable; and so on.

To illustrate how the path lookup works, we show two path-lookup examples in sequence diagrams. The network topology of the examples is represented in Figure 4.2 below. In both examples, the requester is in AS *A*. In the first example, the destination is in AS *D*. In the second example, the destination is in AS *G*. ASes *B* and *C* are core ASes in the local ISD, while *E* and *F* are core ASes in a remote ISD. Core AS *B* is a provider of the local AS, but AS *C* is not, i.e., there is no up-segment from *A* to *C*.

For the sequence diagram of the first example, see Figure 4.3; for the sequence diagram of the second example, see Figure 4.4.

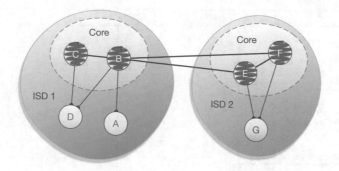

Figure 4.2: Topology used in the path lookup examples.

4.5.2.1 Application and SCION Daemon

For an application to obtain a set of paths to a destination AS, it must send a request to the local SCION daemon, which will proceed as follows[6]:

1. If the destination is not set, invalid, or represents the local AS, then immediately return an error message.

2. Split the path request into segment requests; different strategies can be used, as described above.

3. For each segment request,

   a) return segments from cache if possible;

   b) otherwise, request segments from the local path service, validate the retrieved segments, and add them to cache.

4. Combine all segments into a set of paths.

5. Filter paths with revoked on-path interfaces.

6. Return paths to the application.

4.5.2.2 Local Segment-Request Handler

The segment-request handler of a local path service will proceed as follows:

1. Determine the requested segment type and validate the request.

2. In the case of an up-segment request, load the matching up-segments from the path database and return.

---

[6]The SCION daemon is described in more detail in §12.1.2

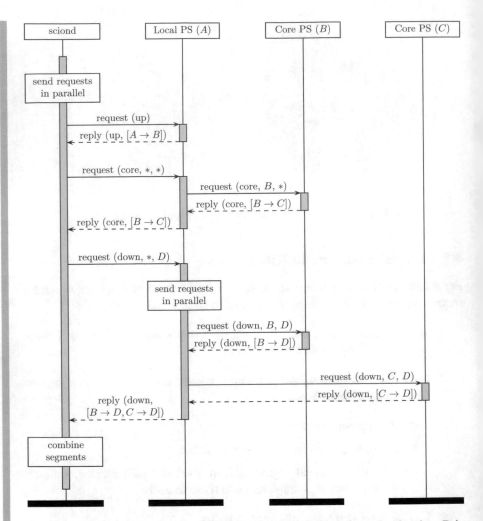

Figure 4.3: Sequence diagram illustrating a path lookup for a destination D in
            the local ISD. The SCION daemon is denoted sciond. The request
            (core, *, *) is for all pairs of core ASes in the local ISD. Similarly,
            (down, *, D) is for down-segments between any core AS in the
            local ISD and destination D.

3. In the case of a core-segment request:

   a) Expand the source wildcard into separate requests for each reach-
      able core AS in the local ISD.

   b) For each segment request,

      i. if the source in the segment request (i.e., the start of the
         path segment) is the local AS, then load the matching core-
         segments from the path database;

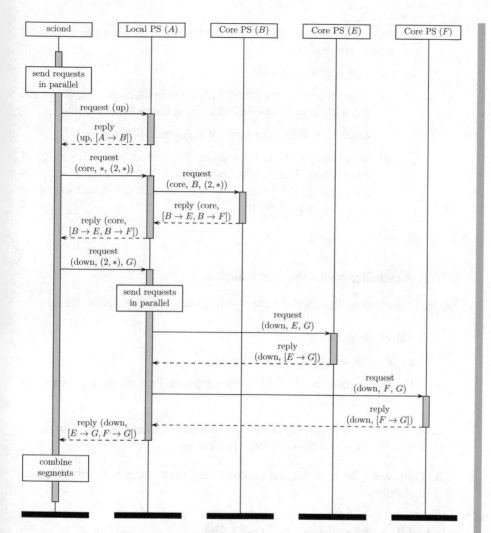

Figure 4.4: Sequence diagram illustrating a path lookup for a destination G in a remote ISD. The SCION daemon is denoted sciond. The request (core, $*$, (2, $*$)) is for all path segments between a core AS in the local ISD and a core AS in ISD 2. Similarly, (down, (2, $*$), G) is for down-segments between any core AS in ISD 2 and destination G.

    ii. otherwise, if possible, return segments from cache;

    iii. otherwise, request the segment from the core path service at the source, validate the retrieved segments, and add them to cache.

4. In the case of a down-segment request:

a) Expand the source wildcard into separate request for every core AS in the source ISD.

b) For each segment request,

    i. if the source in the segment request is the local AS, then load the matching down-segments from the path database;

    ii. otherwise, if possible, return segments from cache;

    iii. otherwise, request the segment from the core path service at the source. Sending the request may require looking up core-segments to the source. Then validate the retrieved segments and add them to cache.

5. Filter revoked segments.

4.5.2.3 Core Segment-Request Handler

The segment-request handler of a core path service will proceed as follows:

1. Validate the request:

    a) The source must be this core AS.

    b) The request must be for a core-segment or a down-segment to an AS in this ISD.

2. If the destination is a core or wildcard address, then load matching core-segments from the path database and return.

3. Otherwise, load the matching down-segments from the path database and return.

4.5.3 Alternative Lookup Strategies

As mentioned earlier, we simplified the path lookup process since the publication of the first SCION book [413], but there are tradeoffs to consider. While our new approach reduces the complexity of the lookup by abandoning recursion and synchronization between core ASes, it might lead to increased latency in some situations. Moreover, recursion would allow core path services to more easily filter traffic and thus mitigate potential DoS attacks.

As opposed to the lookup process we describe above, recursive lookup works as follows: Local path services only communicate with their respective core ASes, which in turn are responsible for contacting core ASes in other ISDs to obtain down-segments. These down-segments are then cached at core ASes. Recursive lookup has the advantage that more caching levels may result, in some cases, in a faster lookup. Furthermore, since the path services of core ASes would only communicate with non-core ASes in their ISD and other core ASes, they could filter traffic more effectively (e.g., by filtering out packets that use more than one path segment).

The two strategies are not incompatible and could, in theory, co-exist in the same network. Core path services could then prioritize messages from other core ASes and path services in their own ISD, in case they receive an excessive number of requests.

This discussion also relates to the question of whether down-segments should be synchronized between all core ASes in the same ISD. Synchronization between core ASes would indeed facilitate the task of obtaining down-segments for non-core path services, as they could then send a request to any of their core ASes without any concerns regarding redundancy or consistency. Unfortunately, consensus between distant servers operated by different entities is difficult to achieve in practice.

4.6 Service Discovery

A SCION AS has three main services that make up the control plane—the beacon, path, and certificate service. Furthermore, several systems that extend the SCION control plane, such as EPIC or COLIBRI (see Chapter 10), register new services in ASes. It is therefore crucial that these control-plane services can be efficiently discovered by local and remote hosts that need to interact with them.

In this section, we describe two means to enable service discovery in a SCION AS.

4.6.1 Service-Level Anycast

In order to facilitate control-plane anycast communication, SCION introduces a dedicated service-addressing scheme: the service (SVC) addresses. For instance, a beacon server that wishes to register segments with a remote AS's path service does not have to know the actual address of a remote path server. Instead, the SCION service address of the path service (SVC_{PS}) suffices.

A service address is essentially a form of wildcard address that will resolve to an instance of the service registered under a given service address. To that end, the ingress SCION border router of an AS will locally resolve service addresses to fully qualified end-host addresses whenever it encounters a packet containing a service address as the destination. Each SCION AS is free to implement any kind of resolution mechanism it sees fit, e.g., DNS-based service resolution.

By replying from its actual address, the destination host can inform the sender about the resolved destination address. The sender can cache that address for future communication. As stream-based communication does not work well with wildcard addressing, hosts first resolve the destination address using datagram-based communication, e.g., UDP/SCION, before initiating the

communication stream with the destination. This interaction led to the introduction of a dedicated discovery service, which we describe in the next section.

4.6.2 Discovery Service

The discovery service has two main responsibilities:

- It keeps track of all the services and the health of their service instances
  in an AS. To that end, service instances register their communication
  endpoints for a specific service address with the discovery service, and
  the discovery service tracks their health through a health-check endpoint.

- It returns the communication endpoints of service instances to local or
  remote requesters.

The discovery service itself is reachable via its own dedicated service address: SVC_{DS}.

To illustrate address resolution involving the discovery service, consider a
beacon server in AS A that intends to register a path segment with the path
service in AS B. The beacon server first resolves the communication endpoint
of the path service by issuing a lookup request containing the service address
SVC_{PS} addressed to SVC_{DS} in AS B. The lookup request reaches the discovery
service in AS B through the service-level anycast mechanism described in the
preceding section. The discovery service in AS B keeps an up-to-date view
of the locally available path-service instances and includes the communication
endpoints of at least one of them in the reply back to the beacon server in AS
A. Note that this exchange is based on UDP/SCION and the reply should fit
in a single datagram. The beacon server in AS A now has the communication
endpoint of at least one path service instance of AS B and can initiate the path-
segment registration exchange.[7]

The first-resolve-then-communicate pattern lends itself naturally to a
discovery-service-like concept. The discovery service can act as a load-
balancer by resolving service addresses to healthy instances with low load.
Clearly, the discovery service needs to be highly available due to its crucial
functionality for SCION control plane interactions. There are various options
how to implement a highly available service, e.g., by combining distributed
consensus and load-balancing to implement the discovery service as a distributed system. The implementation details are up to the operators of a
SCION AS and the SCION technology vendors.

---

[7]The current implementation of the SCION control plane relies on gRPC over QUIC/SCION
for its message exchanges.

4.7 SCION Control Message Protocol (SCMP)

The SCION Control Message Protocol (SCMP) is analogous to ICMP in the current Internet and provides the following functionalities:

- **Network diagnostic:** SCMP enables debugging tools such as the SCION equivalents of Ping or Traceroute.

- **Error messages:** SCMP can be used to signal problems with packet processing or inform end hosts about network-layer problems.

To the best of our knowledge, the SCMP protocol is the first instance of a secure control message protocol for the Internet. The main challenges include scalable Internet-wide key distribution and highly efficient generation of authentication information at line speed. In this section, we describe the design, goals, and use cases of SCMP.

4.7.1 Goals and Design

SCMP must be flexible, as it is used for different purposes in various applications. For instance, (1) some SCMP messages are processed by intermediate routers on the path, while other messages are end-to-end; (2) there are various types of SCMP messages (for various types of diagnostics or network errors); and (3) the messages can influence different parts of the SCION stack (such as the transport protocol or the beacon selection mechanism).

SCMP packets can carry either an error message or an informational message. One basic rule of SCMP is that an error packet should never generate another SCMP packet (to prevent infinite recursion), thus border routers must be able to efficiently check whether a packet is an SCMP error message.

4.7.2 SCMP Message Format

SCMP messages have the following general format:

- Type: Indicates the type of SCMP message. Its value determines the format of the info and data blocks.

- Code: Provides additional granularity to the SCMP type.

- Checksum: Used to detect data corruption.

- InfoBlock and DataBlock: Optional fields of variable lengths whose formats depend on the message type.

Types. There are two types of SCMP messages: error messages and informational messages. Error messages are identified by a zero in the high-order bit of the type value, i.e., error messages have a type value in the range of 0–127. Informational messages have type values in the range of 128–255.

Table 4.2: Examples of SCMP message types.

(a) Error messages.

Type	Definition
1	Destination Unreachable
2	Packet Too Big
3	(not assigned)
4	Parameter Problem
5	External Interface Down
6	Internal Connectivity Down

(b) Informational messages.

Type	Definition
128	Echo Request
129	Echo Reply
130	Traceroute Request
131	Traceroute Reply

4.7.3 SCMP Authentication

SCMP messages provide important information to end hosts. That information may be used to optimize network traffic, switch paths, and modify MTUs, for example. Due to possible far-reaching decisions triggered by these messages, it is crucial to ensure their authenticity. Fortunately, DRKey and the SPAO (see Sections 3.2 and 3.3) provide the appropriate mechanisms for this. The details of SCMP authentication then depend on the type of message.

To prevent DoS attacks, SCMP messages are authenticated using keys where high-speed components such as border routers and responders are on the fast side of DRKey. Those components can thus efficiently derive the keys necessary to authenticate SCMP messages. However, an end host receiving an SCMP message may need to fetch the necessary key from the local certificate service. To further reduce the risks of DoS attacks, rate limiting can be applied to SCMP messages.

4.7.3.1 Authentication of Error Messages

All SCMP error messages must be authenticated with DRKey. Depending on the entity that sends the SCMP error message—border router or receiving end host—second-level "host on the slow side" keys or third-level host-to-host keys are used, respectively. The entity sending the SCMP message is always on the fast side.

A border router in AS A sending a "packet too big" SCMP error message to host H_B in AS B uses the key $K^{SCMP}_{A \to B:H_B}$ to authenticate the message.

4.7.3.2 Authentication of Informational Messages

SCMP informational messages come in pairs of query and response messages (see Table 4.2b). Authentication of the queries (echo request and traceroute request) is optional. Responses (echo reply and traceroute reply) are authenticated if and only if the query is authenticated. Although SCION naturally defends against amplification attacks (see §7.7), the corresponding messages are designed to prevent message amplification: A request always triggers one same-size response (even for traceroute).

This, together with the fact that the respondent is on the fast side, ensures that these queries cannot be used for DoS attacks: Routers can always derive the necessary keys from the secret value for the SCMP protocol. End hosts receiving echo requests can use the second-level key for the sender's AS to derive the necessary third-level key. In addition, both routers and end hosts can employ rate limiting.

Example.   An end host H_A in AS A that sends an "echo request" SCMP message to host H_B in AS B (optionally) uses the key $K^{SCMP}_{B:H_B \to A:H_A}$ to authenticate the message. This key must be fetched explicitly from the local certificate service. Host H_B can derive this key from the second-level key $K^{SCMP}_{B:H_B \to A}$, which it may have cached due to previous communication with end hosts in AS A (otherwise it needs to fetch this key from its certificate service). It verifies the MAC on the request and uses the same key to authenticate its "echo reply" message (if and only if the request was authenticated).

5 Data Plane

LAURENT CHUAT, SAMUEL HITZ, MARKUS LEGNER, ADRIAN PERRIG*

Chapter Contents

5.1	Inter- and Intra-domain Forwarding	94
5.2	Packet Format	95
5.3	Path Authorization	96
5.3.1	Expressiveness–Scalability Trade-off	97
5.3.2	Authenticating Local Information Is Insufficient	98
5.3.3	Authorizing Segments through Chained MACs	99
5.4	The SCION Path Type	101
5.4.1	The One-Hop Path Type	102
5.5	Path Construction (Segment Combinations)	104
5.5.1	Notation	106
5.5.2	Paths Through the Core	106
5.5.3	Paths Outside the Core	109
5.5.4	Efficient Path Construction	112
5.6	Packet Initialization and Forwarding	115
5.6.1	Initialization at Source Host	115
5.6.2	Path Reversal	116
5.6.3	Processing at Routers	116
5.7	Path Revocation	120
5.7.1	Challenges	121
5.7.2	Principles	121
5.7.3	Design	122
5.7.4	Link-Failure Detection	123
5.8	Data-Plane Extensions	124

The ultimate purpose of any network is to deliver data between end hosts, which is the responsibility of the *data plane*. SCION's data plane is fundamentally different from today's IP-based data plane in that it is *path-aware*: In SCION, inter-domain forwarding directives are embedded in the packet header

---

*This chapter reuses content from Pappas, Perrig, Reischuk, Shirley, and Szalachowski [400].

© The Author(s), under exclusive license to Springer Nature Switzerland AG 2022
L. Chuat et al., *The Complete Guide to SCION*, Information Security
and Cryptography, https://doi.org/10.1007/978-3-031-05288-0_5

as a sequence of hop fields (HFs), which encode AS-level hops augmented with ingress and egress interface identifiers.

This design choice has several advantages, see also Chapter 1: Not only does it provide control and transparency over forwarding paths to end hosts, it also simplifies the packet-processing at routers: Instead of having to perform longest-prefix matching on IP addresses, which requires expensive hardware like ternary content-addressable memory (TCAM) and substantial amounts of energy, a router can simply access the next hop from the packet header.

However, this raises a question: If end hosts can choose paths and embed the forwarding directives in packet headers, how can ASes ensure that they can only use the path segments that were created and registered in the control plane (see Sections 4.2 and 4.3)? This property, which we call *path authorization*, ensures that the actual forwarding paths conform to the policies of individual ASes, and prevents end hosts from constructing unauthorized paths or paths containing loops. In SCION, path authorization is achieved through symmetric cryptography, which is both faster and more energy-efficient than today's IP-based forwarding and is thus applicable to high-speed Internet packet processing (see Chapter 16 for an analysis of SCION's energy consumption). This was demonstrated recently by de Ruiter and Schutijser, who implemented SCION forwarding for programmable switches using the P4 language [155].

In this chapter, we first discuss how inter- and intra-domain forwarding fit together in a SCION-based Internet (§5.1) and present the general format of SCION data packets (§5.2). We then discuss in detail the mechanisms through which SCION achieves path authorization (§5.3), the structure of the SCION path header (§5.4), how end hosts can construct end-to-end forwarding paths by combining multiple path segments (§5.5), and how packets are processed at routers (§5.6). Finally, we discuss how failures are communicated to end hosts (§5.7) and SCION's support for extension headers (§5.8).

5.1 Inter- and Intra-domain Forwarding

SCION is an *inter-domain* network architecture and as such does not interfere with intra-domain forwarding. This corresponds to the general practice today where BGP and IP are used for inter-domain routing and forwarding, respectively, but ASes use an intra-domain protocol of their choice, for example OSPF or IS-IS for routing and IP, MPLS, and various layer-2 protocols for forwarding. In fact, even if ASes use IP forwarding internally today, they typically encapsulate the original IP packet they receive at the edge of their network into another IP packet with the destination address set to the egress border router, to avoid full inter-domain forwarding tables at internal routers.

SCION emphasizes this separation, as SCION is used exclusively for inter-domain forwarding, and re-uses the intra-domain network fabric to provide connectivity among all SCION infrastructure services, border routers, and end hosts. This has the advantage that minimal change to the infrastructure is required for ISPs when deploying SCION (see also Chapter 13). An additional advantage of this approach is that it facilitates the transition to new network-layer protocols (e.g., IPv6) as the intra-domain network-layer protocol can be chosen independently by each AS.

The full forwarding process for a packet transiting an AS consists of the following steps:

1. The AS's SCION border router receives a SCION packet from the neighboring AS.

2. The border router parses and validates the SCION header and accesses the egress[1] interface from the packet header.

3. The border router consults its *intra-domain* forwarding information and adds an appropriate *IntraProtocol* header (where *IntraProtocol* is the intra-domain forwarding protocol, e.g., MPLS or IP) with the destination address set to the egress BR.

4. The packet is forwarded within the AS by routers and switches based on the *IntraProtocol* header.

5. Upon receiving the packet, the egress border router strips off the *IntraProtocol* header, again validates and updates the SCION header (see §5.6), and forwards the packet to the neighboring SCION border router.

These forwarding steps are illustrated in detail in §13.6.

5.2 Packet Format

SCION packets in the data plane contain a network-layer (layer-3) SCION header between the transport-layer (layer-4) and link-layer (layer-2) or intra-domain network-layer header, see Figure 5.1. The SCION header itself is composed of a *common header*, an *address header*, a *path header*, and an optional *extension header*, see Figure 5.2. Byte-level header structures are available online at https://www.scion.foundation/specification.

The *common header* contains important meta information like a version number, flags, lengths of the header and payload, etc. In particular, it contains flags that control the format of subsequent headers such as the address and path headers. Furthermore, fields for the traffic class and the flow ID mimic the corresponding fields in the IPv6 header [19, 158, 389, 426] and enable consistent traffic shaping and prioritization inside the network.

The *address header* contains the ISD, AS, and end-host addresses of source and destination. The type and length of end-host addresses are variable and can

---

[1] We use the terms *ingress* and *egress* in forwarding direction. When we refer to the beaconing (construction) direction, we continue to use the terms *ConsIngress* and *ConsEgress*.

| Link-layer/intra-domain header |
| **SCION header** |
| Transport-layer header |

Figure 5.1: Position of the SCION header in a data packet.

| **Common header** |
| **Addresses** |
| **Path** |
| **Extensions** (optional) |

Figure 5.2: SCION header structure.

be set independently using flags in the common header. Furthermore, these addresses are not required to be globally unique or globally routable, they can be selected independently by the corresponding ASes. This means, for example, that an end host identified by a link-local IPv6 address in the source AS can directly communicate with an end host identified by a globally routable IPv4 address via SCION. Alternatively, it is possible for two hosts with the same (private) IPv4 address 10.0.0.42 but located in different ASes to communicate with each other via SCION.

The *path header* contains the full AS-level forwarding path of the packet. A *path type* field in the common header specifies the path format used in the path header. We will discuss the "standard" SCION path type in §5.4, and additional path types and extensions in Chapter 10.

Finally, the optional *extension header* contains a variable number of hop-by-hop and end-to-end options, similar to the extensions in the IPv6 header [158].

5.3 Path Authorization

Path authorization is the property that data packets always traverse the network along paths that were authorized by all on-path ASes in the control plane. In contrast to the IP-based Internet, where forwarding decisions are made by routers based on locally stored information, SCION bases the forwarding decisions purely on the packet-carried forwarding information,[2] which can be set by end hosts.

To ensure path authorization, a SCION border router must be able to check that the path contained in the packet header was authorized by this AS. An

---

[2]This concept is often also called "packet-carried forwarding state" (PCFS).

obvious approach to achieve this would be a local store of valid paths, which, unfortunately, has several disadvantages: Routers would still need to keep a substantial amount of local information, which either requires expensive hardware or significantly slows down packet processing. Furthermore, all this information needs to be distributed to all border routers of an AS, which can cause scalability issues and inconsistencies.

This is why, instead of relying on local state at routers, SCION uses cryptographic mechanisms to efficiently verify path authorization. Unfortunately, asymmetric cryptography, which SCION uses in its control plane (see Chapter 4), is orders of magnitude too slow to be used for this purpose and creates excessive overhead. Instead, SCION uses *symmetric* cryptography in the form of message-authentication codes (MACs) to secure the forwarding information encoded in hop fields.

5.3.1 Expressiveness–Scalability Trade-off

The precise definition of path authorization is a design choice with an inherent trade-off between the expressiveness of policies that ASes can have and the availability and scalability of the system: On the one end of the spectrum are systems like ICING [381], where all on-path ASes authorize the full end-to-end path; this system gives ASes very precise control over allowed paths but introduces high overhead. In addition, situations may arise, where not a single path between a pair of ASes is authorized by all ASes (even if every AS authorizes at least one path). On the other end of the spectrum are systems that authenticate only local forwarding information; an AS can authorize some set of local ingress/egress interface pairs without restricting them further.

SCION strives for a compromise in this trade-off that is (1) sufficiently scalable and allows paths to be authorized as they are discovered, i.e., in a single communication round, and (2) expressive enough to support practically relevant AS policies. In particular, it gives ASes comparable levels of control over paths as today's BGP/IP-based Internet; the path policies that are expressible in SCION are analyzed in further detail in §6.2. The COLIBRI extension, which will be presented in §10.2, adds more fine-grained control for ASes when granting bandwidth reservations.

To that end, SCION authorizes individual *path segments* and allows end hosts to combine them according to certain rules (see §5.5). For the authorization of segments, an AS can choose which beacons to extend and to forward to which neighboring ASes. An AS can thus base its authorization decision on the full upstream part of the segment and its own forwarding information. This means that in Figure 5.3a, AS F can authorize different upstream segments for ASes G and H.

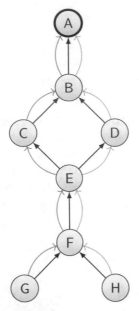 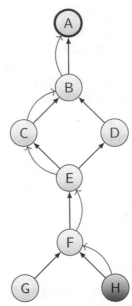

(a) The colored arrows represent authorized up-segments. AS F chose
to selectively authorize the left upstream segment for AS G and the
right one for AS H.

(b) If only local information were authenticated, a malicious source in
AS H, who gains access to hop
fields of AS G could splice the two
segments to obtain the red path.

Figure 5.3: Simple intra-ISD topology where circles represent ASes and the
black arrows denote customer→provider relationships. AS A is a
core AS (marked by the thick circle).

5.3.2 Authenticating Local Information Is Insufficient (Path-Splicing Attacks)

Before presenting the mechanisms that SCION uses for path authorization, we
will discuss why authorizing and authenticating purely local hop information
is insufficient based on the two examples shown in Figures 5.3 and 5.4.

We first consider the intra-ISD topology of Figure 5.3. Here, AS F has received two beacons, which both extended the same original beacon from AS A,
via two different paths; it forwards and thus authorizes the beacon received via
the left (orange) path to AS G and the beacon received via the right (green)
path to AS H. An end host in AS H is only supposed to be able to use the
green path via AS D. However, if a malicious end host gains access to hop
fields (HFs) from the orange path (e.g., through collusion with a host in AS G),
it can combine HFs of ASes H and F from the green path with HFs of ASes E,
C, B, and A from the orange path. As each AS would only check its own hop
information, none of them could detect this *path-splicing attack*.

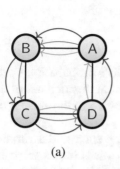
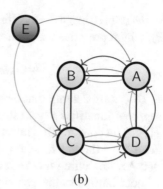

(a)   (b)

Figure 5.4: Simple core topology where the thick circles represent core
ASes and the orange and green arrows represent authorized core-
segments. An attacker in any of the four ASes can combine HFs of
ASes A and D from the green path with HFs of ASes B and C from
the orange path to create a forwarding loop (red arrows).
The simple example in a can be prevented by binding each HF
to a per-segment nonce (timestamp or random value); however, a
malicious AS outside the loop (AS E in b) may create segments
with equal nonces which would then be exploitable.

A similar attack is possible for core-segments, see Figure 5.4: In contrast to
intra-ISD beaconing, which follows parent–child links in one direction, core
beaconing does not have this directionality as core links represent peering-like
connections. It is therefore possible to have at the same time authorized core-
segments $A{\rightarrow}B{\rightarrow}C{\rightarrow}D$ and $C{\rightarrow}D{\rightarrow}A{\rightarrow}B$. Here, an attacker with access to
both these segments could combine HFs of ASes A and D from the green path
with HFs of ASes B and C from the orange path. While no infinite loop can
be constructed this way (as every AS increments the HF pointer of the packet),
the attacker *can* include these HFs several times in the path header of a packet,
which then travels several times around the loop (where the number of hops is
only limited by the maximal size of the path header).

In the next section we show how chained MACs prevent such splicing attacks.

5.3.3 Authorizing Segments through Chained MACs

In the authorization of SCION path segments, an AS authenticates not only
its own hop information but also an aggregation of all upstream hops. In this
section, we describe how path authorization is achieved during the beaconing
process.

As described in §4.1, the first component of every PCB is the info field (INF), which provides basic information about the PCB:

$$INF = \langle\, Flags_{INF} \parallel SegID \parallel TS \,\rangle. \tag{4.2 revisited}$$

Here, TS is a timestamp that denotes when the PCB's propagation started, and the segment identifier $SegID$ stores an aggregate of authenticators of all previous HFs. The core AS that initiates the PCB sets the TS to the current time and the $SegID$ to a unique value for that timestamp.[3]

Each AS (including the initiating AS) adds a hop entry and a variable number of peer entries to the beacon, signs it, and forwards it to a selected set of neighbors. Both hop entries and peer entries contain HFs of the format

$$HF = \langle\, Flags_{HF} \parallel ExpTime \parallel ConsIngress \parallel ConsEgress \parallel HFAuth \,\rangle. \tag{4.8 revisited}$$

In the following, we will describe how the HF authenticator, $HFAuth$, is calculated and how path authorization is achieved. Consider a beacon traversing ASes $AS_0, \ldots, AS_n$ with secret keys $K_0, \ldots, K_n$ in this order. AS_0 is the core AS that initiates the beaconing and creates a random initial segment identifier

$$\beta_0 = \mathrm{RND}(). \tag{5.1}$$

This value is added to the $SegID$ field of the PCB. Based on this and the contents of the HFs, each AS AS_i for $i \in \{0,\ldots,n\}$ can calculate the HF authenticator and an updated segment identifier:

$$\sigma_i = \mathrm{MAC}_{K_i}(\beta_i, TS, ExpTime_i, ConsIngress_i, ConsEgress_i), \tag{5.2a}$$

$$\beta_{i+1} = \beta_i \oplus \sigma_i\left[0{:}l_{SegID}\right]. \tag{5.2b}$$

Here, $\mathrm{MAC}_K(\cdot)$ denotes the calculation of a message-authentication code over "$\cdot$" using key K, $\oplus$ denotes the bitwise XOR operation, and $X[a{:}b]$ is the substring from byte a (inclusive) to b (exclusive) of X. To limit communication overhead, the HF authenticator is also truncated to l_{HFAuth} bytes:[4][5]

$$HFAuth_i = \sigma_i\left[0{:}l_{HFAuth}\right]. \tag{5.3}$$

During beaconing, only the initial random value β_0 is stored in the PCB. The subsequent segment identifiers are not added to the PCB as they can be re-computed based on β_0 and the $HFAuth$ fields of previous hop entries. In the data plane, the $SegID$ field is updated at each hop based on the $HFAuth$ and

---

[3]This can be achieved by using a (pseudo-)random value or a counter.

[4]The default length for the HF authenticator is 6 B, i.e., $l_{HFAuth} = 6$.

[5]Note that this requires using a MAC algorithm that can be truncated safely, such as MACs that also fulfill the properties of a pseudorandom function (PRF). See §17.4 for further details.

must contain the appropriate value at each hop, see §5.6.3 for further details. This requires that $l_{SegID} \leqslant l_{HFAuth}$.

Equation (5.2) applies to "normal" HFs, i.e., core and parent–child links. The computation for peering links is slightly different in that the HF authenticator of a peering HF of AS_i is computed based on β_{i+1} (instead of β_i) and does not affect the updated segment identifier:

$$\sigma_i^P = MAC_{K_i}\left(\beta_{i+1}, TS, ExpTime_i^P, ConsIngress_i^P, ConsEgress_i^P\right). \quad (5.4)$$

"Chaining" an AS's HF authenticator to the previous HFs by including an aggregate of all previous HF authenticators in the MAC computation through the segment identifier rules out the path-splicing attacks described in §5.3.2. Even if two segments have the same info-field timestamp (for example because they have been created from the same initial PCB), a malicious sender cannot splice them to create a new (unauthorized) segment as the segment identifiers at the splicing AS do not match. In principle, in case there is a collision of a segment identifier, splicing multiple segments would become possible. However, as segment identifiers are not under the control of end hosts, they cannot be brute-forced; instead, end hosts would need to wait for accidental collisions. In addition, the probability of an "accidental" match decreases exponentially with the length of the segment identifier, l_{SegID}. Overall, path splicing is prevented in SCION. A formalization of the path-authorization property and a formal verification will be presented in Chapter 22.

5.4 The SCION Path Type

Whenever an end host establishes a connection with another end host, they need to construct the forwarding path. After receiving multiple path segments[6] from the path server during the *path-lookup* process (see §4.5), the end host can combine multiple such segments according to simple rules (which will be discussed in the next section) to construct a forwarding path, which can be embedded in the packet header.

The structure of the final SCION path header is shown in Figure 5.5 and consists of up to three *info fields* and an arbitrary number of *hop fields*.[7] In addition, the path header contains a *path meta header*,

$$PathMetaHdr = \langle\, CurrINF \parallel CurrHF \parallel Seg0Len \parallel Seg1Len \parallel Seg2Len \,\rangle, \quad (5.5)$$

which contains meta information, namely the index of the current info field, *CurrINF*, the index of the current hop field, *CurrHF*, and the lengths of the

---

[6]We use the term "path segment" for the complete PCB including all metadata. The combination of the info field and HFs extracted from a PCB is called a "stripped segment".

[7]The number of hop fields is limited by the maximum header size.

Path meta header
Info field (up)
Info field (core)
Info field (down)
Hop field
. . .
Hop field

Figure 5.5: SCION path type. Not all info fields must be present (at least one, at most three), but their order is fixed.

different segments. The info field contains two single-bit flags, *ConsDir* and *Peering*: *ConsDir* is true if the segment corresponding to that info field is traversed in construction (i.e., beaconing) direction, *Peering* is true if the first or last HF is a peering HF.

The process of extracting info fields and HFs from the segments containing additional information is illustrated in Figure 5.6. Note that ASes at the joints of multiple segments are represented by two hop fields.

5.4.1 The One-Hop Path Type

The one-hop path type is a special case of the SCION path type. It is used to handle communication between two entities from neighboring ASes that do not have a forwarding path. Currently, it is used for bootstrapping beaconing between neighboring ASes.

A one-hop path has exactly one info field and two hop fields with the specialty that the second hop field is not known a priori, but is instead created by the ingress SCION border router of the neighboring AS while processing the one-hop path. Any entity with access to the AS forwarding key of the source AS can create a valid info and hop field as described in Sections 4.1.1.1 and 4.1.1.5, respectively. Upon receiving a packet containing a one-hop path, the ingress border router of the destination AS fills in the second hop field of the one-hop path with the ingress and egress interface information[8] and calculates the appropriate MAC for the hop field. Note, reversing a one-hop path will create a full-fledged SCION path (see §5.6.2).

Because of its special structure, no path meta header is needed. There is only a single info field and the appropriate hop field can be processed by a border router based on the source and destination ISD-AS.

---

[8]The egress interface is empty indicating that the path cannot be used beyond this AS.

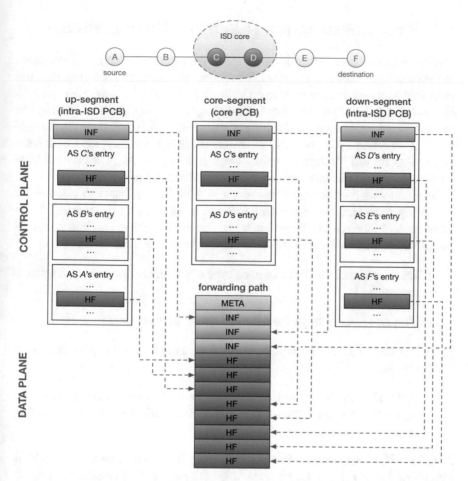

Figure 5.6: Example of path construction where three path segments are combined into a forwarding path.

Info field
Hop field
Hop field

Figure 5.7: SCION one-hop path type. It is a special path type that can be used by two entities in neighboring ASes to communicate without the need for a forwarding path.

5.5 Path Construction (Segment Combinations)

The beacons that core ASes disseminate to their customer ASes are confined to their isolation domain. This isolation property allows SCION to protect the intra-ISD beaconing process from outside adversaries and configuration errors. However, as a consequence, forwarding along single segments is insufficient for global connectivity across different ISDs. Therefore, SCION allows segments of different ISDs (but also of the same ISD) to be combined with each other to obtain full forwarding paths. Segments can furthermore be used bidirectionally, i.e., in the same or in the opposite direction of construction.

Depending on the network topology, a SCION forwarding path can consist of 1–3 segments with corresponding info fields. To protect the economic interest of ASes, "valley routes"[9] need to be prevented. Therefore, segments cannot be combined arbitrarily but have to fulfill the following conditions:

- There can be at most one of each type of segment (up-, core-, and down-segment).[10]

- If an up-segment is present, it must be the first segment in the path.

- If a core-segment is present, it must come before the (optional) down-segment.

- If the *Peering* flag is set in any info field, there must be exactly two segments (up- and down-segments), which both have the *Peering* flag set.

Additionally, all segments without the *Peering* flag need to consist of at least two HFs. These rules must be obeyed by the source host constructing a forwarding path. Note that the type of a particular segment is known to the end host but is not visible in the path header of data packets. A SCION router therefore must verify that these rules were followed correctly, see §5.6.3.

In the following, we present the different allowed segment combinations, assuming that the source and destination end hosts are in different ASes (as end hosts from the same AS use an empty forwarding path to communicate with each other). We stress that although end hosts enjoy freedom in composing their forwarding paths, the possible combinations (i.e., the provided path segments) follow AS routing policies and rules such as the valley-free property (discussed in §6.2). The possible combinations are illustrated in Figure 5.8 and summarized in the following:

---

[9] A valley route contains ASes that do not profit economically from traffic on this route, such as an AS that forwards traffic between parents or peers. The name comes from the fact that such routes go "down" (following parent–child links) before going "up" (following child–parent links).

[10] Strictly speaking, this is not necessary to prevent valley routing. However, allowing multiple up- or down-segments decreases efficiency and the ability of ASes to enforce path policies.

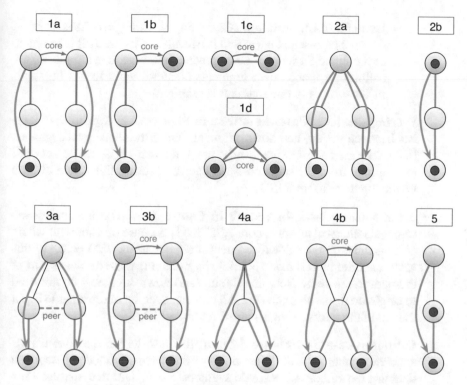

Figure 5.8: Illustration of possible path-segment combinations (i.e., how end
hosts may obtain a forwarding path). Circles represent ASes, gray-
shaded circles represent core ASes. Straight lines represent links.
Orange arrows represent the dissemination of PCBs. Core beacons
are labelled "core" (they can contain more than two ASes and can
traverse multiple ISDs). The links used in the forwarding path are
marked in blue. Peering links are dashed. The small gray circles
represent the end hosts at either end. Paths are symmetric, i.e., they
can be used in both directions, thus there is no distinction between
source and destination.

- **Communication through core ASes:**

  - **Core-segment combination (Cases 1a, 1b, 1c, 1d in Figure 5.8):**
    The up- and down-segments of source and destination do not have
    an AS in common. In this case, a core-segment is required to con-
    nect the source's up- and the destination's down-segment (Case 1a).
    If either the source or the destination AS is a core AS (Case 1b)
    or both are core ASes (Cases 1c and 1d), then no up- or down-
    segment(s) are required to connect the respective AS(es) to the
    core-segment.

- **Immediate combination (Cases 2a, 2b in Figure 5.8):** The last AS on the up-segment (which is necessarily a core AS) is the same as the first AS on the down-segment. In this case, a simple combination of up- and down-segments creates a valid forwarding path. In Case 2b, only one segment is required.

- **Peering shortcut (Cases 3a and 3b in Figure 5.8):** A peering link exists between the up- and down-segment. The extraneous path segments to the core are cut off. Note that the up- and down-segments do not need to originate from the same core AS and the peering link could also be traversing to a different ISD.

- **AS shortcut (Cases 4a and 4b in Figure 5.8):** The up- and down-segments intersect at a non-core AS. This is the case of a shortcut where an up-segment and a down-segment meet below the ISD core. In this case, a shorter path is made possible by removing the extraneous part of the path to the core. Note that the up- and down-segments do not need to originate from the same core AS and can even be in different ISDs (if the AS at the intersection is part of multiple ISDs).

- **On-path (Case 5 in Figure 5.8):** In the case where the source's up-segment contains the destination AS or the destination's down-segment contains the source AS, a single segment is sufficient to construct a forwarding path. Again, no core AS is on the final path.

In the following sections, we describe these cases in detail, following the same structure. Note that we choose one particular direction for each case, but all of them are bidirectional.

5.5.1 Notation

We use the following notation in this section: $HF_{B\underline{A}C}$ stands for a hop field that was generated by AS A, and that can be used for packet forwarding through AS A, with ASes B and C as preceding or succeeding ASes (resulting in the AS sequence BAC or CAB). Note that this is a simplification, as two neighboring ASes can have multiple different connections and even a HF with the same set of interfaces can occur in different segments. If A is the first or last AS in a path segment, the absence of either the preceding or succeeding AS is indicated by the symbol •. Arrows in the figures are oriented according to the direction of beaconing.

5.5.2 Paths Through the Core

It is always possible to construct forwarding paths that traverse one or multiple core ASes. If one or both end hosts are located in core ASes, this is the only option. However, if both end hosts are in non-core ASes, this option is not

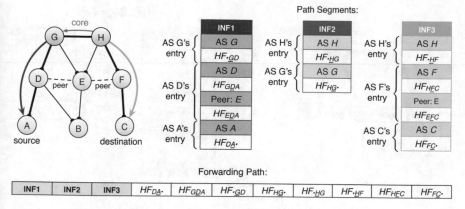

Figure 5.9: An example of a path traversing core ASes.

preferred and is used as a last resort since such a path is usually longer than its alternatives (i.e., peering or shortcut paths), which will be described in §5.5.3.

While less efficient in many cases, creating forwarding paths through the core is less complex than the shortcut options since it simply requires the combination of entire path segments.

5.5.2.1 End Hosts in Non-core ASes

In order to construct a path between two non-core ASes that traverses a core AS, the source requires an up-segment, a core-segment, and a down-segment, which are *connecting*, i.e., the up-segment starts where the core-segment starts or ends, and the core-segment ends or starts where the down-segment starts. This corresponds to Case 1a in Figure 5.8 and a more detailed example is presented in Figure 5.9. In the case where up- and down-segments originate from the same AS, a core-segment is not required (Case 2a). Note that a successful path lookup guarantees that there exists at least one connecting set of path segments.

The destination end host can reverse the received forwarding path as described in §5.6.2 on page 116.

An example of a forwarding path traversing core ASes is presented in Figure 5.9. Combining the path segments is straightforward in this case: the source embeds a series of up-segment hop fields (from its own AS *A* to core AS *G*), two core-segment hop fields (from AS *G* to AS *H*), and three down-segment hop fields (from core AS *H* to the destination AS *C*) as a forwarding path.

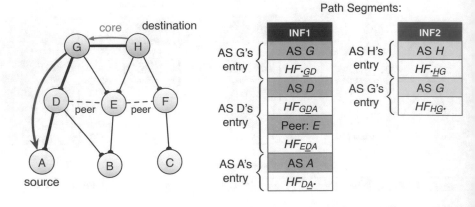

Figure 5.10: An example of path joining when the source is in a non-core AS and the destination is in a core AS.

5.5.2.2 One End Host in the ISD Core

In the case where the source is in a non-core AS, but the destination is in a core AS,[11] the following two options are possible.

First, the source can have a direct up-segment that originates from the destination AS (Case 2b in Figure 5.8). This is possible when the destination AS is a direct or indirect parent of the source AS, and the destination AS has propagated a beacon that was chosen by the source AS as an up-segment.

Second, a direct up-segment to the destination does not exist (Case 1b in Figure 5.8). An example is shown in Figure 5.10 with source in AS A and destination in AS H. The source must obtain an up-segment to a core AS (e.g., to AS G), and a core-segment between that core AS and the destination AS H.

The forwarding path in this case is constructed as a special case of the path through the core from §5.5.2.1. The source just joins all hop fields of the up- and core-segments.

5.5.2.3 Both End Hosts in the ISD Core

Path construction between two core ASes (Cases 1c and 1d in Figure 5.8) is straightforward since path propagation in the core guarantees that there is always a core-segment that connects two core ASes. Thus, the source simply obtains a core-segment and uses it directly.

---

[11]The reverse case works analogously.

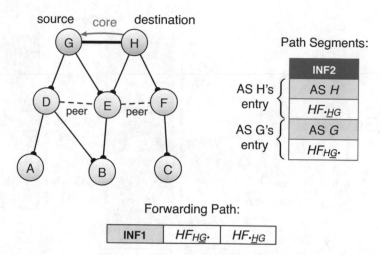

Figure 5.11: An example of a forwarding path between two core ASes.

An example of such a case is presented in Figure 5.11. The source just extracts the hop fields from the core-segment and concatenates them to construct a forwarding path.

5.5.3 Paths Outside the Core

If both end hosts are in non-core ASes, there are several ways to avoid passing through the core (§5.5.2.1) by taking shortcuts. The three variants are described in the following. We assume that after the path lookup process, an end host is provided with sets of up-segments, core-segments, and down-segments.

5.5.3.1 Peering Path

With the proliferation of Internet exchange points (IXPs), communication through peering links is becoming more common. To check whether the destination can be reached via a peering link, the source tries to find a common peering link between the obtained up-segments and down-segments. If such a link exists, a peering forwarding path can be constructed (Cases 3a and 3b in Figure 5.8).

Figure 5.12 presents an example of a path crossing over a peering link. In this example, the two path segments contain the hop fields for the peering link between ASes E and F (i.e., HF_{FEB} and HF_{EFC}).

To minimize traffic overhead, only HFs that are relevant for the forwarding process are part of the forwarding path. In particular, a core-segment and its hop fields are not used to construct a peering path. In §5.5.4, we show an efficient algorithm to find a common peering link in two path segments.

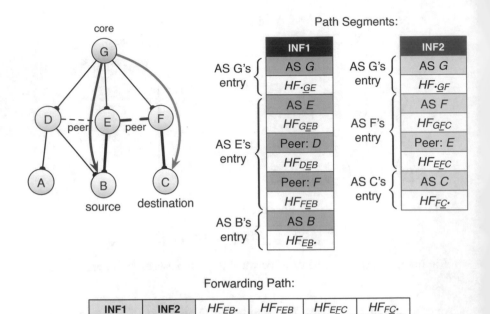

Figure 5.12: Path composition with a peering link.

In the example presented in Figure 5.12, the source specifies its intent to use the shortcut by setting the *Peering* flag on the info fields. It also sets the direction (i.e., *ConsDir*) flag of the hop-field segments. Border routers verify that hop fields are created correctly, i.e., whether their ASes permitted the use of the mentioned peering link. The verification process is described in §5.6.3.

Peering links are allowed between ASes from different ISDs and peering paths with such links are created in the same way as presented. Furthermore, peering links can also exist between a core AS and a non-core AS (but not between two core ASes). In such a case, a peering path may also traverse the core (but no core links).

Peering paths always contain exactly two segments, which both need to have the *Peering* flag. Peering segments are the only ones which are allowed to consist of a single hop field, namely in case the source and destination are located inside one of the ASes connected by the peering link.

5.5.3.2 Shortcut Path (Common AS on Paths)

The up- and down-segments obtained may contain a common upstream AS (Cases 4a and 4b in Figure 5.8). The example in Figure 5.13 shows such a case, where an end-to-end shortcut path can be constructed through AS *D*. In this case, packets do not need to traverse the ISD core, but can be directly forwarded from the common AS *D* to destination AS *B*.

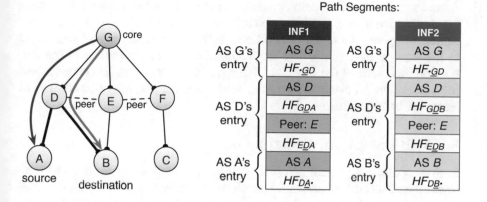

Figure 5.13: Path composition through a common AS.

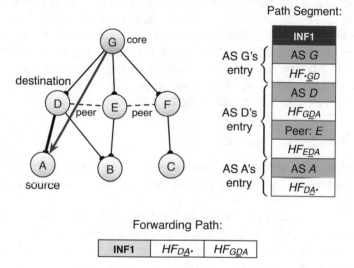

Figure 5.14: Path composition when the destination is contained in the up-segment.

5.5.3.3 Destination AS on Path

In the last case, the destination is an AS on the up-segment (Case 5 in Figure 5.8). No forwarding path for this case should go through a core AS. Also, the up-segment (on which the destination is placed) alone is sufficient to forward packets to the destination.

An example of such a path construction is presented in Figure 5.14, where the destination is within AS D, which is on the up-segment. In this case, the source composes the series of hop fields up to the destination and sets AS D as the destination address. The last router, i.e., the ingress router of the destination AS, verifies the hop field HF_{GDA}. If the verification succeeds and the destination AS matches the router's AS, then the packet is also allowed to terminate at the destination AS.

5.5.4 Efficient Path Construction

In this section, we present an efficient path-construction algorithm, which allows an end host to find and build the shortest forwarding path towards a destination AS. The algorithm is executed by the end host once it has obtained the corresponding up-, core-, and down-segments to the destination AS.

A path segment consists of a list of ASes, where each AS is identified by its ISD and AS identifiers (ISD ID, AS ID) and a list of entries with ingress and egress interfaces. Thus, even if two paths contain the same list of (ISD:AS) tuples, they can differ at the granularity of interfaces (e.g., when two ASes share more than one link). Naïvely, an end host can try to join all possible combinations of up-, core-, and down-segments and select the best end-to-end path among them. However, exhaustive exploration is inefficient if path segments contain many peering links, or if there are many path segments available. We resolve this issue by designing an efficient path-construction algorithm that finds the shortest path(s) in terms of AS hops. The algorithm operates in two steps: the graph-construction step and the path-construction step.

5.5.4.1 Graph Construction

In the first step, the source host creates a weighted and directed graph, based on the up-, core-, and down-segments. The graph is then constructed as follows:

1. For each up-segment of the source host's AS, the algorithm traverses the ASes starting from the source AS and creates a node in the graph for every new AS encountered in the up-segment. The algorithm adds a directed edge from the source AS to the encountered AS, annotated with the hop distance from the source AS. Furthermore, a path identifier is also used to annotate the edge.

   If the encountered AS already has a node in the graph, a new edge is added from the source AS node to the encountered AS node, and the edge is annotated with the same information as described earlier. Thus, multiple edges can exist between two nodes.

   If an AS has a peering link with another peer AS, a new edge is similarly added between the nodes for the source and the peer AS. In this case, the edge is additionally annotated as a peering link.

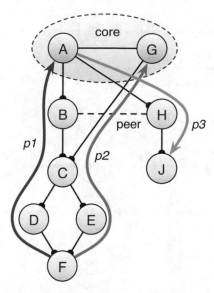

Figure 5.15: An example topology, with up-segments (p_1, p_2) and down-
segments (p_3) obtained by the source (placed in AS F) towards
the destination (placed in AS J).

2. The same procedure is followed for each down-segment of the destination, with two differences. First, the direction of the edge is reversed, so
   that the edge points from each AS on the segment towards the destination AS. Second, peering links are not added, since a valid end-to-end
   path can traverse at most one peering link; if a peering link is traversed,
   it has already been added in step 1 of the algorithm.

3. For core-segments, the algorithm complements the graph as follows.
   First, it selects only the segments that connect the core ASes of the up-
   segments to the core ASes of the down-segments. Then, it traverses
   every AS in the selected core-segments and adds an edge from the core
   AS of the up-segment to the encountered core AS, similarly to the previous two steps. Also, it annotates the edge with the hop distance and
   with a core path flag.

Example. We provide an example to explain how the graph is constructed.
Figure 5.15 shows an example topology with the up-segments (p_1 and p_2) of
the source AS F, and the down-segments of the destination AS J; for simplicity, we omit the core-segments. The source host in AS F obtains the path
segments p_1, p_2, and p_3 and constructs the graph in Figure 5.16, according to
the procedure described earlier.

   All outgoing edges from F point to an AS that is either on an up-segment of
F or has a peering relationship with an AS that is on an up-segment of F (e.g.,

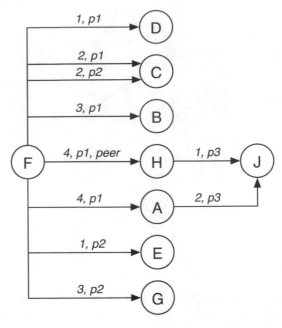

Figure 5.16: Constructed graph based on the up- and down-segments. Edges
          are annotated with the hop distance between the corresponding
          ASes, a path identifier, and a peering-link flag in case a peering
          link is traversed.

AS H). Note that the edges do not correspond to physical links (as shown in
Figure 5.15) and that the weight of each edge denotes the hop distance from the
source AS to the corresponding AS. The reason the graph is constructed in this
way is that the source host should not combine valid path segments to create a
new path segment. For instance, the combination $F \rightarrow E \rightarrow C \rightarrow B \rightarrow A$ is not
valid, as it is a combination of p_1 and p_2.

The same procedure is followed for the down-segments, but the direction of
the edges is reversed. Thus, all edges point towards the destination AS (AS J
in Figure 5.16).

The algorithmic complexity of the graph-construction step is linear with
respect to the number of received path segments (whether up-, down-, or core-
segments), since the algorithm processes every received path segment once.
The intermediate nodes in the graph of Figure 5.16 can be added to a hash
table for efficient lookup, i.e., lookup in constant amortized time, when they
have to be looked up in order to add the corresponding edges in the graph; this
is possible as the number of paths is relatively small.

5.5.4.2 Path Construction

The construction of the graph turns path construction into a simple graph traversal problem. The source host can use existing algorithms to discover the shortest path, all shortest paths, or all paths to the destination. For example, using Dijkstra on the graph in Figure 5.16 yields the shortest path $F \rightarrow H \rightarrow J$, with a total cost of 5.

The paths that are discovered on the constructed graph do not correspond directly to physical paths. However, the edges of the constructed graph are annotated with all the required information so that the source host can discover the actual path and set up the hop fields. For example, the edges of the path $F \rightarrow H \rightarrow J$ inform the source host that the end-to-end path is formed by combining p_1 and p_3 and that it traverses a peering link.

Using Dijkstra for path construction yields an algorithmic complexity of $\mathcal{O}(|L| + |V| \log |V|)$, where $|V|$ is the number of distinct AS nodes in the received path segments and $|L|$ is the number of distinct links in the received path segments; recall that a link is uniquely identified by the interface identifiers and not by the AS identifiers. However, note that the graph-construction step generates a directed acyclic graph (DAG), which enables an even faster shortest-path algorithm: With topological sorting on the DAG, the single-source shortest distances can be calculated in $\mathcal{O}(|L| + |V|)$.

5.6 Packet Initialization and Forwarding

In §5.5, we have described how forwarding paths can be constructed by combining different up-, core-, and down-segments. However, the path header of data-plane packets (§5.4) only contains a sequence of info fields and hop fields without any additional data contained in PCBs. In particular, the path does not contain AS IDs (except for source and destination) and there is no field explicitly defining the type of each segment (up, core, or down).

In this section, we describe the steps required for the source end host and each SCION border router to ensure that only authorized segments can be traversed (§5.3) and the conditions for segment combinations are obeyed (§5.5).

5.6.1 Initialization at Source Host

In addition to extracting and combining the info fields and HFs from the different path segments, the sender must set the *ConsDir* and *Peering* flags and the *SegID* field of all info fields. The *ConsDir* flag is set to true whenever the corresponding segment is traversed in construction direction, i.e., for down-segments and potentially core-segments. It is false for up-segments and "reversed" core-segments. The *Peering* flag is set for both segments if and only if the path is a peering path (Cases 3a and 3b in Figure 5.8). As already stated above, the type of a particular segment is not visible directly in the forwarding path but must be inferred from flags and other information (see §5.6.3).

The *SegID* field of a segment starting at AS_i (always in direction of the final path) must be set to

$$SegID = \begin{cases} \beta_i, & \text{if } Peering = 0, \\ \beta_{i+1}, & \text{otherwise.} \end{cases} \tag{5.6}$$

5.6.2 Path Reversal

When an end host receives a SCION packet, it can use the path directly for sending reply packets using the following steps:

1. Reverse the order of info fields;
2. Reverse the order of hop fields;
3. For each info field, negate the *ConsDir* flag; and
4. Set *CurrINF* and *CurrHF* to 0 and reverse the order of the non-zero *SegLen* fields.

The *SegID* fields remain unchanged.

A similar mechanism is possible for on-path routers, for example to send SCMP messages to the sender of the original packet, see §4.7.

5.6.3 Processing at Routers

SCION border routers need to verify the packet header, update header information, and forward packets through the intra-domain network (ingress routers) or through inter-domain links to the neighboring border router (egress routers). Depending on the type of forwarding path and the location of a particular AS on this path, border routers need to perform different actions. Overall, the position of an AS can be one of the following:

P1 *transit AS* in the middle of a segment with a single HF;

P2 *transfer AS* at the joint of two segments with two HFs;

P3 *source AS*;

P4 *destination AS*;

P5 *pre-peering AS* (before the peering link) with one peering HF:
  P5a not source AS;
  P5b source AS; and

P6 *post-peering AS* (after the peering link) with one peering HF:

P6a not destination AS;

P6b destination AS.

In cases P1–P4, there are additional sub-cases depending on the direction in which the corresponding segments are traversed. In all of these cases, it must be guaranteed that (1) all HFs are verified, (2) all border routers have access to the necessary forwarding information and verification fields, and (3) that the segment combination adheres to the rules defined in §5.5. To simplify the reasoning about necessary verification steps and updates, we formulate several invariants for SCION packets. At a high level, a SCION packet can be in one of two places when it is not being processed at a border router: (1) inside an AS's network being forwarded between border routers and/or end hosts, and (2) on the link between border routers of different ASes.

The following invariants describe the *CurrHF*, *CurrINF*, and *SegID* header fields and the verification status of the HFs during normal forwarding (in the absence of an adversary):

- Inside an AS:
  - The *CurrHF* and *CurrINF* point to the fields relevant for forwarding by the egress border router (if there is one, otherwise to the last HF);
  - The *SegID* field contains the value that is required for verification of the MAC in the current HF; and
  - The current and previous HFs belonging to this AS have been verified (unless the packet originated inside the AS).[12] Strictly speaking, verifying the current HF at the ingress border router is only necessary in the destination AS; however, verifying all HFs before forwarding it through the intra-domain network makes economic sense for all other on-path ASes as well.

- Between ASes AS_i and AS_{i+1} on a particular segment (irrespective of *ConsDir*):
  - The *CurrHF* and *CurrINF* point to the fields relevant for the next ingress border router;
  - The *SegID* contains the value β_{i+1};
  - All HFs of the previous AS have been verified; and
  - All segment combinations until the current segment have been checked.

These invariants ensure that all HFs of a path are verified and that every border router is able to easily access the relevant forwarding information and verification fields. A simplified overview of the invariants and processing at border routers is shown in Figure 5.17.

---

[12]This additional clause is required for the case of the source AS.

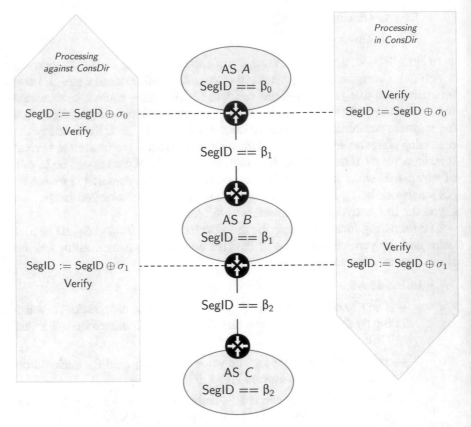

Figure 5.17: Verification of HF authenticator and update of segment identifier
for processing in or against construction direction.

The ingress border router has the following tasks:

1. Check that the interface through which the packet was received is equal
   to the interface in the current HF;

2. Check that the hop field is not expired and within its validity period;

3. (Implicitly) determine the case of the current AS (P1, P2, P4, P5a, or
   P6) based on header fields;

4. (Perform general integrity checks on the packet header);

5. In cases P1, P2, and P4, and if $ConsDir = 0$, XOR $SegID$ of the current
   info field with $HFAuth$ of the current HF, $SegID := SegID \oplus HFAuth$;

6. Compute σ' according to Equation (5.2a) on page 100 using $SegID$ of
   the current info field as the segment identifier and check that it is equal
   to $HFAuth$ of the current HF;

Table 5.1: Table for router at the joint of two segments to decide if segment combination is allowed.

Ingress interface	Egress interface	Decision
CORE	CHILD	
CHILD	CORE	
CHILD	CHILD	OK
CHILD	PEER	
PEER	CHILD	
CORE	CORE	Drop, core loop
PARENT	PARENT	
PEER	PARENT	Drop, valley routing
PARENT	PEER	
CHILD	PARENT	
PARENT	CHILD	Drop, path splicing
other		Invalid configuration

7. In case P2, check that the segment combination is allowed (see rules in Table 5.1);

8. In case P6, check that $ConsDir = 1$ and $Peering = 1$;

9. In case P2, increment the *CurrHF* and *CurrINF* and verify the second HF (steps 2 and 6);

10. In cases P4 and P6b, check that the destination ISD and AS in the packet header are the correct ones (i.e., the border router is located in that AS); and

11. Encapsulate the packet and forward it to the egress border router (based on the interface of the current HF, in cases P1, P2, P5a, and P6a) or the destination host (based on the destination host address, in cases P4 and P6b).

The egress border router has the following tasks:

1. Decapsulate the SCION packet;

2. (Implicitly) determine the case of the current AS (P1–P3, P5, or P6a) based on header fields;

3. (Perform general integrity checks on the packet header);

4. Verify *HFAuth* of the current HF (see steps 2 and 6 for the ingress border router);

5. In cases P1–P3, and if $ConsDir = 1$, XOR *SegID* of the current info field with *HFAuth* of the current HF, $SegID := SegID \oplus HFAuth$;

6. Increment *CurrHF*; and

7. Forward packet to the neighbor AS.

If any step in the verification fails at either the ingress or the egress border router, the router drops the packet and should reply with a "parameter problem" SCMP message, see §4.7, specifying the location and type of problem.

Note that in cases P1, P2, P5a, and P6a, a HF gets verified twice, once at the ingress border router and again at the egress border router. In principle, it would suffice to do it either at the ingress or egress border router. However, parsing the HF is necessary at both border routers and the cryptographic verification creates little overhead. Furthermore, both validations have their own advantages: Validation at the ingress border router has the advantage that incorrect packets can be dropped before using resources to forward them; validation at the egress border router can detect any modification within the network or maliciously crafted packets by end hosts.

5.7 Path Revocation

In contrast to the IP Internet, SCION packets are not dynamically rerouted in the network in case of failures or policy changes due to the packet-carried forwarding state. This feature is a major security advantage, as it provides control and transparency to end hosts and prevents hijacking attacks. However, it comes with new challenges. Path segments are generally valid for several hours and can be stored in many locations throughout the network due to the multi-level caching approach (see §4.5). The question is, how do we ensure that end hosts are notified of failures and have access to up-to-date paths even when links fail or policies change?

Concretely, end hosts must be informed about updated path segments in two cases: (1) proactively in case of changes in routing policies, and (2) reactively in case of link or router failures on the path.

The first case is usually not time-critical and can be addressed through expiration timestamps on path segments in conjunction with ASes ceasing to advertise these paths. However, cached copies will still be usable for as long as the path is valid. We assume that an AS is committed to a path segment it registers for the entirety of its lifetime.

For reactively revoking a path segment due to a link failure on the path, time plays a critical role; the faster end hosts are informed about a faulty path segment, the fewer will attempt to use that segment. This directly translates into a lower amount of wasted resources and better quality of experience. For the rest of this section, we focus on this case of reactively revoking paths.

5.7.1 Challenges

There are complex challenges and potential corner cases related to revocations, which we will discuss in the following.

Time Granularity. The duration for which paths are revoked is a crucial parameter for the efficiency of the system with a fundamental trade-off: Increasing the duration of revocations (or even permanently removing path segments from caches) reduces the number of revocations that need to be processed; however, this can also cause unnecessarily long nonuse of certain links in case of short-term or intermittent failures and cause other overhead as it requires re-registering paths.

Scalability. To ensure scalability and also to prevent denial-of-service (DoS) attacks by malicious entities in the network infrastructure, it is critical to achieve low computational, storage, and bandwidth overhead. Thus, a revocation must not require involved network elements to keep an excessive amount of state or to generate a large number of additional messages within the network. Additionally, a revocation must be efficiently verifiable to prevent overwhelming verifiers through many (possibly forged) revocations. Finally, revocations should be short, to minimize communication overhead.

Security. Path revocation directly affects the paths used by end hosts. Therefore, the system needs to prevent unauthorized or malicious parties from revoking path segments. The system must thus ensure that revocations are authentic, i.e., only the operator of an interface should be able to revoke that interface. It must also be impossible to replay a recorded revocation with the effect of removing a valid path segment (resistance against replay attacks).

5.7.2 Principles

As any revocation system is necessarily imperfect and (short) delays are unavoidable, SCION is designed such that it remains functional even in the face of failures *without* a dedicated revocation system. Even with a revocation system in place, applications should not rely on the system and should continue to function without it. Specifically, applications should have state machines that switch from a broken path even if no revocation was received on that path (e.g., by relying on transport-layer acknowledgments).

Most fundamentally, a SCION network with revocations should work strictly better than without the system. If a revocation feature does not bring a strict improvement (e.g., improves certain cases but deteriorates others), it should not be implemented.

5.7.3 Design

There are two main approaches to handling path revocations:

- **Control plane**: Disseminate revocations through the path infrastructure including border routers, beacon services, and path services.

- **Data plane**: Directly inform end hosts about faulty paths when they attempt to use them.

5.7.3.1 Path Revocation in the Control Plane

The first SCION book describes the control-plane approach to path revocation [413, §7.3]. This approach has the advantage that faulty paths are removed directly from path services such that end hosts do not even attempt to use these paths. Unfortunately, this system has major disadvantages as well.

Control-plane path revocation comes with high complexity and thus opens up attack vectors for DoS attacks. For example, a single revocation can trigger a large amount of additional processing, and keeping track of already handled revocations requires complex logic.

Furthermore, as already discussed above, removing paths from path services can cause issues for short-term failures. Most notably, an application would not even be informed about such a path and would not attempt to use it even after the failure has been resolved and even if it provides substantially better service than the alternative.

In certain cases, the control-plane approach can even cause a temporary loss of connectivity even though a functioning path exists at all times. Consider two ASes that are connected through two different paths. Both of them experience short-lived but non-overlapping outages within a short period. With a control-plane revocation system, both paths would be removed from path services and no paths could be served to end hosts until they had been registered again.

In summary, control-plane path revocations suffer from subtle issues that overall violate the principle of "strict improvement" stated above. Therefore, the current SCION implementation does not include control-plane path revocations (anymore).

5.7.3.2 Path Revocation in the Data Plane

Instead of control-plane path revocations, the current SCION implementation implements path revocation in the data plane. The data-plane approach to path revocations is substantially simpler than the control-plane and fulfills the "strict improvement" principle. Path revocations are not propagated to the control-plane services at all. Instead, routers that cannot forward a packet because of a failure send an authenticated SCMP message to the source host reporting either an "external interface down" or "internal connectivity down" issue. The end host can then directly switch to a backup-path or fetch additional paths from the local path service. Additionally, it is up to the end host whether it wants to remove the revoked path from its local path store or keep it to use it again after the revocation period has expired.

The disadvantage of this approach is that faulty paths can remain in the storage of path services for several hours until their eventual expiration. However, this is only an issue for long-term failures, which can be resolved in an automated way similarly to proactive revocations due to policy changes.

5.7.3.3 Optimizations

There are several optimizations that end hosts can make to improve their quality of experience without a revocation system and with a data-plane-only revocation system.

Backup Paths. End hosts should fetch multiple (ideally disjoint) paths from the local path service to any destination such that they can switch immediately if their preferred path fails.

Proactive Fetching of DRKeys. End hosts can proactively fetch SCMP-DRKeys for all on-path ASes and the destination when starting to use a path. This allows the end host to authenticate any SCMP error messages they might receive without any additional delay.

Probing Traffic. If a failure has been reported for a preferred path, the end host can continue to probe that path (e.g., with echo requests) to quickly detect when that path becomes available again.

Simultaneous Requests for Latency-Critical Applications. To prevent additional delay for connection setup in case of a path failure, the initial request can be sent over multiple paths simultaneously. However, as this causes additional overhead, this should be reserved to latency-critical applications.

5.7.4 Link-Failure Detection

SCION border routers keep track of the state of inter-domain links (i.e., on each of their external interfaces) and the state of intra-AS connectivity to sibling border routers (i.e., between internal interfaces). This connectivity information is used when creating SCMP External Interface Down and Internal Connectivity Down messages (see §4.7.2) to signal connectivity issues on a path.

To that end, SCION uses Bidirectional Forwarding Detection (BFD) [283]. BFD is a protocol intended to detect faults in the bidirectional path between two forwarding engines, including interfaces, data link(s), and, to the extent possible, the forwarding engines themselves, with typically very low latency. It operates independently of media, data protocols, and routing protocols.

BFD packets, as they are described in the relevant RFCs, are placed directly into a SCION payload with no additional intermediate protocol. The *NextHdr* field in the SCION common header indicates that this is a SCION/BFD packet.

A SCION router creates one "external" BFD session for each SCION interface that it owns. Its BFD peer is the SCION router in the neighboring AS. The associated BFD packets use the SCION one-hop path type (see §5.4.1). Furthermore, a SCION router creates one "internal" BFD session for every other SCION router instance within the same AS. The associated BFD packets use an empty SCION path. While it would have been possible to only rely on the intra-domain routing and forwarding protocol for the internal BFD sessions, including a SCION header ensures that the SCION forwarding engine of the BFD peers is operational.[13]

Using BFD, operators have the possibility to fine-tune associated time-out timers to strike a balance for failure-detection time while also ensuring that the number of false positives is minimized.

5.8 Data-Plane Extensions

SCION supports extensions similar to IPv6 [158]. At the time of writing, two extension headers are defined: a hop-by-hop (HBH) options header and an end-to-end (E2E) options header, similar to the hop-by-hop options and destination options of IPv6, respectively. Also similar to IPv6, extension headers can be chained via the "next header" field. Each extension header can be present at most once and, if both are present, the HBH options must come before the E2E options. Both options headers can carry a variable number of type-length-value encoded options.

Hop-by-Hop (HBH) Options. The HBH options are used to carry optional information that may be examined and processed by every router along a packet's delivery path.

End-to-End (E2E) Options. The E2E options are intended for optional information that is only processed at the source and/or destination host. One notable application of the E2E options header is the SCION Packet Authenticator Option described in §3.3.

Header Handling Revision. Please note that extension header handling has been revisited since the publication of the first SCION book [413], where any

---

[13]Without the SCION header, the internal BFD sessions would only test for connectivity on the intra-domain network stack, which could lead to false negatives in case the intra-domain network stack is working properly, but the SCION stack encounters issues.

number of chained hop-by-hop (HBH) and end-to-end (E2E) extensions were permitted. The new design bundles individual extension options into at most two top-level extension header containers, HBH and E2E types, which can be skipped whole. This simplifies the packet parsing and handling logic, as at most two extensions need to be skipped before accessing the L4 payload. In addition, the extension header alignment was changed from 8 to 4 bytes.

Part II

Analysis of the Core Components

6 Functional Properties and Scalability

CYRILL KRÄHENBÜHL, MARKUS LEGNER, ADRIAN PERRIG, SIMON
SCHERRER, SEYEDALI TABAEIAGHDAEI, JOEL WANNER*

Chapter Contents

6.1   **Dependency Analysis** **130**

    6.1.1   SCION Defense Mechanisms against Circular Dependencies   .   131

    6.1.2   Assumptions   132

    6.1.3   Analysis Methodology   132

    6.1.4   Resolution of Apparent Cycles with Finer Granularity   132

    6.1.5   Results   133

6.2   **SCION Path Policy** **135**

    6.2.1   Path Policies among Core ASes   135

    6.2.2   Differences between SCION and BGP Policies   136

    6.2.3   Approaches to Implementing Path Policies in SCION   138

    6.2.4   Gao–Rexford Model, Interconnection Agreements, Convergence   139

    6.2.5   Sample Path Policies   142

    6.2.6   Secrecy of Routing Policies   145

    6.2.7   Influencing End-Host Behavior   145

6.3   **Scalability Analysis** **148**

    6.3.1   High Impact on Scalability: Core Beaconing   148

    6.3.2   Medium Impact on Scalability   149

    6.3.3   Low Impact on Scalability   150

6.4   **Beaconing Overhead and Path Quality** **150**

    6.4.1   Data Sources and Simulation Setup   151

    6.4.2   Overhead Comparison of SCION, BGP, and BGPsec   153

    6.4.3   Path-Quality Evaluation   154

---

*This chapter reuses content from Hu, Klausmann, Perrig, Reischuk, Shirley, Szalachowski, and Ucan [250], Krähenbühl, Tabaeiaghdaei, Gloor, Kwon, Perrig, Hausheer, and Roos [305], Scherrer, Legner, Perrig, and Schmid [453].

In Chapters 2–5, we have presented the main components of the SCION architecture: beacon service, certificate service, and path service; the concept of ISDs and their trust root configuration (TRC); the CP-PKI and symmetric-key infrastructure; the path-exploration (beaconing), -registration, and -lookup mechanisms; and the SCION forwarding process.[1]

These are only the main components—name resolution and end-entity authentication will be presented in Chapters 18 and 19, transition mechanisms like the SCION–IP gateway (SIG) will be discussed in Chapter 13, and various extensions to achieve stronger availability guarantees will be presented in Chapters 8 and 10. Nevertheless, even the subset of main components represents a complex system with numerous interaction points between subsystems. It thus requires a careful analysis of both functional and non-functional (scalability, efficiency) properties.

Such an analysis is presented in this chapter. The security properties achieved by SCION are discussed in Chapter 7.

6.1 Dependency Analysis

The SCION architecture is a large-scale distributed system, designed to be highly reliable. Traditionally, large-scale systems evolve incrementally, leading to a complex set of dependencies between its subsystems. Without active prevention efforts, the resulting system will likely exhibit circular dependencies that threaten system reliability in two ways: During continuous operation, a dependency loop can amplify an outage of a component and cause a large-scale service interruption. Moreover, when re-starting the system or part of it, a dependency loop can at best delay the starting process and at worst completely prevent it.

This problem is not unique to computer networking: In electricity systems, power plants often rely on external power supply to start their operation, but this power supply draws on other power plants. Such a circular dependency becomes problematic in the case of a large-scale outage, where it could prevent the power grid from restarting its operations. Therefore, special measures are taken in the system to ensure that a so-called "black start" is possible [217] using hydroelectric power plants that can resume power production without an external supply. Our goal in SCION is to achieve a similar property: That the Internet can start up even after large outages or attacks, in addition to avoiding cascades of outages caused by fragile interdependencies.

Today's Internet infrastructure has grown incrementally over decades and consists of a complex network of dependencies, which remain poorly under-

---

[1]For the beacon-, certificate-, and path service, see §2.1; for the concept of ISDs and their trust root configuration (TRC), see Chapter 2 and §3.1; for the CP-PKI and symmetric-key infrastructure, see Sections 3.1 and 3.2; for the path-exploration, -registration, and -lookup mechanisms, see Chapter 4; and for the SCION forwarding process, see Chapter 5.

stood due to a lack of rigorous analysis. Moreover, the structure of the Internet ecosystem amplifies the impact of outages due to widespread reliance on third-party services such as DNS or CA providers [282]. Several high-profile outages have demonstrated how devastating this effect can be: For instance, the 2021 Facebook outage showed how an erroneous BGP configuration change could take down large parts of the company's network. Network engineers struggled to revert the change for a prolonged period of time, as they were locked out from their configuration interface, which relied on the network connectivity that was blocked by the outage [364]. This ultimately resulted in one of the largest outages ever recorded.

A secure and reliable routing architecture must be designed specifically to avoid circular dependencies. Operating SCION networks in the real world for several years with multiple bootstrapping phases has provided experimental evidence for the absence of circular dependencies.

However, experimental evidence is clearly insufficient for a system as important as the Internet. For this reason, we present a dependency analysis for SCION in this section: We demonstrate what concepts SCION uses to prevent circular dependencies and describe our methodology to construct and analyze the dependency graph of the whole system. Based on this methodology, we show that none of the SCION systems can be blocked by circular dependencies.

6.1.1 SCION Defense Mechanisms against Circular Dependencies

SCION uses a number of design principles that prevent common sources of circular dependencies:

- **Neighbor-based path discovery:** Path discovery in SCION is performed by the beaconing mechanism. In order to participate in this process, an AS only needs to be aware of its direct neighbors. As long as no path segments are available, communicating with the neighboring ASes is possible with the one-hop path type (§5.4.1), which does not rely on any path information.

- **Path segment types:** SCION uses different types of path segments to compose end-to-end paths. Notably, a single path segment already enables intra-ISD communication. For example, a non-core AS can reach the core of the local ISD simply by using an up-segment fetched from the local path storage, which is populated during the beaconing process.

- **Path reversal:** In SCION, every path is reversible—i.e., the receiver of a packet can reverse the path in the packet header to send back a reply packet without having to perform a path lookup.

- **Availability of certificates:** In SCION, every entity is required to be in possession of all cryptographic material (including TRCs and certificates) that is required to verify any message it sends, see also §3.1.9. This (together with the path reversal) means that the receiver of a message can always obtain all this necessary material by contacting the sender.

6.1.2 Assumptions

In practice, it is not possible to initiate networked communication from a setting in which participants have no initial knowledge at all. In the process that is usually known as *bootstrapping*, some information is configured by the operator or exchanged via out-of-band channels.

In SCION, the following assumptions are required to bring an AS fully online:

- **Local configuration:** Every AS must have a configuration for all links to its neighboring ASes and the corresponding border routers.

- **Base TRC:** The base TRC of the local ISD must be available (see §3.1.8). It contains the information (certificates, core AS numbers) that is required to set up initial communication.

- **Coarse time synchronization:** The AS infrastructure must be given a rough initial time, and each AS must be approximately synchronized with its neighbor ASes. This can be achieved by configuring the time manually or with some other out-of-band mechanism.

- **Intra-AS communication:** As SCION is inherently an inter-domain architecture, we assume that intra-domain communication (i.e., communication between end hosts and infrastructure nodes within an AS) is functional.

6.1.3 Analysis Methodology

In order to verify that SCION does not contain circular dependencies given our assumptions, we have built a model of its subsystems represented in a directed graph. An edge from node u to node v expresses the statement "component u depends on component v". A circular dependency can then simply be identified by a cycle in this graph.

6.1.4 Resolution of Apparent Cycles with Finer Granularity

In order to model the SCION architecture with such a representation, we need to choose an appropriately fine level of granularity. The following example demonstrates the false positives that can initially arise with a corse-grained

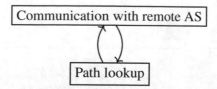

(a) A less detailed representation that contains a circular dependency.

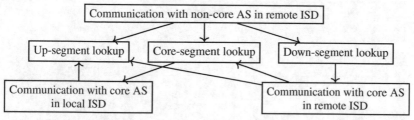

(b) A more detailed representation, distinguishing between different types of communication and path lookups.

Figure 6.1: An excerpt of the SCION dependency graph, showing two different choices for the level of granularity to model the same part of the system.

model, and how they are resolved by refining it. In SCION, communicating with a host in a remote AS requires a path. However, in order to obtain the path it is necessary to communicate with a remote AS. This leads to an apparent cycle in the dependency graph shown in Figure 6.1a.

But if we increase the level of detail in the dependency analysis, it becomes apparent that this is not the case for the actual system. By applying a case distinction on the nodes by type of path segment, we obtain the dependency graph in Figure 6.1b.

Concretely, we highlight that SCION paths consist of different types of segments that are available at different path services: Up-segments can be fetched from the local path service based solely on intra-domain communication—i.e., without requiring any inter-domain SCION paths. These up-segments can then be used to communicate with core path services in the local ISD, which enables fetching core-segments. Up- and core-segments can be combined to fetch down-segments from remote core ASes. Finally, the three types of segments can be combined to communicate with any remote non-core AS.

6.1.5 Results

The full dependency graph corresponding to our model is displayed in Figure 6.2. On this graph, we run an algorithm to search for cycles. This process has determined that the graph representation of SCION's subsystems is

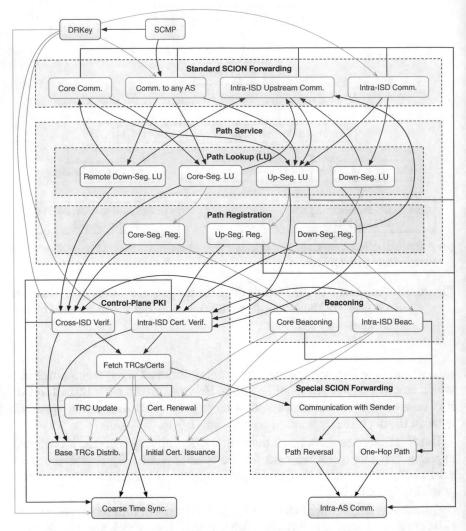

Figure 6.2: Graph representing dependencies between the main components of
           SCION. Black arrows indicate permanent dependencies. Gray ar-
           rows indicate periodic dependencies. Red nodes represent axioms.
           Blue and green nodes refer to the control plane and the data plane,
           respectively. The purple node is a supporting system.

a-cyclical, from which we can follow that the architecture does not suffer from
circular dependencies.

6.2 SCION Path Policy

SCION can support a rich set of path policies, providing ISPs with fine-grained control over permissible paths. This is an important property, as ISPs need mechanisms that implement their traffic policies to match their business model. In today's Internet, ISPs define their routing policies through BGP; in this section, we will highlight the differences between BGP routing policies and SCION path policies.

Before we discuss the differences in more detail, we offer an observation. BGP is sometimes described as the gold standard for routing policies. However, BGP is mainly able to express destination-based policies, but no general source-based policy. It might be concluded from this that current Internet routing policies have evolved towards what is expressible by BGP, and not that BGP has evolved to accommodate the most-desired routing policies (as it has remained virtually unchanged since the late 1990s). In fact, we anticipate that familiarity with SCION path policies may raise awareness of BGP's inadequate expressiveness.

Three fundamental points complicate the definition of SCION path policies. First, because SCION's path exploration is fundamentally different from BGP, policy construction differs from today's Internet. Second, announcing multiple paths creates a challenge (compared to the current situation where ISPs only need to approve and provide a single path to each destination). To our knowledge, there are currently no multipath policy definitions available. Third, while client-based path selection allows better satisfaction of diverse client preferences, it can also pose challenges to an ISP's business model. For example, a client may select a more expensive path, incurring a higher cost for the ISP.

We will first compare SCION's path policy and BGP's routing policy among *core ASes* and show that any BGP policy is also expressible in SCION. We will then focus on non-core ASes and first explain the fundamental differences between BGP routing policy and SCION path policy in terms of expressiveness. Subsequently, we will describe the SCION path-policy framework, illustrate with specific examples how BGP policies can be translated, and what policies SCION can naturally express that BGP cannot. Finally, we discuss how client-based path selection interacts with ISP-based traffic engineering.

Note that extensions like COLIBRI, which will be discussed in §10.2, provide ISPs with additional and more fine-grained ways of shaping traffic through the management of bandwidth reservations.

6.2.1 Path Policies among Core ASes

Among core ASes, SCION's path exploration, registration, and lookup correspond closely to the BGP routing process (except for the cryptographic verification and the multipath capabilities): Every core AS learns the full path to every other core AS via core PCBs, similar to BGP update messages, and can independently decide which ones to use (i.e., register in the local path service) and forward. Any BGP import and export policy can thus be mapped directly to corresponding SCION path policies. However, in contrast to BGP where the actual forwarding path can differ from the one announced, SCION ensures that packets traverse the network along the paths explored by PCBs. Moreover, SCION path-exploration messages do not relate to specific IP prefixes, which are the central element of BGP announcements. As a result, the path discovery in SCION does not involve the merging of multiple BGP update messages by means of prefix aggregation.

For the remainder of this section, we consider policies of non-core ASes unless explicitly stated.

6.2.2 Differences between SCION and BGP Policies

Differences Between SCION Path Policy and BGP Routing Policy

Intra-ISD path exploration in SCION starts from core ASes and extends paths towards the leaf ASes—whereas paths in BGP are constructed from leaf ASes towards all other ASes. This difference suggests that SCION can express a different set of path policies from BGP (but as we describe below, additional mechanisms such as beacon extensions and hop-field encryption can enable a full set of policies).

Example. To illustrate the differences between BGP routing policies and SCION path policies, we present an example in Figure 6.3. In BGP, the routing updates originate at the destination and are flooded through the network. For instance, AS E sends a BGP update message to its provider D, which further disseminates it upstream to A and C. Similarly, C further disseminates the update to A and B, and B sends it on to A. At this point, A can decide how E will be reached. Traffic follows the reverse path of the updates, so traffic destined for E can traverse path A-D-E, or path A-C-D-E, or path A-B-C-D-E. This example demonstrates the limitation that an AS can only control over which downstream path traffic is sent, but has no upstream control. Specifically, traffic destined for E cannot be controlled by E; it will flow over the link that A selects. Moreover, BGP cannot express source-based policies (except the next hop to which an update is sent) as we are going to illustrate later in this section.

6.2.2.1 Control of Non-core ASes

In SCION, ASes can register different segments as up- and down-segments and thus influence their visibility to local and remote end hosts. However, SCION paths are bidirectional, and therefore ASes cannot implement neither source- nor destination-based policies out of the box. Instead, the simplest SCION policies relate to the paths towards the ISD core.

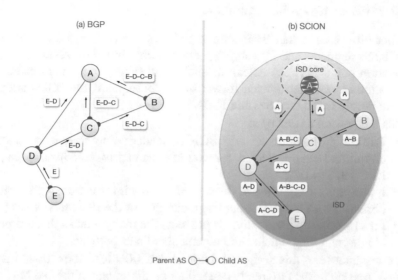

Figure 6.3: Simple network topology and two approaches to path exploration: BGP (where the destination emits BGP updates) and SCION (where core ASes emit PCBs).

Example. Consider that AS A in Figure 6.3 is part of the ISD core and initiates PCBs that it sends to B, C, and D. B and C continue to send the PCB to D. D can now decide which PCB(s) to send on to E based on its path policy. Since SCION is a multipath architecture, D would most likely forward the three PCBs, but to illustrate the policy options we assume it only sends a single PCB. This PCB represents both the path towards the destination (used as a down-segment), as well as the path from the source (used as an up-segment). Thus, it is challenging to compare the policy expressiveness of SCION directly with that of BGP.

This example also illustrates the gain in control for Tier-2 and leaf ASes. In BGP, they can rely on a few techniques to steer inbound traffic, i.e., using BGP Multi-Exit Discriminator (MED), AS path prepending, or selectively advertising longer prefixes. However, these techniques are in practice often filtered by upstream ASes [152], having limited effect. In addition, AS path pre-pending is will not be available in a secure version of BGP, for instance in BGPsec [330, 331]. Single-homed ASes (e.g., AS E in Figure 6.3), fully depend on their upstream ASes to perform path control. Overall, decisions are in the hands of their (direct and indirect) providers. In contrast, SCION enables these ASes to choose over which paths they want to send and receive packets from a set of candidate paths. Of course, these choices can still be restricted by Tier-1 ASes through pricing and other contractual clauses. Overall, SCION enables a more balanced path choice among ASes.

6.2.2.2 AS or IP-Prefix Granularity

Another difference is that BGP routing policies can be based on IP prefixes. In SCION, segments and forwarding paths are only based on ASes (and interconnection points between ASes); as such, neither the source nor destination address influences inter-domain forwarding. Consequently, SCION path policies are purely based on ASes and ISDs.

This has several advantages:

- The end host addresses in the SCION header do not need to be globally unique and can be selected, for example, from private address space (i.e., RFC 1918 [431]);
- It prevents an explosion of the size of routing information, which can be observed with the increasing fragmentation of the IP address space; and
- From a security perspective, it prevents the many attacks that are possible through illegitimate announcements of sub-prefixes.

A consequence of this design decision is that ISPs lose some flexibility in terms of routing traffic differently depending on the subnet of the end host. For example, an ISP in a large country may want to send and receive traffic through external interfaces at an IXP close to the individual end hosts. A solution to enable such decisions in SCION is to split a large AS into several smaller ones, each corresponding to an IP prefix with specific routing policies in BGP. Such policy-defined ASes pose no scalability issue because the hierarchical structure of ISDs allows highly scalable path discovery among non-core ASes.

6.2.3 Approaches to Implementing Path Policies in SCION

The main mechanism to implement path policies in SCION is *beaconing control*. Each AS can decide which PCBs to send on, which peering links to add to the PCB, and which PCBs to register as up- and down-segments. This enables implementation of a first basic level of path policy. An AS can decide which upstream ASes should be avoided when propagating PCBs to downstream ASes, or which path properties are preferred (see §4.4).

In addition, two other approaches can be added to enable more fine-grained path policies:

- **Explicit path policy transmitted as beacon extension:** An AS adds information to the PCB to explicitly indicate which paths are permissible. This can include a list of downstream ASes that are (or are not) permitted to use the PCB's path. The granularity of path policy can be fine-grained to the level of per-link policies, so even peering links can be annotated with a policy in the PCB (see §4.1.2). Downstream ASes that violate a usage policy are accountable for their actions as they sign the PCB and a policy violation is detectable when the path is registered at the core path service.

When the path is only used against the explicit path policy but not registered, detection is more challenging. To detect such misuse, an AS can monitor hop fields (HFs) used in traffic and, in the case of HFs that were not registered by any of the downstream ASes, it can verify whether the source or destination AS is allowed to use the path. Furthermore, violation by an intermediate AS can be detected by tracing the intermediate ASes in a sequence of HFs and verifying compliance with the explicit path policy. Although detection requires operational effort, it is likely to be a sufficient deterrent for misbehavior.

- **Hop field encryption with explicit path activation:** An AS that intends to encode more sophisticated policies can encrypt the hop field in the PCB to make it unavailable unless it is activated by the leaf AS. Activation requires sending a special packet through the network with the entire end-to-end path, so that on-path ASes can inspect and activate the path by decrypting the hop field if they permit the path. A unique policy identifier can be added to the PCB to enable leaf ASes to optimize which paths are attempted to be activated.

At the time of writing, the implementation of SCION supports basic beaconing control; explicit policies and explicit path activation will be included in a future release.

We observe that if an ISP only announces encrypted hop fields that require explicit path activation, end-to-end path setup is slowed down and in the worst case requires several attempts to find a working path. We thus require that each AS must make available at least one upstream path, called the *default path*, that supports arbitrary end-to-end paths. This ensures quick establishment of an end-to-end path that is supported by all upstream ASes. To achieve high availability and rapid failover, two disjoint paths that support arbitrary end-to-end paths should be made available. The multipath system will then continue to seek additional paths as the connection progresses, finding new paths that optimize latency, bandwidth, loss rate, ASes traversed, etc.

6.2.4 Gao–Rexford Model, Interconnection Agreements, and Convergence

As a path-aware networking architecture, SCION enables new types of interconnection agreements that are not possible in today's Internet. Nowadays, interconnection agreements are heavily influenced by the *Gao–Rexford* (GR) model [121, 195], which prescribes that traffic from peers and providers not be forwarded to other peers or providers. More precisely, a policy compliant with the GR model must implement the following two sub-policies capturing two separate aspects of the GR model:

- **GR preference:** The preference policy captures the *economic aspect* of the GR model and is based on the observation that ISPs want to maximize profits: When an ISP has a choice of where to send traffic to, then the preferred order is first towards a customer, second over a peering link, or finally to a transit provider. The reason is that traffic sent to a customer earns a profit from the customer, traffic sent over a peering link has zero marginal cost, but traffic sent to a transit provider incurs a cost. In BGP, the way these preferences are expressed is through the LocalPref setting, which is assigned a different value depending on whether the update was received from a customer, a peering link, or a provider. Gill et al. report that over 85% of ISPs utilize this policy [206].

• **GR export:** The export policy captures the *stability aspect* of the GR model. It dictates that customer routes be announced everywhere, but routes learnt from peering and provider links only be announced to customers. This policy ensures *valley-free* routing [195], where traffic never flows "down" the AS hierarchy to a customer and back "up" towards the destination. By following this simple policy, Gao and Rexford were able to show that BGP converges [195]. Gill et al. find that over 70% of ISPs apply this policy.

The fundamental issue with convergence in BGP is the next-hop principle: ASes can only select a next-hop AS for their traffic and thus rely on that AS to forward the traffic along the route that was originally communicated via BGP. If this assumption is violated—even temporarily—routing loops can arise. Put differently, in a BGP/IP-based Internet, all ASes need to share a common view of the used forwarding paths.

Several examples of topologies and policies have been constructed which cause convergence issues [221]. Some examples, known as "BGP wedgies" [220], converge non-deterministically, such as the DISAGREE example. While this non-determinism is clearly undesirable, it does not constitute a fundamental problem for convergence in BGP. However, other slightly more complicated topologies like the BAD GADGET do not converge at all and cause persistent route oscillations [221]. Even worse, seemingly benign topologies and policies may easily reduce to the BAD GADGET in case one network link fails [221].

Adherence to the GR model is a sufficient but not necessary condition to prevent these issues. However, the susceptibility to oscillations shows that GR-violating policies need to be implemented very carefully and with coordination among all involved parties to ensure routing stability. As a consequence, "sibling" agreements in which two ASes provide each other access to their respective providers and peers (as presented above) generally only exist between ASes controlled by a single organization.

In contrast, SCION does not require all ASes to converge to a uniform set of forwarding paths to use, due to the packet-carried forwarding state. The stability aspect of the GR model is therefore no longer relevant for SCION, and SCION thus presents the exciting opportunity to create and use paths violating

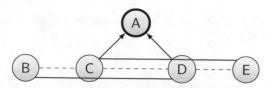

Figure 6.4: Example AS topology with mutuality-based interconnection agreements (the red lines). Peering links are shown as dashed lines, parent–child links as "parent → child".

the GR model without the risk of instability—provided that such paths can be made economically viable. In particular, we observe that while SCION may no longer require the GR conditions for reasons of stability, it must still respect the economic logic that makes the GR model a rational forwarding policy.[2]

Example. Consider the topology in Figure 6.4: While creating the path B–C–D may not lead to convergence problems in SCION, the path is still economically undesirable for AS C, because it would incur internal forwarding costs but does not earn any revenue from its peers B and D. This can be resolved through interconnection agreements based on *mutuality*, a concept that is already present in peering agreements today, but can be leveraged to set up more complex and flexible agreements. Concretely, mutuality means that the mentioned example path B–C–D could be rendered economically viable for C by requiring a *quid pro quo* from D, the main beneficiary of the path. For example, D could offer the path C–D–E to C such that both C and D could save transit cost for accessing ASes E and B, respectively, but incur additional internal forwarding cost.

If the flows over the new path segments are properly balanced and especially if the new path segments allow ASes C and D to attract additional revenue-generating traffic from their customers, such unconventional agreements can be mutually beneficial. Hence, SCION offers opportunities for profit maximization which are not present in today's Internet.

These novel interconnection agreements can be implemented easily in the core, where PCBs are created and forwarded by all core ASes. While all interconnections are abstracted as "core links", they are still based on underlying agreements between ASes. Non-core ASes could forward some PCBs to their peers, potentially with explicit path policies restricting their use.

We have also investigated conditions for the economic feasibility and negotiation mechanisms for these novel agreements as well as their impact on path diversity and quality. Our results show that through these agreements, ASes can reach many destinations through shorter paths and select among a substan-

---

[2]The valley-freedom is enforced through the fact that intra-ISD PCBs only follow parent–child links (see §4.2) and the restrictions on segment combinations (see §5.5).

tially larger set of possible paths. Details can be found in Scherrer, Legner, Perrig, and Schmid [453].

6.2.5 Sample Path Policies

To demonstrate how one can express path policies in SCION, we will consider some other popular routing policies that are used in BGP, and present additional policies that cannot be expressed in BGP. For the current BGP policies, we discuss the policies presented by Gill et al. [206] beyond the GR model: next-hop routing, consistent export, and most stable path. We also discuss hot-potato routing, an example of a complex BGP policy, and finally a source-based policy that BGP cannot express.

6.2.5.1 Next-Hop Routing Policy

Gill et al. conducted a study where they investigate the various BGP routing policies used in practice [206]. It was found that the majority of ISPs use simple routing policies that are easy to configure and maintain. For instance, around 60% of ISPs use a next-hop routing policy, which implies that the BGP LocalPref setting is solely based on the next hop (i.e., the incoming link of the BGP update) and the destination, and not on the intermediate path. This creates a predictable, simple routing policy, and enables implementation of the GR preference policy.

SCION enables control over which PCBs are sent to which next-hop ASes, allowing for policies that take the previous and next hop into account. As SCION paths are bidirectional, both source- and destination-based policies can be expressed, rather than only destination-based policies as in BGP.

6.2.5.2 Consistent Export Routing

Gill et al. also report that consistent export routing is a popular policy, with 65% of the ISPs deploying it [206]. In this policy, if a route with LocalPref $= \ell$ is exported, then routes with LocalPref $\geqslant \ell$ are also exported. This represents a monotonicity property, which again leads to predictable behavior.

The reason why policies in BGP need to be simple and predictable lies in the danger of connectivity loss in case a link fails, or another ISP changes its policy and withdraws a route. With complex policies, outages and loss of connectivity are common [118, 222, 360, 516]. SCION does not need such a rule—given its path exploration mechanism and the default path, which guarantees at least one working path.

6.2.5.3 Route along the Most Stable Path

BGP supports a preference for more stable paths, which is expressed based on the age of a path. Such an approach is not critical in SCION as several paths can be active simultaneously, and one of them can be a path that has been stable over an extended time period.

6.2.5.4 Hot-Potato Routing

Hot-potato routing denotes the strategy of sending a packet off as quickly as possible to the next AS, limiting the amount of resources consumed within the AS. This strategy can lead to asymmetric paths, as packets traveling from A to B may thus take a different path than packets traveling from B to A, and can cause detours [404]. Since SCION defines the specific ingress and egress links, hot-potato routing cannot be achieved by standard SCION. However, if two neighboring ISPs do want to perform hot-potato routing, they can assign the same interface identifier to two different links, and send packets across the closer link. This approach would sacrifice opportunities for multipath communication, so we do not expect it to be used in practice.

6.2.5.5 Example of a Complex BGP Policy

Consider the example depicted in Figure 6.5, where B is an educational network that only permits traffic destined for another educational network. As F is also an educational network, traffic destined for it can traverse B. In BGP, B learns the path originated by F (i.e., F–E–D), and the path originated by G (i.e., G–E–D). Since F is an educational network, B propagates its update with the path F–E–D, whereas the updates of corporate network G are filtered out by B. Traffic follows the reverse path of the updates, so traffic destined for F can come to D from B or C, but traffic destined for G cannot flow across link B–D.

Such a policy can be expressed in SCION, although the path-exploration process is conducted from core to leaf ASes. Considering A is in the ISD core, D receives PCBs with paths A–B and A–C, extends them (by adding information about itself), and propagates them downstream to E. The PCBs sent contain a path policy extension, where either B or D states that only F can use the path A–B–D. This policy is enforced at several points:

- E, on learning D's statement, does not send the PCB containing B to G.
- Core path services refuse path registrations attempted by entities other than permitted ASes (i.e., other than F in our example).
- D can sample traffic to check whether the policy is violated (since SCION addresses contain AS identifiers).

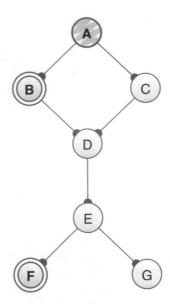

Figure 6.5: Network topology, where AS B represents an educational network
(e.g., GÉANT or Internet2) that provides transit only to educational
entities. AS F is an educational institution, which is allowed to
receive traffic through B. All other entities are commercial ASes.
In the case of SCION, we consider AS A to be a core AS.

6.2.5.6 Source-Based Downstream Path Policy

For a given destination, BGP can only announce a single update, preventing a
diversified routing policy based on the source of the packet. In SCION, path
exploration is realized from the core ASes towards leaf ASes and an AS receives several PCBs with diverse paths. Moreover, in SCION, a provider has
more control over which paths are provided to its customers.

For instance, consider again the topology in Figure 6.5. Using BGP, D must
select which path to A it will forward to E, either D–B–A or D–C–A. Even if
E knew of both paths, all traffic destined for A would either traverse B or C.
Thus, E has no choice but to forward the relevant BGP update for A on to F or
G—it cannot have F use path E–D–B–A and G use path E–D–C–A.

With SCION, both F and G can use paths through B or C, as long as D has
sent one PCB with the path A–B–D and another with A–C–D to E, and E in
turn has extended the PCBs and passed them to F and G. However, if desired,
D can decide not to reveal the connection with B (if, for instance, it is a backup
link), and can keep sending to E only PCBs with the path A–C–D. Similarly,
if E has PCBs with the two paths, it can send both PCBs (i.e., traversing B and
C) to F, and send only one PCB to G, the one traversing C.

6.2.6 Secrecy of Routing Policies

Routing policies are often sensitive information for an ISP's business. ISPs therefore guard their policies, even though the actual routing decisions leak some information about their policy. An important question is whether SCION leaks more policy information than BGP. In the case of explicit path policies, the policy is directly published in the PCB. We anticipate that non-sensitive policies will be published this way. Standard beaconing also discloses policy information, and, as a natural consequence of multipath path discovery, SCION discloses more information than BGP—as any multipath routing protocol would naturally disclose more information. If ISPs were to use encrypted hop fields to hide their policy (even though encrypted hop fields require a higher overhead for path setup), then extensive probing of paths with encrypted hop fields may again reveal the policy. However, in that case an ISP can monitor how much probing is performed and which paths it intends to permit.

So in summary, despite SCION revealing more policy information than BGP, ISPs can monitor and control the amount of disclosed information. Much of the revealed information is a fundamental consequence of multipath communication. We believe that this is a worthwhile tradeoff to make, given the advantages offered by multipath communication.

6.2.7 Influencing End-Host Behavior

With SCION, a path-aware network architecture, end hosts have substantially more control over paths than in today's Internet. This is a design choice with many benefits for reliability, sovereignty, and efficiency. However, this can represent a problem for ISPs, as clients may select communication paths that incur a higher cost than a default path.

6.2.7.1 Conflicting Interests

We first consider situations where an end host's interests conflict with those of its ISP. Such a situation can arise if a path is attractive from a performance perspective (relevant to end hosts), but unattractive from a financial perspective (relevant to ISPs):

1. Peering links are cheaper than parent links for an ISP—in many cases they incur no cost for an ISP. However, an end host can decide not to take a peering shortcut and instead use the longer path through the core because this connection has lower latency, higher bandwidth, or other desirable properties.

2. Low-Earth-orbit (LEO) satellite networks promise to provide substantially reduced latency for long-distance communication if it is integrated into the global Internet [208]. However, connectivity through these satellite networks is also expected to be much more expensive than

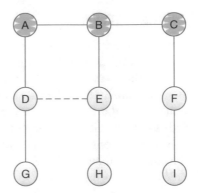

Figure 6.6: The vertical links indicate provider–customer relationships (represented by parent–child links in SCION), and the horizontal links indicate peering relationships (represented by peering links). ASes A, B, and C are part of the ISD core.

standard terrestrial connectivity. Thus, the interests of end hosts (low-latency paths) may conflict with those of an ISP (low-cost paths). Note that integrating LEO satellite networks and other specialized network infrastructure into a BGP-based Internet is very challenging or even impossible (beyond last-mile access).

For the first case, consider the example topology in Figure 6.6, in which a host g in AS G desires to communicate with host h in AS H. If host g makes use of the peering link D–E, then ASes D and E save money as the traffic does not flow through their respective providers A and B. SCION's path control, however, enables hosts g and h to select the path G–D–A–B–E–H for their communication, which would incur a cost for D and E. D and E cannot easily force the sender to use the peering link, as host g needs to be able to use the path D–A as a backup path in case the peering link fails, or also if it wants to communicate with host i in AS I.

Even though the validity of hop fields is mainly checked through their authenticators, an ISP has several options for limiting the use of particular interdomain links by its own end hosts:

1. Providing access to certain (more expensive) links only to end hosts willing to pay an extra charge.

2. Charging end hosts based on the traffic on certain "premium" links or limiting the traffic they can send on them.

3. Limiting the available bandwidth on more expensive links.

Options 1 and 2 (individually or in conjunction) allow an ISP to implement host-specific policies, and naturally lend themselves to the case of satellite networks described above. They can be implemented by filtering packets of end hosts at their access gateways based on the egress interface ID. Additional bandwidth-reservation systems like COLIBRI (see §10.2) provide even more fine-grained control.

Option 3 is a general approach that affects all end hosts equally. It is intended to "guide" flows towards less costly paths by providing more available bandwidth. The multipath path-exploration mechanism will continuously optimize the set of paths used for the communication, and thus the paths and traffic will naturally migrate towards paths with more available bandwidth. This bandwidth information can also be added as metadata to PCBs (via beaconing extensions, see §4.1.2). Also, limiting the total usage of expensive links may be the only means to limit costs if the methods of price-based access control mentioned above are undesirable because of overhead or unavailable for regulatory reasons.

In addition to retaining partial control over their end hosts' path choices, these mechanisms thus enable new business models for ISPs with premium connectivity. The price for access to high-performance paths could even be flexibly determined based on demand, e.g., via spot markets [96].

6.2.7.2 Malicious End-Host Path Selection

In addition to cases where end hosts' interests conflict with those of their ISP, we must also consider malicious end hosts that intentionally want to cause harm to their ISP or other ASes. We will consider more cases in the security analysis (Chapter 7), but discuss a particular case of malicious path selection here.

For example, an end host may intentionally *not* construct a shortcut path to cause additional costs for the AS where the shortcut could occur: Instead of forwarding traffic between two customer ASes, that AS needs to send and receive traffic to and from a provider. In contrast to peering links and special ASes discussed in the previous section, it is highly unlikely that this is of any benefit to the end host (there may be rare cases, where intra-domain connectivity is lost at the shortcut AS), so this behavior constitutes a malicious action with high probability.

For example, consider the topology in Figure 6.5 on page 144 and a host in AS F creating a path to a host in AS G, merging up- and down-segments. A non-malicious host would create path F–E–G, but a malicious host can create path F–E–D–E–G to harm E (or help D) by incurring additional cost for E.

Unfortunately, such malicious path selection cannot be detected easily by the attacked AS. However, note that this mechanism does not provide additional attack surface to a malicious end host: Also in a BGP-based Internet, a malicious end host can always trigger a double charge, by contacting a destination in D and receiving a reply (i.e., using the paths F–E–D and D–E–F).

In conclusion, while there is a conflict between end-host path control and ISPs' traffic flow policies in some cases, the situation in SCION is not worse than in a BGP-based Internet. On the contrary, SCION offers several mechanisms to mitigate the conflict and even enables integrating novel types of network infrastructure into the Internet, and creates new business opportunities for ISPs.

6.3 Scalability Analysis

A multipath routing architecture disseminates multiple paths per destination and thus potentially increases the number of sent control-plane messages and the memory requirements for routing tables. For SCION, scalability was one of the most fundamental design goals;[3] it achieves scalability to an Internet-wide deployment through three properties:

P1 Packet-carried forwarding state (PCFS).

P2 Organizing ASes into sovereign ISDs.

P3 Different control-plane strategies on each of the two levels of routing hierarchy: selective flooding among core ASes (core beaconing) and unidirectional intra-ISD forwarding (intra-ISD beaconing).

We consider all control-plane operations, in turn, to determine which aspect to focus on for the scalability analysis. Table 6.1 shows an overview of various SCION components and their impact on scalability. Since core beaconing has the largest impact on scalability, we present a detailed evaluation in §6.4.2. The remaining operations have a lesser impact on scalability, as we argue below.

6.3.1 High Impact on Scalability: Core Beaconing

In core beaconing, path segments between core ASes are disseminated through *selective flooding*, i.e., an AS selects for each outgoing interface a subset of received PCBs, signs and forwards them. Core beaconing potentially has the highest impact on scalability among the control-plane components. In contrast to intra-ISD beaconing, core beaconing discovers paths between any pair of core ASes and is not restricted by link directionality. Furthermore, the large number of interconnections between core ASes (typically Tier-1 or Tier-2 ISPs) can result in an exponential number of paths.

Consequently, the beaconing must be carefully tuned to avoid the exponential explosion, yet allow for high-quality paths to be found. We analyze the scalability and path quality achieved by the core beaconing in detail in §6.4.

---

[3]The first letter of the SCION acronym stands for "scalability."

Table 6.1: Path-management overhead of various SCION components.

SCION component	Scope			Frequency			Impact		
	AS	ISD	Global	Hours	Minutes	Seconds	Low	Medium	High
Core beaconing			●			●			●
Intra-ISD beaconing	●				●			●	
Core-segment lookup	●				●			●	
Down-segment lookup			●			●		●	
End-host path lookup	●					●	●		
Path (de-)registration	●			●			●		
Path revocation			●	●			●		

6.3.2 Medium Impact on Scalability

There are several mechanisms in SCION that only affect a single ISD, occur rarely, or have additional mechanisms like caching; these have a medium impact on the scalability of a SCION network.

Intra-ISD Beaconing. Intra-ISD beaconing is initiated by the core ASes which disseminate PCBs unidirectionally to the leaf ASes (see property P3). PCBs are forwarded along parent–child links, limiting the number of PCBs and leading to a linear overhead in the number of interfaces. Since the number of PCBs received by non-core ASes in an ISD only depends on the topology of that ISD, regardless of the size and topology of the whole network, intra-ISD beaconing is scalable to any global network (see property P2) as long as the size of ISDs does not increase. In particular, the overhead of intra-ISD beaconing is an order of magnitude lower than core beaconing, as we show in §6.4.2.

Down-Segment Lookup. To fetch down-segments, path services in non-core ASes need to contact core path services in other ISDs. Fetching path segments is a unicast operation to the destination AS's core path service with a small overhead compared to typical data-plane traffic. To further reduce overhead, path services and end hosts cache path segments to directly serve subsequent requests for a given destination AS during the lifetime of the segment (on the order of several hours). This is particularly effective for popular destination ASes, such as CDN providers.

Core-Segment Lookup. To reach other ISDs, core-segments are fetched by non-core path services from core path services in the same ISD. In addition to the points mentioned for down-segment lookup, core-segment lookup requires only intra-ISD communication and thus has a limited impact on scalability (similar to intra-ISD beaconing).

6.3.3 Low Impact on Scalability

Intra-domain mechanisms as well as optional or relatively infrequently mechanisms have a low impact on the scalability of a SCION network.

End-Host Path Lookup. End-host path lookup consists of end hosts fetching path segments from their local path service, which is an intra-AS operation and thus not materially influenced by the size of the global network.

Path Registration and Deregistration. Path (de-)registration is performed every tens of minutes and consists of sending around 10 KiB to the core path service in the same ISD.

Path Revocations. Link failures trigger SCMP error messages sent by SCION border routers to end hosts. The SCMP messages typically produce much less traffic than the data-plane traffic prior to the link failure, and hosts switch to a different path as soon as the SCMP message is received.

Additionally, ASes can proactively de-register path segments from local and core path services. As discussed above, this has a low impact on the scalability of SCION.

6.4 Beaconing Overhead and Path Quality

Routing mechanisms like BGP in today's Internet and SCION's beaconing mechanism have several partially conflicting objectives:

1. Discover high-quality paths according to various metrics (including latency and bandwidth).

2. Discover a diverse set of paths (in the case of multipath routing protocols).

3. Ensure that routing messages cannot be spoofed.

4. Create as little processing and communication overhead as possible.

Clearly, trying to explore all path combinations in a multipath architecture like SCION results in huge overhead in terms of communication and computation as the number of paths can be very large in highly-connected topologies. Therefore, a viable path-dissemination algorithm must select a set of high-quality paths to each destination, while keeping the cost in terms of message overhead as small as possible. As we discussed in §4.4, each AS defines, independently of other ASes, a path-selection algorithm, which selects a set of paths to forward on each outgoing (egress) interface to a neighbor AS.

In this section, we provide insights into the routing overhead and achieved path quality of two sample path-selection algorithms and compare them to BGP and BGPsec.[4] These results are a summary of Krähenbühl, Tabaeiaghdaei, Gloor, Kwon, Perrig, Hausheer, and Roos [305].

For SCION, we consider a simple baseline path-selection algorithm, which optimizes paths for the same metric as BGP—i.e., the AS-level path length—and disseminates the best PCBs in regular intervals. Clearly, this algorithm has several shortcomings:

- The disseminated paths lack diversity, i.e., link disjointness, since the same number of paths are disseminated at every beaconing period.

- The algorithm sends a set of paths irrespective of previously sent paths. Paths that are sent multiple times on the same egress interface cause unnecessary redundancy and waste bandwidth.

For these reasons, we developed a more sophisticated path-selection algorithm (diversity-based path selection), which is described in §4.4.6. This algorithm attempts to maximize the disjointness of paths while reducing the overhead. It achieves this by keeping track of recently sent PCBs on each egress interface, and inhibiting redundant path retransmissions.

6.4.1 Data Sources and Simulation Setup

To compare the overhead and achieved path quality of BGP, BGPsec, and SCION, we combine several publicly available data sets [101, 102, 394] and our own simulations [393, 481]. Simulations are necessary as only BGP is currently deployed on an Internet scale. Although SCION is already deployed on a global scale, the current production networks are not yet large enough for Internet-scale inferences. Similarly, BGPsec is not yet deployed globally.

In the following, we give an overview of the various data sources and simulations we use. Further details—e.g., how we ensure comparability for different data sources and topologies—can be found in our research paper [305].

6.4.1.1 BGP

To obtain the ground truth for BGP, we leverage data from the Route Views [394] update messages data set. The Route Views project has 34 collectors that peer with monitors around the world. Each monitor is a BGP speaker that relays its received update messages as well as its own routing

---

[4]BGP and BGPsec only differ in the overhead and security properties but not in the quality of the paths they discover.

table to the connected collectors. To compare the control-plane signaling cost
of BGP, BGPsec, and SCION, we analyze the received control-plane traffic in
the same ASes and during the same time period. We consider update messages
from RouteViews2 collector during May 2020. To evaluate the communication overhead of BGP, we calculate the size of update messages based on the
individual field sizes defined in RFC 4271 [432].

6.4.1.2 BGPsec

We simulate BGPsec on the entire CAIDA AS Relationships topology using
the SimBGP simulator [481]. We organize the border routers of each AS in a
star topology. Thus, each border router has two interfaces, one connected to
the border router of the neighboring AS and the other to the internal BGPsec
speaker of its own AS. In our configuration, each BGPsec speaker has a minimum route advertisement interval timer of 15 s and a processing delay of 5 ms
for each incoming update message.

We measure the BGPsec churn by using SimBGP to monitor update messages received by the central BGPsec speaker of each AS. Then, we derive
the BGPsec churn per destination prefix based on the size of update messages
destined to that prefix according to RFC 8205 [329]. As every AS in the simulation only announces one prefix, we multiply the churn for each destination
prefix by the number of prefixes this AS announces to obtain the churn per
destination AS.

6.4.1.3 SCION

For SCION, we developed a scalable control-plane simulator using the ns-3
network simulation framework [393]. We simulate core and intra-ISD beaconing on large-scale topologies derived from CAIDA's AS Relationships
Geo [101] data set, which contains the relationships between 12,000 ASes
as well as their interconnection locations. This dataset allows us to infer the
relationships and number of links between neighboring ASes, giving us a
realistic view of the Internet's core AS-level topology. We use the results
collected from simulating SCION beaconing on this topology to approximate
an Internet-scale deployment of SCION.

Core Beaconing.   To simulate core beaconing, we use the subset of the 2000
highest-degree ASes from the topologies mentioned above, by incrementally
pruning the 10,000 lowest-degree ASes. We simulate SCION core beaconing
using both the baseline path-selection algorithm and the diversity-based path
selection. To measure the amount of traffic used for core and intra-ISD beaconing in SCION, we observe the amount of PCB traffic sent on each inter-domain
interface.

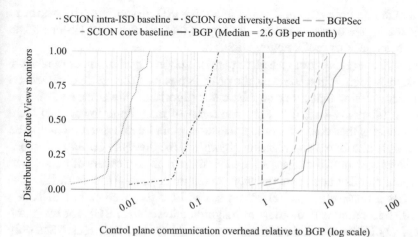

Figure 6.7: Distribution of Route Views monitors based on the link utilization of BGPsec, SCION core beaconing using the baseline and the diversity-based algorithm, and SCION intra-ISD beaconing using the baseline algorithm relative to BGP. This cumulative distribution function (CDF) plot indicates the fraction of monitors (value on the y-axis) that have at most a given overhead (value on the x-axis). For instance, 50% of ASes using the diversity-based SCION core beaconing algorithm have an overhead of less than 0.2 of the overhead of BGP, despite finding 50+ different paths (the number of found paths is not shown on this graph).

Intra-ISD Beaconing. To simulate the intra-ISD beaconing, we first construct a single large ISD[5] by picking the 11 highest-rank American ASes (by customer-cone size) from CAIDA's AS Rank [102] data set. Then, we add their direct or indirect customers to the ISD by iterating down the Internet hierarchy starting with the core ASes. The result is a large ISD with 11 core ASes and 7017 non-core ASes. In a global-scale deployment of SCION, the intra-ISD beaconing process is expected to run on similar sized topologies. For the intra-ISD beaconing simulation we only employ the baseline path-selection algorithm as intra-domain beaconing is less relevant for scalability as discussed in §6.3.

6.4.2 Overhead Comparison of SCION, BGP, and BGPsec

Figure 6.7 shows the overhead, relative to BGP, of BGPsec, the diversity-based path selection applied on core beaconing, and the baseline path-selection algo-

---

[5]Since SCION ISDs provide routing isolation, the intra-ISD beaconing process of each ISD is mutually independent, rendering simulations of multiple, connected ISDs superfluous.

rithm applied on both core beaconing and intra-ISD beaconing. The overhead of BGPsec is one order of magnitude higher than BGP due to larger update messages and lack of prefix aggregation.

While the overhead of the baseline path-selection algorithm in core beaconing is slightly higher than BGPsec, the order is reversed for the diversity-based path selection: For that more sophisticated algorithm, the overhead of core beaconing (diversity-based) is even one order of magnitude lower than BGP, which indicates that core beaconing scales to global deployment. Compared to the simple baseline path-selection algorithm, the diversity-based path selection reduces the control-plane overhead by more than two orders of magnitude. Finally, our results underscore our assessment made in §6.3 that intra-ISD beaconing has a relatively small effect on scalability: the overhead of SCION's intra-ISD beaconing is two orders of magnitude lower than BGP, for two main reasons: (1) PCBs are initiated only by core ASes, and thus leaf ASes only receive PCBs without propagating them; and (2) SCION only announces ASes in contrast to BGP and BGPsec, where an AS sends out BGP update messages for each of its prefixes, which potentially flood the entire Internet and consume additional resources during the convergence process.

Note that the lower overhead of SCION compared to BGP and BGPsec is *despite* the fact that SCION discovers *multiple* (typically 10–100) paths between every pair of ASes where BGP(sec) only finds a single path. Comparing per-path overheads, the relative efficiency of SCION compared to BGP and BGPsec thus increases by another 1–2 orders of magnitude such that SCION core beaconing using the diversity-based path selection is about 1000× more efficient than BGPsec.

6.4.3 Path-Quality Evaluation

We evaluate the quality of paths provided by SCION based on the criteria of link-failure resilience, latency, and achievable bandwidth. The effect of SCION on metrics like energy efficiency and emissions is discussed in Chapter 16. Failure resilience is defined as the minimum number of inter-domain links whose failures disconnect two previously connected ASes. By comparing it to BGP,[6] we observe that the path quality increases in SCION—both with the baseline path-selection algorithm and the diversity-based path selection. A comparison with the optimal path quality achievable in the given topology enables evaluating SCION's diversity-based algorithm, and demonstrates its increased resilience.

We consider the best possible case for BGP by choosing the best path present in Route Views, and by assuming full BGP multipath support between every AS pair, which allows AS pairs to use all available links between them for fast failover and bandwidth aggregation. Additionally, we assume no intra-AS

---

[6]BGPsec does not differ from BGP in terms of path quality.

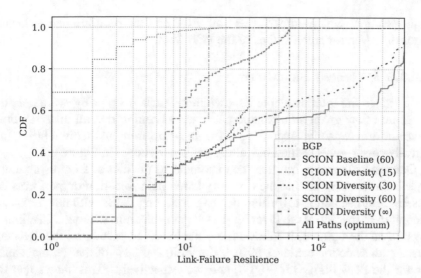

Figure 6.8: Cumulative distribution function (CDF) of the number of link failures that cause the disconnection of pairs of ASes. The different
lines correspond to different routing algorithms; for a particular
value of the link-failure resilience (x axis), the line indicates what
share of AS pairs are disconnected by that many or fewer links.
According to the max-flow min-cut theorem, this also corresponds
to the maximum flow between these ASes (assuming uniform link
capacities). The value in the brackets indicates the PCB limit for
SCION beaconing algorithms, i.e., up to how many paths are disseminated between each AS pair.

bandwidth limitations and consider only propagation delay (no queuing delay)
for the minimum latency.

For SCION, we evaluate the performance with different limits on the number of routes stored in beacon and path service per destination AS, which we
call *PCB limit*. This is an important parameter that affects both the storage
overhead in the infrastructure and (potentially) the quality of discovered paths.

6.4.3.1 Link-Failure Resilience

We find the optimal link failure resilience by computing the min-cut for each
AS pair in the topology, and comparing the different algorithms with the computed value. Even when using BGP multipath, around two thirds of AS pairs
are disconnected (temporarily) when a single link failure occurs, which can
be seen in Figure 6.8. Already the baseline path-selection algorithm performs
significantly better: 80% of AS pairs are only disconnected when 4 or more
link failures occur. The figure clearly shows the benefits of the diversity-based

path selection, which outperforms both BGP and the baseline path-selection algorithm even for small values of the PCB limit.

6.4.3.2 Maximum Bandwidth

We estimate the total available bandwidth for each AS pair by evaluating the maximum flow given all available paths and assuming that all inter-AS links have uniform capacity. According to the max-flow min-cut theorem [189], this directly corresponds to the link-failure resilience shown in Figure 6.8.

Our results thus confirm that the maximum bandwidth of BGP using multipath is the lowest and that the diversity-based path selection outperforms the baseline path-selection algorithm due to prioritizing PCBs with new links. Although the capacity of the diversity-based path selection without PCB limit is lower than the optimal, it follows the same trend. In particular, the diversity-based path selection achieves 99%, 97%, 95%, and 82% of the optimal capacity for the PCB limits 15, 30, 60, and ∞, respectively. This shows that the algorithm effectively finds paths with a diverse set of links and performs close to optimal for small numbers of PCBs sent per source AS.

6.4.3.3 Minimum Latency

We estimate the latency of paths by computing the great-circle distance between routers whose locations are extracted from the CAIDA AS Relationships Geo [101] data set. Our results show that both BGP and SCION typically find a low-latency path close to the optimum since they prefer short AS-level paths which often represent low-latency paths. SCION discovers paths with slightly lower minimum latencies than BGP (2.5 ms difference on average) and the baseline path-selection algorithm performs the best since it is more likely to add short paths that do not increase resilience but reduce latency. To optimize the path-selection algorithm for latency, additional topology information must be disseminated in beacon extensions to consistently pick the lowest-latency path (see Sections 4.1.2 and 4.4).

7 Security Analysis

TOBIAS KLENZE, MARKUS LEGNER, ADRIAN PERRIG, BENJAMIN ROTHENBERGER*

Chapter Contents

7.1   **Security Goals and Properties** **158**
    7.1.1   Overall Security Properties 158
    7.1.2   Security Properties of ASes 159
    7.1.3   Security Properties of End Hosts 160
7.2   **Threat Model** **161**
7.3   **Overview** **162**
    7.3.1   Baseline: Security in Today's Internet 163
    7.3.2   Security Properties Achieved by SCION 164
7.4   **Control-Plane Security** **165**
    7.4.1   Path Hijacking through Interposition 166
    7.4.2   Creation of Spurious ASes 167
    7.4.3   Peering Link Misuse 167
    7.4.4   Manipulation of the Path-Selection Process 168
7.5   **Path Authorization** **170**
    7.5.1   Crafting Unauthorized Hop Fields 171
    7.5.2   Protection of Validity Period 171
    7.5.3   Path Splicing 171
7.6   **Data-Plane Security** **172**
    7.6.1   Modification of Path Header 173
    7.6.2   Detectability 173
    7.6.3   Modification of Address Header 174
7.7   **Source Authentication** **174**
    7.7.1   Reflection Attacks on Local End Hosts 175
    7.7.2   Reflection Attacks on End Hosts in Other ASes 175
    7.7.3   Defenses against Address Spoofing 176
7.8   **Absence of Kill Switches** **176**
    7.8.1   Local ISD Kill Switch 177

---

*This chapter reuses content from Barrera, Klenze, Perrig, Reischuk, Rothenberger, and Szalachowski [52].

© The Author(s), under exclusive license to Springer Nature Switzerland AG 2022
L. Chuat et al., *The Complete Guide to SCION*, Information Security
and Cryptography, https://doi.org/10.1007/978-3-031-05288-0_7

7.8.2   Remote ISD/AS Kill Switch   178
7.8.3   Recovery from Kill Switches   178
7.9   Other Security Properties **179**
7.9.1   Confidentiality (P13)   179
7.9.2   Resilience against Censorship (P14)   180
7.9.3   Anonymity (P15)   180
7.10   Summary **181**

One of the fundamental objectives that guided the design of SCION is security, in particular network security. In this chapter, we state the precise security goals of various network participants and how they are achieved in the presence of different types of adversaries. By contrasting SCION's security properties with the state of security in the current Internet, we emphasize the achieved improvements.

7.1 Security Goals and Properties

The central task of the networking layer of an Internet architecture is to provide connectivity between end hosts. Upholding this connectivity in the presence of faulty networking components and end hosts is hence crucial. However, in the presence of adversaries, fault-tolerance by itself is insufficient: the Internet must also be designed to protect the connectivity between honest agents against malicious or compromised end hosts and ASes. Availability is therefore the central security goal of the networking layer.

In this section, we describe the security properties that SCION aims to provide for two types of network entities: end hosts and ASes. Availability is the overarching security goal for both, as end hosts depend on connectivity, and ASes economically depend on their end-host or AS customers. We dedicate a whole part of this book, Part III, to describing systems and tools to achieve global availability guarantees and protect against denial-of-service attacks. Still, some weaker forms of availability (connectivity, absence of kill switches) are already discussed in this chapter.

Besides availability there are a number of other goals. These include the enforcement of routing policies of ASes, path transparency and control for end hosts, quality of service, traffic attribution, backdoor-freedom, and support for a heterogeneous but global trust environment.

7.1.1 Overall Security Properties

The following security properties are crucial for both ASes and end hosts.

P1 Global connectivity: There is a single connected global network, in which any two entities that wish to communicate can do so, even in the presence of faulty or malicious components.

P2 Routing security: Routing information can be unambiguously attributed to an AS. In particular, the routing information an AS propagates cannot be altered by malicious entities, and both ASes and end hosts can authenticate routing information they receive.

P3 Absence of kill switches: No single actor (or small number of actors) is able to cause global outages. Concretely, the effect of any malicious actions related to routing or PKIs should be *isolated*.

P4 Weak and strong detectability: An on-path data plane attacker, as an intermediate node, cannot effectively disguise their own presence on the forwarding path to subsequent ASes and the destination, even when changing the path information in the packet's header. The property is called *strong* if the entities can detect whether or not an attack occurred. Localization of the attacker is not required under this definition in either case.

Detectability is a data plane property that states that the attacker's AS is a visible part of the forwarding path, as the path is embedded in each packet's header. This property rules out certain attacks, but it does not provide guarantees as strong as the source authentication property defined below.

7.1.2 Security Properties of ASes

ASes need to provide connectivity to their subscribers and customers, but they also have other requirements. In particular, they formulate routing policies, which are used to rule out impractical or uneconomic paths that violate business interests. The enforcement of these policies can be split between the control and the data plane.

P5 Beacon authorization: Only segments that are in accordance with the routing policies of all on-path ASes are created in the beaconing process of the control plane. We call these segments *authorized*.

P6 Path authorization: Packets are only forwarded along authorized segments.

Optionally, ASes can authenticate the source of packets in the data plane. While the source is typically authenticated by the destination end host, doing so at ASes is useful as it allows filtering of spoofed packets before they reach bottleneck links.

P7 Source authentication: The on-path AS authenticates the sender of a packet.

7.1.3 Security Properties of End Hosts

The security goals of end hosts are diverse and manifold, many of which should be addressed in other layers of the network stack. Here, we focus on security properties of end hosts at the network layer. The first property is fundamental to path-aware networking:

P8 Path transparency and control: The sender of packets should know and be able to (partially) control the path taken by its packets. The receiver should also be able to infer the path.

The following properties describe authentication and relate three types of paths: the *intended path*, which the source embeds into the packet header, the *observed path*, which is the path contained in the header when the packet is received by the destination, and the *real path*, which is the path that was actually traversed.

P9 Truthful forwarding: The real path of a packet corresponds to the intended path. This is particularly relevant when path control by the end host is used to avoid certain ASes or ISDs (e.g., for compliance reasons), or when an end host pays for more expensive paths (e.g., through a satellite network).

P10 Source authentication: The receiving host authenticates the sender of a packet.

P11 Packet integrity: Receiving hosts verify that a packet, including its path, is the same as the one sent by the source.

P12 Path validation: End hosts verify that the observed path corresponds to the real path, i.e., that the packet actually traversed all the ASes on the path contained in the packet header. To this end, each AS adds a proof of traversal to the packet header.

Truthful forwarding, i.e., the equivalence of real and intended path is not shown directly, but follows from the other properties, as we illustrate in Figure 7.1. Source authentication, packet integrity and path validation together imply the equivalence of intended, observed, and real path, and hence imply truthful forwarding. However, path validation is not necessarily required. In case that there is at most one on-path adversary, source authentication and packet integrity suffice to imply the same equivalence. This is because the destination authenticates a packet, including the embedded path, and this authentication fails if an on-path attacker changed the path embedded by the sender.

There are several other goals of end hosts, which are more advanced and/or addressed by other layers in the networking stack:

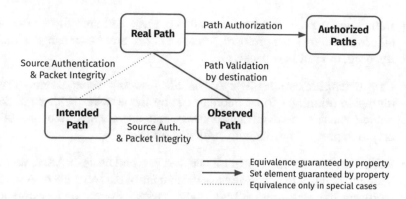

Figure 7.1: Relationship between different types of paths. We consider a scenario, where a packet has reached its destination, which is assumed to be honest, and to have validated the packet successfully. The dotted property edge is only sufficient to guarantee equivalence when there is at most one on-path attacker.

P13 Confidentiality: Communication over the network remains confidential. This is normally achieved at higher layers in the networking stack (e.g., through TLS) but also partially addressed in SCION.

P14 Censorship resilience: As a special case of property P1, end hosts want to communicate irrespective of the presence of censorship mechanisms.

P15 Anonymity: End hosts can communicate anonymously, i.e., observers are not able to infer which end hosts communicate with each other.

7.2 Threat Model

We use different threat models to analyze SCION, as we study security for different entities. For instance, properties that protect ASes against malicious end hosts require different honesty assumptions than properties that protect end hosts from malicious network components. We categorize adversaries along the following dimensions:

- **Type of entity:** Does the adversary control (1) end hosts, (2) networking equipment such as routers, (3) infrastructure services, and/or (4) entire ASes?

- **Number of adversaries:** How many entities are under the control of the adversary?

- **Collusion:** Do malicious or compromised entities collude?

- **Behavior:** A *passive* adversary only observes and analyzes traffic but otherwise follows the protocol, while an *active* adversary can also modify, drop, or craft new packets.

- **Cryptographic capabilities:** A *symbolic attacker model* assumes cryptographic primitives to be unbreakable by the attacker. Other attacker models consider more advanced adversaries that are able to break some cryptographic primitives with sufficient effort.

- **Location:** For a communication between two end hosts or ASes, we can distinguish *on-path* adversaries—which control components or ASes directly on the path—and *off-path* adversaries. Source and destination ASes are special cases of on-path adversaries.

Typically, we consider a distributed, colluding, active, symbolic attacker that controls several entire ASes that we call *compromised*. By compromising an AS, the adversary learns all of its secrets (e.g., cryptographic keys) and has complete control over its networking equipment, infrastructure components, and end hosts. When an AS is compromised, we assume that the adversary can eavesdrop on all control and data messages traversing the AS and inject arbitrary messages. Furthermore, all malicious or compromised entities are assumed to collude and share a channel for information exchange potentially outside of the SCION network.

Our focus is the security of protocols, not of well-established cryptographic primitives. We therefore adopt a variant of the symbolic Dolev–Yao adversary model [168], which assumes the security of the underlying primitives. For instance, for symmetric encryption we assume that the attacker cannot decrypt a ciphertext without the corresponding secret key. We consider both on-path and off-path adversaries, but note that in the presence of on-path adversaries, some security properties are not achieved in SCION, or are fundamentally unachievable. In particular, connectivity and availability between two entities cannot be guaranteed if no attacker-free path between them exists.

Regarding SCION-specific capabilities, we assume the adversary is able to register as an AS with the ISD core and perform regular operations. However, we assume registration operations in the core to be throttled and visible to other nodes within the ISD. In particular, we assume a mechanism that prevents large numbers of malicious ASes from joining the ISD rapidly or automatically. For increased security, we assume that for each new AS registration, the necessary amount of due diligence and verification is performed by core ASes, as we explain in Chapter 3.

7.3 Overview

In this section, we provide an overview of the security properties achieved by SCION and contrast them with those of today's Internet. We focus on the

Table 7.1: Comparison of security properties achieved by BGP, BGPsec, and SCION. The symbols denote whether a property is achieved (✓), partially/conditionally achieved ((✓)), or not achieved (✗). The symbol "—" indicates that the property is not applicable for that system.

	P1	P2	P3	P5	P6	P8	P9	P4	P10	P11
BGP	(✓)	✗	✓	—	—	✗	✗	✗	(✓)	(✓)
BGPsec	(✓)	✓	✗	—	—	✗	✗	✗	(✓)	(✓)
SCION	✓	✓	✓	✓	✓	✓	(✓)	✓	✓	✓

main properties at the network layer—properties P1–P6 and P8–P11. Properties P13–P15 will only be described briefly in §7.9. We describe mechanisms that achieve properties P7 and P12 in §10.1 (page 228).

We start by discussing the security of today's Internet as a baseline and then give an overview of the security properties achieved by SCION and the mechanisms used for that. An overview of all security properties is provided in Table 7.1.

7.3.1 Baseline: Security in Today's Internet

One of the motivations for the design of SCION was the complete lack of routing security in the current Internet (property P2): BGP does not authenticate any routing messages, which enables any AS in the Internet to perform hijacking attacks. The situation has improved in recent years with the standardization of RPKI and BGPsec [328, 329]. Unfortunately, these extensions introduce circular dependencies [137] and global kill switches [441] and thus violate property P3. They have also suffered from a lack of adoption, as they provide little incentive for deployment.

Global connectivity (property P1) is generally achieved by the BGP/IP-based Internet as long as ASes use appropriate routing policies. However, re-convergence of the routing protocol after failures or reconfigurations temporarily violates this property (in addition to the ability of an attacker to perform BGP hijacks, and the presence of circular dependencies with RPKI).

As today's Internet is not a path-aware architecture, and every router locally stores and thus controls forwarding decisions, property P6 is not applicable to BGP. Unfortunately this fact also implies that end hosts have neither control nor transparency over forwarding paths (property P8). Even if end hosts attempt to gain insights into forwarding paths—e.g., via routing messages or tools like traceroute—there is no way of ensuring that data packets follow the same paths (property P9). There are no mechanisms to achieve detectability (property P4).

In today's Internet, source-address spoofing is a widespread issue. With extensions like IPsec [288], end hosts can achieve source authentication and packet integrity properties P10 and P11; however, this requires complicated key management and multiple round-trip times (RTTs) for handshakes, which limits the applicability to certain use cases like virtual private networks (VPNs).

7.3.2 Security Properties Achieved by SCION

We provide an overview of the security properties achieved by SCION in this section and provide pointers to sections with further details on selected properties. In our analysis, we consider all protocols and mechanisms that have been presented in Chapters 2–5.

7.3.2.1 Global Connectivity (P1)

In any protocol, connectivity breaks down when a large enough part of the network is faulty or compromised. Hence, property P1 inevitably requires assumptions on the network topology and the attacker. While SCION is no exception to this observation, it is designed to be as resilient to network failures and attacks as possible. For instance, to guarantee connectivity between two entities, SCION only requires an attacker-free path between them, and that this path has sufficiently good path metrics to be discovered in the control plane.

Between any two ASes, a path can be discovered through SCION's beaconing process. The path-registration and -lookup mechanisms provide access to these paths to end hosts. SCION does not rely on convergence as each path is valid irrespective of other paths; as a result, SCION does not suffer from temporary outages due to path changes. Link failures are resolved within a round-trip time (RTT) thanks to signaling via SCMP.

An on-path adversary can always interfere with end-host communication by simply dropping packets or by not disseminating beacons. However, as a multipath architecture, SCION allows end hosts to choose paths (property P8) and thus avoid the adversary as long as a benign path exists and is discovered in the control plane. We further discuss SCION's connectivity guarantees in the context of censorship in §7.9.

7.3.2.2 Routing Security and Kill Switches (P2 and P3)

Through its flexible control-plane PKI (CP-PKI), SCION supports heterogeneous trust relationships in the Internet without global roots of trust. Consequently, there is no key whose compromise would disrupt global communication. The absence of kill switches is further discussed in §7.8.

The CP-PKI provides each AS with a certified key pair. These keys enable the authentication of all routing messages, which rules out the manipulation of information contained in PCBs. Every AS and end host can verify all routing messages by following the certificate chain. Various routing attacks and how SCION defends against them are discussed in §7.4.

7.3.2.3 Beacon and Path Authorization (P5 and P6)

Beacon authorization is achieved through the authentication of control-plane messages (property P2). A more detailed analysis of path policies that are possible with SCION is provided in §6.2.

We have discussed in §5.3 how ASes achieve path authorization—i.e., ensure that data packets always traverse authorized paths—through the hop authenticators, segment identifiers, and various other checks at border routers (including the interface checks that ensure weak detectability).

Property P6 will be further discussed in §7.5. Both path authorization and weak detectability have been proven formally as we show in Chapter 22.

7.3.2.4 Path Control, Truthful Forwarding, Weak Detectability and Path Validation (P4, P8, P9, and P12)

With SCION, end hosts can select the path their packets should take from a set of authorized paths. As this path information is authenticated using the CP-PKI, end hosts can verify its correctness. However, embedding a path in the packet header by itself does not guarantee that these directives are followed by on-path ASes. Indeed, malicious on-path ASes can modify the path header in various ways as we describe in §7.6. However, the adversary needs to ensure that the interface check of the next AS succeeds, which ensures weak detectability.

By using the SPAO, end hosts can defend against manipulation of the path header for non-colluding adversaries. To defend against colluding adversaries, extensions providing path validation (property P12) are necessary. We describe such an extension, EPIC, in §10.1.

7.3.2.5 Source Authentication and Packet Integrity (P10 and P11)

Both source authentication and packet integrity are achieved in SCION using DRKey and the SPAO. We discuss this further in §7.7.

7.4 Control-Plane Security

In today's Internet, researchers and network operators are noticing that BGP has numerous shortcomings [61, 256, 447, 460] and especially lacks integrity protection for routing update messages. Maliciously acting routers can advertise IP prefixes from address spaces that are unused or that belong to other ASes, and effectively redirect traffic to hosts under the control of the adversary. To address these problems, specific improvements (such as BGPsec) have been proposed [11, 89, 216, 248, 330, 331, 522], although these improved protocols have not seen widespread deployment so far.

Also in SCION, an adversary can attempt to attract traffic by selectively disseminating PCBs or by forging new PCBs. However, SCION control-plane

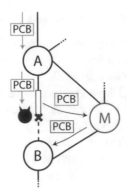

Figure 7.2: Interposition attack.

messages, unlike those of BGP, are authenticated and integrity protected, paving the way for solving issues such as path hijacking. Moreover, SCION's control plane does not rely on convergence (see §6.2.4) as each discovered path can be used independently—in stark contrast to BGP whose iterative refinement of routes can result in slow convergence and intermittent outages. As such, recovery and path-discovery times are shorter and more predictable in SCION.

In the following, we examine several approaches to manipulating paths in the SCION control plane, and show for each case how SCION's design prevents the corresponding attack or helps to mitigate it. Launching a successful attack would require breaking the cryptographic primitives used in SCION. Thus we demonstrate how the SCION beaconing process achieves property P2.

7.4.1 Path Hijacking through Interposition

To attract traffic and thus become on-path, an adversary might try to manipulate the beaconing process. More precisely, as illustrated in Figure 7.2, provider AS *A* sends two PCBs to customer ASes *B* and *M*. Suppose that AS *M* is malicious and can eavesdrop on links between AS *A* and AS *B*; then *M* could try to intercept and disseminate another PCB meant for *B* by injecting its own hop fields into the PCB toward downstream ASes. This could offer *B* an attractive up-segment traversing *M* to the core.

The attack is detectable by downstream ASes, because the PCBs disseminated by *A* towards *B* contain *B* as an egress AS identifier. Therefore, verification of inbound PCBs will fail, because the adversary's PCBs cannot contain *A*'s correct signature.

Assuming that an adversary wants to interpose an AS by modifying an already existing path, they would need to modify the corresponding hop fields. As hop fields are integrity protected and chained together through the segment identifier, malicious modifications of hop fields are prevented (see §7.5.1)

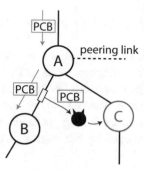

Figure 7.3: PCB theft. AS *A* wants to selectively share access to the peering link with *B*, but not with AS *C*. An eavesdropping adversary reads the PCB intended for AS *B* and re-injects it at their own AS *C* to gain access to the peering link.

However, if the adversary can block the traffic between *A* and *B*, then indeed they can force traffic redirection through *M*. This attack is fundamental and generally cannot be prevented.

7.4.2 Creation of Spurious ASes

An adversary could try to spoof other ASes by introducing nonexistent entities. This would enable the adversary to send traffic with the spoofed entity as a source and thus allows them to plausibly deny the misbehavior and complicate detection of this attack.

However, spoofing a new AS requires a registration of that AS with the ISD core to obtain a valid AS certificate; otherwise the adversary cannot construct valid PCBs. As this registration includes a thorough check and authentication by a CA, this cannot be done stealthily, which defeats the original purpose.

Similarly to creating a fake AS, if an adversary wanted to introduce a new ISD (possibly spoofed), it would need to generate its own TRC, find core ASes to peer with, and convince other ISDs of its legitimacy (see §3.1.8).

7.4.3 Peering Link Misuse

Downstream PCBs may be recorded by an on-path adversary (e.g., by eavesdropping on a link between ASes). By re-injecting that PCB into another link, the adversary can extend paths as long as the PCB is correctly forwarded.

Consider the example in Figure 7.3. AS *A* wants to share its peering link only with one of its downstream neighbors, *B*, and therefore decides to selectively include the peering link in PCBs sent to *B*. The monitoring adversary misuses this PCB to gain access to the peering link by prepending it to their own path. Apart from eavesdropping on the link, the adversary is able to obtain the necessary hop fields by querying a path service and extracting them from registered paths.

SCION successfully mitigates this attack by including specific "next hop" information in the PCB before disseminating it further downstream (see Equation (4.7)). Furthermore, each hop field contains an egress interface. If a malicious entity tries to misuse a stolen PCB by adding it to its own segments, verification will fail upstream as the egress interface mismatches. Therefore, the peering link can only be used by the intended AS.

7.4.4 Manipulation of the Path-Selection Process

Path selection is one of the main benefits of SCION compared to the current Internet, where hosts have no control over the forwarding paths that their packets traverse. With the benefits of freedom regarding path selection, however, comes the risk for hosts to choose non-optimal paths, which we have already discussed in §6.2.7. In this section, we discuss some mechanisms with which an adversary can attempt to trick hosts downstream (in the direction of beaconing) into choosing non-optimal paths.

Still, compared to the current Internet, the path selection in SCION offers higher security overall since (1) *path transparency* enables a path-selecting end host to identify the potentially malicious ASes on the path, and (2) *path control* enables the host to avoid such malicious ASes, even if the cost of such adversary-free paths appears higher.

In SCION, path selection is used in three cases. First, a beacon service selects which PCB to forward to its neighbors. Second, the beacon service chooses which paths it wants to register at the local path service as up-segments and at the core path service as down-segments. Third, the end host performs path selection from all available path segments. We now describe attacks that aim at influencing the path-selection process in SCION. The goal of such attacks is to make paths that are controlled by the adversary more attractive than other available paths. A simple example is a low or even negative price in a pricing system, or announcing high bandwidth and low latency for a path.

The following attacks are only successful if the adversary is located within the same ISD and upstream relative to the victim AS. It is not possible to attract traffic away from the core as traffic travels upstream towards the core. Furthermore, the attack may be discovered downstream (e.g., by seeing large numbers of paths become available), but also during path registrations. After detection, paths traversing the adversary AS can be identified and avoided by regular ASes.

Announcing Seemingly Desirable Path Segments. An adversary who wants to attract traffic from its customers can forward PCBs received from its providers to its customers enriched with metadata (see Sections 4.1.2 and 8.3). While this metadata is signed and thus attributable to the adversary, the

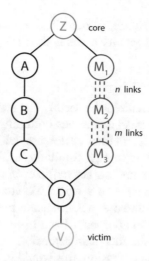

Figure 7.4: An adversary controlling ASes M_i advertises $m \times n$ path segments to increase the chance of selection by victim V.

recipient cannot directly verify the accuracy of the data itself. Therefore, the adversary can add fake metadata which makes the PCB seem desirable. If such an adversary-announced path complies with the policy of the customer ASes, the corresponding PCB may be added as one of the k paths available to the end host. At this point, the end host may select such a path traversing the adversary for communication.

However, if the metadata in the PCBs does not match measurements in the data plane, or high loss rates are experienced, end hosts or customer ASes can easily move to other paths with better quality.

Announcing Large Numbers of Path Segments. A variant of the previous attack is possible if the adversary controls multiple (at least two) ASes. The adversary can create a large number of links between the ASes under its control, which do not necessarily correspond to physical links. In the example shown in Figure 7.4, the adversary controls three ASes with n virtual links between ASes M_1 and M_2 and m virtual links between ASes M_2 and M_3.

For each received PCB, the adversary can thus forward $m \times n$ different PCBs to the customer D. The large number of PCBs increases the chance that one or several of them are selected by D or the intended victim V. In general, the number of PCBs that an adversary can announce that way scales exponentially with the number of consecutive ASes.

While such bogus paths can have some desirable properties, they will need to traverse multiple ASes controlled by the adversary. As there might exist other shorter paths, this decreases their chance of being chosen by a downstream beacon service for PCB dissemination or by a host for construction of a forwarding path. Furthermore, as described above, both end hosts and customer ASes can detect bad-quality paths in the data plane and switch to better paths.

Wormhole Attack. A malicious node M_1 can send a PCB not only to their customers, but also out-of-band to another, colluding malicious node M_2. This creates new segments to M_2 and their customers, which may not correspond to actual paths in the network topology. Similarly, a fake path can be announced through a fake peering link and attract traffic even across ISDs. Without specific prevention mechanisms, such a wormhole attack is unavoidable in routing [249]. To detect wormhole attacks, latency measurements with per-link timestamps are one potential approach. Each ISP would announce the latency of links in the PCB using the metadata extension described in §8.3. In combination with a timestamping extension, this would help reveal the wormhole.

Fake Peering Link Announcement. As an instance of a wormhole attack, an adversary advertises fake peering links, thus offering short routes to many different destination ASes within and outside its own ISD. Downstream ASes will likely have a policy of preferring paths with many peering links and thus are more likely to disseminate PCBs from the adversary. Similarly, hosts are more likely to choose short routes that make use of peering links. However a peering link can only be used if the neighboring AS also announces it. If the adversary is colluding with an external AS, a wormhole becomes possible. In the data plane, whenever a packet containing a fake peering link is received by the adversary, they can transparently exchange the fake peering hop fields with valid hop fields to the colluding AS (see §7.6). To avoid detection of the path alteration by the receiver, the colluding AS can replace the added hop fields with the fake peering link hop fields the sender inserted.

To defend against this attack, methods to detect the wormhole attack are needed [249]. As discussed above, link-latency measurements can help reveal the wormhole and render the fake peering link suspicious or unattractive.

7.5 Path Authorization

Path authorization (P6) is the central property of a path-aware data plane. As we show in Chapter 22, this property has been formally proven for SCION [295, 297]. Here, we discuss how an adversary may attempt to violate the path-authorization property by either crafting completely new hop fields, constructing an unauthorized segment by combining hop fields of different valid segments (path splicing), or modifying the validity period of a segment. In the following we describe why each of these approaches fail in SCION.

7.5.1 Crafting Unauthorized Hop Fields

Hop fields are protected with MACs and, if the corresponding key is unknown, the adversary can at best attempt to perform a brute-force attack to determine the key. Candidate keys can be validated by checking the MAC contained in sample hop fields. As SCION uses 128-bit keys by default, such an off-line attack is computationally infeasible in practice. Furthermore, the keys for the MAC computation are short-lived, with a validity period of 24 hours.

MAC schemes are not generally specified by SCION and may thus be individually chosen by each AS. A hop field's MAC is only checked by the AS that created it, which enables algorithm agility for the MAC scheme. If a certain MAC algorithm is discovered to be weak or insecure, ASes can quickly switch to a secure algorithm without the need for coordination with other ASes. This property is known as cryptographic agility and is further discussed Chapter 17.

The adversary might also attempt to directly brute-force a MAC (instead of the MAC's key). However, this requires an *online attack*: One packet would need to be sent to verify each guess. For an ℓ-bit MAC, the adversary needs to generate $2^{\ell-1}$ packets on average to forge one correct MAC. For the 6-byte MACs, which are proposed at the time of writing, the adversary would need to try $2^{47} \approx 140$ trillion different MACs to successfully forge the MAC of a single hop field. For each incorrect hop field, the corresponding AS returns an SCMP packet. Even though an adversary can observe whether a hop field has been accepted, each incorrect guess is visible to a monitoring entity and thus the attack can be easily detected.

Unfortunately, in the unlikely case where such an online brute-force attack succeeds, the obtained hop field can be reused until its eventual expiration. The EPIC path type presented in §10.1 resolves this issue by introducing per-packet unique MACs.

7.5.2 Protection of Validity Period

The metadata for each path segment stored in the *info field* inside the path in the SCION header (see §5.4 on page 101) includes a timestamp, which is set by the initiator of the PCB. This timestamp cannot be modified by an adversary as it is included in the calculation of the MAC for each hop field. This ensures that the timestamp cannot be set to a later date to extend the validity of the path.

7.5.3 Path Splicing

In a path-splicing attack, an adversary (either controlling an AS or an end host) takes valid hop fields of multiple path segments and splices them together to obtain a new valid path. We have already described such attacks in §5.3.2 and used them to motivate SCION's MAC-chaining mechanism: By including the info field of a segment in the input to each hop field's MAC computation and keeping track of upstream hop fields via a segment identifier in the info field, it is impossible to combine hop fields of multiple segments.

7.6 Data-Plane Security

Besides manipulating the routing decisions in the control plane, adversaries can also attempt to influence forwarding in the data plane and thus violate properties P8 and P9. Because the forwarding path selection has already been made in the control plane, an off-path adversary is severely limited in their abilities. The adversary can merely attempt to disrupt the connectivity of the chosen path and force the host to select a new path. As SCMP error messages are authenticated (see §4.7), an adversary cannot simply trigger a path switch this way. Instead, they would need to flood a link on the path with excessive traffic. We discuss such denial-of-service (DoS) attacks in Chapter 11 and concentrate on the case of an on-path adversary in this section.

To differentiate these attacks from path-manipulation attacks in the control plane, we assume a static control plane. This means that path services have a constant set of paths available, and an adversary is restricted to engaging in attacks by receiving, manipulating, and sending data traffic as opposed to sending control messages. We further assume that path services are available— DoS protections of these services will be discussed in Part III.

Adversaries may try to attract or divert traffic from certain points in the network, craft new hop fields and segments, or combine existing ones in order to create new paths to influence the way outgoing data packets are forwarded, to manipulate the routing history of the packet, and to cover up their own actions.

While some of these attacks seem to be quite severe, they are under the strong assumption of an *on-path* adversary, limiting the possible location of an adversary from 75,000 to a handful of ASes. This is in stark contrast to most attacks considered in protocols such as BGP, where this restriction does not apply. In addition, we will show that the attacks presented here are, if not entirely preventable, at least *detectable*.

In particular, the SCION Packet Authenticator Option (SPAO) described in §3.3 plays an important role in achieving properties P4 and P9–P11. To briefly recapitulate, SPAO enables the sender to protect the header (and optionally the payload) of a SCION packet with either a MAC or a signature, which can be checked by the receiver. All fields of the header are protected, except for mutable fields. Thus, the destination can verify that the destination-observed path is equal to the intended path.

Note that an on-path adversary can always simply drop packets. Therefore, we do not consider any attacks that cause packets to be dropped at a later hop on the path.

7.6.1 Modification of Path Header

An on-path adversary can modify the path header of a packet, and replace one or more path segments or parts of segments with different segments. The only restriction is that the path after the adversary needs to follow authorized segments, as the packet would be simply dropped at some point otherwise. The already traversed portion of the current segment and past segments can be modified by the adversary in a completely arbitrary manner (for instance, deleting and adding valid and invalid hop fields).

The adversary is able to transparently revert any changes to the path header on replies by the destination. For instance, if an adversary M is an intermediate AS on the path of a packet from A to B, then M can replace the packet's past path (leading up to, but not including M). The new path may not be a valid end-to-end path. However, when B reverses the path and sends a new packet, that packet would reach M, who can then transparently change the invalid path back to the valid path to A.

Modifications of the path header can be discovered by the destination if the packet is integrity protected (property P11), e.g., by using SPAO. If there are two colluding adversaries on the path, then they can forward the packet between them using a different path than selected by the source: The first on-path attacker changes the packet header arbitrarily, and the second on-path attacker changes the path back to the original source-selected path, such that the integrity check by the destination succeeds. To defend against multiple on-path adversaries and to prevent this attack, an extension providing path validation (property P12) is required, which is presented in §10.1.

7.6.2 Detectability

For all path-header modifications, the ingress interface of the hop field directly after the adversary must identify the link to the adversary as otherwise the interface check would fail (see §5.6.3). This crucial check ensures the weak-detectability property (property P4): An adversary cannot manipulate a packet's path information to disguise their own presence on the path. While hop fields do not contain globally valid AS identifiers, publicly available segments at path services can be used to map hop fields to AS names, and thus to extract the sequence of ASes that the packet claims to have traversed. This property is considered to be weak, since the receiving host does not know *which* of the hops on the manipulated path is the adversary, and cannot even tell *whether* a segment-replacement attack has taken place.

We can guarantee a stronger detectability property by using the SPAO, which provides end-to-end path-integrity protection. Thus, any path splicing or segment replacement will be noticed at the receiving end host. Strong

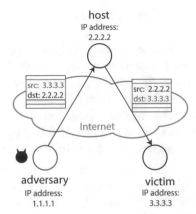

Figure 7.5: Reflection attack using source-address spoofing in today's Internet:
A malicious entity spoofs the source address of the packet, which
will be reflected towards the victim by an arbitrary host, or even
amplified by a vulnerable service.

detectability in this sense does not imply the ability to tell *which* of the nodes
were acting maliciously, only that *some* of them were.[1]

Note, however, that in case of colluding ASes, the second AS could revert
the path modification of the first AS such that the receiving end host could not
detect any modification even with the SPAO. Such attacks can be prevented by
using the EPIC path type presented in §10.1

7.6.3 Modification of Address Header

An on-path adversary can modify the source and destination addresses in the
address header. In case the destination AS is modified, the future path needs
to be modified accordingly; otherwise the packet would simply be dropped by
the last AS on the path.

All these attacks can be detected by the destination when using the SPAO
as the source and destination addresses influence the key used for the MAC
computation.

7.7 Source Authentication

Source-address spoofing refers to an attack where the source address of a data
packet is set not to the sender's actual address but some other address. The
most common defense against this is *source address validation* (SAV), also
known as BCP 38 [186], where an AS filters outbound traffic and drops packets
that claim to have originated in another network or AS.

---

[1]Full-fledged fault localization requires substantially more complicated systems [54].

Unfortunately, far from all ASes employ SAV: The *Spoofer project* [93] has been tracking the global deployment of SAV since 2005. In 2013, more than 40% of all global ASes allowed some level of IP address spoofing [69]. The situation has improved since, but in 2019 still more than 20% of ASes were affected [346]. Current results are published online [94].

This susceptibility of today's Internet to source-address spoofing is often used to hide the true origin of a packet. Services built on connectionless protocols like UDP, for example DNS or NTP, can then be abused for amplification attacks, a form of DoS attack. As shown in Figure 7.5, spoofing the origin of a packet allows an adversary, say 1.1.1.1, to redirect (or even amplify) traffic to the victim 3.3.3.3 by first sending a request to an arbitrary host (or vulnerable service), say 2.2.2.2, whose answer is then sent to the *alleged* originator of the request 3.3.3.3 (instead of to the *true* originator 1.1.1.1).

In the following, we discuss several scenarios of attacks and how they can be prevented in SCION.

7.7.1 Reflection Attacks on Local End Hosts

The address fields in a SCION packet header can be arbitrarily picked by an adversary as they are not integrity protected. However, as SCION does not forward the packet based on the destination address until the packet reaches the destination AS, the effect of address spoofing is limited. Since SCION packet forwarding follows the hop fields, a packet needs a correct sequence of hop fields to be delivered to the destination.

In the simplest source address-spoofing attack in SCION, the adversary simply embeds the source address of a different host v that is located in the same AS. The adversary embeds the hop fields required to reach the destination host h and sends off the packet. Since in SCION the receiver reverses the path for the response packet, v will receive the response. This attack is limited to attacking other hosts within the adversary's AS.

7.7.2 Reflection Attacks on End Hosts in Other ASes

In a more complicated case, the adversary wants the victim (who happens to be in a different AS) to obtain the response packet. Because of the way SCION forwarding operates, the victim host needs to be located in an AS on the path traversed by the hop fields, either before or after the adversary's AS. If the adversary is located in AS A, the victim in AS V, and the host h in AS H, then the adversary can select a path V–A–H (potentially with additional intermediate ASes), set the current HF index to the hop field corresponding to AS A, and send the packet to the local egress border router which will forward the packet on the path toward AS H. When h inverts the path, the response will be delivered to v. This attack can work if the adversary is located in an AS on the up-segment of AS V, in an AS on a core-segment, or in an AS on the down-segment of AS H.

Since SCION border routers check whether the destination ISD and AS numbers match their own ISD and AS numbers to determine if the packet should be locally delivered, an adversary can also create a path $A–V–H$ to mount this attack, which causes the response packet to be delivered to v. Furthermore, to hide their traces and frame an innocent AS F, the adversary can use a path of the form $F–A–V–H$.

7.7.3 Defenses against Address Spoofing

All of these attacks can be prevented if an AS performs filtering similar to SAV and enforces that packets originating from its end hosts have a correct source address (including ISD and AS number) and their hop-field index is set to the first hop field. As this directly improves the security for its own customers, the AS itself has a much stronger incentive to perform SAV on its customers than in today's Internet. Furthermore, the fact that the adversary has to be located in an AS encoded by the hop fields in the packet ensures weak detectability (property P4) and facilitates adversary localization.

Finally, the SPAO enables even the first packet sent to a destination to be authenticated with DRKey, thus achieving source authentication (property P10). This directly prevents all these attacks. In case a malicious AS creates many fake hosts to overwhelm a destination server, source authentication enables the server to perform per-AS load balancing (see also §9.2).

In a special case where the destination needs to perform a path lookup to return a packet (e.g., if a path has just expired or if a unidirectional path was used), the destination can insist on source authentication via the SPAO to prevent reflection attacks as described in the beginning of this section.

Armed with these countermeasures, we conclude that source-address spoofing is not effective, and SCION achieves property P10.

7.8 Absence of Kill Switches

Monopolistic trust root architectures such as DNSSEC and RPKI/BGPsec enable entities in possession of private keys to shut down portions of the namespace controlled by those keys. The introduction of these *kill switches* into DNS and BGP has created skepticism and concern over the potential outages that could arise should private keys be misused or fall into the wrong hands [441].

Briefly, monopolistic trust architectures work by delegating trust from a top-level key to lower-level keys. Due to this hierarchical trust structure, each key has full control over keys beneath it in the hierarchy. Kill switches work by revoking or maliciously substituting a public-key certificate at a specific point of the hierarchy (e.g., the certificate for the root zone in DNSSEC). The incorrect certificate will cause downstream signature validations to fail. A compromise of the top-level zone is the worst-case scenario; the entire namespace can be shut down. However, targeting specific zones lower in the hierarchy is also possible.

In DNSSEC and BGPsec, root keys are secured through multiple layers of physical security accompanied by a key rollover schedule. However, the key-management processes must still be performed in a known jurisdiction; for DNSSEC, key-management ceremonies are held quarterly at Verisign facilities in the United States. Even though the key-rollover ceremony is monitored and logged, it may still be possible that a state-level adversary can gain access to root-level keys.

We note that private keys can be leaked through several means. Attackers might exploit a software vulnerability in a system with access to the key, or use social engineering for access to that system. Employees may go rogue or be threatened/extorted to reveal the keys. Short asymmetric cryptographic keys (e.g., RSA keys shorter than 1024 bits) are likely breakable today by well-sponsored nation states actors, or can be broken in the future as computing speeds continue to increase or quantum computers become practical. The broad attack surface makes it challenging to implement comprehensive protection mechanisms.

Even without key compromise, centralized architectures risk being unavailable if one or more of their critical services becomes unavailable. For example, a DoS attack on services publishing revocation lists will cause clients to receive an incorrect view of currently valid certificates.

SCION's trust architecture is fundamentally different from that in the cases described above. In SCION, each ISD manages its own trust roots instead of a single global entity providing those roots. This structure gives each ISD autonomy in terms of key management (i.e., all key management operations can take place without contacting a parent authority) and in terms of trust. All entities inside the ISD already subscribe to the ISD's policies. What SCION enables is trust transparency for entities to know what additional roots need to be trusted for a given communication.

Despite not having centrally controlled trust, local kill switches are to some extent possible in SCION. The following sections explain these cases and possible countermeasures.

7.8.1 Local ISD Kill Switch

As in the case of DNSSEC and BGPsec, executing a kill switch inside a local ISD can be done at different levels of the AS-level hierarchy. One difference in SCION is that core ASes cannot be switched off by a parent authority since they manage their own cryptographic trust roots. Another difference is that the attack vector of intra-ISD kill switches has only two entry levels; all ASes obtain certificates directly from the CAs included in the TRC.

If one of the core's root keys is compromised, an adversary could issue illegitimate AS certificates, which may be used in further attacks. However, multiple different voting keys (defined by the voting quorum) would be required to maliciously change the TRC through a TRC update.

Moreover, the core might stop propagating PCBs, precluding the discovery of new paths. In this case, downstream ASes will notice that PCBs are no longer being propagated, but all previously discovered (and still valid) paths are still usable for data-plane forwarding until they expire.

Perhaps a more stealthy kill switch would be to shut down path services in victim ASes. While this cannot be done remotely, an adversarial entity controlling an ISD (e.g., a government) might compel core and non-core ASes to stop replying to path requests. Alternatively, the compelled ASes might return only a subset of all available paths. If this attack were used in conjunction with blackholing, senders in the ISD would have difficulty getting traffic out of the ISD. We would like to emphasize, however, that such attacks are even easier to perpetrate in today's Internet. In SCION, existing paths can continue to be used in the data plane as long as the traversed ASes allow the forwarding.

7.8.2 Remote ISD/AS Kill Switch

Since SCION ISDs independently manage their own cryptographic keys and namespace, it is not possible for a remote attacker (outside the target victim's ISD) to cause a kill switch in a different ISD. That is, without access to the private keys forming the trust root in the remote ISD, the attacker is limited to data-plane attacks. Even if private keys became available to a remote attacker, they would need access to an AS inside the remote ISD to inject faulty information.

7.8.3 Recovery from Kill Switches

In the event of a key compromise of a non-core AS, the impacted AS needs to obtain a new certificate from the core. This process will vary depending on internal issuance protocols. If any of the root keys or voting keys contained in the TRC are compromised, the TRC must be updated as described in §3.1.6. Only in the case of a catastrophic compromise of multiple voting keys at the same time must a trust reset be triggered.

If the core AS has not been compromised, but is instead acting maliciously (e.g., by not propagating PCBs downstream or tampering with responses for paths or certificates), one way to recover is for downstream ASes to self- organize and form a new ISD. By now operating autonomously, the new ISD can begin path discovery and traffic forwarding.

SCION, unlike BGP, has no notion of routing convergence. Instead, the flooding of PCBs disseminates topology information. This means that in the worst case, if all paths must be re-created, fresh paths are established after a single flood has reached all ASes.

7.9 Other Security Properties

In this section, we discuss mechanisms to achieve properties P13–P15.

7.9.1 Confidentiality (P13)

If a malicious AS resides on the path between source and destination, it is able to inspect the traffic between the communicating end hosts. On-path traffic inspection is often referred to as *surveillance*. Mass surveillance has increased in scope and awareness worldwide over the past decade. As document leaks have repeatedly shown, state-level adversaries are interested in monitoring, recording, and analyzing user data wherever possible, both passively and actively [218].

On-path adversarial traffic inspection is inevitable in any network infrastructure, including today's Internet. Therefore, end hosts need to use encryption to hide sensitive data. Systems at higher layers in the networking stack like TLS 1.3 achieve this securely. SCION supports these systems through DRKey and other mechanisms that enable lightweight end-host authentication to protect key-exchange messages (see Chapter 20).

However, even encrypted data traffic can leak information through traffic patterns (such as traffic volumes and inter-packet delays). Research has shown that traffic analysis is highly effective in identifying the online activities of users [565] and can even be used to uncover spoken phrases in voice-over-IP traffic [549, 550]. SCION has a number of built-in features that make analysis more difficult compared to today's Internet.

First, SCION's ISDs prevent entities in remote ISDs from manipulating the local control plane, thus preventing adversaries from tricking other networks into sending traffic through them. This reduces the attack surface for traffic analysis to *on-path* attackers. Within ISDs, the core ASes may appear to be a good vantage point for performing surveillance. Oppressive states may collude with or compel their local core ASes to perform surveillance. However, peering links offer an alternative path out of the ISD without traversing the core. Furthermore, due to source-selected paths and path transparency, senders can select paths through ASes they trust, and thanks to the secure routing process and packet-carried forwarding state, data packets cannot be redirected by off-path adversaries.

Second, SCION's native multipath communication hampers surveillance as an adversary only has access to a subset of packets that traverse its infrastructure. To obtain full information about a user's traffic,[2] the adversary would have to be concurrently present on *all* paths that a traffic flow is using. This results in a costly effort since path selection in SCION can be

---

[2]The effectiveness of splitting traffic over multiple paths strongly depends on the scheduling strategy of the sender. Certain strategies like round-robin do not substantially increase the difficulty of traffic analysis [154].

made dynamic and unpredictable; that is, an end host can select paths based on an arbitrary selection process and change this selection at any time. Moreover, even if the adversary could eavesdrop on all the paths, it would need to correlate the packets that traversed different links but originate from the same user, which causes higher overhead for the surveillance.

Besides traffic patterns, metadata included in headers (such as source and destination addresses) allows identifying and tracking Internet users. We discuss privacy-enhancing technologies facilitated by SCION in §7.9.3.

7.9.2 Resilience against Censorship (P14)

The goal of censorship is different than that of surveillance: Instead of observing the communication of end hosts, a censor attempts to interrupt certain communications altogether. However, despite this difference, censorship is enabled in similar situations, namely when the adversary is on the communication path. Consequently, the same defense mechanisms can be used: Through path selection, end hosts can circumvent ASes that attempt censorship; and through peering links, they can even avoid core ASes.

An adversary controlling core ASes can interfere with communication by withholding path segments to prevent end hosts from using peering links and force them to use paths that include the adversary's AS. However, paths can be shared out-of-band, enabling end hosts to circumvent this attack. Furthermore, SCION's path-lookup system is extensible. For example, downsegments could also be stored at path services in non-core ASes; in that case, paths traversing peering links could be fetched without relying on core ASes but only involving on-path ASes.

These mechanisms are only effective as long as an adversary-free path exists; if an adversary controls all ASes in the user's ISD, the mechanisms are insufficient to avoid censorship. In such a case, more sophisticated censorship-evasion techniques offered by anonymity systems (see §7.9.3) are necessary [507]. We stress, however, that the isolation properties provided by ISDs do not facilitate censorship.

7.9.3 Anonymity (P15)

Communicating on the Internet inevitably leaks information. In particular, network headers reveal information (e.g., source address, flow information), which might threaten anonymous communication and privacy. Based on this information, a state-level adversary could be able to enforce censorship. To counteract identification of users and hosts, SCION's path awareness and packet-carried forwarding state as well as inter-domain routing based on AS numbers (but not end-host addresses) enabled the development of several privacy-enhancing technologies including HORNET [105], TARANET [107], and OTA [325].

HORNET is a low-latency onion-routing system that operates at the network layer and makes use of symmetric cryptography for data forwarding. It offers payload protection by default using a shared secret key between end hosts and routers and can defend against attacks that exploit multiple network observation points. Instead of keeping state at each relay, connection state is carried within packet headers, allowing intermediate nodes to quickly forward traffic without per-packet lookup. Its highly efficient and largely stateless design is facilitated by SCION's packet-carried forwarding state.

TARANET is an anonymity system that implements protection against traffic analysis at the network layer, and limits the incurred latency and overhead. In the setup phase of TARANET, traffic analysis is thwarted by mixing. In the data transmission phase, end hosts and ASes coordinate to shape traffic into constant-rate transmission using packet splitting. TARANET relies on a network architecture that does not leak any extra information about the end hosts or the path before or after the forwarding AS, except the ingress and egress interfaces that are needed as forwarding information through the AS. This property is satisfied by SCION.

OTA uses per-packet *one-time addresses*, which are issued by ASes to their customer hosts. Each one-time address is only used once as either a source or a destination address. This eliminates flow information from packet headers— implicitly (e.g., the standard 5-tuple in TCP/UDP packets) and explicitly (e.g., flow identifier)—while still allowing demultiplexing of seemingly unrelated packets to flows. This system fundamentally relies on the fact that ASes can unilaterally distribute end-host addresses because inter-domain routing and forwarding only depends on the source and destination ASes.

7.10 Summary

In this chapter and the previous one, we have analyzed various functional and non-functional properties of SCION and compared them to those of today's Internet. We have shown that SCION scales better and, at the same time, provides built-in security defenses against many well-known network attacks that plague today's Internet operators and users.

SCION's basic components cannot prevent all attacks. To achieve availability guarantees and rule out some particular types of path-manipulation attacks, extensions are needed, which will be discussed in Part III.

We finally refer our readers to external evaluations of future Internet architectures (FIAs), for instance to a recent study by Ding et al. [166], which also demonstrate that the security properties achieved by SCION are stronger than those of other FIAs.

Part III

Achieving Global Availability
Guarantees

8 Extensions for the Control Plane

MARC FREI, MATTHIAS FREI, SAMUEL HITZ, JONGHOON KWON, CHRISTOPH LENZEN, ADRIAN PERRIG

Chapter Contents

8.1 **Hidden Paths** **185**
   8.1.1   Overview of Hidden-Path Communication 186
   8.1.2   Detachable Extension for PCBs 187
   8.1.3   Segment Registration 188
   8.1.4   Path Lookup 188
   8.1.5   Security 189
8.2 **Time Synchronization** **190**
   8.2.1   Design Principles 190
   8.2.2   Solution Overview 192
   8.2.3   System Architecture 194
8.3 **Path Metadata in PCBs** **197**
   8.3.1   Supporting All Types of Segment Combinations 198
   8.3.2   Types of Metadata 200

SCION's extensible architecture enables new systems that can take advantage of the novel properties and mechanisms provided. As compared to the current Internet, most of the benefits are achieved through the use of packet-carried forwarding state (PCFS), path transparency, and control. In this chapter, we describe three control-plane systems to supplement the SCION architecture: Hidden paths enable the selective announcement of path segment information to chosen recipients; the SCION time synchronization provides highly secure, available, and scalable global time; and finally, we highlight the power of adding metadata about paths to PCBs.

8.1 Hidden Paths

In Chapter 4 we discussed how path segments are registered at path services by ASes. All these segments are *public*—every AS and all end hosts in the SCION network can obtain them and use them to send data packets. This is crucial to achieving global connectivity as discussed in Chapter 7.

© The Author(s), under exclusive license to Springer Nature Switzerland AG 2022
L. Chuat et al., *The Complete Guide to SCION*, Information Security
and Cryptography, https://doi.org/10.1007/978-3-031-05288-0_8

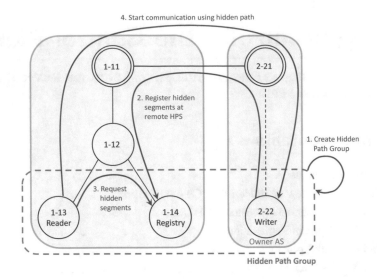

Figure 8.1: Hidden-path communication scheme.

However, this also means that any host in the Internet can attempt to flood these links with excessive amounts of traffic in a (D)DoS attack. While it is very difficult to overload inter-domain links in the core of the Internet, access links of small ASes generally do not have sufficient bandwidth to withstand such attacks.

An AS may therefore want to keep secondary *hidden* access links that are not publicly registered but only shared with specific authorized ASes or end hosts that belong to the same company or select customers. These hidden segments are not registered at standard public path services but instead at a hidden-path service, which enforces access control to authorized entities.

Thus, even if the public access link of the AS is unavailable during a DDoS attack, critical communication with branch offices and important customers would most likely still be possible through the hidden segment.

In this section, we present the control-plane components that are necessary for hidden-path communication. To ensure that hop fields of hidden segments cannot be reused by entities observing packets in the data plane, additional mechanisms are required; these will be presented in §10.1.

8.1.1 Overview of Hidden-Path Communication

Hidden-path communication works in hidden-path groups as illustrated in Figure 8.1. In general, four types of ASes are involved in the communication:

- **The owner AS** is responsible for maintaining the hidden-path group configuration and distributing it to all servers that require it. The owner has read and write access to the hidden-paths registry.

- **Writer ASes** are allowed to register hidden segments; they have read and write access to the hidden-paths registry.

- **Reader ASes** are allowed to access hidden segments; they have read access to the hidden-paths registry.

- **Registry ASes** run the hidden-path service at which writers register hidden segments and readers fetch them. Even though they can write data to the registry storage they maintain, they effectively only have read access: As the path segments themselves contain signatures of on-path ASes, these cannot be forged by the registry ASes.

Basic hidden-path communication works as follows:

1. The owner AS creates a hidden-path group and shares the configuration through secure (out-of-band) channels with reader and registry ASes. It is the group owner's responsibility to disseminate updated versions to all members.

2. Writer ASes register down-segments at registry ASes of their hidden-path group.

3. Reader ASes query registry ASes for hidden segments.

4. Reader ASes can communicate with writer ASes using hidden path segments.

8.1.2 Detachable Extension for PCBs

Communication over hidden paths must use the EPIC path type to avoid leaking hop authenticators in the data plane (see §10.1). EPIC itself requires longer authenticators for each hop field. These additional authenticators are added to PCBs as detachable extensions (§4.1.2). This approach has multiple advantages:

- No separate beaconing process is necessary for hidden segments.

- A single PCB can be registered both as a hidden segment and a standard public SCION segment (after removing the additional information in the detachable extension).

In particular, a PCB can be used as a hidden and public segment at the same time by being registered both at the hidden-path service of the hidden-path group and at the public (local and core) path services. Although the hidden segment is no longer technically *"hidden"*—non-members of the hidden-path group are able to forward data packets through the same access link using the public segment retrieved from the public path services—the benefit of the hidden segment on the high-availability guarantee for communication is intact;

prioritization of the EPIC traffic at on-path border routers on the hidden access link, which can only be sent using the additional authenticators included in the detachable extension, still provides defense against DDoS attacks, as we discuss in §10.1.4.

8.1.3 Segment Registration

Each AS needs to configure which segments should be publicly registered with the standard path services and which ones should be registered as hidden segments. These decisions are based on a policy that is defined for each SCION interface ID in the local AS. If an interface is marked as "public", then down-segments going through the interface will be registered via the normal SCION down-segment registration mechanism. For segments that should be hidden, one or multiple hidden-path groups can be specified; for each group, the segment is registered at the hidden-path service of all corresponding registry ASes.

Communication with the hidden-path service to register segments is similar to the communication with the core path service for standard down-segment registration. Importantly, the hidden-path service needs to verify that the sender of the segment is a writer in the hidden-path group in which it tries to register the segment.

8.1.4 Path Lookup

Looking up hidden segments is similar to fetching standard SCION segments. In addition to the SCION daemon of the end host, this involves one or multiple hidden-path services, which take the role of path services for public SCION segments.

8.1.4.1 SCION Daemon

In addition to up-, core-, and down-segments, the SCION daemon of an end host is responsible for fetching hidden down-segments. The daemon is configured with the hidden-path group IDs it should query. Using the configured hidden-path group IDs the daemon queries the local hidden-path service for the given destination. Once the daemon has all segments collected it combines the segments to paths and returns the paths to the requester.

8.1.4.2 Hidden-Path Service

The hidden-path service can be queried for hidden segments to a given destination. The request includes a set of hidden-path group IDs and a destination ISD–AS identifier. Upon receiving a request, the service must verify that the requester is allowed to access the requested hidden-path groups: If the requester is a local end host, the service checks if it is authorized according to internal policies; if the requester is a remote hidden-path service, the service checks that the requesting AS is a reader in the corresponding hidden-path group.

Once validation is completed, the hidden-path service can take one of the following actions for each group ID in the request:

- Look up segments in the database for hidden-path groups where the local AS is a registry and thus the service is authoritative.

- Forward the request to a hidden-path service in an AS that is a registry for the requested hidden-path groups. Forwarding is only allowed if the request comes from an end host within the AS.

Note that multiple forward requests might be required depending on the hidden-path groups that are requested. These requests can be bundled in case multiple hidden-path groups have the same registry ASes. This is an optimization and therefore not required.

Hidden-path services in remote ASes can be discovered through the service-discovery mechanism described in §4.6.

8.1.5 Security

Various operations in this design are security-critical and require client/server authentication:

- For the creation of hidden-path groups, we assume that the chosen out-of-band mechanism is safe, e.g., using an encrypted connection through a secondary interface that is physically separated from the primary network connection.

- For segment registrations from a beacon service to the hidden-path service, the registry and writer AS must be authenticated. This can be done using TLS based on AS certificates from the CP-PKI. Even though the segments themselves contain signatures, the writer AS is also authenticated through the CP-PKI, preventing unauthorized addition of hidden paths.

- The SCION daemon querying paths from the local hidden-path service is secured by AS-internal policies or PKIs.

- For inter-AS hidden-segment lookups, clients are authenticated using TLS client certificates that are created with the AS's secret key from the CP-PKI, similar to PILA (see Chapter 20).

8.2 Time Synchronization

Due to unavoidable clock drift, computing devices have to correct their clocks periodically to keep the relative offsets among clocks in an acceptable range. This is exactly the goal of time synchronization. Extensive studies have been performed on the fundamental problem of time synchronization in distributed systems to improve accuracy, fault tolerance, and security. However, no existing system can satisfy all of these qualities. In the Network Time Protocol (NTP [368]) architecture, achieving high accuracy of synchronization for every host on the Internet would hardly scale. Sophisticated time synchronization protocols such as Precision Time Protocol (PTP [259]) or Datacenter Time Protocol (DTP [323]) require special hardware support, suffering from the lack of incremental deployability [323, 564]. The security and resilience to extreme conditions of the current time synchronization systems can also be improved.

In this section, we introduce a clean-slate redesign of a global time synchronization infrastructure operating on SCION. Our vision is to achieve reliable time synchronization for any host on the Internet with an accuracy of at least 100 ms. We want to achieve this through a system that is scalable, fault-tolerant, and resilient to attacks by malicious system components and entities. The system operates in a decentralized fashion without a global root of trust, and does not depend on the perpetual availability of external time references such as Global Navigation Satellite Systems (GNSSes).

8.2.1 Design Principles

With the ambition of a secure and highly available Internet architecture, SCION also aspires to support infrastructure components and applications with a scalable and fault-tolerant time synchronization architecture that is robust against attacks on availability and accuracy. The core challenge in achieving global time synchronization in SCION is the tension between the sovereign operation of ISDs and the requirement to arrive at a globally agreed-upon synchronized time across ISDs.

8.2.1.1 Challenges

Despite the large body of work on clock synchronization for distributed systems, the following challenges still provide room for improvements: precision stability and fault tolerance.

- **Precision stability**: Synchronization precision in most NTP-like protocols relies heavily on network latency symmetry. A client queries a remote time server to learn the server's clock value and estimate the network delay; one-way delay is estimated by halving the round-trip time and compensating for the measured time offset. It is an intrinsic assumption that the communication is perfectly symmetric, i.e., the propagation delay is the same in both directions. In a real system, however, this assumption will most likely not hold. Routing asymmetry in the Internet is a widely observed phenomenon [156, 228, 403]. Even on a symmetric path, messages will encounter different buffer times [184, 333]. Random wire delay introduced by other traffic is another source of latency asymmetry [280].

- **Fault tolerance**: Routing failures or misconfiguration can significantly impact the clock synchronization architecture. Most clock synchronization protocols allow the Internet routing architecture, such as BGP-IP, to manage packet transmission. The underlying architecture thus achieves agility to network failures. However, considering the slow convergence after network failures in the current Internet [241, 313], reliable clock synchronization is elusive. Furthermore, active network attackers might attempt to delay, drop, replay, and forge clock synchronization packets in order to desynchronize target systems. Botnet-size adversaries might also compromise a fraction of time servers to disrupt target systems or services [161, 414].

8.2.1.2 Requirements

The fundamental goal of this work is for collaborative networks without complete mutual trust to introduce a reliable time synchronization architecture, which achieves a maximum clock offset between any two nodes of $\leq 100\,\text{ms}$ on a global scale. More precisely, the bounds should be at least as good as in more traditional time synchronization architectures under ideal external conditions. Furthermore, synchronization must be reliable even in the occurrence of extreme events, like a massive failure of reference clocks, being disconnected from parts of the network, or the presence of active attackers. The impact of such events should be limited with regard to the offset between any two non-faulty clock values in the system.

8.2.1.3 Constraints

Based on the system's goal, a number of key constraints can be inferred for the design.

- We need a decentralized architecture, which cannot depend on a global root of trust.

- We need to avoid reliance on other (single or co-dependent) external systems that do not or might not offer the same degree of resilience to attacks and extreme conditions.

- In particular, we cannot depend on the perpetual availability of GNSSes, because they might be incapacitated by extreme events such as Coronal Mass Ejection (CME) [276].

- We must ensure resilience against component failures and attacks by malicious components or entities.

8.2.1.4 Threat Model

We assume that the adversary can compromise several network entities. The adversary has full control over its territory (i.e., compromised entities and corresponding links) and can eavesdrop, inject, intercept, delay, and alter on-path packets with negligible latency inflation. Besides the presence of an active attacker, we also consider network failures due to, e.g., link congestion, misconfiguration, and physical errors that might hamper reliable clock synchronization. However, we restrict the threat model up to $f < n/3$ Byzantine faulty nodes in the system, where n refers to the number of core ASes.

8.2.2 Solution Overview

Given the heterogeneous trust relationships in the Internet, we seek a globally distributed infrastructure design that does not place trust in any single entity, and can tolerate a fraction of faulty or malicious entities. An interesting property of this setup is that despite the absence of a global leader that determines the time, the globally agreed time is an emergent property of the distributed time synchronization protocol. To achieve high utility for practical applications, we attempt to keep the global SCION time synchronized with Coordinated Universal Time (UTC) provided by reference clocks like, e.g., GNSSes, as long as the received value for UTC is within a close enough range to the globally approximated SCION time.

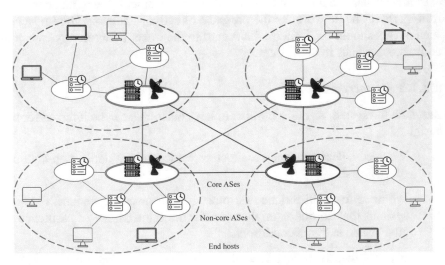

Figure 8.2: Global time synchronization topology.

8.2.2.1 Structure Design

Hierarchical Synchronization. The core idea of our approach is to construct a hierarchical time synchronization topology, where each AS contributes with a time service, similar to the DNS infrastructure. Figure 8.2 illustrates the basic architecture.

We leverage the structure of AS tiers (e.g., Tier-1, Tier-2, and Tier-3) by observing that higher-tier ASes can contribute more significantly as they have larger budgets and more personnel.

For instance, we assume that core ASes operate reliable infrastructure and have local atomic clocks deployed to guarantee long-term time stability in case of a significant global event. Our approach constructs a two-level hierarchy, where the core ASes run a global peer-to-peer synchronization algorithm (see §8.2.3.1). All the other ASes then synchronize their time with their upstream providers. As each AS operates at least a minimal time server infrastructure, end hosts can then synchronize their time with the local time server, ensuring scalability.

GNSS Independency. We avoid fully trusting any external system, in particular GNSSes such as GPS, Galileo, GLONASS, or BeiDou. This is because GNSSes are vulnerable to manipulation by nation states and also to natural phenomena like, e.g., Coronal Mass Ejection (CME) events, which would likely paralyze the global GNSS infrastructure for an extended time period [276].[1] The typically high precision of GNSS time, however, can nevertheless be used to improve precision of local clocks as long as the GNSS time remains consistent across core ASes. Using GNSSes under such normal conditions keeps the globally synchronized SCION time as close as possible to UTC. With this approach, we increase the precision and usefulness of our system in ideal conditions without relying on GNSSes in situations where they are unavailable or cannot be trusted.

8.2.2.2 Communication Design

Path-Aware Synchronization. We make use of SCION's fine-grained path control for the time synchronization communication. By selecting paths based on the topology and metadata provided by the SCION network infrastructure, the system can improve security and fault tolerance while also striving to achieve delays between two nodes that are as constant and symmetric as possible, which optimizes the accuracy of the overall synchronization. A constant delay means that the time required to deliver a packet between two nodes in a network shows relatively low variation over time. A symmetrical delay means that the time required to forward a packet from node A to node B in a network

---

[1]In case of CME, terrestrial cables and much of our IT infrastructure would be affected. A weak CME could affect satellites, but not terrestrial infrastructure [276].

is as close as possible to the time required to return the packet from B to A. It therefore can be advantageous to base the path selection on a metadata framework that associates individual path segments with information like, e.g., the expected latency or the type of the links along a candidate path.

Byzantine Fault Tolerance.   An AS's time server communicates with other ASes' time servers, gathering multiple time offsets simultaneously. The goal of having these multiple external time sources is to maintain sovereign operation of ISDs. This means that non-faulty ASes are able to maintain internal time synchronization within an ISD and across ISD boundaries even if external entities (like, e.g., GNSSes) are supplying erroneous values, experience outages, or attempt to maliciously manipulate clock values (provided that enough network paths through correctly operating ASes exist).

8.2.3 System Architecture

As a global time synchronization architecture, the system is not limited to any particular protocol. By default, we leverage NTP message exchanges to measure clock offsets, but any other existing or newly emerging time synchronization protocol can potentially be adopted by network entities. The following section describes the inter-ISD and intra-ISD time synchronization shown in Figure 8.2.

8.2.3.1 Core Synchronization

Each core AS operates a time server (TS) that acts as a stratum 1 server similar to the traditional NTP architecture. The core TSes typically use one or more GNSSes as their external time sources (stratum 0). We assume that the core TSes are equipped with local clocks driven by high quality oscillators with an autonomous free run accuracy of at least ± 10 ms after one year. To achieve global scale and robustness against faulty external time references, core TSes continuously synchronize their local clock also with every other reachable core TS.

Algorithm.   The general approach is to periodically compute an approximate agreement on clock correction values at each TS based on the relative clock offsets to its peer TSes. Each TS corrects its local clock towards the approximate global clock value. To avoid large discontinuities, the resulting corrections are capped by the maximum time drift of the local clock (scaled by a configurable coefficient) over a given time interval as specified by the clock manufacturers (under the assumption of controlled ambient conditions like, e.g., temperature).

In addition to these periodic global synchronizations, each TS synchronizes the local clock with its local reference clocks. These local clock corrections are also limited to the expected maximum drift of the local clock scaled by a configurable coefficient, but they are applied more frequently than the global clock corrections.

Byzantine fault tolerance properties of the approximate agreement on the global clock value among all core TSes are achieved by deriving the algorithm from the extensively studied clock synchronization algorithm by Welch and Lynch [536]. The core idea is that in each round every participating node collects an array of relative clock offsets to each peer. Based on this information, a fault-tolerant midpoint (or averaging) function is computed, resulting in a global approximate agreement on the relative time differences among the core TSes [169]. With this algorithm, the system is able to tolerate faults or malicious behavior of up to one third of the nodes in the set of participating core TSes.

A requirement for the core synchronization algorithm is that all core TSes agree on the same set of core TSes to sync with. In SCION it is possible to provide TSes in the core network with a consistent enough view of all core ASes in the network to satisfy this requirement in practice. TRCs are disseminated among all ISDs as part of the core beaconing. Each TRC consists of a signed collection of cryptographic information entries, including a list of the core ASes in a given ISD. By leveraging the precisely defined processes for verification, update, revocation, and recovery from catastrophic events, the set of all core ASes can therefore be maintained based on TRCs without introducing additional mechanisms besides what is already provided by the CP-PKI.

Leap Seconds. The unique aspect of UTC compared to its base time, International Atomic Time (TAI), a high-precision atomic coordinate time standard derived from a few hundred atomic clocks distributed over 50 national laboratories, is the leap second. By occasionally applying a one-second adjustment, UTC accommodates the irregular rotation of the Earth. With an announcement from the International Earth Rotation and Reference Systems Service (IERS) in advance of 6 months, an adjustment for a leap second is scheduled. A leap second is applied at the end of a UTC month, most likely June or December. When it is scheduled, a leap second is inserted or deleted between 23:59:59 and 00:00:00.

For practicality, we attempt to keep SCION time synchronized with UTC, so the leap seconds also need to be applied to SCION time if mandated. Unlike UTC, however, SCION time is a global clock value derived from all core ASes. It requires consensus on applying leap seconds in SCION time from a large fraction of participants. For simplicity, we mandate that all core ASes running TSes apply leap seconds by default. When IERS announces the next leap second, AS administrators are responsible for scheduling the leap second adjustment. For those that do not apply leap seconds as scheduled, gradual convergence to the leaped SCION time through the global clock correction will be achieved.

8.2.3.2 Networking

A central tenet of the SCION networking architecture is comprehensive path transparency and control that enables senders to simultaneously select multiple paths to carry packets towards the destination. In general, this multipath communication capability can be used to optimize bandwidth and latency as well as to enhance overall availability due to increased resiliency against link failures, or to avoid untrusted infrastructure along the way. In the specific application of global time synchronization, multipath communication enables designing a system that is able to approximate optimal accuracy while also improving security and fault tolerance of the synchronization even over a public Internet. Another benefit of using a path-aware networking substrate is that end hosts gain the ability to assess networking conditions in terms of the number of available paths, the trustworthiness of these paths, and also the advertised quality metrics. These quality attributes could be associated with a given clock value providing additional QoS information for applications and even end users with demanding timing requirements.

Path Selection.   When a sender is creating a packet to be sent over the network, it first queries a set of paths to the target end host. These paths are discovered and disseminated by the SCION control plane based on individual path segments at the level of ASes. Infrastructure services in the control plane also provide the functionality to combine path segments into actual end-to-end paths. For typical network topologies it is to be expected that the result set for a path query to a given target end host will consist of up to a few dozens of paths, especially between core ASes. The sender will select one or more paths out of the set of available paths, which can be used simultaneously even from endpoints connected by a single link. For each selected path, the sender will create a separate packet that includes the selected path in the packet header as so called "packet-carried forwarding state" before it gets sent out.

There are two categories of information that SCION can provide to support the end-to-end path selection process.

The first category relates to the network topology: Each path is specified by a cryptographically secured list of ASes and ISDs through which the path leads. This defining structural property can be used by applications in the path selection process to optimize architectural qualities like fault tolerance (against faulty or malicious actors) by picking paths that are as diverse and disjoint as possible. Certain routes can also be excluded based on commercial or legal criteria. This property is the basis for security arguments and effective fault tolerance in the time synchronization network.

The core time synchronization algorithm described in §8.2.3.1 thus selects among the available paths towards a given peer a subset of paths with maximum disjointness exactly to increase fault tolerance and to minimize the impact of on-path attackers attempting to drop or delay packets. The resulting

NTP-based time offset measurements over the selected paths are then collected and the current time offset to the given peer is computed based on these measurements.

In addition, it is useful to have a second category of information about the actual properties of the available paths, such as information about the expected latency or the type of the traversed links. This metadata is collected and published by individual ASes and disseminated via PCB extensions, see §8.3. Investigating how SCION's path metadata system can be integrated into the path selection process in order to also optimize the accuracy and precision of network-based offset measurements would be an interesting direction of future research.

8.2.3.3 Intra-ISD Synchronization

Within an ISD, time synchronization conceptually follows the provider-customer relationship: Each AS synchronizes with its provider ASes. However, considering that multiple tiers inside an ISD may exist, recursive time synchronization from a core AS to a leaf AS by strictly synchronizing with the respective provider ASes will introduce additional inaccuracies. For this reason, the intra-ISD synchronization architecture allows downstream ASes to directly synchronize with their core ASes while end hosts synchronize with their local TS. Relaxing the layering in this way enables us to minimize the potential amplification of errors in precision caused by recursive synchronization processes.

All non-core ASes therefore run TSes that operate as stratum 2 servers in the traditional NTP sense. By default, these TSes synchronize with their core TSes. Ideally, core ASes operate multiple TSes for fault tolerance, such that their customers can benefit from the highly reliable time synchronization architecture. As in the case of the core synchronization between core ASes, each regular AS might have multiple paths to reach its core ASes, achieving similar fault tolerance and resilience against on-path attackers also within an ISD.

Under normal conditions, end hosts are synchronizing their clock with TSes in the local AS. Nonetheless, it is also possible to synchronize with one or multiple upstream TSes if a local TS fails.

8.3 Path Metadata in PCBs

As we discussed in §4.1.2, SCION's routing messages—PCBs—are extensible; in addition to basic routing information, additional data can be added in extensions (signed, unsigned, or detachable). Metadata about segments and individual ASes is highly valuable both for the beaconing process itself (i.e., when ASes decide which PCBs to register and extend) and the path construction and selection by end hosts.

Metadata can be classified as *static*, which refers to properties that remain unchanged over the lifetime of a segment (i.e., minutes to hours), and *dynamic*, referring to properties that change at shorter timescales. For example, the installed bandwidth on each link of a path is static, as it will not change unless the underlying network infrastructure is modified. On the other hand, the bandwidth available to an application over this path is dynamic, as it depends on the current load on the network infrastructure along the path.

In this section, we describe how static metadata can be added in a *static info extension*, which is a signed extension and thus protected by each on-path AS's signature.

8.3.1 Supporting All Types of Segment Combinations

To describe the properties of all hops of a path, it is necessary to include information about the links between the ASes (inter-AS hops), and also for the path from the ingress to the egress inside an AS (intra-AS hops). As an end host can combine segments in various ways (§5.5), it is insufficient to include only information about the hop between the PCB ingress and egress interfaces: At any AS where two path segments can be joined together, the information about the resulting intra-AS hop must be contained in (at least) one of the AS entries.

We make the assumption that the link properties are symmetric, that is, they are the same in both directions. This assumption simplifies the encoding and reduces the communication overhead.[2] The following scheme ensures that each hop is described by (at least) one AS entry and avoids redundancy of information that is added to PCBs.

Core Beaconing. In core beaconing, an AS includes metadata for

- its own intra-AS hop between the PCB's ingress and egress interface; and

- the inter-AS hop at the egress interface.

Intra-ISD Beaconing. In intra-ISD beaconing, an AS includes metadata

- for its own intra-AS hop between the PCB's ingress and egress interface;

- for the intra-AS hop between egress and any interface to child links with ID smaller than the egress interface (for shortcut combinations);

- for the intra-AS hop between egress and any interface to core links (for combining up- and core-segments or core- and down-segments);

---

[2]If a property is asymmetric for a particular link, the more conservative value can be added to the PCB.

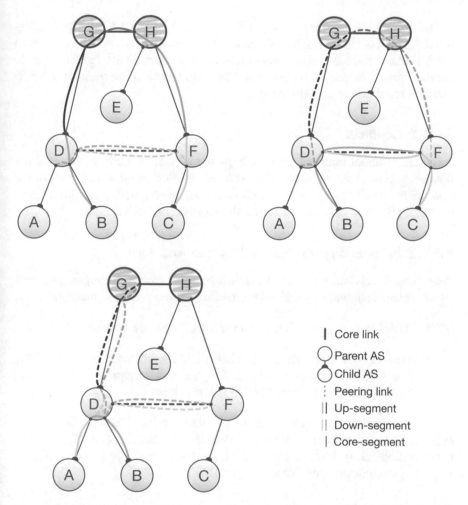

Figure 8.3: Illustration of the information included in the individual AS entries
          in path segments and how this information is used in three types of
          segment combinations. Each color represents an AS entry. Dotted
          colored lines represent information about links that is included in
          the PCBs, but not used in this path segment combination.

- for the intra-AS hop between egress and any interface to peering links
  and the corresponding inter-AS hop at that interface (for peering combinations); and

- for the inter-AS hop at the egress interface.

Information about the intra-AS path from/to the end hosts in source and destination ASes cannot be included easily as the location of end hosts is unknown
at beaconing time.

For example, when propagating a PCB from AS G to AS B, AS D in Figure 8.3 must also include intra-AS metadata for the path between interfaces to ASes B and A such that end hosts obtain full end-to-end path information for shortcut paths. Similarly, metadata for the path between interfaces to ASes B and F is required for peering paths.

8.3.1.1 Conflicts

Information about inter-AS peering hops is included in both the up- and the down-segment. Here it is possible to have conflicting information from the two ASes, which must be considered when combining path segments. These conflicts can be resolved by selecting the more conservative value.

8.3.1.2 Authenticity, Accountability, and Accuracy

Both signed and detachable PCB extensions are protected through signatures, which ensures authenticity and accountability for the provided metadata:

- No AS can tamper with the metadata included in a beacon.

- The source of the metadata is visible and cannot be repudiated. Thus, if an AS was detected to be including false information, it can be held accountable by, e.g., being added to a block list.

The responsibility for providing all of this metadata lies with the individual ASes. Accuracy (and completeness) of the information could be certified (cryptographically) through trusted third parties or could be legally enforced as part of the requirements for joining an ISD.

8.3.2 Types of Metadata

Various types of metadata can be included in the static info extension.

- **Latency:** The propagation delay between two border routers is a static property as long as intra-domain paths remain stable or predictable over the lifetime of the segment. This represents the delay under ideal networking conditions and does not include queuing delays. Latency information can be used by both ASes and end hosts to optimize their path-selection algorithm for low-latency paths.

- **Bandwidth:** The installed bandwidth between any pair of border routers is a static property. It represents the available bandwidth under ideal networking conditions (without congestion). Like latency information, bandwidth information can be used by ASes and end hosts to construct paths with consistently high bandwidth.

- **Geographic information:** Geographic information in the form of coordinates (latitude and longitude), and optionally a civic address, can be provided for border routers and/or intermediate routers. This information can be used by end hosts to ensure geo fencing, that is, ensuring that paths do not leave a particular area or, alternatively, that they never cross territory that is considered "undesirable" by the end host. In addition, geographic information can serve as a proxy for latency information (if that is not available).

- **Link type:** Link Type information gives a broad classification of the different underlying infrastructure used by inter-AS links. It can distinguish between *direct* links (direct physical connections), *multi-hop* links (connections with local routing/switching), and *overlay* links (connections that are laid over the legacy Internet).

- **Internal hops:** The Number of AS-internal hops counts the internal hops (e.g., internal IP or MPLS routers) between the ingress and egress routers. This information can serve as a proxy for latency information (if that is not available) and can be used to exclude undesirable paths from the selection.

- **Power consumption & emissions:** An AS can include power consumption and CO_2 emissions of intra-AS hops, which enables the construction of energy-efficient and/or low-emission paths. Similar to other information, this could be certified by independent entities. The details of how this metadata is computed and used are described in Chapter 16.

- **Note:** Simple plain-text notes can be attached as a tool for network engineers to communicate interesting or important information to their other ASes and end hosts.

9 Monitoring and Filtering

MARKUS LEGNER, ADRIAN PERRIG, BENJAMIN ROTHENBERGER, SIMON SCHERRER*

Chapter Contents

9.1   Replay Suppression **204**
    9.1.1   Filtering Outdated Packets Based on Timestamp   204
    9.1.2   Duplicate Detection Using Bloom Filters   205
    9.1.3   Unique Packet Identification   206
    9.1.4   Security Considerations   206
    9.1.5   Prerequisites for Replay Suppression   207
9.2   High-Speed Traffic Filtering with LightningFilter **207**
    9.2.1   Problem Definition   208
    9.2.2   Overview of LightningFilter   209
    9.2.3   Deployment Scenarios   211
    9.2.4   Details of the LightningFilter Pipeline   212
    9.2.5   History-Based Resource Allocation   215
    9.2.6   Traffic Filtering   215
    9.2.7   Best-Effort Traffic   216
    9.2.8   Evaluation Results   217
9.3   Probabilistic Traffic Monitoring with LOFT **217**
    9.3.1   System Overview   218
    9.3.2   Algorithm   220
    9.3.3   Evaluation   223

An important aspect of achieving security for SCION is not only to ensure correct operations by *preventing* malicious activities, but also to *detect* misbehavior. For instance, a MAC or digital signature prevents adversaries from forging or altering authenticated information. However, an adversary may still duplicate legitimate packets to perform a DoS attack by overflowing links (§9.1), sending excessive numbers of packets to overwhelm end systems (§9.2), or simply sending too much traffic, thus causing congestion in the network

---

*This chapter reuses content from Rothenberger, Legner, Frei, Gude, Jacky, Sprenger, and Perrig [443], Scherrer, Wu, Chiang, Rothenberger, Asoni, Sateesan, Vliegen, Mentens, Hsiao, and Perrig [455].

© The Author(s), under exclusive license to Springer Nature Switzerland AG 2022   203
L. Chuat et al., *The Complete Guide to SCION*, Information Security
and Cryptography, https://doi.org/10.1007/978-3-031-05288-0_9

(§9.3). This chapter presents SCION's countermeasures, which all can be implemented efficiently on high-speed routers and end systems.

9.1 Replay Suppression

Authentication of packets—e.g., via the SPAO—ensures that packets can be attributed to the sender and cannot be changed in transit. However, an on-path attacker could still record authenticated packets and replay them (potentially many times) to flood the network or overload the receiver. This is an issue for several of the systems we describe in this part of the book, including Lightning-Filter (§9.2), EPIC (§10.1), and COLIBRI (§10.2).

Existing end-to-end security mechanisms like IPsec and TLS defend against packet replay through sequence numbers. Unfortunately, this requires keeping per-connection state, which is impractical for in-network replay suppression (required for EPIC and COLIBRI) and provides additional attack vectors for end-to-end applications (LightningFilter).

In this section, we describe an efficient approach for performing replay suppression based on timestamps and Bloom filters. This requires global time synchronization with a precision on the order of 100 ms, which we achieve through SCION's time synchronization system described in §8.2.[1]

9.1.1 Filtering Outdated Packets Based on Timestamp

To limit the time period for which the replay-suppression system must keep track of every single packet—we call this the *monitoring period*—it must be able to filter out outdated packets. This can be achieved by adding precise timestamps to packets, which in turn requires globally synchronized clocks.

The maximum propagation delay in the Internet provides a lower bound for the monitoring period. This can be estimated based on maximally distant communication endpoints, the speed of light in fiber, and the maximum latency inflation due to detours and buffering. According to a study by Singla et al. [475], the median latency inflation of network paths over the minimal latency given the great-circle distance between two points and the speed of light in vacuum is 2.3 (including the effect of the lower speed of light in fiber) with few outliers reaching a latency inflation of 10. Assuming two points on Earth with the maximally possible great-circle distance of 20,000 km, this extreme worst-case value yields a maximal (one-way) propagation delay of 700 ms.

Given a time-synchronization system achieving a precision on the order of 100 ms (e.g., §8.2), the maximum time difference between two entities is 200 ms. Together with the maximum propagation delay, this means that any packet with a timestamp of more than 900 ms in the past can be considered

---

[1]A system based on per-AS sequence numbers that only requires *local* time synchronization was designed by Lee et al. [326].

outdated and dropped or forwarded with lower priority. Thus, the monitoring period can be limited to approximately 1 s. This calculation also demonstrates that clock synchronization with a precision of around 100 ms is sufficient as even better precision does not significantly reduce the monitoring window.

9.1.2 Duplicate Detection Using Bloom Filters

Bloom filters [78] are a space-efficient probabilistic data structure for set-membership queries, where false positives can occur, but false negatives cannot. Querying a Bloom filter containing a set of elements either returns that it (1) possibly is, or (2) definitely is not an element of the set. They are often used in packet-processing environments due to their space efficiency compared to traditional hash maps. However, since regular Bloom filters do not support the deletion of entries, we make use of a data structure consisting of multiple Bloom filters that are periodically rotated [326].

In this data structure, a packet hash (see §9.1.3) is inserted only into one of the filters (the currently active Bloom filter); however, when establishing whether the packet has been previously observed, all filters are checked. A positive response from any of the filters indicates a packet replay. The filters are periodically rotated in a round-robin fashion, such that they cover a sliding time window approximating the length of the monitoring period. In order to delete entries for packets older than the monitoring period, the oldest filter is reset and reinitialized to zero, before being reused as the next active filter.

The use of Bloom filters as a method of suppressing duplicates raises the question of optimal choice of parameters. The number of filters N and the filter rotation period R are determined by the monitoring period length M. After inserting packets in the active filter, this filter must still be checked for duration M. During this time, the remaining $N - 1$ filters are used. Thus, the rotation period must satisfy $(N - 1) \cdot R \geqslant M$, and, crucially, we must also reserve some time for resetting the next active filter. Additionally, we can choose the filter size m, in number of bits, and the number of hash functions k for the Bloom filters.

For example, if the monitoring period is $M = 1$ s and we use $N = 2$ Bloom filters, we can choose the filter rotation period as $R = 1.1$ s. A traffic volume of 10 Gbps of minimum-sized packets results in a maximum number of expected entries for each filter of $n = R \cdot \frac{10\,\text{Gbps}}{64\,\text{B}} = 21.5 \cdot 10^6$. We choose the acceptable false-positive rate p as 0.01%. Now the optimal values for m and k can be determined from n and p with well-known formulas for Bloom filters. The number of hash functions is chosen as $k = 13$, and the size of the filter is chosen as $m = -\frac{n \ln(p)}{\ln(2)^2} = 4.12 \cdot 10^8$ bit $= 51.5$ MB.

In conclusion, this data structure in combination with timestamp-based filtering effectively removes duplicated or replayed packets. In other words, packets that are stored and replayed significantly after their observation time will be caught due to an outdated timestamp. Packets that are replayed shortly after their observation will be caught by the Bloom filter-based data structure, whereas packets that are replayed with a modified timestamp will be filtered by the source-authentication module.

9.1.3 Unique Packet Identification

To prevent false positives beyond what is unavoidable due to the use of Bloom filters, we must ensure that any two packets can be distinguished. Concretely, we require packets to explicitly or implicitly contain a unique packet identifier (ID). This unique ID can either be used directly as the packet hash for inserting into the Bloom filters, or indirectly—e.g., through a MAC that depends on the unique ID (see next section).

Given a sufficiently precise timestamp, the combination of source address (including AS and ISD) can serve as such a unique ID. If the timestamp is not precise enough, the payload hash can be used in addition. To enable fast processing, it is important that the payload hash is encoded in the packet header.

9.1.4 Security Considerations

The replay-suppression system is used to prevent replay attacks. However, we must ensure that the system (1) cannot be evaded by malicious actors, and (2) cannot be attacked itself through poisoning attacks.

9.1.4.1 Preventing Evasion

If the adversary could modify the unique packet ID of packets they want to replay, they could trivially evade detection of these packets. This means that replay suppression is only meaningful in combination with packet authentication. More concretely, both the unique packet ID and the timestamp (if it is not included in the packet ID) must be authenticated by the source.

9.1.4.2 Preventing Bloom-Filter Poisoning

If an adversary could predict the hash values for their packets, they could poison the Bloom filters to increase the false-positive rate and cause benign packets to be dropped. This could severely deteriorate performance for honest senders and would thus represent an attack surface for DoS attacks.

In general, Bloom filters are vulnerable to algorithmic-complexity and poisoning attacks [31, 128, 203]. These can be prevented by using an appropriate hash-function family like SipHash [39]—which is designed and evaluated to be a cryptographically strong pseudorandom function—and prepending a local secret value, which is rotated regularly, to the input of each hash computation.[2]

---

[2]Even when using cryptographic hash functions, prepending a secret value is necessary to prevent offline brute-forcing attacks to craft poisoning packets.

9.1.5 Prerequisites for Replay Suppression

In summary, any protocol that should be protected by our replay-suppression system must provide the following in each packet:

1. A timestamp with a precision of at least 100 ms (and a corresponding global time synchronization) to filter out long-outdated packets and limit the monitoring period;

2. A unique packet ID to be able to distinguish any two packets; and

3. Authentication of at least the timestamp and the unique packet ID.

The hash used to insert the packet into the Bloom filter can be either the unique packet ID or the MAC with which the packet is authenticated: Given sufficiently long MACs, collisions—while in principle possible—are highly unlikely. The false-positive rate caused by these collisions is much lower than that of the Bloom filters themselves.

9.2 High-Speed Traffic Filtering with LightningFilter

Intrusion-detection systems and firewalls have become indispensable for detecting and preventing a range of attacks in today's Internet. Unfortunately, far from being a panacea, these defense systems suffer from several shortcomings: First, traffic filtering is hindered by the ever more ubiquitous use of end-to-end encryption. Indeed, deep packet inspection is impossible without terminating encryption (and thus breaking end-to-end secrecy). Therefore, firewalls are often demoted to filtering based on header attributes and packet metadata. As these attributes are typically not authenticated, adversaries can spoof their IP address and thus render filters ineffective in many cases. When a firewall also incorporates VPN functionality, spoofed IP packets can have significant impact, as VPNs have been shown to be extremely susceptible to stateless flooding attacks [494]. Second, the complex filtering rules of modern firewalls are computationally expensive to enforce. As a result, enterprise-grade firewalls with a throughput beyond 100 Gbps can cost several hundred thousand USD [40]. Furthermore, the advertised performance of firewalls is often much lower if an adversary sends worst-case traffic in a denial-of-service (DoS) attack [144]. Finally, despite the computational effort spent per packet, firewalls and intrusion-detection systems suffer from substantial false-positive and false-negative rates.

To remedy these issues, we developed LightningFilter, a high-speed traffic-filtering mechanism that leverages DRKey to enable authenticated traffic shaping based on the AS number of the source host. This provides the basis for meaningful header-based filtering rules and reduces the load on traditional fire-walls during DoS attacks. LightningFilter is orthogonal to (and can thus com-plement) both traditional firewalls and firewalls based on programmable hard-ware. Our open-source prototype implemented purely in software achieves line-rate for bandwidths up to 160 Gbps running on commodity hardware.

9.2.1 Problem Definition

LightningFilter's goal is to pre-filter traffic to an amount that is manageable by the network infrastructure receiving that traffic, to prevent spoofing of header fields (including the source address), and to ensure that a client located in a non-compromised domain[3] is not affected by DoS attacks.

9.2.1.1 Adversary Model

For the design and analysis of LightningFilter, we consider the general adver-sary model described in §7.2. We consider both malicious ASes and end hosts that can reside at arbitrary locations in the network but not in the AS where LightningFilter is deployed (the destination AS). In addition, the Lightning-Filter system itself and the protected entities within the local network are as-sumed to be benign.

The goal of the adversary is to disrupt access to a specific domain for legiti-mate clients by consuming available resources at the leaf AS (e.g., at a firewall). Concretely, this can be done in the following ways:

- The adversary may send an excessive amount of legitimate or arbitrary (potentially invalid) traffic from all entities under its control (which can be a botnet in a distributed DoS attack).

- The adversary may hide its identity using source-address spoofing and magnify the impact of the attack using traffic reflection and amplifica-tion.

- An on-path adversary may carry out a replay attack (i.e., by re-sending observed packets).

Since this work focuses on DoS attacks targeted at end systems, we assume that the adversary is not able to saturate the bandwidth of network links to the domain, which can be prevented by using systems based on bandwidth reservation (see §10.2).

---

[3] In principle, a "domain" can be any clearly defined network segment such as an AS or an IP prefix. In the context of SCION, we consider SCION ASes.

9.2.1.2 Design Goals

LightningFilter is designed to achieve the following objectives:

- **Guaranteed access for legitimate users within traffic profile:** The system must ensure that a client in a non-compromised domain (i.e., a domain without an adversary) has a guarantee to reach a target domain even in the presence of adversaries in other domains. We define a traffic profile as a sequence of measurements over a specific period of time (profiling window) on a per-flow basis (flow count). As long as the traffic of a flow is within such a traffic profile, its packets are guaranteed to be processed.[4]

- **Enabling traditional firewalls to filter packets using metadata:** The system should enable traditional firewalls to employ meaningful rule-based packet filtering using packet metadata (such as the 5-tuple in the packet header). Without LightningFilter, these filtering rules can be circumvented by spoofing attacks due to the lack of authentication.

- **Elimination of collateral damage across domains:** The system should guarantee that compromised domains cannot introduce collateral damage on non-compromised domains by consuming all available resources. Legitimate clients within a compromised domain, however, may be affected by an adversary consuming excessive resources at a target domain. This provides an incentive for domain owners to eliminate attack traffic sent by their end hosts.

- **Non-goal:** Guaranteed traffic delivery to the domain is not a goal of this system, but can be achieved by a complementary system in SCION.

9.2.2 Overview of LightningFilter

Considering our threat model, the adversary's goal is to consume all available processing resources to prevent legitimate clients from reaching a target service, e.g., by sending an excessive number of requests. To prevent a single entity from achieving this goal, the available processing resources should be subdivided and distributed among all clients. However, allocating an equal share of resources to each entity inhibits high utilization and potentially punishes benign traffic. As a consequence, researchers have suggested the use of more dynamic approaches, such as history-based filtering [213, 407] or binning of requests [470]. The potentially huge number of clients poses a challenge to the former approaches, as storing a traffic history (e.g., packet counters) per client is impractical. Instead, we propose to aggregate and store traffic profiles at the level of domains, i.e., ASes. These traffic profiles denote a sequence

---

[4]The replay-suppression system causes a negligible number of packets to be dropped due to false positives; however, end hosts must be able to handle packet loss anyway.

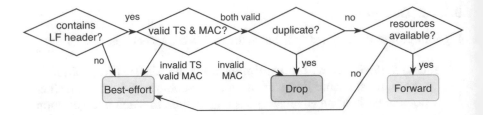

Figure 9.1: Illustration of the LightningFilter pipeline. We use the abbrevia-
          tions LF for LightningFilter and TS for time stamp. "Forward" and
          "Best-effort" represent two different processing pipelines, whereas
          "Drop" means that the packet is discarded.

of flow counts measured within a profiling window. However, since an adversary can employ source-address spoofing to circumvent history-based filtering,
these approaches are only meaningful in combination with source-address authentication. Defense systems based on traffic history are susceptible to replay
attacks, for instance by an on-path adversary.

We designed the LightningFilter service to authenticate, pre-filter, and rate-
limit traffic such that only an upper bound of traffic is forwarded. For this
purpose, LightningFilter keeps aggregates and stores long-term traffic profiles
using probabilistic counting on an AS-level (although other domain granulari-
ties would also be possible). To determine the limit of manageable traffic, an
administrator of the protected entities would conduct an internal capacity assessment and then configure the receive limit (specified in packets or bytes per
time interval, e.g., Mbps) in LightningFilter. If the receive limit is exceeded,
LightningFilter starts filtering packets by computing a per-AS resource allocation based on the traffic profile and a prediction mechanism that accounts for
recent packet arrival rates. Thus, the infrastructure behind LightningFilter only
receives as much traffic as it can handle.

To avoid source-address spoofing and replay attacks, LightningFilter contains modules for checking the source authenticity and payload integrity of
packets. It leverages DRKey (see §3.2) and the SCION Packet Authenticator
Option (SPAO; see §3.3) for filtering out replayed or duplicated packets. If the
traffic does not contain sufficient information to be processed by Lightning-
Filter or only partially complies with the filtering rules, it is processed in a
best-effort manner. For this purpose, LightningFilter employs two different
processing pipelines for fully pre-filtered, authenticated traffic and for best-
effort traffic. Furthermore, even if the receive limit is not reached, traffic that
has successfully traversed the modular packet check pipeline will be forwarded
with higher priority than traffic of the best-effort pipeline. An overview of
LightningFilter's processing pipeline is illustrated in Figure 9.1.

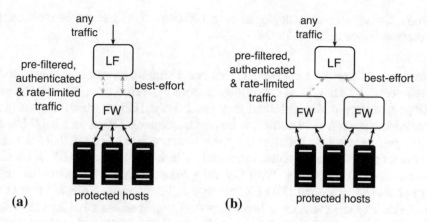

Figure 9.2: Deployment scenarios with LightningFilter upstream of (a) a single firewall and (b) multiple firewalls.

9.2.3 Deployment Scenarios

LightningFilter can be placed directly behind border routers to filter ingress traffic for an entire domain, or, more typically, it can be used to protect any smaller subnetwork of end hosts. LightningFilter is intended to complement, not replace standard firewalls; it is deployed upstream of one or multiple standard firewalls. Figure 9.2 illustrates two potential deployment scenarios: Fully authenticated and rate-limited traffic (green dashed arrow) can be marked and processed with simpler rules by a standard firewall, while (unauthenticated) best-effort traffic (blue dotted arrow) is processed normally including potentially expensive packet inspection. Differentiated treatment of authenticated and best-effort traffic can be either done within a single firewall (a), or across multiple firewalls (b). In both scenarios, LightningFilter can set a flag in the IP header indicating successful authentication and thus preferential treatment to the firewall.

LightningFilter's deployment has several advantages for the protected network. The authentication of packets can enable simplified processing if domains are trusted (e.g., simple firewall rules based on header fields). The deployed firewall(s) only need to be able to cope with the maximally expected amount of legitimate traffic instead of the maximally expected attack traffic due to the pre-filtering and rate limiting of LightningFilter. This translates to substantial cost savings, as there can be a difference of several orders of magnitude between legitimate traffic and attack traffic [542].

In this proposed deployment, the LightningFilter system must be powerful enough to cope with the maximal amount of attack traffic. However, as our prototype implementation demonstrates, LightningFilter can filter up to 160 Gbps of traffic on commodity hardware (see §9.2.8). Consequently, an inexpensive commodity server can be deployed on a 100 Gbps link, and can be replicated to process larger traffic volumes.

Example. Consider a company C that hosts various services for clients and employees. Under typical conditions, incoming traffic rates do not exceed 5 Gbps and even under high load they reach only 8 Gbps, which is what the internal servers can handle. Still, to be resilient against DDoS attacks, C has a 100 Gbps Internet link. Unfortunately, this also requires a firewall that can handle the worst-case traffic volumes including attack traffic, i.e., 100 Gbps, which costs over USD 300K [40]. With LightningFilter, which requires commodity hardware costing around USD 6K, company C only needs a firewall that can handle the expected amount of legitimate traffic and can use a 10 Gbps firewall with a cost of around USD 45K [41]. Thus, C drastically reduces expenses while achieving an even higher level of DDoS protection due to the additional source authentication.

9.2.4 Details of the LightningFilter Pipeline

The design of LightningFilter is split into control-plane and data-plane elements, where all data-plane elements consist of specialized data structures and efficient operations to keep the computational effort that is spent for each individual packet as low as possible. Thus, the technical design of LightningFilter focuses on efficiency, which in turn enables the execution of per-packet operations even in environments with high traffic rates. For example, to enable per-packet counting on a multi-core system, compact data structures that store a partial view of the current state are used for each processing core. Later, the stored values are aggregated to gain a system-wide view of all packet counts. The control-plane part is responsible for managing the systems and interacting with peripheral systems (e.g., a downstream firewall).

In the following, we will discuss the specialized data-plane modules that comprise LightningFilter. A central design principle followed by all those modules is that the computational effort spent per packet must be kept minimal to enable per-packet counting in environments with high packet rates. For example, to process 5 Gbps of 64-byte-sized packets on one core, the processing budget is approximately 100 ns per packet.

9.2.4.1 Module 1: Packet Authentication

Checking the authenticity of a packet is not only fundamental for detecting changes to a packet in-flight, but also essential for preventing source-address spoofing, which would allow an adversary to circumvent IP-based defense mechanisms (e.g., filtering based on per-IP-prefix history [407]). However, to employ authentication checks on a per-packet basis only very efficient operations are eligible. While the use of asymmetric cryptography scales well in the number of hosts, the computation overhead is significant and thus not suited for a per-packet usage. On the other hand, using symmetric cryptography would traditionally require the filtering service to store a key for each packet source. To avoid per-host state while still using symmetric cryptography, we use DRKey.

Key Establishment. As described in §3.2, DRKey enables the creation of a hierarchy of symmetric keys. These keys are dynamically derived using a pseudorandom function (PRF) on a deterministic input, avoiding per-host state. In DRKey, each AS has a local secret value for standard protocols, for instance SV_A^{LF} in AS A, which can be shared with trusted internal infrastructure elements (e.g., border routers, servers) but not with external entities. Based on the local secret value, external AS-level keys can be derived:

$$K_{A \to B}^{LF} = \mathsf{PRF}_{SV_A^{LF}}(B).$$ (9.1)

The certificate servers deployed in each AS exchange the AS-level keys, e.g., AS A provides $K_{A \to B}^{LF}$ to AS B, and AS B provides $K_{B \to A}^{LF}$ to AS A. By applying the PRF again, both ASes A and B can independently derive host-level keys $K_{A:H_A \to B:H_B}^{LF}$, see §3.2 for details.

An end host H_B located in AS B can query B's certificate server to obtain its host-level key. The host can then use this key to calculate a MAC for each packet to allow the destination AS or host to verify the packet's source and check its integrity; as the destination can dynamically derive the key on demand, it does not need to keep per-source state and can already authenticate the first packet received from any host. The advantage of DRKey is that it allows on-the-fly key derivation and enables high-speed packet verification, together in less than 100 ns on commodity hardware.

Depending on the location where LightningFilter is deployed, there are two scenarios:

- If the entity controlling the LightningFilter instance is trusted by the AS, it can obtain the secret value SV_A^{LF} directly. From this, it can independently derive host-level keys.

- If the entity is not fully trusted, it needs to proactively fetch and store second-level keys $K_{A:H_A \to B}^{LF}$ for all protected hosts H_A and all ASes B from which it expects traffic. From these keys, it can then again derive the host-level keys to perform packet authentication.

Header and Payload Authentication. Header and payload authentication is performed via the SPAO. To be able to quickly perform source authentication and perform payload authentication in parallel, LightningFilter uses an algorithm based on a combination of a cryptographic hash function and a MAC.

First, the module checks the authenticity of the packet header (including the packet source) based on the per-packet MAC. Second, the integrity of the payload is checked based on a cryptographic hash, which is included in the MAC computation. Both MAC and hash are included in the SPAO header. If either of the checks fails, the packet is discarded. Authentication based on DRKey allows authenticity checks within 100 ns and enables Lightning-Filter to rapidly determine whether a packet header has been tampered with. The secondary check of the packet payload based on a cryptographic hash is performed in parallel and only needs to be completed if the MAC is correct. The per-packet MAC is computed as

$$tag = \text{MAC}_{K^{LF}_{A:H_A \rightarrow B:H_B}}\left(\overline{hdr}\right), \tag{9.2}$$

where $K^{LF}_{A:H_A \rightarrow B:H_B}$ is a symmetric key between host H_A in AS A and host H_B in AS B and $\overline{hdr}$ is the immutable part of the SCION header (including the payload hash). The hash value is computed over the packet payload, $h = \text{H}\left(\overline{pld}\right)$.

9.2.4.2 Module 2–3: Filtering Duplicated and Outdated Packets

In addition to the previous checks, we assume that an on-path adversary can collect packets and replay them from any location under their control at any time. Thus, in addition to the previous checks, LightningFilter requires a module that removes duplicated and outdated packets.

This is achieved with the system described in §9.1. The prerequisites (§9.1.5) are fulfilled through the timestamp from the SPAO, the source address and timestamp as packet IDs, and the authentication provided by the SPAO. As input to the Bloom filter, we use the MAC in the SPAO header, which is dependent on (the immutable part of) the packet header and the payload.

Traffic older than one second (i.e., the monitoring period) is not dropped but simply assigned to best-effort traffic. This ensures that deploying Lightning-Filter does not have disadvantages in case of unsynchronized clocks (see §9.2.7).

9.2.4.3 Module 4: Filtering Based on Traffic Profile

Now that all received packets can be attributed to their source and all duplicated and replayed packets are filtered, the available resources should be subdivided to prevent a single entity from consuming all resources by sending an excessive number of requests. For this purpose, we take advantage of the observation that traffic loads in the Internet typically exhibit recurring patterns on a daily, weekly, and even monthly basis [21]. This means that the traffic load of a service on a specific weekday at a certain time is similar to the preceding day or the same weekday a week before. This observation enables us to achieve high-speed filtering of malicious traffic while introducing a minimal latency increase and avoiding excessive use of resources.

To efficiently keep track of a per-AS traffic profile, we introduce a novel data structure that consists of two types of hash tables. The first type of hash table, which we call *cardinality table*, is designed to store counters (e.g., number of packets, number of bytes) on a per-AS basis. It is designed with a focus on memory efficiency and should fit into CPU caches to allow for fast entry access. As a trade-off, the use of this hash table results in an approximate counting process, such that packets of previously unobserved ASes will not be counted if the location in the hash table is already occupied. As a consequence, the hash table is frequently reset and its content is aggregated into a second hash table. The purpose of the second hash table, the *aggregation table*, is to store slices of the traffic profile, which represent specific time intervals (e.g., 30 minutes).

9.2.5 History-Based Resource Allocation

If the receive limit is exceeded, LightningFilter starts filtering based on a per-AS resource allocation, which is computed based on the traffic profile that has been collected over an extended period of time. To improve the stability of this prediction in the presence of ongoing attacks, traffic profile slices at the same time in previously recorded weeks and on the previous day are combined with immediately preceding slices to compute the allocation. The slices from weeks ago represent the general trend, while more recent data account for variations in the popularity of the service. While an attack is ongoing (and the packet limit is exceeded), the traffic profile is not updated, such that the attack traffic does not affect the allocation. Apart from attacks, the resource allocation should also be robust to sudden, benign increases of popularity (e.g., flash crowds).

A simple but efficient variant to predict the per-AS resource allocation is using linear regression of the values for the same AS at previous points in time. Instead of linear regression, more complex functions can be used.

After the resource allocation for each AS has been computed, it is scaled such that the sum of all packet counts per time interval is equal to the receive limit. Alternatively, the allocated resources for an AS could also be set manually, for instance to express a service level agreement (SLA).

9.2.6 Traffic Filtering

If an entity becomes the target of a DoS attack and the number of packets exceeds its receive limit, LightningFilter filters traffic based on the per-AS resource allocation. LightningFilter checks for each packet if the source AS still has resources available in the current time interval. If the AS has not exhausted its allocations, the service starts processing the packet using all active modules and decrements the respective AS's packet allocation for each forwarded packet. If no more resources are available for a specific AS, its traffic

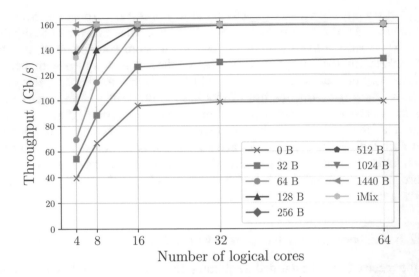

Figure 9.3: Throughput of LightningFilter for 4 × 40 Gbps traffic with a varying number of cores and payload sizes. iMix [274] refers to a mix
of packet sizes consisting of 7 × 84 B, 4 × 576 B, and 1 × 1500 B,
which results in an average packet size of 366 B.

is treated as best-effort traffic (and dropped if no resources are available). Similarly, packets that contain no SPAO header or an outdated timestamp are also
allocated to best-effort traffic. Packets with invalid MACs are discarded.

After the filtering time interval has ended, the table containing the resource
allocation is reset to the previously computed allocation.

9.2.7 Best-Effort Traffic

As using time synchronization or packet authentication is not mandatory,
clients that send packets with an invalid timestamp or no SPAO header should
still be able to eventually reach the service, even if the service is under a DoS
attack. For this purpose LightningFilter allocates a share of the receive limit
(e.g., 10%) to best-effort traffic that is processed at a lower priority. As a
consequence, best-effort traffic will be placed in a processing chain separate
from authentic traffic and will be processed when resources are available.

Allocating a fixed amount of traffic to be processed as best-effort also helps
to cope with the case in which an AS is not yet present in the traffic profile. However, to provide an incentive for deploying time synchronization and
source authentication, traffic that contains a valid timestamp and packet authentication will be processed with higher priority and will always bypass traffic
without SPAO header or an invalid timestamp.

9.2.8 Evaluation Results

We have implemented and evaluated a prototype of the LightningFilter system, the details of which we describe in an upcoming research paper. We achieved high-speed packet processing through the use of the Intel DPDK framework [421] and specialized hardware instructions for the AES block cipher [13].

Figure 9.3 shows the throughput of legitimate traffic achieved by our prototype, running on a commodity server with four 40 Gbps Ethernet interfaces, for a varying number of processing cores and payload sizes. With an increasing number of cores and packet size, the overall throughput of the system increases. For packet sizes of 64 B and 16 processing cores, the system achieves line-rate saturation.

We have evaluated LightningFilter's performance under different attack scenarios, with varying amounts of legitimate traffic together with artificially generated attack traffic of different types. In all experiments, with traffic volumes of up to 160 Gbps, the prototype system successfully filtered attack traffic and forwarded legitimate traffic[5].

9.3 Probabilistic Traffic Monitoring with LOFT

Flow-monitoring algorithms aim at measuring quantitative properties of network flows (i.e., packet sequences associated with a connection between two endpoints), such as the total volume sent by a flow. Flow monitoring is thus a key enabler of numerous applications targeted at improving the security and efficiency of networks, such as bandwidth-reservation systems like COLIBRI (see §10.2), anomaly detection, and flow-size–aware forwarding. Given the central role of flow-monitoring systems in these applications, flow-monitoring algorithms have to fulfill stringent requirements, most importantly *processing efficiency* and *estimation accuracy*.

First, flow-monitoring algorithms have to enable *efficient processing* because these algorithms are often employed on inter-domain border routers and therefore are supposed to handle the data rates observable on inter-AS links, which can reach up to multiple Tbps. Hence, the acceptable per-packet overhead induced by a flow-monitoring algorithm is strongly limited. This constraint on the per-packet overhead directly translates into a memory-consumption constraint: Keeping up with Tbps link speeds is only possible if the monitoring state fits into fast caches and main-memory look-ups can therefore be avoided.

---

[5]The high throughput volume is achieved by distributing flows across different processing cores. In case a processing core is overwhelmed (by an attack or a very large flow), the system can fall back to randomly distribute packets across the processing cores, potentially compromising on packet ordering for large-volume flows.

Second, flow-monitoring algorithms need to produce highly *accurate estimates* of flow properties because the effectiveness of the applications based on flow monitoring is directly determined by the monitoring accuracy. As an example, consider the policing of bandwidth reservations, which corresponds to identifying all flows that send above a given rate, i.e., *overuse flows*. If a flow-monitoring algorithm is inaccurate, e.g., only detects overuse flows that are considerably larger than the given threshold, an attacker could exhaust link bandwidth by creating a large number of reservations and moderately overusing each reservation, undermining the guarantees provided by the bandwidth-reservation system.

Existing approaches to flow monitoring can be broadly divided into three categories. The first category is given by the straightforward approach of universal individual-flow monitoring (e.g., NetFlow [123]). However, monitoring every flow with an individual counter quickly exhausts the available memory resources on high-speed routers that handle millions of concurrent flows. Hence, more practical approaches rely on *probabilistic* techniques that trade off accuracy against resource efficiency. The probabilistic techniques encompass the other two categories of flow-monitoring algorithms, namely selective individual-flow monitoring (e.g., EARDet [553]) and sketch-based monitoring (e.g., Count-Min Sketch [139]). In our research on overuse-flow detection, we found that all previous algorithms, even probabilistic algorithms, only deliver the desired accuracy given untenable memory consumption and, as a result, prohibitively high per-packet overhead. Therefore, we developed LOFT [455], a sketch-based probabilistic flow-monitoring algorithm specifically targeted at detecting low-rate overuse flows. In contrast to previous algorithms, LOFT is able to reliably detect flows that only slightly overuse their reservation, while conforming to strict limits on computational complexity and memory consumption.

In this section, we present LOFT both in a high-level overview (cf. §9.3.1) and in a more detailed algorithm analysis (cf. §9.3.2). We also briefly discuss experimental results that corroborate the superior performance of LOFT (cf. §9.3.3).

9.3.1 System Overview

On actual systems, the overuse-detection capabilities of LOFT are leveraged by a *flow-policing mechanism*, which is typically composed of four basic building blocks as illustrated in Figure 9.4: (1) a flow *classifier* that extracts each flow's ID and determines its permitted bandwidth; (2) a *blacklist* that filters out blacklisted flows; (3) a *probabilistic overuse flow detector* (here given by LOFT) tasked with finding suspicious flows that could potentially be overusing; and (4) a *precise monitoring* component, which analyzes individual suspicious flows to determine which ones are actually misbehaving and should be added to the blacklist. The precise-monitoring component can access a limited

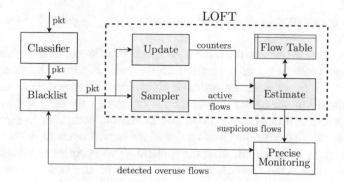

Figure 9.4: Architecture of flow-policing mechanism based on LOFT.

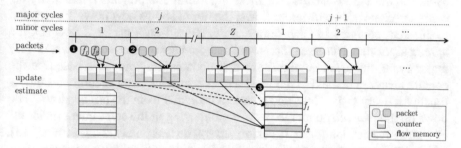

Figure 9.5: LOFT Timeline. ❶ A hash value is computed on the flow ID of the incoming packet, and the corresponding counter is increased by the packet size. ❷ A different set of counters and hash function are used when the minor cycle changes. ❸ After Z minor cycles, the *estimate* algorithm aggregates the counter values and attempts to identify a group of overuse flows.

amount of fast memory (e.g., on the order of the amount used in the overuse-flow detector). This limited fast memory restricts the number of suspicious flows that can be simultaneously monitored by the detector, leading to false negatives.

LOFT itself contains four components: the *update* algorithm, the *estimate* algorithm, the *sampler*, and the *flow table*. Packets not rejected by the blacklist are forwarded to the *update* algorithm and the *sampler*. The *estimate* algorithm consumes their output, updates the *flow table* and creates a list of suspicious flows for precise monitoring.

The LOFT *update* algorithm targets one short time interval at a time, which we call a *minor cycle* (e.g., 12.5 ms). For each minor cycle, the *update* algorithm collects aggregated traffic information over groups of flows by using a single counter array. For every packet, LOFT maps the packet flow ID to one counter in the current counter array and increases that counter by the packet size. At the end of the minor cycle, the counter array is passed to the *estimate* algorithm.

The *estimate* algorithm operates on a larger time scale, at intervals that we call *major cycles* (e.g., with duration 250 ms; see Figure 9.5): every time a major cycle is concluded, the *estimate* algorithm analyzes the counter arrays stored from all minor cycles during the major cycle, extracts estimates for the bandwidth utilization of every flow, and stores the estimates in a *flow table*. From this, the algorithm produces a list of suspected overuse flows, which is handed to the precise-monitoring component. The *sampler* provides a list of active flows which the *estimate* algorithm uses in its analysis.

In summary, the *estimate* algorithm uses the sequence of counter arrays generated by the *update* algorithm to create a final flow estimate in each major cycle. Figure 9.5 visualizes how these algorithm components interact. By depicting packets of two different flows, the figure shows how flows are mapped to different counters in each minor cycle. Based on a packet's flow ID, the *update* algorithm increases the associated counter by the packet size. Every major cycle, the *estimate* algorithm first updates the *flow table* based on the counter arrays generated in the most recent Z minor cycles, recomputes the estimate for every active flow, and creates a *watchlist* from the largest flows. The flows on the *watchlist* undergo precise monitoring in the subsequent major cycle. This precise monitoring is performed by the leaky-bucket algorithm [513], which detects any violation of a flow specification without false positives. Any flow that is found misbehaving by precise monitoring can thus be blocked by inserting it into the *blacklist*.

9.3.2 Algorithm

After the overview of the LOFT algorithm presented in the previous section, we discuss the building blocks of the algorithm in more detail in this section.

9.3.2.1 Update Algorithm

Similar to a sketch, the *update* algorithm maps each flow to one counter $ctr_{j,k}$ in a counter array in minor cycle k of major cycle j. Each counter tracks the aggregate bandwidth of all the flows mapped to that counter during a minor cycle. In each minor cycle, the association of flows to counters is randomized by changing the hash function $H_{j,k}$ for every minor cycle. When a minor cycle ends, the counter array is moved to main memory and an empty counter array is initialized in fast memory.

9.3.2.2 Estimate Algorithm

At the end of a major cycle, which contains a certain number Z of minor cycles, the *estimate* algorithm performs a flow-size estimation for every flow asynchronously (e.g., in the userspace of the router), while the *update* algorithm continues to aggregate traffic information. The flow-size estimate builds on two values: the *volume sum* and the *cardinality sum*.

For the volume sum, the *estimate* algorithm sums up the values of all the counters to which a flow was mapped. This aggregation over time reduces the counter noise in the sense of uneven flow size within the same counter: Intuitively, an overuse flow will be consistently associated with large counter values, which results in a large volume sum for that flow.

Although using multiple counters can reduce counter noise, it is neither sufficient nor innovative, as the Count-Min Sketch [139] uses the same idea (although by applying different hash functions concurrently instead of sequentially) and delivers insufficient accuracy. Indeed, the key to reducing counter noise lies in the cardinality sum. In order to compute this sum, an active-flow list constructed by the *sampler* (see §9.3.2.3) is consulted to compute the cardinality of each counter, i.e., how many flows are associated with each counter within each minor cycle. For every flow, the *estimate* algorithm sums up the cardinality values of all counters associated with the flow. The cardinality sum reduces the distortion created by the varying cardinalities of counters: Intuitively, an overuse flow will be associated with a large counter value even when the number of flows in that counter is small.

When dividing the volume sum by the cardinality sum, the strongest increases in a flow estimate are produced when the flow is mapped to high-value counters that contain a small number of flows. Indeed, flows with these characteristics are highly likely to be the largest flows among the investigated flows and are therefore candidates for more precise monitoring.

Formally, after major cycle j, we define the estimate of a flow f to be

$$\overline{U}_j(f) = A_j(f)/C_j(f) \qquad \text{where} \tag{9.3a}$$

$$A_j(f) = \sum_{j' \in J_j(f)} \sum_{k=1}^{Z} ctr_{j',k}[H_{j',k}(f)], \tag{9.3b}$$

$$C_j(f) = \sum_{j' \in J_j(f)} \sum_{k=1}^{Z} |ctr_{j',k}[H_{j',k}(f)]|, \tag{9.3c}$$

and $J_j(f)$ contains all major cycles $j' \leqslant j$ in which flow f was active. The term $|ctr_{j',k}[x]|$ denotes the number of flows that have been mapped to counter x in minor cycle k of major cycle j' (counter cardinality). $A_j(f)$ is the value aggregate of the counters that flow f has been mapped to (volume sum) and $C_j(f)$ is the summed count of the flows in these counters (cardinality sum) up to and including major cycle j. In order to avoid preserving counter arrays from past major cycles, the terms $A(f)$ and $C(f)$ are kept in the flow table and updated after every major cycle, i.e.,

$$table[f].A \leftarrow table[f].A + \sum_{k=1}^{z} ctr_{j,k}[H_{j,k}(f)], \qquad (9.4)$$

and analogously for $C_j(f)$. These updates are made for all flows f that were active in the most recent major cycle and are thus in the active-flow list generated by the sampler (cf. §9.3.2.3).

These estimates need to be adjusted when some flows send intermittently. For example, suppose flow f_1 sends x GB in the first and the third major cycle and nothing in the second major cycle, and flow f_2 sends x GB from the first to the third major cycle, i.e., $J_3(1) = \{1,3\}$ and $J_3(2) = \{1,2,3\}$. Then $A_3(1)/C_3(1)$ and $A_3(2)/C_3(2)$ will be almost the same. Suppose all counters contain exactly y flows, then $A_3(1)/C_3(1) = \frac{x+x}{y+y} = \frac{x}{y} = \frac{x+x+x}{y+y+y} = A_3(2)/C_3(2)$. However, the total traffic sent by flow f_2 in these three cycles is actually 1.5 times larger than flow f_1 and should result in a higher flow-size estimate.

To fix this problem, we reduce a flow-size estimate $\overline{U}_j(f)$ relative to the number of major cycles where flow f was not active, i.e.,

$$U_j(f) = \frac{|J_j(f)|}{j} \cdot \frac{A_j(f)}{C_j(f)}. \qquad (9.5)$$

To enable this computation, the flow table must track $|J_j(f)|$ for every flow f.

Another issue is that as $A_j(f)$ and $C_j(f)$ are accumulated, the *estimate* algorithm is actually computing their average over time. An attacker can take advantage of this approach by sending low-rate traffic in the beginning for a period of time, and then start sending bursty traffic. It may not be detected by our system as its long-term average looks the same as a non-overuse flow, so old values must be discarded at some point. Therefore, we define the reset cycle θ, and clear all the data every θ minor cycles. This reset may sound risky, as an overuse flow could send its traffic around the reset point so that its estimated size is reset before being detected by our system. However, an attacker does not know the reset point. Moreover, even if an attacker could infer the reset point, the overuse traffic sent by a flow with such a strategy is bounded, as we show in the mathematical analysis of the full-paper version [456].

9.3.2.3 Sampler

LOFT requires a list of active flows for which an estimate must be computed. In order to generate such an active-flow list, we use sampling and limit the number of sampled packets per second to be λ. LOFT considers a randomized sampling period, which is a random variable of an exponential distribution with mean $\frac{1}{\lambda}$. This randomization prohibits an attacker flow from circumventing the sampling by sending at the appropriate moments. Having an active-flow list for a major cycle j also allows the cardinality $|ctr_{j,k}[x]|$ of counters in the *estimate* algorithm to be computed, namely by counting how many active flows were mapped to each counter with the respective hash function $H_{j,k}$ for any minor cycle k.

9.3.3 Evaluation

In this section, we provide evidence of LOFT's desirable performance characteristics. In particular, we focus on evidence showing that LOFT fulfills the requirements of processing efficiency and detection accuracy, which are the two main requirements outlined at the beginning of this chapter.

9.3.3.1 Processing Efficiency

To measure the processing efficiency of LOFT, we performed scalability experiments with two prototypes: one implementation based on the DPDK framework and one implementation based on a Xilinx FPGA.

DPDK. For the scalability experiments with DPDK, we implemented LOFT in C on the Intel DPDK framework [421]. The application uses n worker threads that execute the *update* algorithm and a separate thread running the *estimate* algorithm every major cycle. The major and minor cycle indices are computed based on a monotonic clock with nanosecond resolution.

To understand the scalability of LOFT in a DPDK environment, we evaluate the maximum packet rate with respect to the packet size and number of cores that execute the *update* algorithm concurrently. Figure 9.6a shows that LOFT is able to achieve line-rate of 160 Gbps for iMix-distributed traffic using 16 cores that execute the *update* algorithm and one core that runs the *estimate* algorithm. With fewer cores, line rate can only be achieved for larger packet sizes (1024 B).

Since traffic flows with small packets perform considerably worse than flows with large packet sizes, we additionally evaluate the overhead introduced by LOFT by comparing it to regular L3 packet forwarding in DPDK. Figure 9.6b shows the throughput of regular L3 forwarding and the throughput of forwarding with additional LOFT processing for different packet sizes. Moreover, the figure gives the LOFT throughput as a percentage of the corresponding base throughput. As can be seen in the figure, even regular L3 packet forwarding using eight processing cores cannot achieve line-rate for packet sizes smaller than 1024 B. Compared to regular packet forwarding, LOFT introduces overhead for small packet sizes, which results in a maximum packet rate of $\sim$50 million packets per second (Mpps) using eight processing cores, i.e., $\sim$6 Mpps per core. As for larger packet sizes this maximum packet rate does not become exhausted, this effect is diminished. However, LOFT still achieves a much higher packet rate than the alternative schemes with the best accuracy, i.e., HashPipe and HeavyKeeper: Prior work has shown a packet rate of $\sim$2 Mpps

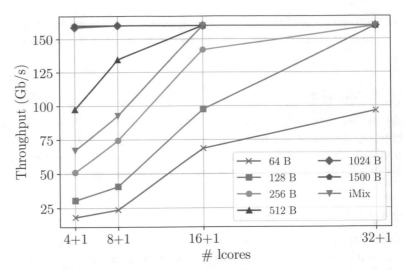

(a) DPDK throughput for a given number of logical cores and packet sizes.

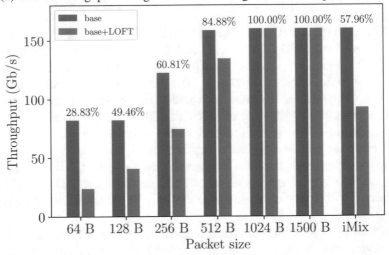

(b) DPDK overhead compared to regular L3 forwarding for 8+1 cores.

Figure 9.6: DPDK throughput and overhead.

per core for HashPipe [556] and a packet rate of ∼2.5 Mpps per core for Heavy-Keeper [557].

FPGA. For further scalability experiments, we also implemented the *update* algorithm of LOFT (i.e., the *update* algorithm using 16,384 counters) on an FPGA. The used Xilinx Virtex UltraScale+ FPGA is programmable via the Netcope NFB 200G2QL platform [385] and possesses two 100 Gbps NICs and an operating frequency of 200 MHz. The LOFT implementation can process a packet in every cycle. For minimum-size packets of 64 B, each NIC manages to transfer one packet per cycle to the corresponding instance of the LOFT implementation, which allows a packet rate of 200 Mpps per NIC to be achieved. As the FPGA platform contains two NICs, it achieves a total packet rate of ~400 Mpps. This high throughput demonstrates that LOFT is suitable for high-speed packet processing if implemented on programmable NICs. Details of the FPGA implementation design can be found in Sateesan et al. [451].

9.3.3.2 Detection Accuracy

In order to compare LOFT to other flow-monitoring algorithms in terms of detection accuracy, we performed an extensive simulation-based analysis, which can be found in the LOFT paper [455]. The central metric of this analysis is *detection delay*, i.e., the time between the first violation of the given rate threshold by a flow and the detection of that flow. Our analysis shows that low-rate overuse flows (i.e., with overuse ratios ranging from 1.5 to 3) are detected more than 300 times faster by LOFT than by other algorithms.

10 Extensions for the Data Plane

GIACOMO GIULIARI, MARKUS LEGNER, SI LIU, ADRIAN PERRIG, THILO WEGHORN, MARC WYSS*

Chapter Contents

10.1 Source Authentication and Path Validation with EPIC ... **228**

10.1.1 Level 1: Improved Path Authorization 228

10.1.2 Level 2: Source Authentication for Routers 229

10.1.3 Level 3: End-Host Path Validation 230

10.1.4 Protecting Hidden Paths with EPIC-HP 231

10.1.5 Checking Every Packet Everywhere with EPIC-SAPV 234

10.1.6 Performance of EPIC 235

10.2 Bandwidth Reservations with COLIBRI **237**

10.2.1 Introduction 237

10.2.2 Background: Enabling Technologies 239

10.2.3 Overview of COLIBRI 241

10.2.4 The COLIBRI Architecture in Detail 246

10.2.5 Fair Bandwidth Allocation with N-Tube 256

10.2.6 Security Analysis 263

10.2.7 Discussion 265

10.2.8 COLIBRI in the SCION Ecosystem 266

Two of the most surprising properties SCION enables are that—for the first time—minimum bandwidth guarantees and highly efficient per-packet source authentication can be achieved on global communication networks. Such properties were so far considered elusive at best, and impossible at worst. In this chapter, we describe the technical innovations and systems that lead to the accomplishment of these two properties, while the following chapter discusses how the interaction between all elements within the entire SCION ecosystem, including the here-mentioned innovations, ensures the availability and authenticity of communications end-to-end.

---

*This chapter reuses content from Giuliari, Roos, Wyss, García-Pardo, Legner, and Perrig [209], Legner, Klenze, Wyss, Sprenger, and Perrig [327], Weghorn [534].

© The Author(s), under exclusive license to Springer Nature Switzerland AG 2022

L. Chuat et al., *The Complete Guide to SCION*, Information Security and Cryptography, https://doi.org/10.1007/978-3-031-05288-0_10

10.1 Source Authentication and Path Validation with EPIC

EPIC [327] is a series of protocols providing improved security properties for the SCION data plane: achieving a stronger path-authorization property, enabling on-path routers to perform source authentication, and allowing end hosts to verify that the packet was processed by all intended on-path ASes (i.e., light-weight path validation). EPIC was designed with efficiency as a core goal and achieves these properties with substantially lower communication and processing overhead than previous systems like ICING [381] or OPT [411].

To achieve this, data-plane packets include a hop validation field (HVF) for each AS on the path. A HVF is a short authenticator calculated by the packet source that can be subsequently verified by one specific on-path AS. These packet-specific HVFs of all data-plane packets inspired the name EPIC, which stands for "every packet is checked."

We have designed different "levels" of EPIC with different prerequisites and achieved properties. In Sections 10.1.1–10.1.3 we describe EPIC levels 1–3 of the protocol series and the properties they achieve.[1] Those chapters introduce the theoretical framework behind EPIC. How those three EPIC levels are implemented in SCION is explained in Sections 10.1.4 and 10.1.5.

10.1.1 Level 1: Improved Path Authorization

We discussed in Chapters 5 and 7 how the HF authenticators included in SCION data-plane packets achieve path authorization: Data-plane packets always follow path segments authorized by the control plane. This property is achieved under two assumptions: (1) The authenticators cannot be forged by attackers, and (2) all authorized paths are publicly known and their authenticators thus do not need to be kept secret.

The first assumption depends on the length of authenticators, l_{HFAuth}: For 6-byte-long authenticators, the assumption is generally satisfied, as it is infeasible to conduct an online brute-force attack that requires sending $2^{47} \approx 140$ trillion packets. The second assumption holds for standard SCION segments that are publicly registered at path services; unfortunately it does *not* hold in the context of hidden paths (see §8.1): By capturing SCION traffic destined to an AS behind a hidden path, an attacker can copy the HF authenticators and use them to send traffic over the hidden path itself.

In both cases, the central issue is that a HF authenticator that is once brute-forced or observed in the data plane can be reused for sending an arbitrary number of packets until the HF expires.

For those reasons, EPIC level 1 replaces the *static* HF authenticators of the standard SCION path type by *per-packet* HVFs. In the following we consider

---

[1]The presented algorithms are slightly modified compared to the publication by Legner et al. [327] and adapted to the current SCION packet structure.

a source host H_S that sends a packet over a path consisting of ASes $A_1,\ldots,A_\ell$. The source host then computes the HVF V_i for on-path AS A_i as follows:

$$V_i = \text{MAC}_{\sigma_i}\big(Src, TS_{\text{INF}}, TS_{\text{pkt}}, \text{len}(P)\big)\,[0{:}l_{\text{HVF}}]. \tag{10.1}$$

Here, σ_i denotes the non-truncated version of the HF authenticator (see Equations (5.2a) and (5.4) in §5.3.3), and Src refers to the combination of H_S and the ISD–AS pair of A_1. TS_{pkt} is a current high-precision timestamp relative to the timestamp of the first info field, TS_{INF}. The length of this timestamp is chosen such that the tuple $\langle Src, TS_{\text{INF}}, TS_{\text{pkt}} \rangle$ uniquely identifies a packet; for example, given a maximum validity period of one day for path segments, a 6-byte timestamp is sufficient to uniquely identify one packet every 0.3 ns, which corresponds to a maximum throughput of 39 Tbps when using 1500-byte packets. The last input to the MAC function is the length of the payload, $\text{len}(P)$. When creating a SCION packet, H_S truncates all the MACs ($[0{:}l_{\text{HVF}}]$) and appends the resulting HVFs together with TS_{pkt} to the packet. The additional space for the HVFs is recuperated by shortening the HF authenticators to a length of l_{SegID} bytes (instead of l_{HFAuth}) as their main purpose now is to enable the HF chaining.

In addition to the standard SCION packet processing as described in §5.6.3, the border routers of A_i check that TS_{pkt} is fresh and verify the validity of V_i by recalculating and comparing it to the one provided in the packet. Note that the freshness check requires synchronized time among all on-path ASes (§8.2).

With such per-packet HVFs, hidden paths can be protected effectively as HVFs discovered in packets traversing a hidden path do not enable a malicious entity to send additional traffic on its own anymore. To prevent an attacker from replaying observed packets, all EPIC protocol levels rely on a replay-suppression system to filter duplicate packets (§9.1). The replay-suppression system requires a sufficiently precise timestamp, the possibility to uniquely identify packets, and authentication of this information (§9.1.5). This is provided in EPIC through the precise timestamp (which together with Src and TS_{INF} uniquely identifies a packet) and the per-AS MACs in the packet header.

10.1.2 Level 2: Source Authentication for Routers

The second EPIC level extends the first one by enabling all on-path ASes to verify the packet source (source authentication) and the destination host to additionally validate the authenticity of the payload (packet authentication). The destination host can therefore detect and discard spoofed packets efficiently using LightningFilter (§9.2). Source authentication at on-path ASes ensures that spoofed packets are detected as early as possible and thus helps mitigate DDoS attacks.

To achieve this, H_S obtains a DRKey (§3.2) shared with each on-path AS (K_i^S) and one DRKey shared with the end host H_D (K_D^S):

$$K_i^S := K_{A_i \to A_1 : H_S}^{EPIC}, \tag{10.2a}$$

$$K_D^S := K_{A_\ell : H_D \to A_1 : H_S}^{EPIC}. \tag{10.2b}$$

In EPIC level 2, the source host calculates the HVFs as follows:

$$V_i = \text{MAC}_{K_i^S}\left(Src, TS_{\text{INF}}, TS_{\text{pkt}}, \text{len}(P), \sigma_i\right)[0:l_{\text{HVF}}]. \tag{10.3}$$

Note that apart from the symmetric keys, the inputs contributing to the HVFs are the same as in EPIC level 1. For packet authentication, EPIC level 2 mandates including the SCION packet authenticator option (SPAO, see §3.3), where it uses symmetric cryptography based on K_D^S to authenticate all immutable and predictable fields.

As in EPIC level 1, the source host sends a packet including TS_{pkt} and the HVFs, and each border router on the path checks the freshness of the timestamp and verifies the corresponding HVF. In addition, a border router first needs to derive the shared key with the source host to compute the HVF. The receiving end host must derive the DRKey from the second-level key $K_{A_\ell : H_D \to A_1}^{EPIC}$ or, if that key is not cached, fetch it from the certificate service to verify the SCION packet authenticator.

10.1.3 Level 3: End-Host Path Validation

In Chapter 7, we discussed how packet authentication prevents path modifications by non-colluding attackers. However, two colluding attackers could still modify the path and then change it back to the original one. Therefore we have designed EPIC level 3 to further extend the security properties of EPIC level 2 by enabling the source and destination hosts to perform path validation: This is a stronger property, namely that the end hosts can verify that every AS on the intended path[2] has processed the packet.

Path validation is particularly important if the source intends packets to traverse particular ASes because they offer a desired quality of service (e.g., satellite networks, which offer lower latency) or because they perform important in-network functions (e.g., filtering). In many cases, paths are selected for compliance reasons (packets are not allowed to leave a certain jurisdiction); unfortunately, it is impossible to prevent on-path ASes from simply duplicating packets, so the paths need to be selected carefully to consist only of trusted ASes in such a case.

The calculation of a HVF in EPIC level 3 is exactly the same as Equation (10.3) for EPIC level 2; however, provided we use a MAC algorithm that

---

[2] See Figure 7.1 on page 161 for the terminology.

fulfills the requirements of a pseudorandom function, and its output is at least $2l_{HVF}$ bytes long[3], we obtain a second MAC at no additional processing cost:

$$\widetilde{V}_i = \mathrm{MAC}_{K_i^S}\left(Src, TS_{INF}, TS_{pkt}, \mathsf{len}(P), \sigma_i\right)[l_{HVF}{:}2l_{HVF}]. \qquad (10.4)$$

This enables an on-path router to derive V_i and $\widetilde{V}_i$ with a single cryptographic operation. Upon receiving a packet, the border router first validates it in the same way as in EPIC level 2 by comparing the HVF with V_i. If the validation is successful, the router replaces the HVF in the packet with $\widetilde{V}_i$, thus proving that the packet was processed by the corresponding AS.

To perform path validation, the source and destination hosts must know the expected values in the HVFs after processing by on-path routers (i.e., $\widetilde{V}_i$). The source host, H_S, achieves this by storing all the $\widetilde{V}_i$ in a key–value store under the (unique) key (TS_{INF}, TS_{pkt}). By including these predicted values of the HVFs in the input to the SPAO, the destination host, H_D, can check that they are correctly updated and thus verify that the packet was processed by all on-path ASes (otherwise at least one HVF would not have been updated and the SPAO validation would fail).

After verifying the packet, H_D replies to H_S with an authenticated packet containing $\widetilde{V}_1, \ldots, \widetilde{V}_\ell$, TS_{INF}, and TS_{pkt} to enable H_S to perform path validation. H_S subsequently compares the HVFs previously stored under (TS_{INF}, TS_{pkt}) to the ones contained in the response packet. If they match, also H_S is assured that the EPIC level 3 packet followed the intended path.

10.1.4 Protecting Hidden Paths with EPIC-HP

The concrete implementation of EPIC level 1 in SCION, called EPIC-HP, does not strictly follow the description in §10.1.3. Instead, the implementation is a lightweight variant that is optimized for the most common use case of hidden-path communication (§8.1): A leaf AS wants to keep one access link to its provider hidden. In this case, only the last two hop fields require special protection, all other hop fields can use standard SCION MACs.

EPIC-HP is a separate path type (§5.2); in addition to the full SCION path type header, it contains a timestamp (TS_{pkt}) and HVFs for the penultimate and last HF (PHVF and LHVF), which are supposed to remain "hidden" and therefore require protection through the EPIC mechanism.

This approach allows for standard SCION forwarding, with the sole difference that the two last hops additionally verify the PHVF and LHVF, respectively. The included SCION path header ensures that border routers can always reply with SCMP packets, and that the destination behind a hidden link can respond with packets using the regular SCION path type. The router or

---

[3]See Chapter 17 for further details on the cryptographic function.

destination host only has to extract the SCION path header from the EPIC-HP header and reverse the path.[4]

Importantly, the AS protected by the hidden path is the *last AS*: The PCB defining the hidden path ends in this AS, meaning that the AS does not forward the PCB to further downstream ASes, as otherwise those neighbors will inevitably leak authenticators of the hidden path when they publicly announce their own path segment.

A standard PCB only contains the l_{HFAuth} bytes of the truncated HF authenticator (§4.1); the remaining $l_{HVF} - l_{HFAuth}$ bytes of the authenticator are added as detachable extension as explained in §8.1.2. Through this approach a single PCB can be used for both standard path registration (after removing the detachable extension) and hidden path registration.

There are two main scenarios, further illustrated in Figure 10.1, for the use of EPIC-HP in SCION: The first option is a "fully hidden access link", which is not registered publicly but only shared with authorized entities. To remain publicly reachable, the protected AS must have additional access links for which segments are publicly registered. The second option is to register a downsegment both publicly as a standard SCION segment and at a hidden-path service as an EPIC-HP segment. This is possible as the additional authenticators are included in detachable extensions in the PCB (see §8.1.2). In that case, the access link can be used by anybody, but authorized entities can use the EPIC-HP path type for their traffic, which is prioritized.

We describe both scenarios in further detail in the following two sections.

10.1.4.1 Fully Hidden Access Link

The last and penultimate ASes on the hidden path only allow incoming traffic using the EPIC-HP path type on the interface pairs that affect the hidden path. With such a setup, it is not possible for unauthorized sources to reach the services in the last AS. Thus, EPIC-HP effectively prevents adversaries from running attacks like denial of service or attack preparations such as scanning the services for vulnerabilities.

If some host inside an AS with such a setup wants to communicate with a host inside another AS that is also behind a hidden path, both hosts need to have valid authenticators to send traffic over the corresponding hidden paths— i.e., the hosts can exclusively communicate using EPIC-HP. Note that hosts behind a hidden path can send packets using the SCION path type towards hosts in other ASes, but that those hosts cannot send a response back if they do not have the necessary authenticators.

---

[4]If the source is protected by a hidden path itself, sending back SCMP packets by on-path border routers is generally not possible, as the routers would need to send back EPIC-HP packets, for which they do not have the necessary authenticators. The destination host in such a case also has to answer with EPIC-HP packets and is responsible itself for configuring or fetching the necessary authenticators.

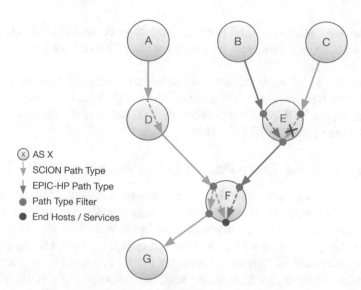

Figure 10.1: Example scenario where EPIC-HP is used to protect hidden-path communication. AS *F* is the AS protected by the hidden path (blue). The hidden path terminates at AS *F* (represented by the black dot), so AS *F* did not forward the PCB that defines the hidden path further down to AS *G*. This is however still allowed for traffic using the SCION path type (green): There are SCION paths that enter AS *F* from AS *D*. One of the two paths ends in AS *F*, while the other one is extended further to AS *G*.

The border routers of AS *F* and AS *E*—the last and penultimate ASes on the hidden path—further implement path-type filtering (orange dots). For example, AS *E* will block (red "X" in the figure) traffic using the SCION path type from AS *C* that is destined towards AS *F*, as it would affect the hidden path. Additionally to blocking non-EPIC-HP path type traffic, ASes must also prioritize EPIC-HP traffic.

The ASes can always decide to be more restrictive, for example AS *F* could additionally disallow traffic using the SCION path type from AS *D*, so that it is reachable through the hidden path only.

10.1.4.2 Prioritization at Access Link

The last and penultimate ASes on the hidden path allow EPIC-HP and other path types simultaneously, but prioritize traffic using the EPIC-HP path type over the SCION path type. This has the advantage that this link can still be used as a normal SCION access link. DoS attacks are not possible in this case, because an adversary is limited to sending low-priority SCION path type packets. Due to the strict prioritization of EPIC-HP path type traffic at the routers, the authorized sources will still benefit from high communication guarantees despite DoS attacks. However, an adversary can still reach the services behind the hidden link.

In this scenario, hosts behind a hidden path can send packets using the SCION path type towards hosts in other ASes, and those hosts can reply with traffic using the SCION path type when they do not have the necessary authenticators to send back EPIC-HP traffic.

10.1.5 Checking Every Packet Everywhere with EPIC-SAPV

EPIC levels 2 and 3 are planned to be implemented in SCION based on an additional path type, called EPIC-SAPV. The abbreviation refers to the provided security properties source authentication and path validation (SAPV).

This path type contains TS_{pkt}, a reverse flag (R), a validation flag (V),[5] and a slightly modified SCION path, where each HF authenticator is shortened to l_{SegID} bytes and the rest is replaced with the corresponding HVF. This embedding of the HVFs allows the EPIC-SAPV header to achieve a constant overhead in terms of header size compared to the regular SCION path type. The unaltered l_{SegID} bytes of the HF authenticator are necessary to be able to recompute the *SegID* when processing a HF against the beacon-construction direction. As the full HF authenticator contributes to the HVF (see Equation (10.3)), this shortening of *HFAuth* does not weaken the path-authorization property; on the contrary, due to the packet-specific MACs, the HVF cannot be reused to send additional traffic as explained above. The border routers verify and update the HVFs as defined in EPIC level 3.

10.1.5.1 Path Validation for the Source

After verifying the SPAO, the destination host examines the validation-flag, which determines whether the source host also wants to validate the packet path, meaning that it expects to receive the $\tilde{V}_1, \ldots, \tilde{V}_\ell, TS_{INF}$, and TS_{pkt}. If this is the case (V=1), the destination host prepares an answer packet, but it is free in the choice of the packet's path type:

- As parts of the HF authenticators were overwritten by the HVFs, it is not possible to invert the modified SCION path inside the EPIC-SAPV packet and directly return a standard SCION packet over the same path. Instead, a new path has to be retrieved for response packets.

- To achieve the best security guarantees, the response should also use the EPIC-SAPV path type. But also here, a new path has to be retrieved, and additionally the corresponding DRKeys need to be fetched. To prevent circular confirmations, the validation-flag must not be set (V=0) in the EPIC-SAPV response packet.

---

[5]Both flags are included in the input to the SPAO but not the HVFs.

- EPIC-SAPV supports direct reply packets that are forwarded as best-effort traffic, as described below. This is the default way of sending confirmations to the source. To prevent circular confirmations, the validation-flag must not be set (V=0).

If the validation-flag was not set, the source host is not interested in validating the path, and therefore the destination host does not send back a reply packet.

10.1.5.2 Direct Reply Packets

SCION adheres to the principle that any entity receiving or forwarding a packet is able to send a reply to the source of that packet.[6] This is necessary for SCMP as on-path routers are unable to request path segments back to the source for efficiency reasons. Furthermore, it means that the destination can also send acknowledgments or other replies without having to fetch new path segments or DRKeys.

Therefore, also EPIC-SAPV supports direct reply packets. The hop fields in the EPIC-SAPV path header cannot be simply inverted as in the regular SCION header however, as the HVFs are computed based on the length of the original packet and the DRKeys shared by the on-path ASes with the original packet's source. Furthermore, the HVFs have already been modified by the on-path routers on the forward path.

Hence, the recipient or an on-path router can send a reply packet by inverting the hop fields in the EPIC-SAPV path header, and further adding an *Original Packet Size* field to the path meta header (which contains the packet size of the original packet), and setting the reverse-flag (R=1). This flag indicates to the border routers on the backwards path that the HVF in the packet was already updated and that the source and destination in the address header were switched; correspondingly, border routers use the original source's DRKey (identified via the *destination address* field in the packet header) and the *Original Packet Size* field to re-compute the updated HVF as in Equation (10.4).

While EPIC-SAPV packets in the forward direction can be prioritized, packets with the reverse-flag set must be treated as best-effort traffic and do not pass through the replay-suppression system.

10.1.6 Performance of EPIC

To show that EPIC is practically feasible, we implemented and evaluated EPIC level 3 prototypes for the source, the routers, and the destination using the Intel DPDK framework [421].[7] The prototype is evaluated on a commodity server with an 18-core Intel Xeon 2.1 GHz processor executing the component to be

---

[6]There is only one exception when using hidden segments (§10.1.4).

[7]As other EPIC levels have a strict subset of processing steps, they would achieve strictly better performance.

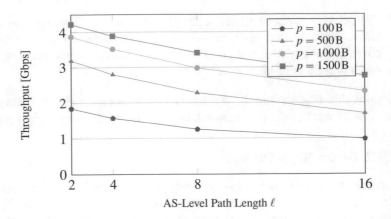

Figure 10.2: EPIC level 3 packet throughput generated by the source on a single core for different payload sizes.

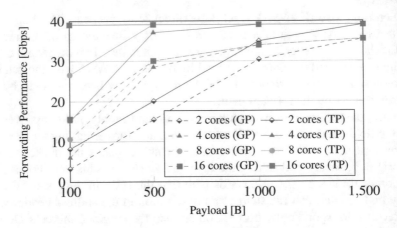

Figure 10.3: Throughput (TP) and goodput (GP) of a router plotted against the payload for 2, 4, 8, and 16 cores and $\ell = 8$.

tested, i.e., the source or router. This machine is connected with a packet generator and bandwidth monitor via a 40 Gbps Ethernet link.

The current average path length in the Internet is less than 4 AS-level hops [251, 350, 526, 552]. However, as we expect that number to increase due to the benefits of being an AS in a path-aware Internet, we evaluate to an extreme path length of up to 16 AS-level hops.

The throughput achieved by the source (using a single CPU core) is shown in Figure 10.2. For packets of $p \geqslant 500\,\mathrm{B}$ and path lengths of $\ell \leqslant 8$, the prototype implementation consistently achieves throughput above 2 Gbps.

Figure 10.3 shows the forwarding performance of an EPIC level 3 router for a path of length $\ell = 8$. In these measurements, we assume no cached hop authenticators or DRKeys; they are always recalculated on the fly. For packets with a payload $p \geqslant 500$ B, the 40 Gbps link is saturated for all path lengths using only 4 cores; using 16 cores, the link is even saturated for small packets ($p = 100$ B). As the implementation is easily parallelizable, it can be used even on 100 Gbps or 400 Gbps links by adding more processing cores or dedicated hardware. An important observation is that the processing time of the router is 445–460 ns independent of both payload size and path length. The forwarding performance in terms of Mpps (million packets per second) is thus also independent of these parameters and amounts to approximately 2 Mpps per processing core.

A more detailed evaluation of EPIC is included in our research paper [327]; it shows that the system can be implemented efficiently on commodity hardware, it is parallelizable and scales well to core links on the Internet, has significantly lower communication overhead compared to existing systems, requires virtually no state on routers, and limits additional control-plane overhead.

10.2 Bandwidth Reservations with COLIBRI

In this section, we present a collaborative lightweight inter-domain bandwidth-reservation infrastructure: COLIBRI. Our system overcomes the limitations of previous reservation systems and can provide worst-case minimum bandwidth guarantees. This is made possible through path choice and stability provided by the path awareness of SCION, and the confluence of several additional technologies developed in the context of SCION: the global symmetric-key distribution system DRKey enabling efficient per-packet authentication (§3.2); an efficient replay-suppression system (§9.1); and an overuse-flow-detection system (§9.3). COLIBRI is an evolution of the Scalable Internet Bandwidth Reservation Architecture (SIBRA) [55], and thus many design principles are shared between the two.

10.2.1 Introduction

Systems that attempt to achieve global quality-of-service (QoS) guarantees require scalability in three different areas given an increase in network size, number of flows, and traffic volume:

- **Control-plane scalability** concerns the communication, processing, and storage overhead for admission control and resource allocation.

- **Data-plane scalability** concerns the processing and storage overhead for packet forwarding, including authentication, monitoring, and policing to detect or prevent adversarial actions.

- **Management scalability** concerns the human effort required to configure and administer ASes.

Past efforts fall short in achieving scalability in one or several of these three areas, or cannot handle adversarial settings. Of the many systems proposed to achieve guarantees for global communication, the two archetypal and most widely deployed architectures are Integrated Services (IntServ) [551] and Differentiated Services (DiffServ) [389]. They constitute the two extreme points in the trade-off spectrum between scalability and strength of offered guarantees, and we will therefore use them as representatives of many other systems that employ the same ideas and suffer from the same shortcomings.

IntServ provides very strict guarantees on the communication parameters through end-to-end reservations, but is known to scale poorly in all three areas because of the complex decisions that have to be made during reservation requests through the Resource Reservation Protocol (RSVP) and the amount of per-flow state that on-path routers have to keep.

DiffServ, on the other hand, provides hosts with a way to divide their traffic into a number of classes according to the application's requirements, indicated in the IP packet's TOS header field. The task to schedule and prioritize the packet in order to enforce the desired service level is then delegated to the on-path routers. DiffServ scales well with respect to the three areas, as the only information needed for the traffic differentiation is carried in the IP packet header. Unfortunately, DiffServ does not provide any guarantees in an inter-domain or adversarial setting. It is thus mainly applicable to intra-domain contexts, for example to achieve traffic isolation and prioritization (see §11.4).

In COLIBRI, scalability is achieved through hierarchical decomposition at several levels:

- **Topological reservation hierarchy:** Reservations follow SCION's ISDs and segment types, reducing the global coordination effort.

- **Temporal reservation hierarchy:** Intermediate-term AS-to-AS reservations carry most of the computational overhead, which is thus amortized over time. Starting from these, short-term reservations are inexpensively set up for host-to-host communications.

- **Monitoring hierarchy:** Stateful reservation monitoring is performed at the edge. At transit and core ASes—where the number of flows is too high to be processed statefully—neighbor-based and probabilistic monitoring are used.

- **Policing hierarchy:** Each AS is responsible for enforcing that none of its hosts overuses a reservation. In the inter-domain, ASes check that others fulfill this responsibility: Because of per-packet authentication and duplicate suppression, misbehavior can be unambiguously attributed to the offending AS and can thus be punished (e.g., by canceling existing reservations and declining future ones).

The core property we seek in COLIBRI is to provide minimum bandwidth guarantees between any pair of ASes on a given path, irrespective of DDoS attacks or other allocations. Naturally, the actual amount of the minimum bandwidth guarantee depends on the network topology, and diminishes with increasing path length. In the common case, allocations will be much higher, but worst-case guarantees enable Internet-scale QoS. Our implementation of COLIBRI shows that it achieves scalability in three areas—control plane, data plane, and management—and can operate securely in a federated Internet in the presence of adversaries.

10.2.2 Background: Enabling Technologies

Traditionally, inter-domain bandwidth reservation systems were fundamentally limited by the following challenges, each of which needs to be resolved to achieve a viable system: (1) per-flow state in the data plane, (2) lack of path stability, (3) lack of path diversity, (4) flows overusing their reservation, (5) framing attacks by on-path routers through packet replay or alteration, (6) framing attacks by off-path entities using source-address spoofing, and (7) DoS attacks on the packet-authentication system on routers. In the remainder of this section, we describe the developments in networking that are at the basis of COLIBRI and allow both performance and security challenges to be overcome. Table 10.1 lists a summary of these points.

10.2.2.1 Path-Aware Networking

SCION is a path-aware architecture that uses packet-carried state to encode forwarding paths, reducing the state that must be kept at intermediate routers. SCION also provides two other essential properties for a resource-reservation architecture: path stability and path choice.

Path Stability. Since routing decisions are decoupled from the dissemination of path information, SCION does not suffer from the long convergence times that affect path-vector protocols such as BGP. Reservation guarantees are hard to achieve and maintain if communication has to wait for re-convergence after every routing event. Moreover, the packet-carried forwarding state makes forwarding immune against routing attacks attempted by off-path adversaries, preventing the denial of reservations by means of BGP hijacking or similar attacks. Together, these properties ensure that AS-level paths, and any reservations on them, are stable in time and cannot be affected by off-path entities.

Path Choice. In SCION, hosts can choose to use one (or several) of the paths discovered through the beaconing process. This feature enables multiple options for reservation architectures: If the reservation request cannot be met

Table 10.1: Challenges faced by attack-resistant resource-reservation systems, and the technical solutions used in COLIBRI.

Challenge	Enabling technology	
Per-flow state in the fast path	Packet-carried state	§2.4
Re-convergence changes reservation path Path-hijack invalidates reservation*	Path stability	§2.4
No reservation space available on path On-path adversary*	Path choice	§2.4
Large number of reservations	ISDs and segment types	§2.3
Authentication overhead of signatures	Symmetric-key auth.	§3.2
Framing attack with spoofed packets*	Per-packet source auth.	§3.3
Framing or DoS through packet replay*	Duplicate suppression	§9.1
Over-use of legitimate reservation*	Probabilistic monitoring	§9.3
Coordination of short-lived reservations	Time synchronization	§8.2
Unused reservations waste bandwidth	Traffic prioritization	§11.4
Difficult admission decisions	Reservation hierarchy	§10.2.3

*Adversarial action.

on the first path, COLIBRI can attempt to make a reservation on the alternative paths, which increases the probability of a successful reservation.

Isolation Domains and Path Segments. Through its ISDs and different types of segments, SCION provides the ideal structure to decompose reservations and thus increase scalability.

10.2.2.2 Reservation Protection

In order to assign, enforce, and bill reservations correctly without per-flow state on routers, a data packet must carry cryptographically protected information that allows an on-path AS to (1) verify that the packet is sent over a valid reservation, and (2) attribute the packet to the source. As every packet must be checked individually, efficient cryptographic mechanisms are of paramount importance. In addition, the authentication of control-plane packets must also be very efficient to avoid additional attack vectors (e.g., signature flooding), and prevent denial-of-capability (DoC) attacks [35].

Authentication of data packets can be achieved by computing per-packet MACs for each on-path AS, similar to the mechanisms in EPIC. DRKey is the fundamental building block that allows COLIBRI to achieve highly efficient authentication of control-plane packets: The source computes MACs for all on-path ASes, which can then be checked by on-path ASes without per-source state and using only highly efficient symmetric cryptography.

Duplicate Suppression. Even if packets are source-authenticated, the reservation architecture is still susceptible to replay attacks: An on-path adversary can capture an authenticated packet and send it repeatedly at a higher rate than what the reservation allows, thus causing congestion and framing the honest source. Because of this, an efficient duplicate-suppression system with minimal state requirements is needed.

Monitoring and Policing. Cryptographic tags enable stateless checking of the authenticity of reservations but are insufficient for enforcing bandwidth limits. To prevent overuse of reservations, additional monitoring and policing systems need to be used, i.e., systems the ASes use in order to make sure that flows originating from, traversing, or ending inside their networks do not use more bandwidth than allocated. ASes can thus identify misbehaving flows, and take action to ensure a fair use of resources.

Stateful flow monitoring (e.g., using a standard token-bucket algorithm [362]) enables precise bandwidth measurements, but requires per-flow state. Therefore, it can only be deployed at the edge of the network, where the number of flows is small, or to monitor a small set of suspicious flows. In contrast, probabilistic monitoring using LOFT (§9.3) can be performed even in the Internet core with very large numbers of flows.

10.2.2.3 Time Synchronization

Scheduling reservations across AS boundaries also requires a shared notion of time. Routers on the path agree to provide the required bandwidth for an interval of time, and therefore have to be synchronized for the system to work properly. Time synchronization between ASes is also important for duplicate detection and traffic monitoring. SCION's time synchronization provides a sufficient level of synchronization (§8.2).

10.2.3 Overview of COLIBRI

In this section, we provide a high-level description of the COLIBRI system. The core ideas are summarized in Figure 10.4.

10.2.3.1 General Concepts and Intuition

With COLIBRI, hosts can make short-term bandwidth reservations to protect end-to-end communications. However, as the Internet core could serve an enormous number of such end-to-end reservations (EERs), the decision whether or not to allocate resources must be very efficient.

To simplify and speed up this process, COLIBRI bases the allocation of EERs on a second type of bandwidth reservations called segment reservations (SegRs). SegRs have longer validity periods than EERs and are fewer in number, as they are established between ASes instead of end hosts. To increase efficiency, we offload the burden of computationally expensive operations to SegR allocations; hosts can then use the pre-computed SegRs to quickly set up EERs, whereby ASes enforce that the total bandwidth of EERs does not exceed the capacity of the underlying SegR. We further restrict the number of SegRs by leveraging the decomposition of paths into different segment types in SCION (§2.3.1). Forwarding is not impacted by these control-plane operations, as they take place in the dedicated COLIBRI service (CServ), outside the routers' fast path.

End hosts can then request SegRs from their AS (§10.2.3.2) and assemble them to cover the complete path to the destination's AS. The resulting SegRs can be used to request EERs (§10.2.3.3). Based on the capacity of SegRs, ASes on the path of EERs request can immediately grant or refuse the request.

To avoid per-flow state in the fast path, and to verify the legitimacy of a packet in the data plane, COLIBRI protects the packet-carried forwarding state through cryptographic tags. The keys used to compute these tags, called hop authenticators (similar to those in EPIC), are pseudorandomly generated by on-path ASes and sent—in encrypted form—to the initiator during the reservation process. At forwarding time, each on-path AS dynamically re-creates the authentication key and verifies the tag, efficiently authenticating the packet and the reservation. Thus, we overcome the other major issue of previous bandwidth-reservation systems: the need to maintain per-flow state on routers.

10.2.3.2 Infrastructure

ASes that use COLIBRI must perform three main tasks: (1) control the admission procedure, (2) manage SegRs, and (3) rate-limit the EERs of their end hosts. Tasks (1–2) are handled by the CServ, task (3) by the COLIBRI gateway.

COLIBRI Service. Every AS runs a CServ in addition to the standard SCION services (see §2.1), which handles all control-plane tasks related to COLIBRI within an AS:

- The CServ requests and renews SegRs according to expected traffic requirements. Since link utilization often exhibits repeating patterns over time, an AS can make predictions on future requirements and reserve appropriate bandwidth for segments in advance.

- It provides previously established and registered SegRs to end hosts and remote CServs. Non-core ASes store up-SegRs, core ASes store core- and down-SegRs.

- It handles all reservation setup and renewal requests and performs the necessary admission calculations.

To improve scalability, the CServ can be distributed: A central service is used to keep track of SegRs, while the handling of EERs is delegated to sub-services close to or at the border routers.

COLIBRI Gateway. An AS is held accountable by other ASes for misbehavior of its end-host customers. Therefore, it is crucial that all COLIBRI traffic originating from end hosts within an AS pass through a COLIBRI gateway, which (1) performs stateful monitoring and rate limiting for all EERs, and (2) embeds cryptographic tags into packet headers, allowing on-path border routers to authenticate the source and data in packet headers. To ensure consistent processing and prevent attacks by end hosts, every EER needs to be tied to a specific gateway.

End-Host Networking Stack. COLIBRI requires a modification of the SCION daemon to enable an application to explicitly request and renew EERs.

In principle, any transport protocol can be used with COLIBRI, as the gateway drops packets if the guaranteed bandwidth is exceeded and thus provides feedback to the congestion-control algorithm. Still, a tighter integration with the transport protocol is necessary to reap the full benefits of COLIBRI. For example, in QUIC [267], it is straightforward to disable congestion control and set the sending rate to the reserved bandwidth.

10.2.3.3 Control Plane

COLIBRI's control plane manages the selection, creation, and renewal of SegRs and EERs, which we discuss individually. Both SegRs and EERs are unidirectional, which reflects traffic demand in the Internet: while some ASes mainly send traffic (e.g., CDNs), others predominantly receive data (often called "eyeball networks"). Renewals are discussed in detail in §10.2.4.3.

Segment Reservations. SegRs are intermediate-term reservations made between two ASes and are valid for approximately five minutes. This duration has been chosen as a compromise between excessive overhead for short intervals and insufficient flexibility for long intervals. To improve the scalability of the SegRs and avoid setting up SegRs between any pair of the currently over 75,000 ASes [59], we leverage SCION's ISDs. Following this structure among ASes, COLIBRI distinguishes three types of SegRs mimicking the segment types in SCION: up-SegRs (from non-core ASes towards core ASes inside one ISD); down-SegRs (from core ASes towards non-core ASes inside one ISD); and core-SegRs (among core ASes, potentially in different ISDs).

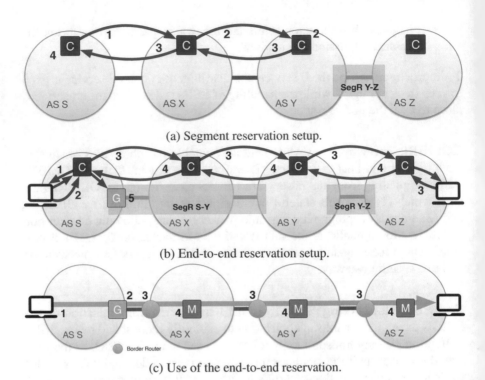

(a) Segment reservation setup.

(b) End-to-end reservation setup.

(c) Use of the end-to-end reservation.

Figure 10.4: Overview of the COLIBRI system. C are CServs, G is the
COLIBRI gateway, the yellow-orange circles are border routers,
and M are traffic monitors. A description is provided in Sections 10.2.3.3 and 10.2.3.4.

SegRs are always initiated by the first AS on the segment. For down-SegRs,
the first AS only sets up a SegR upon an explicit request by the last AS.

An AS can decide which SegRs to request based on historical data or traffic predictions. It selects segments generated by the SCION control plane and
sends a SegR request (**1** in Figure 10.4a) specifying requirements for the reservation such as requested minimum bandwidth. Each AS on the segment can
calculate how much bandwidth can be granted; if this is higher than the requested minimum, it records this reservation locally. It then updates the request with the granted amount of bandwidth and forwards it to the next AS
on the segment (**2** in Figure 10.4a). The last AS sends a reply via the same
segment to the request initiator (**3** in Figure 10.4a).

If the request was successful (each AS granted more than the minimum
amount of bandwidth), each AS locally stores the final amount of bandwidth
granted and includes a cryptographic token in the response that allows the request initiator to use the segment for COLIBRI traffic (**4** in Figure 10.4a; see
§10.2.4.6 for details). In case of an unsuccessful request, the ASes clean up
their temporary reservations and the initiator can determine the location of po-

tential bottlenecks on the segment. All important information is authenticated using DRKey as discussed in §10.2.4.6.

A central mechanism in this procedure is the admission algorithm that determines how much bandwidth can be granted; this will be discussed in §10.2.5.

End-to-End Reservations. EERs in COLIBRI are short-term reservations between two end hosts, with a validity period of 16 seconds. This interval is chosen by considering the trade-off between the overhead of creating the reservation, the duration a reservation remains idle after the connection is closed, and the need for flexibility to adapt to changing traffic patterns.

An end host H_S intending to communicate with another host H_D must first obtain one or several SegRs that can be connected to form a seamless path between the ASes of H_S and H_D. H_S obtains these reservations from the CServ (**1** in Figure 10.4b), in a similar way as it would otherwise obtain a path to H_D from the path service. It can combine SegRs to obtain a complete path and send an EER request (**2** in Figure 10.4b), which has a similar format to that of the SegR request described above. Again, each AS on the path decides how much bandwidth can be granted and forwards the request packet (**3** in Figure 10.4b). In contrast to the SegR request, this decision is simple: The intended bandwidth is granted if there is sufficient available bandwidth in the underlying SegR (see §10.2.4.10 for further details). It is the responsibility of the AS in which H_S is situated to set a limit on the maximum bandwidth that H_S can request. This intra-AS admission policy can be defined by each AS independently.

In addition to all ASes on the path, the destination H_D also needs to grant the request. A response is generated similarly to the SegR request: Each AS on the path updates the reservation information and (if the request was successful) includes the necessary hop authenticators, in the response (**4** in Figure 10.4b). The hop authenticators are stored at the source AS's COLIBRI gateway (**5** in Figure 10.4b), and the rest of the response is forwarded to H_S, who can then start using the EER to send traffic on that path.

The host can base the amount of requested bandwidth on the expected traffic, e.g., the known bitrate of a video stream. In cases of less regular traffic, the host may need to employ heuristics to determine how much bandwidth to request. Both control and data traffic will be handled by the end-host networking stack.

10.2.3.4 Data Plane

Packet Creation and Forwarding. After setting up an EER, an end host can use the reservation to send traffic to the specified destination. The end host includes the COLIBRI routing information in the packet header and sends it to the COLIBRI gateway of its AS (**1** in Figure 10.4c). There, the AS performs stateful traffic monitoring and uses the hop authenticators that were set up during the request process to calculate per-packet MACs for each AS on the path that at the same time (1) provide source authentication, and (2) prove that the COLIBRI path was previously authorized by that AS (**2** in Figure 10.4c). The details of this are described in §10.2.4.6.

On-path routers check the corresponding MAC by recalculating it locally (**3** in Figure 10.4c). We stress that these routers *do not require per-flow state* for this: All necessary keys can be derived on the fly from a single AS-specific secret value. Furthermore, no forwarding-table lookup is necessary as the forwarding information is already included in the packet header. Thus, after authenticating the packet, the router simply forwards the packet to the destination host or next border router.

Traffic Split.   Even in full deployment of COLIBRI, not all traffic can be sent over EERs: First, COLIBRI control-plane messages (except EER-renewal requests) cannot use an EER. Second, reservations are only useful for flows of some minimum size. Third, in some cases the communication requires replies that are however not large enough to need their own reservation—e.g., the acknowledgments for a video stream. Since reservations are unidirectional, these must be sent as best-effort traffic.

Therefore, COLIBRI reserves a fixed minimum of bandwidth for best-effort traffic. The remaining bandwidth is split further between control traffic on SegRs and traffic over EERs. Note that no bandwidth is wasted: If SegRs or EERs are under-utilized, the remaining bandwidth can be used for best-effort traffic. In practice, queuing techniques such as priority queuing or class-based weighted fair queuing can provide this separation on a shared physical infrastructure, see §11.4 for details.

Monitoring and Policing.   The use of cryptographic MACs in packet headers guarantees that an end host can send traffic only over an authorized reservation. However, a host can over-use a legitimate reservation and exceed the allocated bandwidth. To avoid misuse of reservations, an AS running COLIBRI performs two tasks: (1) It ensures that its own customers respect the bandwidth of their EERs, and (2) it monitors traffic from other ASes to detect overuse of EERs or SegRs (**4** in Figure 10.4c). This second task provides an incentive for ASes to comply and perform monitoring properly, as they are held accountable for the behavior of their customers. Further details are provided in §10.2.4.11.

10.2.4 The COLIBRI Architecture in Detail

10.2.4.1 AS Types

For EERs, we distinguish between four types of ASes depending on their position (they correspond to the ASes in Figure 10.4):

- The source AS is where H_S is located (AS S);

- A transit AS is an on-path AS in the middle of a SegR (AS X);

- A transfer AS is at the joint of two SegRs and, according to the structure of segments, necessarily a core AS (AS Y); and

- The destination AS is where H_D is located (AS Z).

10.2.4.2 AS Operation

The CServ is primarily responsible for two control-plane tasks (see §10.2.3.2): (1) controlling the admission procedure, and (2) managing SegRs. These decisions can be automated by collecting metrics of the network and predicting bandwidth usage heuristically.

Before an AS can provide bandwidth reservations to its customers, it must have a set of SegRs available, from which hosts can construct EERs. It is in the interest of the AS to predict usage patterns for reservations accurately, as insufficient bandwidth on its SegRs leads to suboptimal customer experience, while excessively large reservations may incur additional costs for the AS.

Since link utilization often exhibits repeating patterns over time, an AS can make predictions on future requirements and reserve appropriate bandwidth for segments in advance. Additionally, the AS can use live data on the utilization of SegRs to adjust the bandwidth of reservations on the fly, promptly reacting to unforeseen short-term demands. These prediction mechanisms depend on metrics local to the AS, and the heuristics can be fine-tuned by the operators to achieve optimal results.

10.2.4.3 Reservation Versions and Renewal

Reservations are created with short lifetimes to ensure that they do not occupy bandwidth for longer than necessary. If the initiator of a reservation intends to keep using it beyond its expiration time, it can issue a renewal request to extend the reservation, and possibly adjust the bandwidth to shifting traffic demands. This renewal request can be sent via the existing reservation. Expiration and renewal work slightly differently for EERs and SegRs.

EERs. Since EERs are frequently renewed, it is crucial that they can be renewed without service interruptions. For this purpose, the design allows multiple versions of the same EER to exist simultaneously. The initiator can renew the reservation ahead of time to obtain a new version with a later expiration time, which allows for a seamless transition between two versions of an EER. To avoid having to keep track of too many versions, CServs can rate-limit the amount of renewal requests for an EER (e.g., to one per second).

EERs automatically expire and there is no mechanism to remove them earlier. The gateway generally uses a single version (typically the latest one) to send traffic. However, even using multiple versions simultaneously does not provide more bandwidth, as all versions are mapped to the same underlying reservation ID in the probabilistic traffic monitor (see §10.2.4.11).

The initiator of an EER is not the only entity that could be interested in adjusting the reserved bandwidth. An AS on the path may also wish to reduce an EER's bandwidth, e.g., if it receives an increasing number of contending requests. As in the setup procedure, during a renewal request all on-path ASes can specify the amount of bandwidth they are willing to grant, enabling ASes to quickly adapt to changes in demand without interrupting service over existing reservations.

SegRs. Although there is a less frequent need to renew SegRs, due to their longer lifetimes, COLIBRI's design ensures that EERs are not affected by a version change of their underlying SegR. In contrast to EERs, only a single version of a SegR can exist at any time and a pending version obtained through a renewal request must be activated explicitly using a separate request. Making this switch explicit allows ASes to precisely control the time to change to a new version and ensure that no over-allocation with EERs can occur.

10.2.4.4 Packet Format and Header Fields

COLIBRI is implemented as a separate path type in the SCION architecture. Abstractly, a COLIBRI packet traversing AS_0–AS_ℓ has the following format:

$$\text{PACKET} = (\text{PATH} \parallel \text{RESINFO} \parallel \text{EERINFO}^8 \parallel Ts \parallel \qquad (10.5a)$$
$$V_0 \parallel \ldots \parallel V_\ell \parallel Payload),$$
$$\text{PATH} = ((In_0, Eg_0) \parallel \ldots \parallel (In_\ell, Eg_\ell)), \qquad (10.5b)$$
$$\text{RESINFO} = (SrcAS \parallel ResId \parallel Bw \parallel ExpT \parallel Ver), \qquad (10.5c)$$
$$\text{EERINFO} = (SrcHost \parallel DstHost), \qquad (10.5d)$$

where V_i denotes the hop validation field (HVF) of AS_i, which authenticates parts of the packet header and will be explained in detail in §10.2.4.6; PATH is a list of ingress–egress interface pairs; $SrcAS$ is the source AS; $SrcHost$ and $DstHost$ are the end-host addresses within the SCION address header; Bw, $ExpT$, and Ver denote the reservation bandwidth, expiration time, and version, respectively; and Ts is a high-precision packet timestamp relative to $ExpT$ and uniquely identifies the packet for the particular source.

This packet format is used for all COLIBRI control- and data-plane traffic. In the case of SegRs, AS_0–AS_ℓ denote the ASes that constitute the particular segment, for EERs they correspond to the ASes on the end-to-end path.

---

[8]The EERINFO field is only used for data-plane packets on EERs.

Reservation IDs. The reservation ID (*ResId*) must be unique per source AS, which is achieved by having the CServ increase the reservation ID for every new SegR or EER. Thus, the pair (*SrcAS, ResId*) uniquely identifies every SegR and EER globally.

10.2.4.5 Control-Plane Packets

The control-plane comprises setup and renewal requests for SegRs and EERs. The initiator AS relies on path segments generated through the SCION beaconing process for the PATH and AS IDs.

SegR Setup and Renewal. A SegR setup is initiated by the source AS using best-effort traffic, where the payload consists of the PATH, the RESINFO, and the minimum acceptable bandwidth. All on-path ASes add further data to the payload, which is used on the backwards path to calculate and agree on the allocation size. A SegR can be renewed over the existing SegR. Because SegR renewal packets already contain the PATH, *SrcAS*, and *ResId*, the source AS only needs to specify the new *Bw* (and new minimum bandwidth), *ExpT*, and *Ver* in the payload.

EER Setup and Renewal. Every EER is established over one, two, or three SegRs (similarly to how path segments can be combined in SCION). For an EER setup, the source AS creates a COLIBRI packet for the first SegR, and adds the EER PATH, the EER RESINFO, the EERINFO, plus the *ResId*s of all segments to the payload. If the EER is intended to be created over more than one SegR, each transfer AS copies the payload of the previous SegR's COLIBRI packet to a new COLIBRI packet for the following SegR. This is possible because the transfer AS can look up all the necessary information for the following SegR based on the corresponding *ResId*. This way each AS on the end-to-end path obtains the EER setup information in the payload, based on which they either grant or deny the requested allocation. Additionally, the request is also forwarded using intra-AS communication from the CServ to the destination end host specified in the *DstHost* field, who also has to explicitly accept the EER request.

As for SegR renewals, an EER can be renewed over the existing EER, where the source AS only needs to specify the new EER *Bw*, *ExpT*, and *Ver* in the payload.

10.2.4.6 Packet Authentication

Authentication of Control-Plane Messages. To authenticate the payload of control-plane packets, COLIBRI relies on DRKey (§3.2). The source AS calculates a MAC over the payload for each on-path AS, using the key $K_{AS_i \rightarrow SrcAS}^{COLIBRI}$. AS_i can then efficiently recompute this key on the fly and verify

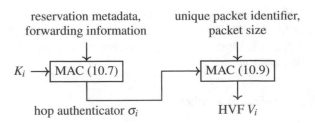

Figure 10.5: Two-step MAC calculation for the ith HVF of a data packet sent
on an EER.

the authenticity of the payload. It also uses the same key to compute a MAC
over the information that it added to the packet payload itself.

Segment Reservations. The only packets that are sent over SegRs are
control-plane packets (SegR renewal and EER setup requests), the payload
of which is authenticated using DRKey as discussed above. Nevertheless, it is
important that routers can statelessly verify the validity of a SegR, and authenticate forwarding information in the header. Therefore, during every SegR setup
and renewal, each on-path AS calculates a cryptographic token in the form of
a truncated MAC over reservation metadata and forwarding information, using
the AS's secret key K_i:

$$V_i^{(S)} = \text{MAC}_{K_i}(\text{RESINFO} \parallel (In_i, Eg_i)) [0{:}l_{\text{HVF}}]. \tag{10.6}$$

Here, l_{HVF} denotes the length of HVFs in bytes and is chosen as a compromise
between security and communication overhead as well as other practical considerations like header alignment. These tokens are returned to the initiator of
the SegR request and stored in its local database. All the information included
in this calculation is explicitly contained in the header of packets sent over this
reservation (see Equation (10.5)), enabling the router to validate the token on
the fly during packet processing without requiring per-reservation state.

While this use of relatively short static MACs in principle enables the reuse
of once-observed or brute-forced tokens, as we discussed in the context of
EPIC (§10.1), this is not problematic in practice due to the short lifetime of
reservations. Also, it is not necessary to "chain" the forwarding information
of different ASes, which is done in SCION and EPIC to prevent path splicing,
as the tokens used in COLIBRI explicitly include the globally unique combination of *SrcAS* and *ResId* (§10.2.4.4).

End-to-End Reservations. The authentication mechanism for data packets
on EERs must, simultaneously,

- be highly efficient, to support rates of hundreds of Gbps;

- avoid any per-reservation state on routers; and

- enable each on-path AS to conclusively attribute the packet to a reservation and the source AS.

We achieve these properties through a two-step authentication mechanism, sketched in Figure 10.5.

First, during every EER setup or renewal, each on-path AS calculates a hop authenticator, similar to the SegR token in Equation (10.6) but without truncation:

$$\sigma_i = \text{MAC}_{K_i}(\text{RESINFO} \parallel \text{EERINFO} \parallel (In_i, Eg_i)).$$ (10.7)

Similar to SegRs, all the information included in this calculation is an explicit part of the data-plane packet header enabling a router to compute the hop authenticator on the fly during packet processing. These hop authenticators are returned to the source AS AS_0 in a secure channel protected through authenticated encryption with associated data (AEAD):

$$AS_i \longrightarrow AS_0: \qquad \text{AEAD}_{K^{COLIBRI}_{AS_i \rightarrow AS_0}}(\sigma_i).$$ (10.8)

The source AS now shares a reservation-specific key with each on-path AS, which is not known to any other entity. These keys are subsequently used to calculate and verify per-packet MACs in the data plane as described in the following section.

10.2.4.7 Processing at Gateway and Border Router

COLIBRI Gateway. The gateway receives COLIBRI EER packets from end hosts, where all the COLIBRI header fields are empty, with the exception of the *ResId* and the *Payload*. The gateway maps the *ResId* of incoming EER packets to the corresponding PATH, RESINFO, EERINFO, and the hop authenticators, which it obtained earlier during an EER setup or renewal, and performs deterministic traffic monitoring as described in §10.2.4.11. It then generates a current high-precision timestamp *Ts*, from which it computes the HVFs for all on-path ASes,

$$V_i^{(E)} = \text{MAC}_{\sigma_i}(Ts \parallel PktSize)[0:l_{\text{HVF}}],$$ (10.9)

where *PktSize* is the size of the packet (including the COLIBRI header). The gateway thus confirms that it has performed the mandatory flow monitoring and authorized this packet. After filling in the missing EER packet contents, the gateway sends the packet to the border router responsible for the egress interface Eg_0.

Router. Upon reception of a COLIBRI packet, the border router of the ith on-path AS validates the packet format, header contents, and packet freshness, and checks whether the reservation has not expired yet. If the packet is a SegR packet, the border router validates the V_i in the packet header by recomputing it using Equation (10.6). If the packet is an EER packet, the border router instead computes the authenticator σ_i using Equation (10.7), from which it derives the V_i as defined in Equation (10.9). In both cases, if this recomputed HVF matches the one in the packet, the packet is forwarded to the next entity. For a SegR this entity is the local CServ and for an EER it is the border router of the next AS according to the path information in the PATH field, except for the border router of the last AS, which forwards the EER packet to the end host specified in *DstHost* instead.

10.2.4.8 Reply Packets on EERs

Similar to EPIC-SAPV (§10.1.5.2), COLIBRI supports direct reply packets, which are unprotected and forwarded as best-effort traffic. This is achieved by reversing the order of hop fields in the COLIBRI path header and setting a "best-effort EER response" flag. Concretely, direct reply packets are created by an intermediate border router or the destination end host as follows:

1. Take the HVFs from the original packet, reverse their order, and add them to the header of the response packet.

2. Reverse source and destination addresses.

3. Set the "best-effort EER response" flag.

4. Add an *Original Packet Size* field to the COLIBRI meta header, containing the packet size of the request packet.

On the reverse path, border routers check the packet's authorization by recomputing the HVF as in Equation (10.7), where they use the *Original Packet Size* field. The packet is forwarded hop-by-hop without prioritization until it reaches the destination (i.e., the source of the original packet). The packet is neither subjected to replay suppression, nor to probabilistic traffic monitoring.

10.2.4.9 Bidirectional Protection of EERs

Standard COLIBRI EERs are unidirectional, i.e., they only reserve and protect bandwidth from source to destination, but not on the reverse path. This is motivated by the observation that traffic—and congestion—is highly directional. Therefore, creating symmetric bidirectional reservations would likely incur a waste of reserved bandwidth in one of the two directions. Despite this inefficiency, bidirectional reservations can be beneficial in some cases. In TRS (described in §11.5.4), for example, both the request and response traffic needs to be protected. We therefore describe a minor modification to COLIBRI that allows construction of bidirectional EERs, which thus also protect response traffic.

To allow border routers to distinguish between these different types of communication, the COLIBRI meta header requires two different flags for best-effort response traffic, and for the reverse-EER traffic.

Overview. When establishing bidirectional SegRs and EERs, on-path ASes reserve a fraction of bandwidth for the reverse path as well—10% of the bandwidth of the forward direction, for example.

The data-plane operation of EERs remains unchanged on the forward direction. In the reverse direction, the COLIBRI gateway at the source AS sends a number of packets containing the encrypted hop validation fields (HVFs) for the reverse path. The destination host can decrypt the reverse-path HVFs, and include them in packets to use the reservation on the reverse path.

These packets are then processed as normal COLIBRI packets, and are subject to monitoring, replay suppression, and prioritization. A flag in the header specifies that the packet is protected by a reverse-path EER.

The source is thus responsible for the traffic on the return paths, and it is held accountable in case the destination exceeds the reserved bandwidth. It can control the maximum amount of traffic the destination can return through the number of authenticators it sends.

Reverse-Path Hop Validation Fields. When the COLIBRI gateway at the source AS receives EER packets from the source host, it also prepares the reverse-path HVFs for the destination. Similarly to Equation (10.9) in §10.2.4.7, for each on-path AS i, these are computed as

$$V_i^{(E)} = \mathrm{MAC}_{\sigma_i}(ReverseFlag \parallel Ts + \Delta T \parallel MaxPktSize)\,[0{:}l_{\mathrm{HVF}}], \quad (10.10)$$

where ΔT defines the point in time at which the reverse-path HVF will be usable by the destination, and *MaxPktSize* is the maximum packet size for the response packets. With these two parameters, the source COLIBRI gateway can effectively limit the bandwidth available for the reservation on the reverse path, for example to 10% of the forward EER bandwidth. The *ReverseFlag* is needed to differentiate these reverse-path HVFs—used to protect traffic—from the best-effort return traffic HVFs discussed in the previous paragraph.

Communication. The COLIBRI gateway generates multiple sets of reverse-path HVFs, one for each packet the destination may want to send on the reverse path. These are then encrypted with authenticated encryption with associated data, using the second-level DRKey $K_{S \to D:H_D}^{\mathrm{COLIBRI}}$ for the destination end host H_D. The gateway then sends the HVFs to the destination host, piggybacked by the source packets in a E2E header option §5.8.

The destination host decrypts the HVFs with the appropriate key, fetched from the local certificate service. To use the reverse-path reservation, the destination creates a new packet with one set of reverse-path HVFs, it sets the reverse-path flag in the COLIBRI meta header, and adds the response payload. Instead of using the packet size in the MAC computation directly (as in Equation (10.9)), on-path routers check that the packet size is smaller or equal to *MaxPktSize*, and then compute the MAC using this latter value.

10.2.4.10 Admission Procedure

Segment Reservations. As a first step, any two neighboring ASes agree on the bandwidth available for COLIBRI traffic on their inter-domain link and negotiate the pricing model. These typically long-term contractual agreements in the order of months are always bilateral to facilitate negotiation and billing. Based on these, each AS can define the local traffic matrix that describes the allocation of COLIBRI traffic between each interface pair.

A fundamental challenge of COLIBRI is to distribute the bandwidth of an ingress–egress interface pair in a fair manner. We have developed an algorithm that achieves "bounded tube fairness"; this algorithm is called N-Tube and is described in §10.2.5.

To perform the SegR admission calculation for a SegR request, the CServ must look up all existing SegRs that use the same egress interface. A careful implementation using memoization techniques still achieves the calculation in constant time on the order of a millisecond.

End-to-End Reservations. The EER admission depends on the type of AS:

- **Source AS:** The source AS has a direct business relationship with the end host. It is free to define which EERs can or must be provided to an individual end host in this contract. Upon receiving an EER request, the source AS checks (1) if there is sufficient bandwidth in the first segment reservation, and (2) whether the request can be granted under its policy.

  The source AS can set aside some bandwidth of the SegR to be able to construct EERs for itself. This is useful for control traffic—for example, for path registrations and lookups (§11.3.5) or key exchanges (§11.5.4).

- **Transit AS:** A transit AS grants the request if the first AS on that segment has approved it. However, to defend against malicious ASes, it again checks if there is sufficient bandwidth in the corresponding segment reservation.

- **Transfer AS:** A transfer AS needs to check if there is sufficient bandwidth in *both* SegRs it connects. Furthermore, the transfer AS between up- and core-SegR needs to distribute the core-SegR's bandwidth between all up-SegRs in case more EER bandwidth is requested than available in the core-SegR. This is done proportionally to the total of all requested EERs (capped at the up-SegR) that compete for the same core-SegR.

- **Destination AS:** The destination AS follows the same decision process as the source AS.

10.2.4.11 Monitoring and Policing

Monitoring and policing in COLIBRI is split into multiple components: deterministic monitoring and traffic shaping at the source AS, probabilistic monitoring at all other ASes, and policing of reservations. Control traffic on SegRs does not need to be processed at line rate, and can therefore be monitored and rate-limited at the CServ.

Deterministic Monitoring at Source AS. Traffic over EERs from end hosts inside an AS is monitored deterministically by the COLIBRI gateway (in parallel with the calculation of the HVFs). An efficient way to limit the transmission rate of the flows from customers while still permitting short-term spikes in traffic is the token bucket algorithm [362], which only needs to keep a timestamp and a counter in memory for each flow. When a flow exceeds the maximum transmission rate for longer than the burst threshold, packets are simply dropped.

Probabilistic Monitoring at Other ASes. The probabilistic overuse flow monitor (OFD) represents the centerpiece of the monitoring architecture in transit and transfer ASes. The main challenge with monitoring traffic that originates from other ASes is that it generally carries a very large number of flows, and in order to sustain line rate, the amount of memory used by the OFD must be small enough to fit in a fast cache. In §9.3 we have described how LOFT achieves very precise monitoring under these conditions.

COLIBRI requires that an OFD be deployed in each AS to monitor EERs. For each packet, the OFD receives as input (1) the normalized packet size,

$$normalized\ packet\ size = \frac{total\ packet\ size}{reservation\ bandwidth} \tag{10.11}$$

and (2) the source AS and the reservation ID, all of which can be read or calculated from the packet. The OFD then tracks bandwidth usage of each EER separately using the pair $(SrcAS, ResId)$ as a flow label; in particular, it combines packets of all versions of an EER in the same flow.

Normalizing the packet size as defined in Equation (10.11) enables monitoring reservations with different bandwidth guarantees using a single OFD and guarantees that a sender using multiple versions of the same EER can obtain at most the maximum bandwidth of all valid versions but not more. As the total packet size (including headers) is authenticated through the HVFs (10.9), framing attacks are prevented. Including the header size in the monitoring ensures that malicious source ASes cannot flood the system with packets with very small or no payload.

Due to the probabilistic nature of the OFD, it may report false positives—legitimate flows that appear to be overusing their bandwidth. For this reason, the suspicious EERs are subjected to deterministic monitoring as described in §9.3, which inspects the reservation precisely—similar to the monitoring at the source AS—to determine with certainty if it is being overused.

Policing.   When a flow is confirmed to be exceeding its EER bandwidth, it can be concluded that the source AS did not perform its monitoring task properly. Typically, the AS that detects the abuse takes two measures: (1) blocking further traffic over the reservation, and (2) penalizing the source AS. Measure (1) is crucial to avoid deteriorating service to legitimate reservations and is achieved by keeping a list of blocked source ASes. As this blocklist is very short—only a tiny share of the 75,000 ASes is expected to misbehave at any point in time—it can be implemented as a simple hash set.

After taking the immediate first step, the border router reports the offense to the local CServ. As misbehavior of the source AS has been established with certainty (due to the cryptographic checks), it is possible for the service to take drastic measures such as completely denying future admission of reservations originating from that AS.

10.2.5 Fair Bandwidth Allocation with N-Tube

The admission algorithm for SegRs must ensure that no AS or group of ASes can reserve excessive amounts of bandwidth. Concretely, for a particular segment, an AS should be able to reserve some positive minimum amount of bandwidth, which only depends on the network topology and link capacities. In particular, this minimum is independent of the number of end hosts or ASes under the control of an adversary; correspondingly, Basescu et al. [55] called this property "botnet-size independence".

We achieve this property through N-Tube, an admission algorithm that guarantees "bounded tube fairness". At a high level, N-Tube distributes the capacity among different competing SegRs proportionally to their adjusted bandwidth demand, which is obtained by

1. limiting the total demand coming from an ingress interface by that interface's capacity;

2. limiting the total demand between an ingress and an egress interface by the egress interface's capacity; and

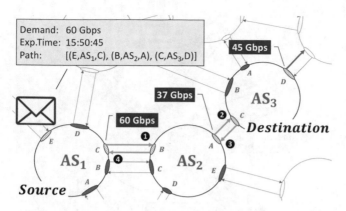

Figure 10.6: The process of making a reservation.

3. limiting the total demand of one particular source AS at a particular egress interface by that interface's capacity.

Note that under network conditions where there exists sufficient bandwidth for all requests, N-Tube is not required and every AS obtains the amount of bandwidth it requests. N-Tube is only required as soon as there are some links where more bandwidth is requested for SegRs than is available. In an extreme situation where every AS requests the maximum possible bandwidth for every path, N-Tube converges to allocations with topology-dependent minimum bandwidth guarantees, thus achieving botnet-size independence.

10.2.5.1 Design Goal and Properties

Our goal is an algorithm that allocates bandwidth according to the demands of source ASes and guarantees a minimum bandwidth allocation even under heavy congestion or flooding attacks. Concretely, N-Tube should satisfy the following properties:

G1 Availability: Any successful reservation request can reserve bandwidth, in spite of network congestion.

G2 Immutability: The allocated bandwidth of any existing reservation stays fixed until it expires.

G3 Stability: In periods of steady and constant demand, the bandwidth allocation in the entire network stabilizes in a predictable period of time.

G4 Minimum Bandwidth Guarantee: After the network stabilizes, there is a lower bound on the allocated bandwidth, i.e., a minimal bandwidth guarantee even with high external demands such as link-flooding attacks.

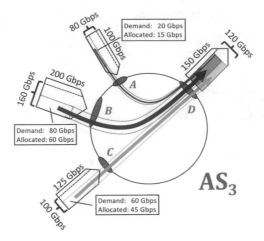

Figure 10.7: N-Tube's bandwidth computation at AS_3 distributes the egress link's (adjusted) bandwidth capacity D proportionally to three ingress demands.

G5 Bounded Tube Fairness: Bandwidth allocation is distributed proportionally to the requested demands, but adjusted to the maximally available bandwidth.

In the following, we present the main ideas and mechanisms of the algorithm through the example shown in Figure 10.6.[9] There, the source AS (AS_1) has chosen a path segment via AS_2 to AS_3 and requests a bandwidth of 60 Gbps. The path and requested bandwidth is included in the authenticated request message and processed by each on-path AS.

To ensure that any request immediately obtains some (potentially very small) amount of bandwidth (G1), N-Tube only reserves a fixed portion δ (which is a globally fixed parameter with $0 < \delta < 1$) of each link's total capacity, called the adjusted capacity. For any new reservation request, N-Tube again initially allocates at most the portion δ of the remaining free capacity, and thereby keeps the rest of the link's capacity available for other new reservations. In the following, we use $\delta = 0.8$.

10.2.5.2 Bounded Tube Fairness (G5)

The main challenge for a resource-allocation algorithm is to treat all reservations fairly, and to provide a lower-bounded bandwidth allocation for honest ASes—even when adversaries try to congest a link by demanding excessive bandwidth.

---

[9]The algorithm is formally specified and analyzed in §22.5 with respect to (G1–G5).

To provide fair bandwidth allocation, N-Tube bounds excessive demands by the links' capacities, and shares the resulting demands proportionally. This is illustrated in Figure 10.7 by the bandwidth-allocation computation at AS_3:

1. AS_3 factors the demands converging at a given egress interface by each ingress interface. These factored demands are called *tubes*.

2. AS_3 bounds the accumulated demand of each tube by its ingress and egress links' adjusted capacities, which we call their *bounded tube demands*.

3. AS_3 proportionally shares the egress link's adjusted capacity between its bounded tube demands.

AS_3 has three interfaces A, B, and C with ingress link capacities 100 Gbps, 200 Gbps, and 125 Gbps, respectively, and an interface D with an egress link capacity of 150 Gbps. The links' adjusted capacities are obtained by multiplying each link's capacity with δ (we use $\delta = 0.8$ in this example), and are indicated by the dotted lines. The three demands of 20 Gbps, 80 Gbps, and 60 Gbps from interfaces A, B, and C are factored into three tubes, and the adjusted capacity of 120 Gbps at interface D is proportionally split among them into 15 Gbps for A, 60 Gbps for B, and 45 Gbps for C.

10.2.5.3 Minimum Bandwidth Guarantee (G4)

By bounding the accumulated demands of tubes in the second step of the bandwidth-allocation computation, we guarantee that each tube obtains a fair share of the egress link's capacity. Whenever we must reduce a tube, we say it has *excessive demands*, and we proportionally reduce all demands inside it.

We illustrate how N-Tube computes bandwidth allocations in the presence of adversaries with three examples. We assume that all ASes on the given path are honest, and any AS off this path may be adversarial. The goal of the adversaries is to reduce as much as possible the allocated bandwidth for the honest ASes between interface C and interface D. Hence, we allow adversaries to demand an arbitrary amount of bandwidth to subsequently congest the egress link at interface D. We then show how the bandwidth-allocation computation still provides a minimum bandwidth guarantee.

Limit Demands on an Ingress Link by Its Adjusted Capacity. In the example of Figure 10.8, two adversarial ASes demand in total 400 Gbps (150 Gbps and 250 Gbps) of bandwidth from B to D. N-Tube bounds these demands by the ingress link's adjusted capacity at interface B of 160 Gbps. Hence, D's adjusted capacity of 120 Gbps is split proportionally between 20 Gbps from A, 60 Gbps from C, and 160 Gbps, instead of 400 Gbps, from B.

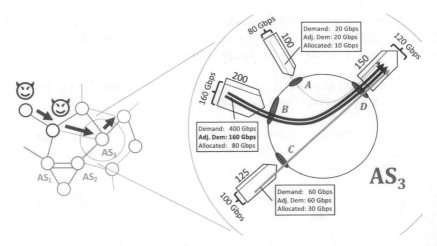

Figure 10.8: Limit demands on the ingress interface.

Limit Each AS's Demands by the Egress Link's Capacity. In Figure 10.9, an adversarial AS demands in total 200 Gbps to interface D: 80 Gbps and 120 Gbps through interfaces A and B, respectively. However, since its combined demand of 200 Gbps exceeds the egress link's capacity, N-Tube reduces both demands proportionally by a scaling factor. The scaling factor is the ratio of the egress link's adjusted capacity to the total adjusted demand of the adversarial AS, i.e., $120/200 = 0.6$. This results in the reduced demands of 48 Gbps ($= 0.6 \cdot 80$ Gbps) and 72 Gbps ($= 0.6 \cdot 120$ Gbps) from interfaces A and B, respectively.

Hence, D's adjusted capacity of 120 Gbps is split proportionally between 60 Gbps from interface C, and the reduced demands of 48 Gbps and 72 Gbps, from interfaces A and B. The computed allocations are therefore 40 Gbps, 32 Gbps, and 48 Gbps, respectively.

In this case, each AS must keep per-source AS state, i.e., how much bandwidth each source AS has reserved through this AS. This is feasible since the number of ASes in a network is much smaller than the number of flows and the admission processing is performed at the dedicated CServ.

Worst Case, Minimum Bandwidth Guarantees. In the example of Figure 10.10, all off-path ASes are adversarial and demand as much as they can on all the ingress links, i.e., a maximum of 80 Gbps on interface A and 160 Gbps (40 Gbps and 120 Gbps, respectively) on interface B. This represents a worst-case attack: Even when more adversarial ASes are present, their bandwidth demands will be adjusted, and therefore limited as described in the two previous examples.

The interface D's adjusted capacity of 120 Gbps is split proportionally between the bounded demands of 80 Gbps from interface A and 160 Gbps from

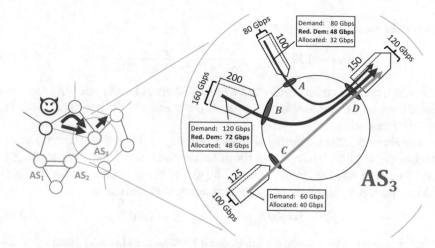

Figure 10.9: Limit demands on the egress interface.

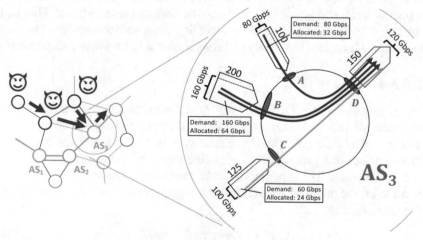

Figure 10.10: Worst-case minimum bandwidth guarantee.

interface B, and benign demand of 60 Gbps from interface C. Hence, this benign demand cannot be reduced to less than 24 Gbps by any amount of external demands. This provides the minimum bandwidth guarantee for the honest
reservation at AS_3.

Global Lower Bound.   By applying the idea from the previous example, we
can provide a local lower bound llb_3 for the proportion of D's adjusted link
capacity that can be allocated to the benign demand. In the worst case, all
tubes have excessive demands at interface D. Then llb_3 is given as the ratio of
the benign demand of 60 Gbps and the accumulated adjusted capacities of all

ingress links, i.e.,

$$llb_3 = 0.17 \approx \frac{60\,\text{Gbps}}{100\,\text{Gbps} + 160\,\text{Gbps} + 80\,\text{Gbps}}. \tag{10.12}$$

Likewise, we can compute llb_1 and llb_2 with respect to AS_1 and AS_2's ingress links' capacities. Note that these local lower bounds do not depend on the adversarial demands.

The request ratio $reqRatio_0$ at the source AS_1 is defined as the ratio of the benign demand of 60 Gbps and the total demand of reservations starting at interface E of AS_1 (see Figure 10.6). The global lower bound glb for the bandwidth that can be allocated to the honest demand is derived as

$$glb = reqRatio_0 \cdot llb_1 \cdot llb_2 \cdot llb_3 \cdot 120\,\text{Gbps}. \tag{10.13}$$

Note that glb only depends on the request ratio at the (honest) source and the capacities of the on-path ASes' ingress links. The request ratio is bounded by the number of reservations the source AS_1 starts at its interface E, i.e., under its own control, and is not influenced by (malicious) reservations. This is, intuitively, why N-Tube's bandwidth-allocation computation provides minimum bandwidth guarantees for SegRs and thus achieves botnet-size independence.

10.2.5.4 Stability (G3)

The validity period of COLIBRI's SegRs is limited to a maximum value of approximately 5 minutes. This upper bound, $maxT$, on the expiration time forces ASes to renew their reservations regularly, which allows N-Tube to stabilize the allocations in a predictable time period $stabT$ of constant demands after a burst in demands. The time period $stabT$ needed to stabilize demands can be shown to be the product of $maxT$ and the length of the longest reserved path $\hat{p}$ in the network, i.e.,

$$stabT = length(\hat{p}) \cdot maxT. \tag{10.14}$$

Intuitively, the bandwidth computation at the first AS only depends on requested demand at that AS, and the constant bandwidth allocations are then successively propagated to all ASes on the path at each renewal of the reservation. This provides an informal argument that N-Tube achieves stability. In the paper, we also show that, after the entire network stabilizes, N-Tube's bandwidth allocations also satisfy bounded tube fairness.

10.2.5.5 Immutability (G2) and Availability (G1)

By reserving only a fraction δ of the available bandwidth, N-Tube can always provide a positive (but possibly small) amount of bandwidth for a new reservation, which ensures availability. Immutability is ensured as COLIBRI does not change established reservations until they either expire or are explicitly deleted by the source.

10.2.6 Security Analysis

In this section, we briefly show how the interplay of cryptography, monitoring, and policing—as presented in the previous sections—protects COLIBRI reservations from attacks and thus enables worst-case minimum bandwidth guarantees. In this analysis, we consider the adversary model described in §7.2 and explicitly distinguish between on-path and off-path adversaries. Finally, we provide an analysis of typical attacks and how they are prevented in COLIBRI.

10.2.6.1 Attacks on Reservation Traffic

Volumetric DDoS Attacks. DDoS attacks against COLIBRI traffic can be carried out with (1) best-effort traffic, (2) bogus COLIBRI traffic, or even (3) authentic COLIBRI traffic that is overusing a reservation. The first attack is prevented by traffic isolation (described in §11.4). As the authentication procedure at border routers uses efficient symmetric cryptography (§10.2.4.6), bogus COLIBRI packets are quickly identified and dropped, countering the second attack. Finally, monitoring quickly identifies and blocks all overusing COLIBRI reservations (§10.2.4.11). In the worst case, an attacker that controls an AS can cause very brief congestion but would afterwards be prevented from creating reservations. Note that it is not possible to create congestion with COLIBRI traffic that respects its reservations, as the admission procedure ensures that the sum of all reservations does not exceed the capacity (§10.2.4.10).

Framing DoS Attacks. An adversary could try to turn the monitoring subsystem against benign ASes by (1) spoofing the source AS, or (2) capturing legitimate packets and replaying them to overuse the reserved bandwidth, thus framing the legitimate source. Since overusing ASes are blocked by on-path ASes, this is another form of DoS. COLIBRI avoids the first attack thanks to efficient source authentication (§10.2.4.6), and the second by using in-network duplicate suppression at benign ASes, thus discarding all copies of the same packet (§9.1).

10.2.6.2 Attacks on the Admission Algorithm

Another way to degrade the service of target users is to try and obtain an unfairly large share of bandwidth from the reservation process. For SegRs, the admission algorithm N-Tube guarantees that the total amount of bandwidth that any AS or group of ASes can reserve is limited, thus ensuring that a benign AS can *always* obtain a finite minimum bandwidth.

The distribution of EER bandwidth is the responsibility of source and destination ASes, as described in §10.2.4.10. These are free to define policies on how to allocate the capacity of the SegRs they set up among their customers. As they have direct business relationships with end hosts and control their address space, they can easily define and enforce these rules.

Finally, all on-path ASes check that the total bandwidth of EERs on a particular SegR does not exceed that SegR's capacity. Therefore, a source AS cannot cause over-allocation by requesting an excessive amount of EER bandwidth.

10.2.6.3 Attacks on Reservation Setup

As we have seen, COLIBRI traffic is fully protected against DoS attacks when a reservation is established. The only remaining avenue for malicious actors is to try and prevent legitimate ASes or end hosts from setting up COLIBRI reservations in the first place. In particular, an adversary can attempt to (1) exhaust the resources of the CServ with bogus requests, or (2) block the reservation setup packet before it reaches the CServ by congesting the network with best-effort traffic. These attacks are known as denial-of-capability (DoC) attacks [35], and COLIBRI is designed with multiple mechanisms to defend against them.

Efficient Authentication at the CServ. Every control-plane message is authenticated using symmetric cryptography that can be derived efficiently at the CServ. Thus, the CServ can very efficiently filter unauthentic packets and employ per-AS rate limiting.

Protected Control Traffic and Reservation Renewal. As soon as a SegR or EER exists, renewal requests can be sent over this reservation and are thus isolated from flooding attacks with best-effort traffic. Therefore, ASes that want maximum protection against DoC (e.g., towards business-critical destination ASes) can preemptively set up a low-bandwidth, inexpensive SegR to these destinations; should the need arise, the reserved bandwidth can be flexibly increased through renewal requests that are then protected from DoC attacks. Finally, EER requests are sent as control traffic over existing SegRs and thus also protected from best-effort traffic.

Prioritization of Initial Requests. Since renewals are protected by existing reservations, the only remaining DoC attack surface is the initial SegR request and the DRKey exchange. Keys only need to be exchanged once per day, which is possible ahead of time. Concerning SegRs, besides proactively setting up reservations ahead of time as described above, ASes can use the isolation mechanisms described in §11.4 to forward SegR requests with higher priority than best-effort traffic. As SegR requests are processed, authenticated, and filtered at each AS's CServ, they could only be used to flood the network of neighbor ASes, which is easily detectable.

In §11.5, we present a complete framework for how this last attack surface for DoC attacks can be prevented.

10.2.7 Discussion

The absence of global bandwidth-reservation architectures can be attributed to many factors—excessive overhead, intricate management, negligible benefits over best-effort—that amount to a general lack of incentives for deployment [289]. COLIBRI addresses all of these concerns.

Low Overhead. Protecting performance-sensitive (e.g., low-latency) traffic is one of the main benefits of bandwidth reservation systems. However, if a system's overhead creates effects similar to or worse than congestion, as in many past proposals, this benefit is negated. COLIBRI, on the contrary, is efficient in the data plane and can fill multi-Gbps links. Moreover, unlike in previous systems, best-effort traffic coexists with COLIBRI reservations and can utilize unused bandwidth, thus avoiding waste of network resources.

Simple Management. Coordinating reservation admission and billing in a distributed, global setting poses many challenges. A global transit ISP does not have enough information to properly allocate bandwidth to flows that are not terminating at its direct customers. Further, billing such reservations is difficult, as costs are shared among many entities.

End-to-end information on flows is, however, not necessary when granting COLIBRI reservations, as the admission decision depends on neighbor-based policies (represented by traffic matrices). The admission procedure is fully automated and built into the COLIBRI control plane. Additionally, thanks to the locality of policies, billing can be implemented with scalable neighbor-to-neighbor settlements, similarly to today's AS peering agreements.

A Profitable Service. For a long time now, overprovisioning has been the primary means with which ISPs maintain sufficient levels of QoS [254]. As the Internet has remained successfully operational so far, the common belief is that this measure is largely sufficient. However, congestion is widely observable both at the edges and in the Internet core [164]. Further, routing optimization can create congestion unexpectedly by converging routes to a popular prefix onto links with inadequate capacity [278]. Finally, recent attacks have been able to fill even large inter-domain links [193, 562]. COLIBRI reservations enforce fair sharing during peak traffic, mitigating the impact of these events, and enhancing the survivability and profitability of ASes. These characteristics add value to the service proposition of an ISP, and will therefore be drivers for deployment.

Secure Operation. Past systems did not consider adversarial actions, and could therefore be abused. COLIBRI employs a combination of light-weight cryptography, monitoring, and policing that make it the first system to achieve worst-case bandwidth guarantees on public networks.

10.2.8 COLIBRI in the SCION Ecosystem

Due to COLIBRI's scalability properties, its use is not limited to critical traffic; we envision that virtually *every* flow over the Internet that exceeds a minimal total traffic volume and duration will ultimately use a COLIBRI reservation. Notably, COLIBRI's guaranteed bandwidth is well suited for video streaming, which was responsible for more than 60 percent of all Internet traffic in 2019 [450]. Other predominant applications like gaming and file sharing (8 and 4 percent of global traffic, respectively) are a good fit for COLIBRI as well. Together, these use cases are responsible for almost three quarters of global traffic, and they will likely grow further.

The higher quality of service and predictability of COLIBRI reservations has several additional benefits besides resilience against attacks: Because the sender knows the bandwidth at which it can send, transport protocols can be substantially simplified with constant-bitrate sending instead of the complicated congestion-control mechanisms deployed today. Furthermore, as packet loss is very rare if not completely absent, more lightweight acknowledgment mechanisms can be used, where the frequency of acknowledgments is strongly reduced compared to today's transport protocols.

Only short connections, traffic patterns with high peak-to-average ratio like web browsing, or connectionless protocols like name resolution would still use best-effort traffic, as creating the COLIBRI reservation causes unreasonable overhead in these cases since the reservation would be severely under-utilized.

Overall, COLIBRI is ideal to support the traffic patterns of the most prevalent applications today and likely also in the future. It leads to higher quality of service, better control and predictability, and defends against volumetric DDoS attacks to achieve much higher availability of communications in the public SCION Internet.

11 Availability Guarantees

GIACOMO GIULIARI, ANNIKA GLAUSER, MARKUS LEGNER, ADRIAN
PERRIG, BENJAMIN ROTHENBERGER, JOEL WANNER, MARC WYSS

Chapter Contents

11.1 Availability Goals and Threat Landscape **268**
    11.1.1 Adversary Model 268
    11.1.2 Availability Properties 269
11.2 Overview **270**
11.3 Defense Systems **271**
    11.3.1 Basic Fault and Attack Isolation 272
    11.3.2 Overview of Protection of Data-Plane Traffic 273
    11.3.3 Overview of Protection of Control-Plane Services 274
    11.3.4 Further Supporting Systems 275
    11.3.5 Secure Path Discovery and Dissemination 276
    11.3.6 Data Delivery 276
    11.3.7 Packet Authentication and Filtering 277
    11.3.8 Bootstrapping of Defense Mechanisms 277
11.4 Traffic Prioritization **278**
    11.4.1 Traffic Classes 278
    11.4.2 Setup-Less Neighbor-Based Communication (SNC) 279
    11.4.3 Traffic Marking 281
    11.4.4 Priority Processing Using Queuing Disciplines 282
11.5 Protected DRKey Bootstrapping **283**
    11.5.1 Assumptions 283
    11.5.2 Overview 284
    11.5.3 Protecting COLIBRI SegR Setup 284
    11.5.4 Telescoped Reservation Setup (TRS) 285
    11.5.5 Security Analysis 287
11.6 Protection of Control-Plane Services **288**
    11.6.1 Criticality Criteria for Control-Plane Interactions 288
    11.6.2 Filtering at the Control Service 293
    11.6.3 Attack Resilience of the Control Service 294
11.7 AS Certification **294**

© The Author(s), under exclusive license to Springer Nature Switzerland AG 2022
L. Chuat et al., *The Complete Guide to SCION*, Information Security
and Cryptography, https://doi.org/10.1007/978-3-031-05288-0_11

11.7.1   Certification Overview   295
11.7.2   Level 1: Bronze   295
11.7.3   Level 2: Silver   296
11.7.4   Level 3: Gold   297
11.7.5   Interaction between Different AS Certifications   297
11.8   Security Discussion **297**
11.8.1   Availability Properties for the Control Plane   298
11.8.2   Resilience to Common Attacks   298
11.8.3   Full Dependency Analysis   300

As we argue in Chapter 7, the most fundamental requirement for any networking architecture is to be available: It is trivial to design a perfectly secure but unavailable network by simply cutting all connections or dropping all packets. Unfortunately, the current Internet falls short of achieving availability in an adversarial environment despite a multitude of defense mechanisms; in contrast, availability was a core requirement during the design of the SCION architecture.

This chapter describes how SCION systems and mechanisms are orchestrated to guarantee[1] availability even in the presence of malicious network actors. In essence, the property the SCION network can offer is communication with a non-zero lower bound on the bandwidth for network paths that are operated by non-malicious entities. SCION is the first public network architecture that can achieve this property.

11.1 Availability Goals and Threat Landscape

First, we discuss the threat landscape and describe the availability goals we strive to achieve.

11.1.1 Adversary Model

To achieve high availability, we need to defend against all types of network-level denial-of-service (DoS) attacks. These attacks can either target the network links, or nodes in the network that operate either the control plane or data plane.

Control-plane attacks target the protocols related to topology discovery, name resolution, and other systems that are prerequisites for end-to-end communication. An attacker can try to provide incorrect information

---

[1]We use the term "guarantee" in this section to indicate assurance with very high probability. In practice, a packet can always be lost due to a variety of reasons, for example, an intermittent electrical problem on a link, or a rebooting router.

(e.g., BGP hijacks in today's Internet) or prevent legitimate users from accessing these services.

Volumetric attacks flood network links with large amounts of traffic to exhaust the network capacity, create congestion, and cause high rates of packet loss for legitimate hosts. These attacks are often executed as distributed DoS (DDoS) attacks using bot nets and typically target the access links of end hosts (e.g., companies), but can also focus on core links in the network [279, 495].

End-system attacks exhaust the (computational or memory) resources of servers to limit the processing of legitimate requests. Typical attack targets include web servers, VPN endpoints, and firewalls.

The impact and stealthiness of these attacks are often increased through techniques such as source-address spoofing. We will discuss in detail how the SCION control plane is protected against these attacks (see §11.6), and how a generic client–server interaction can be protected in the data plane. We use the general adversary model described in §7.2. In some cases, we make a distinction between malicious (or compromised) end hosts and malicious (or compromised) ASes. Of particular relevance for this chapter are volumetric attacks on links in other ASes or resource-exhaustion attacks on (SCION infrastructure) servers in other ASes. We generally assume that no on-path ASes are malicious or compromised. This assumption is intimately related to availability, as a malicious on-path AS can always simply drop all packets.

11.1.2 Availability Properties

A SCION-enabled public Internet[2] achieves strong availability guarantees. In particular, it provides the following availability properties:

- **Secure path discovery and dissemination.** If a standard SCION path between two ASes consisting only of benign (i.e., not compromised or malicious) ASes exists, this path can be (1) discovered through SCION's beaconing process, and (2) made available to end hosts. The routing information disseminated by an honest AS cannot be modified by any other AS.

- **Data delivery.** For essential traffic, any AS can obtain a worst-case bandwidth guarantee to any other AS. The source AS can distribute this capacity among its end hosts. Traffic that does not exceed this capacity is isolated from best-effort traffic along the path and is thus assured to reach the destination with very high probability.

---

[2]There will likely be a substantial transition period in which the new SCION-based Internet will coexist with the current IP- and BGP-based Internet. During this transition period, the properties described here only hold for the SCION network.

- **Packet authentication and filtering.** An AS or end host can efficiently authenticate the source and payload of packets using symmetric cryptography and filter traffic based on simple rules.

In the following, we summarize how these three properties are achieved and how their combination ensures global availability guarantees.

11.2 Overview

Attacks on availability, also known as DoS attacks, come in several different forms, namely control-plane attacks, volumetric attacks, and end-system attacks (see §11.1.1). The defense against this diversity of DoS attacks requires the interplay of different defense systems. On a fundamental level, SCION's availability guarantees are achieved through three steps related to the three different classes of DoS attacks:

1. Ensuring that authentic information about network paths is discovered by autonomous systems (ASes) and distributed to end hosts;

2. Providing guarantees that traffic reaches the intended destination; and

3. Enabling the destination to filter unwanted traffic at very high rates with high accuracy.

The first point, secure path discovery, is one of the most fundamental design goals of SCION and is directly built into its secure control-plane protocols and public-key infrastructure (PKI). In particular (and in contrast to today's BGP-based routing) all routing messages are authenticated and no AS can modify another AS's information. End hosts' access to path information is ensured by replicating, distributing, and caching path segments and restricting access to path services to the intended clients.

Delivery guarantees are ensured through different logically isolated traffic classes. For larger flows, end hosts are able to obtain short-lived (16 s) bandwidth reservations through the *COLIBRI* system (§10.2), which allows them to send traffic at a specified rate with guaranteed delivery. This is particularly useful for real-time applications such as video streaming, voice over IP (VoIP), video conferences, and gaming. To ensure scalability, COLIBRI has been designed to require minimal coordination between different ASes and support stateless routers. An additional defense mechanism is SCION's *hidden paths* approach (§8.1), enabling the restricted sharing of path information to selected ASes or end hosts. In the case of a DoS attack targeting a public access link, a company can still use the hidden path for critical communication.

Finally, it is important that the destination is capable of filtering unwanted traffic at the bandwidth speed of its access link, as otherwise the end system or firewall may become overloaded. SCION achieves this through highly efficient

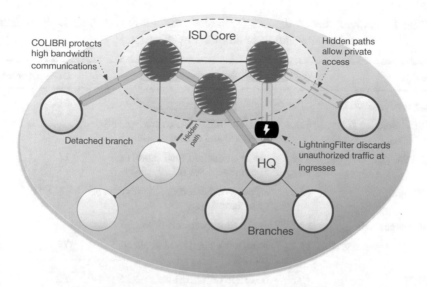

Figure 11.1: Holistic view of defense mechanisms in SCION. Circles represent
           ASes, thick circles indicate the ASes belonging to one particular
           entity (e.g., a company). This entity can protect its communica-
           tions from DoS attacks thanks to the interplay of different SCION
           services: COLIBRI, hidden path, and LightningFilter.

packet authentication based on symmetric cryptography and keys provided by
the DRKey system. This allows the destination to use *LightningFilter* (§9.2)
to filter authenticated traffic through simple rules and only rely on a legacy
firewall for the pre-authenticated remaining traffic.

Figure 11.1 illustrates the various defense systems from the point of view of
a company that consists of multiple branches.

11.3 Defense Systems

To achieve the properties presented above, we require the interplay of several
defense mechanisms described in this book, to be deployed both in the network
and at end hosts. Taking into consideration the different attack classes that
threaten the availability of SCION networks, we provide an overview of the
main defense systems, their dependencies, and their interactions in Table 11.1
and Figure 11.2. These defense systems can be divided into the following three
basic categories:

1. basic fault and attack isolation;

2. protection of data-plane traffic; and

3. protection of control-plane services.

Table 11.1: Overview of the SCION suite containing the main DoS defense mechanisms and their supporting systems.

System	Ref.	Properties	Dependencies
Supporting Systems			
Traffic isolation (TI)	§11.4	different traffic classes are logically separated and cannot interfere	
CP-PKI	§3.1	management and verification of asymmetric keys for all ASes	TI
DRKey	§3.2	efficient derivation and distribution of symmetric keys	CP-PKI
Time sync. (TS)	§8.2	global time synchronization for ASes	
Replay suppression (RS)	§9.1	in-network filtering of duplicated packets	TS
LOFT	§9.3	detection of bandwidth overuse	TS
Defense Systems			
Beaconing	§4.2	secure path discovery through authenticated routing messages	CP-PKI, TI
Hidden paths	§8.1	access links are only shared with selected entities; unauthorized traffic is filtered early	
EPIC-HP	§10.1.4	protection of hidden paths through per-packet MACs	RS
COLIBRI	§10.2	QoS through bandwidth guarantees for ASes and end hosts	DRKey, RS, LOFT, TI, TS
LightningFilter	§9.2	highly efficient packet authentication and filtering	DRKey, RS

This section will only provide a high-level summary of those mechanisms. The details of the systems can be found in Part I and Chapters 8–10.

11.3.1 Basic Fault and Attack Isolation

From the very beginning, SCION was designed to provide fault and attack isolation through several mechanisms (see also Chapter 7):[3]

- Isolation domains (ISDs) group ASes belonging to a common sovereign region (e.g., a country) and ensure that the routing within the ISD cannot be affected by external ASes. A configuration error or control-plane attack can thus only affect other ASes within the same ISD; this is in stark

---

[3]This is even part of the acronym: Scalability, Control, and *Isolation* on Next-Generation Networks.

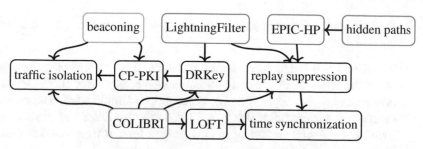

Figure 11.2: Overview of SCION's main DoS defense mechanisms and supplementary systems and their interdependencies ($A \rightarrow B$ indicates that A depends on B). The colors indicate the type of system: systems related to secure path discovery and dissemination (green), packet transmission (blue), packet authentication and filtering (orange), and supplementary systems (black).

contrast to BGP, where an event can cause global outages (e.g., BGP prefix-hijack) [315]. The fact that each ISD can choose its roots of trust independently in its trust root configuration (TRC) prevents external Internet kill switches (§7.8).

- SCION's path awareness—every packet carries the full forwarding path in its header—prevents any off-path entities from interfering in the forwarding process. As long as the ASes on the path are honest (a substantially weaker assumption compared to today's Internet), an end host's traffic is thus guaranteed to follow the intended path.

- Most current router and switch hardware supports queuing disciplines that enable a logical separation of different traffic classes. This ensures that a DoS attack using best-effort traffic cannot affect or disrupt communication in a higher-priority system (see §11.4 for a detailed description).

11.3.2 Overview of Protection of Data-Plane Traffic

A secure control plane based on a secure routing protocol can prevent a substantial portion of attacks but is insufficient by itself; it needs to be complemented by a secure *data plane*. As described in §11.3.1, SCION already protects data traffic from several types of attacks. Additional security systems provide even stronger availability guarantees:

- The DRKey infrastructure (§3.2) provides a symmetric key to any pair of ASes or end hosts, which can be efficiently derived by one of the communication partners. This enables authentication of all packets at the network layer and is used in the SCION packet authenticator option (SPAO, see §3.3) to achieve highly efficient packet authentication.

Furthermore, DRKey is used in several other systems to achieve source authentication.

- The SPAO and DRKey are the foundation of LightningFilter, which enables the receiver of a packet to authenticate its source (both AS and host address) and apply filtering according to simple rules. In contrast to existing systems like IPsec or TLS, LightningFilter can already authenticate the *first* packet of a flow in less than 100 ns on commodity hardware.

- COLIBRI (§10.2) is the first scalable system providing global quality of service (QoS) through bandwidth reservations. These reservations are further differentiated into segment reservations between ASes and end-to-end reservations between end hosts. After obtaining such a reservation (which can be processed efficiently through the use of DRKey), a sender is guaranteed the corresponding bandwidth for its traffic—irrespective of traffic sent by other hosts. COLIBRI requires minimal coordination between ASes and does not require per-flow state on routers.

- Particularly protected communication between direct neighbors can be used to bootstrap COLIBRI reservations and DRKey exchanges (Sections 11.4.2 and 11.5). This ensures that keys can be exchanged, and that segment reservations can be set up even during an ongoing DoS attack.

- By default, all SCION paths are public and can be used by any end host. However, SCION also supports hidden paths (§8.1), which are not registered at public path services but instead only shared with specific ASes or hosts. Through this mechanism, some critical connections can remain hidden and unwanted (DoS) traffic can already be filtered before it reaches the victim AS by its provider (who is likely capable of handling larger amounts of traffic).

- The EPIC system (§10.1) calculates per-packet message authentication codes (MACs) to ensure that using hidden paths does not leak any information to on-path entities who can observe packets.

11.3.3 Overview of Protection of Control-Plane Services

SCION relies on several crucial control-plane services for its operation: beacon services (to discover paths), certificate services (to manage keys and validate signatures), path services (to collect and disseminate path information), and COLIBRI services (to manage reservations). Similar to the Domain Name System (DNS) in today's Internet, these services are potential targets for DoS attacks and need to be properly protected. SCION applies multiple techniques to ensure their availability:

- Most of these services of an AS only need to be accessible for a small number of other ASes or the AS's customers. In this case, the service can whitelist the authorized ASes and ISDs and filter requests using source authentication provided by the SPAO. For example, beacon services should whitelist all neighboring ASes, while requests from other ASes are discarded.

- As a slightly weaker form of source authentication, packets can be filtered based on the SCION path in the packet header. The length of the traversed path contained in a packet indicates how many hops the source is away. For example, packets with an empty path originate from within the same AS, and packets with a path consisting of two hop fields come from neighboring ASes.

- In addition, services can employ rate limiting and history-based filtering to prevent flooding attacks from authorized but compromised or malicious sources. These filtering mechanisms are implemented in LightningFilter (§9.2).

- SCION's highly available control-plane PKI (CP-PKI) provides the keys to enable authentication of all control-plane messages.

11.3.4 Further Supporting Systems

- A potential attack on systems that use source authentication is an on-path replay attack, where an attacker intercepts a legitimate source-authenticated packet from an AS and forwards many copies of it (e.g., by compromising a router). The intent of the attacker is to exhaust allocated resources for the corresponding AS in case of rate limiting or COLIBRI reservations. To prevent packet replay, these defense systems rely on a system detecting and filtering out duplicated packets. Such a *replay-suppression* system is described in §9.1. For COLIBRI, this system needs to be deployed in each AS, e.g., integrated with the border routers. For end-to-end applications, LightningFilter includes a replay-suppression module.

- COLIBRI provides bandwidth shares to ASes or end hosts. To prevent them from exceeding these shares, the bandwidth needs to be monitored. However, due to very high data rates in the Internet core, efficient deterministic monitoring is infeasible. Instead, SCION uses LOFT (§9.3), a probabilistic traffic-monitoring system. As misbehavior can be attributed to the source based on packet authentication, it can be punished harshly (e.g., by blacklisting the source AS) such that even a small chance of detection is sufficient to discourage overuse.

11.3.5 Secure Path Discovery and Dissemination

A fundamental prerequisite for the availability of a network is that an end host
has access to authentic path information. An attacker could thus try to deny
an end host access to the necessary path information or provide incorrect path
information. To ensure that correct path information is discovered, SCION's
routing protocol signs and verifies all its messages using keys provided by
the CP-PKI. By defining a dedicated top-priority traffic class for control-plane
messages, an external attacker can never disrupt path discovery. SCION's concept of ISDs ensures that the routing protocol of one ISD (e.g., one country)
cannot be affected by any other ISD. The authentication of all control-plane
messages based on the CP-PKI prevents the injection of bogus messages and
thus routing hijacks.

After paths are discovered, they are registered at path services from where
they can subsequently be fetched by end hosts. This path lookup by an end
host is the first step of communication and is comparable to a DNS lookup that
is required for most connections in today's Internet. DoS defense is facilitated
by the strict access-control rules specifying which hosts or other servers are
allowed to contact which path service: Non-core path services are only accessible to local end hosts, core path services only to ASes in the same ISD or
other core ASes. Between path services, low-rate COLIBRI reservations can
be used to guarantee communication. In addition to standard techniques like
replication and caching, a path service can be further protected with Lightning-
Filter that applies source authentication, path-length filtering, and rate limiting.

11.3.6 Data Delivery

There are various ways in which an adversary can try to prevent traffic from
reaching the intended destination. We assume that all on-path ASes are benign,
as they could otherwise simply drop all packets. SCION's design choice to embed the forwarding path in packet headers rules out an important attack vector,
namely hijacks by off-path adversaries. Its multipath capabilities also allow
the source to circumvent congested links. However, these mechanisms only
provide relatively weak guarantees, in particular in the face of a determined
and powerful attacker. With access to a sufficiently large botnet, an adversary
can perform Coremelt [495] or Crossfire [279] attacks targeting network links
or overload the victim's access link.

Access links can be protected through the use of hidden paths, which ensure
that traffic from unauthorized sources is already filtered before it reaches this
bottleneck link. Delivery guarantees during DDoS attacks on network links
are provided by COLIBRI: Hosts can obtain bandwidth reservations to any
destination, assuming that all on-path ASes support COLIBRI and have some
remaining capacity. It is crucial to note that while this capacity may be temporarily used for best-effort traffic, it is reserved for COLIBRI as soon as there
is demand. A DoS attack based on best-effort traffic can therefore not cause

any harm to COLIBRI reservations. If no capacity is available, the source host may need to wait for several seconds; for important applications it is therefore necessary to set up reservations sufficiently early and to keep renewing them.

In addition to COLIBRI, the DRKey-bootstrapping protection framework provides two concepts (SNC and TRS) for bootstrapping COLIBRI reservations, described in §11.5.

Overall, we envision an ecosystem where the majority of traffic is sent over COLIBRI bandwidth reservations and only a relatively small amount is sent as best-effort traffic, see §10.2.8. This allows us to remediate the negative effects of congestion and congestion control. Furthermore, a reservation-based system enables more sophisticated routing policies and traffic engineering for ISPs.

11.3.7 Packet Authentication and Filtering

In many cases, the amount of traffic a victim can handle is not restricted by its access link, but by its firewall or by the server's computational resources. In today's Internet, it is easy to spoof the source address of packets, which renders simple firewall rules ineffective [173]. On the other hand, complex heuristics and deep packet inspection are computationally costly and make high-bandwidth firewalls very expensive. In a SCION ecosystem, DRKey provides symmetric cryptographic keys that enable the destination to authenticate the source of all packets at the network layer. With LightningFilter (§9.2), simple rules can define trusted sources and their allowed bandwidth. Authenticated packets from trusted sources can then be rate-limited and bypass the standard firewall, which thus does not need to process all packets. During a DoS attack, the traffic allowed by LightningFilter can always reach the destination even if the firewall is overloaded and needs to drop packets.

The keys provided by DRKey are not just useful in LightningFilter, but can also serve as pre-shared keys for existing network-layer security protocols like IPsec; in these systems key management and distribution are often the most difficult problem for practical deployment. This enables existing high-bandwidth firewalls based on IPsec to achieve similar properties to those of Lightning-Filter, assuming their resilience to flooding attacks is improved [494].

11.3.8 Bootstrapping of Defense Mechanisms

To bootstrap most defense mechanisms in SCION, an AS needs to successfully exchange an AS-level (first-level) DRKey with other ASes it wants to contact. This key can then subsequently be used to bootstrap other defense mechanisms such as COLIBRI, LightningFilter, etc. If an attacker is already present before the initial DRKey key exchange, the exchange might be challenging. However, an AS only needs to get a *single* packet through (in each direction) to successfully set up a first-level DRKey.

A simple approach to tackling this challenge is to continuously retry the first-level DRKey exchange until it is successful. Despite the presence of an attacker, it is likely that a packet will eventually reach the corresponding destination AS, given that SCION already fundamentally provides strong properties to evade volumetric attacks, such as path selection and isolation.

The DRKey-bootstrapping protection framework described in §11.5 ensures that first-level DRKeys can always be exchanged, and that COLIBRI SegRs can be established.

11.4 Traffic Prioritization

To achieve availability guarantees for priority data-plane traffic (e.g., COLIBRI) and control-plane messages, they must have a reserved bandwidth share both on SCION routers and on intra-domain network devices (routers and switches). We thus need to distinguish between individual traffic types, assigning each of them a priority or bandwidth share. For this purpose, all packets with a designated priority are required to be marked and processed with the corresponding priority both at border routers and internal routers of ASes. This ensures that a DoS attack using best-effort traffic cannot affect or disrupt communication in a higher-priority traffic class.

11.4.1 Traffic Classes

We envision the following traffic classes (in decreasing order of priority) that are assigned a certain percentage of the total available bandwidth:

- **Setup-Less Neighbor-Based Communication (SNC):** This class supports critical control-plane traffic between infrastructure services in neighboring ASes; for example, beaconing, setup of COLIBRI segment reservations, or keep-alive messages. SNC is described in more detail in §11.4.2. The bandwidth share of the SNC traffic class can be defined in accordance with the neighboring ASes. We expect a range of 1–2%.

- **COLIBRI traffic:** Traffic with bandwidth reservations should be forwarded with higher priority. According to our vision for the use of COLIBRI (§10.2.8), this will be the predominant traffic class in the long term with a share of 60–80%.

- **EPIC-HP traffic:** In order to guarantee that traffic protected by EPIC-HP is delivered even under DoS situations, this class should be forwarded with higher priority than the best-effort traffic.

- **Best-effort traffic:** Even in the long term, some bandwidth share must remain reserved for best-effort traffic. This is necessary for traffic where setting up reservations causes excessive overhead. Overall, we expect a bandwidth share of 20–40%.

It is crucial that (1) unused capacity of one traffic class is not wasted but made available to the other classes, and (2) lower-priority traffic classes are not threatened by starvation. These two properties can be achieved through the use of appropriate queuing disciplines on both SCION and AS-internal routers, which we discuss in §11.4.4.

11.4.2 Setup-Less Neighbor-Based Communication (SNC)

The Setup-Less Neighbor-Based Communication (SNC) is a traffic class dedicated to achieving highly available communication between neighboring ASes. Since the SNC bandwidth share between two neighbors is established when they first connect, this system does not require a reservation phase. The protected neighbor-to-neighbor communication is used as a highly-available local channel to protect the bootstrapping of control-plane services or other systems.

Crucially, SNC is used in §11.5 to protect the end-to-end establishment of DRKey, which is essential for the availability of COLIBRI and EPIC. SNC is also responsible for the protection and prioritization of beaconing (§11.4.1), and the setup of COLIBRI SegRs (§11.5.3). The protection of SNC traffic by means of queueing disciplines is further discussed in §11.4.4.

11.4.2.1 SNC Implementation

The idea behind SNC is to have a static reservation between two neighboring ASes, which exists without having to explicitly request or renew it. SNC is a contractual obligation of an AS to its neighbor to forward a certain amount of traffic destined to itself with strict priority. Its purpose is to bootstrap a basic level of quality of service (QoS) without already requiring any reservation, i.e., without the need of pre-distributed capabilities.

SNC Data Plane. To implement SNC from some AS A towards its neighboring AS B, AS A marks certain verified outgoing packets destined to AS B. The border routers of AS B that are responsible for the communication with AS A then limit the throughput of marked packets (AS A knows and adheres to this limit), and forward them prioritized to the destination host or infrastructure service within its network. An illustration is shown in Figure 11.3. Packet labeling can be realized without any further encapsulation, for example using the "traffic class" field in the SCION header (see §5.2), or similar fields in IPv4 and IPv6 (as discussed in §11.4.3).

Monitoring and Policing. The monitoring of SNC traffic at the border router of AS B can be implemented efficiently by means of the leaky-bucket or token-bucket algorithm [362]. A crucial point is that an AS is accountable to its neighbor for rate-limiting marked outgoing traffic, and any excess traffic is immediately dropped.

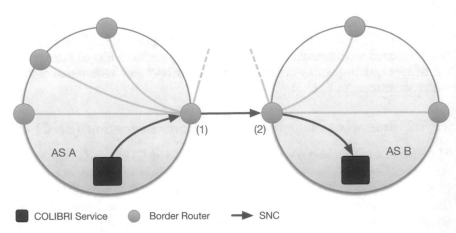

Figure 11.3: Illustration of setup-less neighbor-based communication (SNC)
protecting COLIBRI traffic. The grey lines denote traffic between border routers. To protect SNC from such traffic, border
router (1) forwards marked packets from the COLIBRI service
with strict priority, and border router (2) does the same but only
after policing the traffic. The COLIBRI service is illustrated as an
AS-internal service, but it could also be directly integrated into the
ingress or egress border router.

11.4.2.2 Security Analysis

Inter-Domain Attacks Against SNC. A border router usually handles a
multitude of inter-domain connections. If ingress traffic from more than one
connection is destined to the intra-AS network, congestion can arise at the
corresponding router interface. Due to strict prioritization, SNC packets are
guaranteed to be forwarded irrespective of the traffic pattern of other inter-
domain connections.

This might lead to ASes tagging all their traffic (destined to the intra-AS
network of its neighbors) with the high-priority flag, aiming at better delivery guarantees also for illegitimate best-effort traffic. However, because each
border router limits the throughput of SNC traffic for every neighbor, non-
compliant neighboring ASes cannot abuse this system and can be punished for
misbehavior.

Intra-Domain Attacks Against SNC. As the SNC traffic marker is po-
liced at the access switches—as discussed in §11.4.3—end hosts cannot modify or inject SNC traffic arbitrarily. Only infrastructure services and border
routers, which are controlled and trusted by the network operator, can mark
packets as SNC traffic.

11.4.3 Traffic Marking

For the purpose of marking traffic with different priorities, we suggest using the differentiated services code point (DSCP) contained in the header of IPv4 and IPv6 packets as part of the differentiated services field [389], as well as in the SCION common header (§5.2). Alternatively, MPLS also provides the traffic class field [27], which can be used for traffic prioritization within an AS. The field used to mark traffic for priority forwarding is called *traffic marker* in the following discussion. We distinguish between (1) the SCION traffic marker in the SCION common header, and (2) the intra-domain traffic marker in the IP or MPLS header.

The SCION traffic marker is only necessary to enable an AS to specifically mark outgoing packets as SNC traffic for its neighbor. It thus requires coordination between neighboring ASes to define the value for this traffic class in the SCION common header. All other cases only require intra-domain traffic markers, since they can be identified based on the path type and other header fields by SCION border routers. As the outer IP or MPLS headers are removed at the edges of ASes, their encodings do not require inter-AS coordination. This allows ASes to choose specific and distinguishable encodings for the various traffic classes, which can co-exist with other QoS systems.

To prevent end hosts or entities in other ASes from misusing priority queuing within ASes, the traffic marker of traffic originating from these nodes must be checked, set, or removed at strategic points in the network. For this purpose, we distinguish between nodes belonging to the AS infrastructure (e.g., AS routers and infrastructure services) and other entities (e.g., end hosts or routers of other ASes), see Figure 11.4. Whenever a packet crosses the boundary to the AS infrastructure, the traffic marker needs to be checked and cleared if needed.

For example, an end host could set the traffic marker arbitrarily to get prioritized processing within an AS. Thus, an AS needs to check the field at its network ingress (e.g., access-layer switches at points-of-presence for ISPs). Similarly, in the event that a border router receives a packet from a neighboring AS, it needs to correctly set the traffic marker. Additionally, in specific cases these checks must also consider the directionality of the traffic, because reservations in COLIBRI are unidirectional and packets traveling in the opposite direction of the reservation should be treated as best-effort traffic.

To summarize, the traffic markers must be checked at the following points in the network (see Figure 11.4):

- **Access-layer switches** connect end devices to the network of an AS. They need to reset the traffic marker except for packets destined for an infrastructure service or COLIBRI gateway (see below). For this purpose, an access-layer switch stores a list of these services and compares the destination address of a packet with this list. In addition, access-layer switches should also check that the source address in the IP header

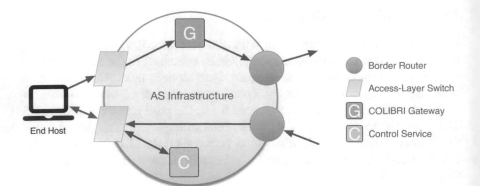

Figure 11.4: Entities involved in the checking of the traffic markers.

and the SCION header are the same and associated with the end host to prevent source-address spoofing (see §7.7). As an alternative to these checks, (untrusted) hosts can be placed into a separate AS.

- **COLIBRI gateways** handle all traffic that is sent on COLIBRI end-to-end reservations. They perform rate limiting for all reservations and then set the appropriate traffic marker for the COLIBRI class. Reservations in COLIBRI are unidirectional; on-path routers and the destination can send packets on the reverse path without involving a COLIBRI gateway, but they are treated as best-effort traffic and must not have a priority traffic marker.

- **Border routers** receive both outgoing traffic to neighbor ASes (egress traffic) and incoming traffic from these ASes (ingress traffic). For egress traffic, the border router simply removes the outer network-layer header and sets the SCION traffic marker only for SNC packets. For ingress traffic, the border router processes the SCION header and then adds an intra-domain header with the appropriate traffic marker based on the path type and the SCION traffic marker: Valid COLIBRI traffic on end-to-end reservations in forward direction is marked accordingly. Packets marked as SNC traffic in the SCION traffic marker obtain the corresponding intra-domain traffic marker if and only if they come from a direct neighbor and do not exceed that neighbor's guaranteed bandwidth.[4]

11.4.4 Priority Processing Using Queuing Disciplines

Differentiated treatment of different traffic types with their corresponding priority can be implemented in practice using *queueing disciplines* on legacy

---

[4]This rate limiting can be performed with a token-bucket algorithm similar to the deterministic traffic monitoring in COLIBRI (§10.2.4.11).

routers. By using this technique, packets are not simply forwarded in a first-in-first-out (FIFO) fashion, but, depending on their type, entered into one of several queues. There are different queuing disciplines that can be used to provide isolation between different traffic classes: rate-controlled priority queueing (PQ), (distributed) weighted fair queueing ((D)WFQ), or weighted round-robin (WRR) queueing [464].

For example, in DWFQ, which is implemented in many off-the-shelf routers (e.g., Cisco routers [120]), packets are assigned to different queues based on the differentiated services field. Each queue has a weight that dictates which share of the total bandwidth is assigned to the packets contained in it.

Strict priority queuing can be used as well but requires additional mechanisms that limit the maximum bandwidth of high-priority traffic classes to prevent the starvation of lower-priority traffic. This is possible for the traffic classes defined in §11.4.1: Control-plane traffic can be rate-limited at infrastructure services and border routers, and COLIBRI traffic is rate-limited at COLIBRI gateways and further monitored inside the network (§10.2.4.11). Access switches can also monitor and police traffic originating from end hosts.

11.5 Protected DRKey Bootstrapping

The availability of many systems in SCION relies on the availability of pre-shared DRKeys. In this section, we describe how SNC (§11.4.2) and COLIBRI SegRs (§10.2) can be combined to protect the bootstrapping of DRKey, thus fully removing all the remaining denial-of-DRKey attack surface.

Although this section is of theoretical importance, it has limited bearing on real-life deployments. In practice, as we argue in previous sections, the long-lasting validity of reservations and keys is sufficient to prevent attacks.

11.5.1 Assumptions

Adversary Model. We consider off-path adversaries, who otherwise have all capabilities described in §7.2. Concretely, they can modify, drop, and inject packets anywhere in the network, except for the path for which a legitimate source wants to create a reservation. The adversary could be an end host, router, a complete AS, or any other entity (or group of entities). The assumption that all the links and routers on the reservation path are not controlled by an adversary is unavoidable and comes from the observation that no availability properties can be provided in such a scenario: An on-path adversary can always maliciously delay or drop packets. The adversary's objective is to prevent the source from obtaining the requested reservations.

Network Requirements. As noted in §10.2, bandwidth-reservation protocols require the underlying network architecture to provide *path stability*:

Paths must not change frequently or unpredictably, as otherwise a reservation protocol cannot adhere to its guarantee. SCION provides path stability by design—unlike the current inter-domain routing protocol, BGP—and it is instantiated with the CP-PKI that can be used to bootstrap trust for key establishment protocols. The source AS discovered the path for which it wants to create a reservation through beaconing.

11.5.2 Overview

There are two main concepts behind our DRKey-bootstrapping protection framework: setup-less neighbor-based communication (SNC) and telescoped reservation setup (TRS).

SNC, which is described in detail in §11.4.2, refers to pre-established bandwidth reservations between two neighboring ASes, and enables bootstrapping of new COLIBRI reservations and extending existing ones. Assembling multiple such neighbor-based reservations yields a protected reservation setup channel for a longer path. This construction constitutes the core idea behind this DRKey-bootstrapping protection framework.

However, there is a fundamental difference between requesting a shared key with DRKey and requesting a COLIBRI SegRs: A DRKey request necessitates the interaction of only two entities—the requesting AS and the destination AS—while in a SegRs request all on-path ASes are involved. For this reason, we cannot assume that intermediate ASes will be able to check the authenticity of key requests, as we do for SegRs requests. This problem is solved with TRS, which is described in detail below. At a high level, the source AS alternatingly extends a partial reservation and fetches keys with the next AS through a relay service, until it has established a secure reservation to the destination.

11.5.3 Protecting COLIBRI SegR Setup

COLIBRI SegR requests and responses are protected by assembling a sequence of SNCs. The COLIBRI service of each AS is responsible for validating the authenticity of each request and its source, where unauthentic requests get dropped and authentic ones are extended with the granted reservation. Valid requests are then forwarded with SNC priority to the next AS on the SegR request path. A visualized description can be found in Figure 11.5.

Rate Limiting. The COLIBRI service performs rate limiting of setup requests based on ingress and egress ASes to avoid the overuse of SNC bandwidth. Specifically, an AS reserves—for each ingress-egress pair—a fair share of the egress link's SNC bandwidth, weighted by the ingress link's SNC bandwidth.

The border routers of an AS only need to enforce that the aggregate SNC traffic incoming from other ASes is less than the maximum SNC bandwidth

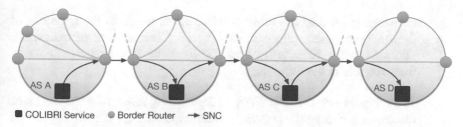

Figure 11.5: Assembling multiple SNCs. Every COLIBRI service checks the authenticity of the requests. Authentic requests are rate-limited, marked, and forwarded to the border router. The response packet is sent back following the same steps in the opposite direction.

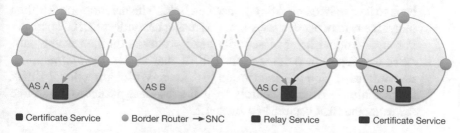

Figure 11.6: Illustration of steps 2 to 4 of one round of the telescoped reservation setup. Here, AS A wants to fetch a symmetric key from AS D, for which it sends a key request over the existing reservation (green line) to AS C, which relays the request protected by SNC to AS D. Step 5 of the TRS round is depicted in Figure 11.5.

of the ingress link. If it is higher, then the neighboring AS must be faulty or malicious, and the excess SNC traffic can be dropped.

11.5.4 Telescoped Reservation Setup (TRS)

Key exchange and reservation setup in our DRKey-bootstrapping protection framework consists of multiple rounds of alternating sub-path reservation setups and SNC extensions for the key exchange. It comprises the following six steps, which are depicted in Figure 11.6.

1. Based on SNC, the source AS (A) exchanges DRKeys and subsequently establishes a SegR with the second AS on the path. The source AS then creates a bidirectional end-to-end reservation (EER)—based on this SegR—with an endpoint co-located with the COLIBRI service, called *relay service*. The purpose of the relay service is to relay DRKey exchange requests.

2. Using the EER thus established (in the example shown in Figure 11.6, there already exists an EER to AS C), the source AS sends a key-

exchange request to the relay service, but destined to the neighbor of the AS hosting the relay service (AS *D*). The amount of bandwidth allocated to the EER is used to rate-limit the key-exchange requests (more on this in the following).

3. The relay service at the last AS of the existing reservation (*C*) receives the request, checks the request's validity, and relays it to the specified neighbor's (*D*'s) certificate service. It does this based on SNC, and therefore the request is guaranteed to arrive at the next AS, which verifies the request and prepares the key.

4. To return the key to the source AS, the last AS (*D*) sends a reply packet back to its neighboring AS (*C*) based on SNC. The neighboring AS then relays the packet back to the source AS using best-effort SCION packets. For an additional layer of security against attacks on the reverse path, ASes can deploy a modification to EER that allows for *bidirectional* reservations, which we describe in §10.2.4.9. With bidirectional EERs, the penultimate AS can securely relay the response packet to the source AS using the EER established in step 2.

5. Based on SNC, the source can create a SegR and subsequently an EER for the prolonged path (now including *D*), as now the source is in possession of all necessary DRKeys.

6. Repeat steps 2 to 5 until the source has received the keys of all on-path ASes and has established an EER to the destination AS.

Rate Limiting. The use of SNC bandwidth for TRS is rate-limited based on the size of the SegRs constructed between ASes. Whenever the relay service of an AS receives an EER setup request, it checks the bandwidth granted to the underlying SegR. The bandwidth granted to the EER is then computed as the fair share of the SNC bandwidth of the egress link, weighted by the bandwidth of the SegR. A minimum bandwidth guarantee, sufficient to relay one request, is always granted with every SegR.

The N-Tube (§10.2.5) algorithm—responsible for the computation of the bandwidth allocated to SegRs—guarantees that no group of malicious ASes can monopolize the available bandwidth. The same property then translates to the SNC bandwidth reserved for TRS. Thus, the fair use of SNC bandwidth can be efficiently enforced using the COLIBRI monitoring and policing system.

Multi-SegR TRS. EERs can be created over a composition of multiple SegRs. Non-core ASes, however, do not have control on core- and down-SegRs, and therefore they cannot run the TRS all the way to the destination. Nonetheless, this does not pose a problem in practice: Non-core ASes can reach core ASes with SegRs using TRS; core ASes then (1) have likely already established SegRs with the destination ISDs' cores, or (2) can set up new SegRs on behalf of the source AS. Even non-core ASes can thus indirectly create SegRs spanning multiple ISDs using TRS, and obtain the same protection guarantees.

One-RTT DRKey Exchange. For TRS to function optimally, the DRKey exchange has to be carried out in one round-trip time (RTT), i.e., using just one request and one response packet: A longer exchange needlessly congests the SNC channel. The DRKey exchange presented in §3.2.4.1—based on TLS and authenticated with the CP-PKI—is therefore suboptimal, as a TLS handshake with client authentication typically requires three RTTs to complete (including the TCP handshake).

As an alternative, we describe here a 1-RTT protocol for DRKey exchange, similar to the TLS 1.3 1-RTT handshake:

- The source sends a packet to the destination, containing a signed Diffie–Hellman public key.

- After authenticating the packet, the destination generates the other Diffie–Hellman public key, and uses the shared secret thus constructed to encrypt the DRKey for the source. Then, it sends back the encrypted DRKey and its public key, both signed.

- Finally, the source authenticates the response message, reconstructs the shared secret, and decrypts the DRKey.

This protocol provides perfect forward secrecy, and does not require the use of public-key cryptography for encryption.

11.5.5 Security Analysis

Attacks against the DRKey-Bootstrapping Protection Framework. Every step of the COLIBRI SegR and DRKey setup is guaranteed to succeed irrespective of any malicious interference. For the COLIBRI SegR setup, every request is protected by the assembled SNCs. For the key exchange, in every round of TRS the key-exchange request is first protected by the previously established EER, and then by SNC for the last inter-domain link. This means that every COLIBRI SegR and DRKey setup request will always reach its destination, as long as the SNC bandwidths are chosen high enough to support the reservation setup traffic.

With TRS, the packet is guaranteed to reach the penultimate AS through the EER, and from that point only needs to traverse one inter-domain link. This is repeated $n - 2$ times for an n-hop path, leading to a linear number of required packets. Note that this computation is for a highly unlikely worst-case scenario in which the source AS has no existing sub-path SegRs and no locally stored DRKeys for on-path ASes.

Sybil Attacks. As the N-Tube algorithm achieves botnet-size independence (§10.2.5), creating additional ASes or ISDs does not allow an adversary to reserve more bandwidth. Thus, Sybil attacks on the DRKey-bootstrapping protection framework are not possible.

Attacks on COLIBRI and SNC. The DRKey-bootstrapping protection framework composes COLIBRI reservations and iteratively extends them using SNC. In the previous paragraphs we have argued that this composition is indeed secure. Regarding the security of COLIBRI and SNC, further information can be found in the relevant sections (§10.2.6 and §11.4.2.2, respectively).

11.6 Protection of Control-Plane Services

The SCION control plane is based on the control service containing different modules such as the beacon service (BS), path service (PS), certificate service (CS), discovery service, and COLIBRI service. As these services are critical for the SCION infrastructure, they must be protected against attacks on their availability. In this section, we analyze the interactions for each of the services by identifying their supported request types, which are classified by criticality. We then specify what defense mechanisms are needed to prevent attacks and discuss how these mechanisms can be bootstrapped.

11.6.1 Criticality Criteria for Control-Plane Interactions

We identify queries to the above services with one or more of the following properties as critical for DoS attacks:

1. *External:* The query can be issued outside of the current AS, which allows the adversary to send requests from multiple different locations.

2. *Computation:* Requests to a service that require a large computation are problematic. For example, if an initial request towards a service contains a signature, this can potentially result in large computational overhead when the service receives invalid forged signatures.

These criteria do not include traffic *amplification*, i.e., requests that trigger a large response from a service, which could be abused to mount reflection-based amplification attacks. As the connections to the control services are based on stateful transport protocols, which prevent reflection of response messages, the control services cannot be leveraged for traffic amplification attacks.

Table 11.2: Requests received by a beacon service.

		Criticality		Filter Configuration			
Request type	Source	Ext.	Comp.	Whitelist	Path len.	Rate limiting	Src. Auth.
Non-core							
PCBs	up neigh. BS	✓	✓	up neigh.	neighbors	interval [AS]	external
Int. TRC / cert.	local PS			infra	empty	–	–
Ext. TRC / cert.	down neigh. BS / core PS in local ISD	✓		ISD	–	profile [AS]	external
Core							
PCBs	core neigh. BS	✓	✓	core neigh.	neighbors	interval [AS]	external
Int. TRC / cert.	local PS			infra	empty	–	–
Ext. TRC / cert.	neigh. BS	✓		neighbors	neighbors	profile [AS]	external

In the following, we analyze requests to control plane services and classify them as critical or not based on the properties mentioned above. An overview of the requests and their criticality is provided in Tables 11.2–11.6.

Beacon Service. The beacon service (BS) receives path-segment construction beacons (PCBs) and queries for TRCs or certificates.

Inter-ISD/intra-ISD PCBs: BSes in both core and non-core ASes receive either inter-ISD or intra-ISD PCBs, which contain a signature for authenticity. BSes in non-core ASes receive PCBs from BSes of upstream ASes, whereas BSes in core ASes receive PCBs from neighboring core ASes potentially located in a different ISD.

TRC / certificate query: After passing a PCB to a neighboring downstream or core AS, the beacon service might receive a request for a missing TRC or certificate from the beacon service in the neighboring AS. Similarly, when the beacon service registers paths at either the local path service or a core path service, these path servers can ask for any missing TRCs or certificates.

Path Service. The path service (PS) receives path lookup requests, path registrations, path revocations, and TRC and certificate queries.

Path lookup: Non-core PSes receive path lookup requests from any host in their local AS. Core path services answer path lookup requests from clients of their local AS and requests from any other path services of core and non-core ASes, possibly located in a remote ISD.

Path registration: After converting PCBs to path segments, beacon services register them at their local path service with a signed path registration request. Path registration at core path services is done by downstream beacon services of the local ISD and the core beacon service within the same AS. To ensure the authenticity of the requests, they contain a signature.

Table 11.3: Requests received by a path service.

	Request type	Source	Ext.	Comp.	Whitelist	Path len.	Rate limiting	Src. Auth.
			Criticality		Filter Configuration			
Non-core	Path lookup	host in local AS			–	empty	interval [IP]	–
	Path reg.	local BS		✓	infra	empty	–	–
	Path revo.	local BS		✓	infra	empty	–	–
	Int. TRC / cert.	host in local AS			–	empty	interval [IP]	–
	Ext. TRC / cert.	core PS	✓		–	segment	profile [AS]	external
Core	Int. path lookup	host in local AS			–	empty	interval [IP]	–
	Ext. path lookup	any PS	✓		–	–	profile [AS]	external
	Int. path reg.	local BS		✓	infra	empty	–	–
	Ext. path reg.	down BS	✓	✓	ISD	segment	profile [AS]	external
	Int. path revo.	local BS		✓	infra	empty	–	–
	Ext. path revo.	down BS or PS / core PS in same ISD	✓	✓	ISD	segment	profile [AS]	external
	Int. TRC / cert.	host in local AS			–	empty	interval [IP]	–
	Ext. TRC / cert.	down or core PS	✓		–	segment	profile[AS]	external

Path revocation: All path services receive signed path revocation messages from their local beacon services. Core path services also receive them from downstream beacon services, downstream path services and neighboring core path services.

TRC / certificate query: Non-core path services get TRC and certificate requests from clients in their local AS, if they miss a TRC or a certificate to verify the path segments they received from the path service. Additionally, non-core PSes receive those queries for the same reason from any core path service that has requested down-segments to the local AS. TRC and certificate requests for core path services can come from the same entities that send path lookup requests: clients of the local AS, downstream path services, and any core path service.

Certificate Service. The certificate service (CS) receives requests for TRCs and certificates, requests corresponding to the DRKey system (both first and second level), and requests related to the CP-PKI.

TRC / certificate query: Certificate services answer TRC or certificate requests from anyone in their AS.

First-level DRKey: As the DRKey values change over time, the keys need to be periodically re-fetched for every AS for which communication is authenticated with DRKey. Certificate services thus receive first-level DRKey requests from any AS in the Internet. Those requests are only issued once per epoch, where an epoch is set to 24 hours in the current SCION implementation. If 10,000 ASes ask for a first-level DRKey, there may be one request every few seconds. To ensure authenticity, these requests are signed.

Table 11.4: Requests received by a certificate service.

	Request type	Source	Criticality		Filter Configuration			
			Ext.	Comp.	Whitelist	Path len.	Rate limiting	Src. Auth.
All	TRC / cert.	host in local AS			–	empty	interval [IP]	–
	First-level DRKey	any CS	✓	✓	–	–	interval [AS]	(external)
	Second-level DRKey	host in local AS			–	empty	interval [IP]	–
	TRC / cert. update	CS in same ISD			ISD	–	interval [AS]	external

Table 11.5: Requests received by a discovery service.

	Request type	Source	Criticality		Filter Configuration			
			Ext.	Comp.	Whitelist	Path len.	Rate limiting	Src. Auth.
	Service reg.	service in local AS			infra	empty	–	–
All	Health info.	service in local AS			infra	empty	–	–
	Service disco.	anyone	✓		–	–	interval [IP]	external

Second-level DRKey: When hosts of an AS need a second-level DRKey, they request it from the local certificate service.

Certificate re-issuance request: Non-core AS certificates are issued by core ASes of their ISD, but are only valid for a limited amount of time. Before their certificate expires, non-core ASes ask the core AS that issued their certificate to re-issue it for the next time period.

Discovery Service. The discovery service keeps track of all the services in an AS (see §4.6.2). Service discovery is another crucial functionality for the SCION control plane interactions and thus also the discovery service needs to be highly available. As a fallback mechanism in case the discovery service is not available, entities can use service addresses to reach control services in other ASes.

Service registration: Service instances within an AS register their communication endpoints for a specific service address with the discovery service.

Health-check information: The discovery service checks the healthiness of registered service instances through a health-check endpoint. To that end, the service instances send health-check information to the discovery service.

Service discovery: The discovery service receives requests for communication endpoints of service instances from local or remote entities. The reply sent by the discovery service normally fits into a single datagram.

COLIBRI Service. The task of the COLIBRI service (CServ) is to manage SegRs and EERs, and to control the admission procedure (for more information, see §10.2.4.10). All requests are source-authenticated with DRKey. This prevents malicious ASes from sending too much control traffic, at the expense of being blacklisted. Furthermore, all requests (except new SegR setups) travel inside existing reservations with defined portions of bandwidth dedicated to control traffic. This prevents the CServ from receiving more requests than the on-path ASes can handle.

SegR new setup: The CServ requests new SegRs using best-effort traffic.

SegR renewal: Renewals of existing SegRs are transported on top of existing SegRs. Since SegRs define a control traffic split, the allowed rate is already established by all on-path ASes.

SegR teardown: This request erases all indices for this SegR and removes the SegR itself. The request travels inside the existing SegR.

SegR index confirmation and activation: These requests set an index state from temporary to confirmed and from confirmed to active. The requests may travel inside an existing SegR or with best effort, depending on whether the temporary index was created from a renewal or a completely new SegR setup.

SegR index cleanup: Removes an existing index for a SegR. The request travels inside an existing SegR.

List SegRs: This request returns the existing SegRs registered in this CServ. It is used by a source CServ to obtain candidate SegRs for the creation of a new EER.

EER setup or renewal: This request creates a new EER or renews an existing one. It travels inside the SegRs that will be stitched to the EER.

EER index cleanup: Removes an existing EER index. The request travels inside the SegRs that are stitched to that EER.

End-host initiated list SegRs: The request is sent by an end host to a CServ in its AS.

End-host initiated EER setup or renewal: The request is sent by an end host to a CServ in its AS.

End-host initiated EER index cleanup: The request is sent by an end host to a CServ in its AS.

Table 11.6: Requests received by a COLIBRI service.

	Request type	Source	Criticality		Filter Configuration			
			Ext.	Comp.	Whitelist	Path len.	Rate limiting	Src. Auth.
All	SegR new setup	CServ	✓		neighbors		profile [AS]	external
	SegR renewal	CServ	✓		COLIBRI only		–	external
	SegR teardown	CServ	✓		COLIBRI only		–	external
	SegR idx confirm & act	CServ	✓		neighbors		profile [AS]	external
	SegR index cleanup	CServ	✓		COLIBRI only		–	external
	List SegRs	CServ	✓		neighbors		profile [AS]	external
	EER setup/renewal	CServ	✓		COLIBRI only		–	external
	EER index cleanup	CServ	✓		COLIBRI only		–	external
	Trigger list SegRs	host in local AS		–		empty	interval [IP]	–
	Trigger EER setup/renew	host in local AS		–		empty	interval [IP]	–
	Trigger EER idx cleanup	host in local AS		–		empty	interval [IP]	–
	Manage end-host admission	host in local AS		–		empty	interval [IP]	–

Manage end-host admission: This request allows the CServ to define which EER setup requests will be approved on behalf of the end host sending the "manage end-host admission" request. The request defines the EER setup or renewal requests concerned, and is AS-internal.

11.6.2 Filtering at the Control Service

In previous sections, we identified critical request types for each module hosted by the control service and described different filtering mechanisms to prevent targeting its availability. In the following, we discuss how these mechanisms are employed and configured to prevent misuse of critical request types.

ISD / AS whitelisting only allows requests from certain categories to pass the filter. All other requests are dropped by default. For the whitelisting categories, we distinguish between (1) *infra* which only accepts requests from infrastructure nodes of the local AS; (2) *ISD* which only accepts requests that originate within the local ISD; and (3) *up/core/– neighbors* which only accepts requests from neighboring ASes that are located either upstream, connected via a core link, or a regular neighboring AS.

Path length filtering filters packets based on the length of the traversed path contained in the packet header. The module filters based on the following categories: (1) *empty* that checks for an empty path (e.g., AS internal traffic); (2) *neighbors* that only accepts requests from one-hop neighbors; and (3) *segment* that filters requests based on the number of path segments (e.g., to filter requests from external ISDs).

Rate limiting allows clients to issue a fixed number of queries in a given time frame. This number is either (1) fixed per *interval*, or (2) based on a traffic *profile*, and can be employed on various scopes (IP address, AS, or ISD identifiers).

Source authentication only lets authenticated requests that contain a valid SPAO pass the filtering module. The module has the following two configurations: (1) *internal* that only accepts AS-internal traffic if it is authenticated; and (2) *external* that only accepts traffic originating outside of the local AS if it is authenticated.

11.6.3 Attack Resilience of the Control Service

Rate Limiting of Control-Plane Traffic on the BR.   Control-plane traffic is forwarded with priority on BRs and legacy switches based on traffic marking and queuing disciplines. To avoid flooding attacks on the control service using control-plane traffic, BRs must employ rate limiting for control-plane traffic that originates in a neighboring AS and is destined to a control service within their AS. To distinguish control-plane traffic from regular traffic, the traffic class field in the SCION header is used. Furthermore, traffic filtering based on the path type and path length is employed to identify traffic from neighboring ASes.

Rate limiting in combination with the separate queue on the BR for control-plane traffic prevents flooding attacks on the control service. At the same time, it makes fundamental control plane mechanisms possible, such as beaconing, service discovery, and certificate and TRC requests from neighbors.

Replication of the Control Service.   As simple but powerful mechanism to increase the reliability, fault tolerance, and availability of a service, is to replicate the service within the AS. Replication is a well-established technique in today's Internet for systems such as DNS. Using SCION's isolation properties and service discovery mechanism, replicated instances can be made available only to specific groups of end hosts. For example, an AS could set up path server replicas such that one replica handles all requests originating from outside the AS, whereas another replica can only be reached by end hosts within an AS. Thus, even if the PS replica handling external requests is targeted by an attack, the other replica will remain available.

11.7 AS Certification

SCION aims to make operation and configuration as simple as possible. While many of the systems and defense mechanisms may themselves be complex, this complexity is hidden from network administrators and end users.[5] Internet service providers (ISPs) and other ASes can largely reuse their existing physical infrastructure and intra-domain networking protocols when deploying SCION.

---

[5]This is similar to cryptography: While cryptographic mechanisms are extremely complicated, no understanding is required to open an encrypted connection to a web server via Transport Layer Security (TLS).

In contrast to today's Internet with only around 75,000 ASes, SCION has support for and benefits from a larger number (potentially millions) of ASes. We expect that companies will operate their own ASes and thus be less dependent on their ISPs. Still, to fully benefit from the more powerful SCION extensions and to obtain security and availability guarantees comparable to dedicated lines, a company relies on some additional support from its service provider:

- An ISP needs to put in place a system to isolate different *traffic classes* in its internal network, e.g., by using queuing disciplines available on standard routers.

- To benefit from the bandwidth guarantees of COLIBRI, each AS on the path between the intended communicating parties needs to support the system. This means that all these networks need to enable COLIBRI on their border routers, deploy the COLIBRI service, and analyze their own networks to provide meaningful and accurate bandwidth guarantees in their own network.

To standardize these support efforts, we envision that ASes can obtain certifications that reflect their availability guarantees provided to customers and peers.

11.7.1 Certification Overview

Bronze. ASes that obtain a Bronze certification guarantee to provide defense mechanisms against basic DoS attacks. For this purpose, the AS deploys basic DoS filtering mechanisms that include traffic prioritization on BRs and legacy routers, as well as basic filtering on the control service for all potential request types (see §11.6.2). In addition to this, ASes and their end hosts can employ hidden paths (see §8.1) as well as EPIC-HP and -SAPV (see §10.1) to defend themselves against malicious attacks.

Silver. Silver-certified ASes provide advanced DDoS filtering by deploying the same defense mechanisms as Bronze, but also secure the control service and discovery service by deploying LightningFilter in front of them.

Gold. ASes with Gold certification also protect against volumetric DDoS attacks and provide a guaranteed bootstrapping of the defense mechanisms by deploying COLIBRI, SNC, and TRS. COLIBRI-related traffic gets priority over standard traffic on SCION BRs and legacy routers.

11.7.2 Level 1: Bronze

ASes with the Bronze certification make use of defense mechanisms against basic DoS attacks that have moderate deployment effort:

1. *Traffic prioritization* on border routers and legacy routers (§11.4). This ensures that DoS attacks using legacy traffic cannot affect or disrupt communication of the SCION control- and data-plane.

2. *Filtering on the Control Service* for all potential request types (§11.6 and §11.6.2). The filtering mechanisms include whitelisting, rate limiting, path-length-based filtering, and source-authentication-based filtering.

Attack Examples Prevented by the Bronze Certification. The following attack examples are prevented by the defense mechanisms in the Bronze certification.

- **Exhausting a switch buffer using legacy (i.e., non-SCION) traffic.** This is not possible because of traffic prioritization and queueing disciplines at border routers and legacy routers. Thus, SCION traffic enjoys priority over legacy traffic.

- **Injecting PCBs from non-neighboring ASes**. PCBs are only accepted from whitelisted neighbors and path-length-based filtering is employed at the control service.

- **Reflection attacks using TRC requests**. Source address spoofing in SCION is thwarted using DRKey-based authentication, preventing reflection attacks for authenticated requests. In addition, the control service employs rate limiting to prevent clients from issuing excessive amounts of requests.

11.7.3 Level 2: Silver

Silver-certified ASes provide advanced DoS filtering by deploying the same defense mechanisms as Bronze, but also secure the control service and other essential control-plane services with LightningFilter. The deployment of LightningFilter in front of the control service drastically reduces its filtering effort as source authentication, duplicate filtering, and traffic profile-based filtering will have already been performed by LightningFilter. However, since most end hosts will use an end-to-end encrypted connection to the control service (e.g., using QUIC), the LightningFilter is not able to see the type and content of requests. Thus, request-type-specific filtering still needs to be performed on the control service.

The combination of LightningFilter and filtering on the control service drastically limits the attack possibilities, and thus prevents the control-plane services from being overwhelmed by excessive amounts of requests.

Attack Examples Prevented by the Silver Certification.

- **Request flooding on control service.** LightningFilter uses a per-AS resource allocation based on traffic profiles. This guarantees access to the control service for legitimate users's requests that are within their traffic profile.

11.7.4 Level 3: Gold

The Gold certification builds on the Silver certification, but also enables protection against volumetric attacks, and guarantees bootstrapping of the defense mechanisms. This requires the deployment of hop-by-hop protocols such as COLIBRI, which provides global quality of service through bandwidth reservations. After obtaining such a reservation, a sender is guaranteed the corresponding bandwidth for its traffic and remains unaffected by traffic sent by other hosts. The COLIBRI system is complemented with LOFT to monitor misbehavior in regards to the bandwidth reservations. Furthermore, SNC and TRS are required in order to obtain protection for the reservation setup and the key exchange.

Attack Examples Prevented by the Gold Certification.

- **Volumetric attacks.** Regardless of the size of an attack, COLIBRI traffic will always receive the guaranteed bandwidth share.

11.7.5 Interaction between Different AS Certifications

AS certifications of the ASes on a path will likely be heterogeneous, and ASes will need to interact with other ASes with different certifications. In this case, the availability guarantees achieved will be those of the lowest certification level in the set of involved ASes. In addition, hop-by-hop protocols such as EPIC and COLIBRI require support from all on-path ASes. As a consequence, we envision that ASes employ beaconing strategies that attempt to construct path segments with uniform certifications, ensuring the highest level of availability guarantees possible along the path segment.

11.8 Security Discussion

This section analyzes the potential damage that can be caused by an attacker under the assumption of the threat model described in §11.1.1. We assume that the bootstrapping of the defense mechanisms is completed, which is reasonable because a cold boot of all the defense mechanisms is extremely unlikely. In the event that the entire system is forced to cold boot, it would be unrealistic for an attacker to be able to exhaust all available resources at this point. Furthermore, affected ASes with a Gold certification can use additional mechanisms to protect their DRKey bootstrapping, which is a further guarantee for a successful boot (for more details, see §11.5).

11.8.1 Availability Properties for the Control Plane

We now discuss how our defense mechanisms ensure availability guarantees for the three fundamental control-plane processes that enable communication on the data plane.

Path Segment Discovery.   BRs process control-plane traffic with a higher priority compared to regular traffic and employ rate limiting for control-plane traffic that originates from their neighbors and is destined to a control service. This prevents flooding of the control service using control-plane traffic and ensures that path discovery using beaconing is possible even in case of an ongoing attack.

Path Segment Registration.   After a path segment has been discovered, it must be registered upstream at the PS of the core AS. Consequently, a BS in a non-core AS must be able to communicate to a path server located in a core AS within the same ISD. For ASes with a Gold certification, this is ensured by maintaining COLIBRI segment reservations, which can be used to bootstrap a low-rate end-to-end reservation to register the path segment at the core PS. Segment reservations in COLIBRI can even be established before the corresponding path segment is registered at the core path server. Bronze- and Silver-certified ASes can rely on EPIC-SAPV to protect path segment registration from DDoS attacks based on standard SCION traffic or packet replay.

Path Segment Fetching.   A PS in a non-core AS must be able to fetch path segments from (1) the PS in the core AS within the same ISD (for core-segments and down-segments in the same ISD), and (2) the core PS in a core AS in a different ISD (for down-segments). For ASes with a Bronze or Silver certification, this communication can be protected by using EPIC-SAPV. ASes that have obtained Gold certification can additionally work with COLIBRI reservations. Similar to path segment registration, a single or a combination of segment reservations can be used to bootstrap low-rate end-to-end reservations. Specifically, the second case occurs if an up-segment reservation and a core-segment reservation need to be combined to obtain a guaranteed connectivity to the core PS in a different ISD.

11.8.2 Resilience to Common Attacks

In addition to the availability properties for the control plane, we discuss how our defense mechanisms prevent common DoS attacks.

Control-Plane Attacks. An attacker can try to provide incorrect information or prevent legitimate users from accessing control-plane services. In §11.6, we analyzed every possible request to the control service and classified them according to their criticality in regards to DoS attacks. For each of the requests, we employ a set of filtering mechanisms to prevent an attacker from misusing them to bring down a control service or misuse the control service to attack other targets (e.g., using a reflection attack). The configuration of the filters is further discussed in §11.6.

Volumetric Attacks. As a specific instance of this attack class, an adversary might be in control of a botnet distributed over multiple ASes. The bots can be used to issue a large amount of requests towards a target, such that a network link towards the target becomes congested.

By design, SCION provides strong properties to evade volumetric attacks, such as path selection and isolation. In addition, ASes that are certified with the Gold standard have a guaranteed defense against such distributed DoS attacks. For example, COLIBRI enables end hosts to create end-to-end reservations with guaranteed bandwidth, which is unaffected by volumetric attacks due to COLIBRI's botnet-size independence property[6].

End-System Attacks. In this type of attack, an adversary attempts to exhaust the (computational or memory) resources of servers to limit the processing of legitimate requests. Typical attack targets include web servers, VPN endpoints, and firewalls. For example, computational resources could be exhausted using request flooding [560], whereas memory could be exhausted using attacks such as slowloris [98].

For the analysis of end-system attacks in SCION, we distinguish between attacks on the AS infrastructure and attacks on other end-systems such as regular end hosts. Entities are considered to belong to the AS infrastructure if they are under administrative control of the AS. This includes border routers, control-plane elements such as the discovery and control services, and legacy switches. Attacks and defenses in the control plane have been discussed above. While BRs are reachable from any ASes and thus can easily be targeted, several mechanisms are used to reduce their attack surface. For example, BRs do not need to maintain any per-flow state, but perform their actions based on packet-carried forwarding state.

Regular end systems can defend against DoS attacks by utilizing the packet header information to filter out attack traffic more precisely, or by deploying hidden paths to ensure strict access control. Moreover, congested paths can be circumvented using path selection.

---

[6]The minimum bandwidth guarantee that can be obtained is independent of the size of the botnet.

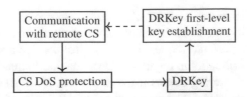

Figure 11.7: An excerpt from the extended dependency graph with a periodic
           dependency, indicated by a dashed arrow. For the purpose of this
           example, we use a slightly simplified representation that does not
           distinguish between different types of communication.

11.8.3 Full Dependency Analysis

For a more formal treatment of the interplay between the various components
that compose availability guarantees, we present an extension to the dependency analysis methodology introduced in §6.1, which focuses on the functional properties provided by the basic components of SCION. We augment
this graph with the additional systems enabling availability guarantees, whose
interdependencies are shown in Figure 11.2. At first sight, it seems impossible to prevent dependency cycles in the extended graph without unrealistic
assumptions: Figure 11.7 shows such an instance.

In this graph, we use a dependency to express that a certificate service can
only be reliably reached if its DoS protection is operational. Otherwise, an
adversary may be able to deny communication. However, advanced DoS protections for the control service include source authentication and therefore rely
on the DRKey system, which in turn requires communication with remote certificate services to establish its first-level keys. Crucially, we have described
in §11.3.8 that this step relies only on a single packet to be sent and is an infrequent operation. We call such a dependency *periodic* and indicate it in the
graph using a dashed line: A system u depends periodically on a system v if u
can operate for prolonged periods of time, even while v is not operational. This
notion is applied in the analysis to similar concepts such as path registration.

The property to verify for the extended dependency graph is that there exists
no cycle of dependencies consisting only of full non-periodic dependencies.
We have shown this property using a simple modified graph search algorithm
and can therefore conclude that even when considering the full set of DoS
defense systems under adversarial conditions, SCION has no problematic dependency loops.

Part IV

SCION in the Real World

12 Host Structure

Sergiu Costea, Matthias Frei, Marten Gartner, David Hausheer, Thorben Krüger, Jordi Subira Nieto, François Wirz*

Chapter Contents

12.1 **Host Components** **303**

   12.1.1 SCION Dispatcher 304

   12.1.2 SCION Daemon 305

   12.1.3 Application Networking Library (snet) 306

   12.1.4 SCION on Android 307

12.2 **Future Approaches** **307**

   12.2.1 Performance-Aware Path Choice 308

   12.2.2 Nesquic 309

   12.2.3 TAPS API 310

   12.2.4 Happy Eyeballs with SCION 310

   12.2.5 SCION Browser Extension 311

   12.2.6 Dispatcher Replacement 312

   12.2.7 SCION Protocol Number 314

In this chapter, we discuss how SCION-enabled end hosts can benefit from SCION properties. For native communication over SCION to be possible, hosts are expected to have a few software components installed and configured. Nevertheless, as we show in Chapter 13, there are mechanisms to ensure that even legacy (i.e., non-updated) hosts can benefit from SCION.

We first introduce how SCION-enabled hosts can communicate over SCION today (§12.1), then we provide an outlook on how we envision SCION-based communication in the future (§12.2).

12.1 Host Components

We introduce the main host software components that enable applications to natively communicate via SCION.

---

*This chapter reuses content from Lee, Perrig, and Szalachowski [322].

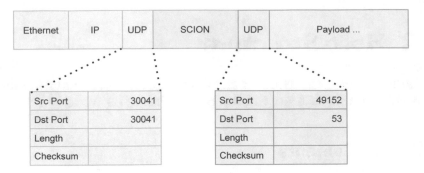

Figure 12.1: The packet layout of a SCION packet containing a UDP payload,
           transported over UDP/IP/Ethernet. This is how SCION packets
           are communicated within an AS. The UDP/IP underlay (yellow)
           shows the fixed port 30041.

The central component of the host structure is the dispatcher (§12.1.1),
which handles all incoming and outgoing SCION packets. The SCION daemon (§12.1.2) handles control-plane messages (e.g., to fetch paths to remote
ASes). Furthermore, to encourage developers to write SCION applications,
the snet library (§12.1.3) provides a simple interface for sending and receiving
SCION packets. Binary packages are available[1] for the main components, so
that SCION can be easily installed with a simple package manager.

We then present the SCION app for Android (§12.1.4), which demonstrates
the portability of software components thanks to their implementation in Go.

12.1.1 SCION Dispatcher

In the current implementation of SCION, we use a UDP/IP underlay to communicate between nodes within an AS: All SCION packets, from end hosts
to routers, routers to end hosts, and between two end hosts of the same AS,
are communicated through a designated UDP port (30041). Figure 12.1 shows
the layout of a SCION packet communicated via UDP/IP as the underlying
network.

The central component of the SCION host stack is the Dispatcher, a single
process within each host that handles all SCION packets on this designated
UDP port. Its main role is to receive incoming packets and deliver them to
individual SCION applications. This setup is shown in Figure 12.2.

When applications on the local system want to send or receive traffic on
a UDP/SCION address, they register the desired local UDP/SCION address
with the dispatcher. The dispatcher maintains a table of all registrations, which
it uses to look up where to forward traffic whose destination contains a local

---

[1]The SCION Debian packages are available here: `https://packages.netsec.inf.ethz.`
`ch/debian/`

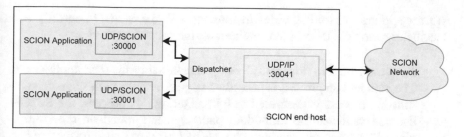

Figure 12.2: SCION Dispatcher overview.

address. Both the registration and future data traffic between the application and the dispatcher go through a UNIX domain socket. Once registered with the dispatcher, applications can start sending and receiving UDP/SCION traffic.

When sending packets, the dispatcher will take care of encapsulating the SCION packet with a UDP/IP header and then send it out on the network towards the destination. The destination is either a SCION Router or a SCION-enabled host in the local AS.

When receiving packets, the dispatcher consults the registration address table to determine to which application the packet is destined, and then delivers the packet to the application. For example, in the case shown in Figure 12.2, if a SCION packet with destination port 30000 arrives, it is sent to the application represented by the upper box in the figure.

When an application closes one of its connections, it closes the UNIX domain socket; the dispatcher removes the associated registration from its tables.

In addition to forwarding, the dispatcher handles some SCMP-specific tasks:

- If an SCMP packet received from the network is for a local application (e.g., a ping reply), it will select the entry in the registration table based on information in the SCMP packet and deliver it. On the application side, snet (§12.1.3) takes care of the packet depending on application policy.

- If an SCMP packet received from the network is for the local host (e.g., a remote system is sending a ping to the local system), then the dispatcher processes the packet itself and, if needed, replies to the sender.

The Dispatcher code is executed in userspace. We envision that in the future, the functionality of the dispatcher will be executed inside a kernel module or as an eBPF program, as discussed in §12.2.6.

12.1.2 SCION Daemon

The SCION daemon is a background process running on end hosts with the goals of (1) handling SCION control-plane messages, and (2) providing an

API for applications and libraries to interact with the SCION control plane. Specifically, the SCION daemon implements the following services:

- **Path lookup:** Provides path lookup functionality for host applications. The path lookup process is described in §4.5, and the path segment combination process is described in §5.5. During this process, the SCION daemon validates the individual path segments based on the control-plane PKI. The path segments are cached here, so that information can be shared across all SCION applications running on a host.

- **Topology information:** Provides information about the topology of the local AS. Topology information includes addresses of border routers (with their interface identifiers) and information on running services (e.g., RHINE (§19.3) or path servers).

- **Extensions:** Various SCION extensions and sub-protocols, such as COLIBRI (§10.2) and EPIC (§10.1), implement their control plane as part of the SCION daemon and extend its API.

12.1.3 Application Networking Library (snet)

Snet is a Go library[2] that hides away the complexity of the underlying SCION stack behind a simple interface for sending and receiving network traffic. The library achieves this by implementing the Go standard library interfaces for network connections: `net.PacketConn` and `net.Conn`. With snet, developers can write SCION network applications quickly, as they only need to focus on the high-level functionality of their application. The compatibility with Go standard library interfaces also enables developers to use SCION network connections in many third-party frameworks such as servers or RPC frameworks. It is also possible to reuse existing networking libraries for SCION. For example, the squic sub-package of snet provides an implementation of QUIC over UDP/SCION by integrating the quic-go library [353].

Currently, the library includes full support for UDP/SCION communication and several utility functions for interacting with SCMP and the SCION daemon. Critical to UDP/SCION support is the SCION dispatcher, the local process that handles the forwarding of data between SCION hosts, described in §12.1.1.

We illustrate snet operations by taking a look at the lifetime of a snet UDP/SCION connection. When an application creates a new UDP/SCION connection object, snet contacts the SCION dispatcher on the local system and informs it of the application's desired local address-port pair. The dispatcher registers this pair, and will later use it to deliver SCION packets.

---

[2]The documentation for snet can be found at `https://pkg.go.dev/github.com/scionproto/scion/go/lib/snet`

If the registration is successful, snet gives the application a handle to the new connection. The application can now use this handle to send or receive packets. Whenever snet needs to send data on the SCION network, it serializes the payload and prepends the necessary headers. It then sends the resulting packet to the SCION dispatcher for forwarding. The dispatcher takes care of sending the packets on the underlying network, and also forwards back return traffic based on the registered local address.

The application can close the connection once it is no longer needed. At this point, snet frees up any resources associated with the connection (e.g., by removing any active registrations from the dispatcher).

In addition to basic packet forwarding, snet also includes support for SCMP, thus allowing applications to explore the network or react to SCION network events. For example, applications can send pings to remote SCION hosts, or can react to SCMP error messages about on-path interfaces being temporarily down. Snet, in fact, does not react directly to a failed path, rather it expects the application on top to change path. We later present Nesquic (§12.2.2), a library that automatically handles network events.

12.1.4 SCION on Android

To demonstrate the portability of the SCION software components, we have developed a SCION app for Android,[3] which enables running an entire SCION AS on an Android smartphone. The SCION Android app makes the setup and configuration of a SCION AS on Android effortless. In its first release, the app supports pinging other SCION ASes and reading data from a SCION sensor server.

The latest app contains the SCION source code compiled into a binary that can be executed on Android. Every SCION component (dispatcher, daemon, etc.) runs in a separate process, which communicates with the SCION Android app via sockets, as is done on other platforms. This approach scales sufficiently well to run an entire SCION AS on an Android device, without the need to depend on an existing AS.

In the future, we anticipate that the SCION host components could be entirely integrated into the kernel.

12.2 Future Approaches

In the following, we present some approaches currently under development, which show how we envision the use of SCION by application developers in the future. First, we discuss the challenges in supporting applications to make optimal path choices in §12.2.1 and present the prototypical implementation

---

[3]The SCION Android app is available from the Google Play store at https://play.google.com/store/apps/details?id=org.scionlab.scion

of a QUIC/SCION socket in §12.2.2. A key question is how SCION should co-exist with the current IPv4 / IPv6 network stacks. A possible approach described in §12.2.3 is based on the TAPS API which provides a protocol-agnostic API for the transport layer. An alternative to the TAPS approach is to augment the Happy Eyeballs mechanisms as discussed in §12.2.4. We also describe how SCION can be supported directly in web browsers (§12.2.5). Finally, we present a potential approach for replacing the dispatcher in order to achieve higher throughput (§12.2.6).

12.2.1 Performance-Aware Path Choice

On the one hand, the path-aware architecture of SCION offers a new world of optimization possibilities for networking applications. On the other hand, an application trying to make optimal use of a path-aware architecture is faced with a wide spectrum of path choices. Path selection can be tackled with different tactics, each one with its own advantages, disadvantages, and implementation challenges.

- **Path racing:** Connection setup can be raced over multiple paths, such that only the connection first established will be used. This approach is useful for quickly selecting a low-latency path without relying on previous measurements and without incurring an increased delay. However, it also raises the question of how many and which paths are selected for connection racing.

- **Path coordination with peer:** When a connection is setup, a return path has to be selected. This opens up additional choices: The two hosts can negotiate both forward and return paths, based on in-band coordination. By default, the reverse path is used.

- **Path health and performance monitoring:** Path performance can be passively monitored on paths in use. Alternatively, paths can be actively probed. The latter approach raises the question of how many and which paths are actively probed.

- **Multipath or single-path transport layer:** Does the transport layer protocol support multipath communication explicitly (similar to MPTCP or proposed multipath extensions for QUIC), or use a more traditional transport layer and attempt to optimize the choice of a single active path?

- **Path switching and fail-over:** When the metrics suggest that performance is deteriorating, the path choice should be revisited. Which metrics are taken into account? At which point should the paths be switched? How does switching paths affect the congestion control, and thus the effective performance, of the transport protocol?

The rich diversity of opportunities for how to select paths spans an exciting research area. Although the design space is already vast, several SCION extensions introduce additional degrees of freedom: A given path can be used in "regular SCION mode" or with extensions such as COLIBRI (§10.2) or EPIC (§10.1). We anticipate that the resulting approaches will introduce an unprecedented quality of communication.

12.2.2 Nesquic

Addressing the challenges raised in §12.2.1, we have implemented a prototype of QUIC/SCION socket with automatic, performance-aware path choice, called Nesquic. The client side implements an active path probing and path-selection behavior. The server uses a simpler mechanism which sends reply packets over the path on which the last packet from the client was received. The client's path-selection logic is implemented as a "sandwich" around the quic-go [353] library:

- **On top:** An interface layer that intercepts the relevant API calls to quic-go.

- **Below:** The snet socket used by QUIC to send UDP packets, so that packets are sent over SCION via a path that can be actively selected by the controller.

- **On the side:** A controller that actively probes available paths and selects a best active path. The latency of the paths is probed by means of SCMP Echo Request packets.

  The controller reevaluates the path choice at regular time intervals, or when the probing observes drastic changes (SCMP error messages or timeout for the active path).

Additionally, the connection establishment is raced over all available paths. This provides a first latency measurement for all paths, before the active probing starts.

The library allows specifying policies for the filtering and preferred ordering of paths. Some basic policies are built-in, but the library also allows defining custom policies relying on additional processing of the path metadata. The built-in policies include the following:

- interactive selection;
- pinned fixed path;
- allowed/disallowed ISDs or ASes;
- latency (low to high);
- latency augmented with geographic data, a heuristic to improve latency estimates by using coordinates of intermediate hops (low to high);

- bandwidth (high to low); and

- total number of hops (low to high).

Policy chaining/composition is supported for sequential policy application.

12.2.3 TAPS API

There is a growing trend [365, 492, 537] to implement network communication in applications using specialized libraries that offer high-level abstractions over the transport layer, saving developers from the tedious task of having to interact directly with the BSD sockets API. While it would be relatively straightforward to include SCION support in popular networking libraries, a more immediate approach exists.

The Transport Services (TAPS) working group of the IETF [261] is currently elaborating a protocol-agnostic general-purpose API [509] for the transport layer, which is intended to replace the traditional BSD sockets API. While this API specification is still in a draft stage (as of 2021), it is sufficiently mature to serve as a framework for exploring potential ways for SCION to co-exist with other networking technologies. Among the proposed features, the dynamic selection of transport protocols and network paths at run time are the most relevant for SCION. When the network protocol is chosen automatically (but based on high-level application requirements) as a function of the operating system's networking stack, the OS can serve as the sole entry-point to add support for any novel networking protocol or technology, including SCION. In other words, using a suitable automated protocol selection strategy (see §12.2.4), SCION support can be seamlessly added to any TAPS-enabled application, without requiring any intervention from application developers, while protocol racing prevents any negative impact on the quality of experience from the user's perspective.

The fundamental compatibility of SCION with an otherwise IPv4/IPv6-based TAPS-like API implementation—including protocol selection at runtime—was demonstrated in a working prototype [544]. We are currently following up on this work with PANAPI [303], a more sophisticated system design that is intended to explore the requirement-based automation (and optimization) of SCION path-selection in the back-end, while retaining compatibility with the TAPS API on the front-end.

12.2.4 Happy Eyeballs with SCION

The Happy Eyeballs method initially developed for dual-stack IPv4/IPv6 hosts [457, 541] can be extended to use SCION as an additional connectivity option. Availability of SCION connectivity to a server can be easily discovered using either RHINE (Chapter 19) or DNS (as in the original Happy Eyeballs approach). In order to facilitate adoption, TXT records can be used to signal a SCION address, as opposed to the A and AAAA records used for IPv4 and IPv6 addresses. On the client side, our end-host bootstrapping procedure (see §13.2) provides the required setup for connecting to the local SCION AS and contacting the local control services. Using these techniques, both servers and clients can seamlessly make a transition to SCION.

An alternative to the TAPS approach proposed above is to augment the Happy Eyeballs implementation provided in some networking libraries such as libcurl [338] to also establish a connection over SCION. This enables support for SCION in the application without any changes other than using the networking library.

12.2.5 SCION Browser Extension

Over the last few years, browsers have gained a number of privacy-enhancing features such as incognito mode, ads and tracker blockers, and anonymous-network integration. Unfortunately, today's Internet does not allow end users to influence how their data is transmitted through the network. In contrast, SCION provides route control, path awareness, and explicit trust information. We introduce a SCION browser extension that empowers users to leverage such properties.

A SCION-extended browser's main goal is enabling users to surf the SCION network and the legacy Internet while providing a smooth experience for them. The browser must be able to seamlessly identify HTTP(S) requests intended for the SCION network, while allowing users to browse the legacy IP web. In addition, the SCION-extended browser provides an intuitive interface that allows the user to check metrics and select paths based on various properties, e.g., latency optimization, CO_2 optimization, or geo-fencing.

Example: Path Selection through Browser Extension. The user browsing from a host in ISD-AS 17-ffaa:1:f1f wants to reach, via SCION, an available HTTP resource (e.g., *https://www.example.com*), which is located in ISD-AS 18-ffaa:0:1206. Upon requesting the resource, the user obtains a graphical representation of the available communication paths through the browser extension.

In Figure 12.3 the north path traverses the AS 17-ffaa:0:1101 while the south path crossing the EU avoids this AS. By way of example, we assume that the user is reluctant to send traffic across the AS 17-ffaa:0:1101. By using the graphical interface the user can easily divert traffic towards the south path.

Example: Geofencing. Another feature example is geofencing. This feature allows the user to select which countries or areas the traffic can go through or, similarly, which areas must be avoided.

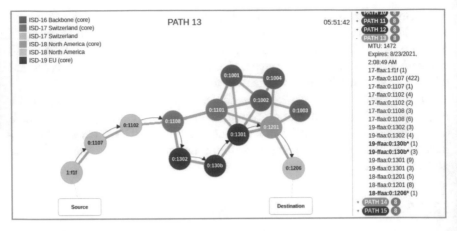

Figure 12.3: Path selection UI in a SCION-enabled browser.

The SCION browser extension demonstrates that end users do benefit from SCION expressive policies and features, and can do so in an effortless manner. Its current implementation relies on SCION support on the end host (i.e., the host must run a dispatcher §12.1.1 and daemon §12.1.2). We envision further integration with browsers, where SCION is packaged directly in the browser, and where the end host relies on bootstrapping mechanisms §13.2 in order to seamlessly obtain SCION connectivity.

12.2.6 Dispatcher Replacement

As presented in §12.1.1, the SCION dispatcher is responsible for handling incoming and outgoing SCION traffic. From the design perspective, having one component handling both directions of traffic reduces complexity. The outgoing traffic always comes from a specific source port. For incoming traffic, the SCION border routers do not need to look up L4 ports before sending packets to end hosts, which reduces routers' workload. Also, firewalls are easier to configure with this setup. Unfortunately, the current design also has drawbacks. First, the achievable bandwidth of high-performance applications is significantly limited because each packet is forwarded by a userspace application and thus traverses the Linux kernel twice before it can be processed by the application. Second, forwarding UDP packets over stream-based UNIX sockets requires additional packet assembling logic. Furthermore, the dependency of applications on the dispatcher over UNIX sockets increases the complexity of deployment, because applications must register again if the dispatcher restarts, and sharing UNIX sockets using container technologies like Docker also has pitfalls. Finally, since the registration of applications is performed by the dispatcher, detecting SCION applications with built-in networking tools is not possible.

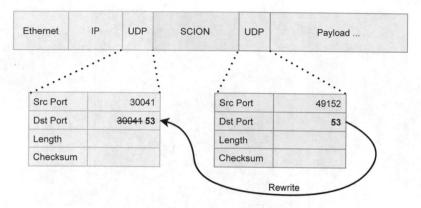

Figure 12.4: Overwrite IP/UDP underlay destination port with SCION/UDP destination port.

To overcome the above drawbacks, we envision a complete re-design of the SCION dispatcher. The dispatcher itself will be replaced by an SCMP daemon [466], which creates and handles SCMP messages. Such daemon forwards SCMP errors to the local SCION apps based on the (partial) quote of the packet causing the error. Furthermore, applications will open native UDP ports instead of registering at the dispatcher to connect to the SCION network. Instead of using a routing table for registered applications, as the dispatcher does at the moment, the forwarding of incoming SCION packets to the applications will be solved by rewriting the L4 UDP port with the L4 SCION/UDP port, which already contains the correct port. This keeps the benefits of having all SCION traffic coming to a specific port on each host. Figure 12.4 shows the port rewriting in SCION packets. The original UDP destination port 30041 is the port on which all incoming SCION traffic enters end hosts.

By removing the dependency on UNIX sockets and therefore also the additional packet assembling logic, a performance increase is expected. However, packet forwarding in userspace applications still limits performance. We consider kernel-based approaches to solve this. The extended Berkeley Packet Filter (eBPF) [518] provides functionalities to attach hooks to a set of events in the Linux kernel. One of these events is the eXpress Data Path (XDP) [239], a hook to perform actions on raw packet data before the packets arrive at the kernel network stack. The port rewriting can then be done before reaching the kernel and packets are passed directly to their receiving applications. Although combining eBPF and XDP is currently targeted at Linux, there is work ongoing for a port on Windows and also userspace versions of eBPF are available on macOS. Due to the design of this approach, it can be deployed as an optional component and the packet forwarding falls back to the userspace forwarder if the required features are not available. Figure 12.5 shows the data flow of incoming and outgoing SCION packets. If eBPF/XDP are available

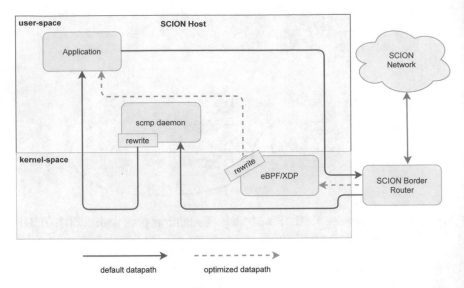

Figure 12.5: Data flow using the SCMP daemon and optionally eBPF/XDP fast
           forwarding.

on the host, ports are rewritten before they are passed to the kernel network
stacks and therefore arrive directly at the respective application. Otherwise, a
forwarding component in the SCMP daemon will perform the rewriting and
forwarding of the packets. For outgoing traffic, a forwarding component is
not required. Since applications know the destination where they send SCION
traffic (normally a SCION border router), they can send the traffic directly.

To summarize, the current design of the SCION dispatcher has several drawbacks in its current implementation. With the proposed conceptual and technical changes, deployment, network tooling and performance are expected to
improve significantly.

12.2.7 SCION Protocol Number

As discussed in §12.1.1, SCION packets are currently encapsulated in UDP/IP,
with a fixed port number for end hosts. This approach allows a relatively
straightforward deployment to end hosts and software routers, as normal UDP
sockets can be used to send and receive SCION packets. Furthermore, from
the viewpoint of legacy network devices and middleboxes in a SCION-AS's local infrastructure, SCION traffic encapsulated in UDP/IP looks like any other
UDP traffic, facilitating a smooth initial deployment of SCION.

While this is a reasonable and pragmatic approach, it has some downsides.
First, the UDP/IP header contains fixed port numbers and a redundant frame
length. The checksum of the encapsulating UDP must be recomputed (or at
least updated) at each SCION router. At the same time, the checksum is also

redundant, as the checksums in the SCION packet payload are sufficient. The encapsulating UDP header, eight bytes in total, thus provides no information but constitutes only wasted payload capacity. Admittedly, this waste amounts to less than one half percent of a typical message size, but at the traffic volumes of today's Internet, this can amount to an energy and monetary overhead.

An alternative future approach would be to obtain an IP protocol number assignment for SCION, allowing SCION packets to be encapsulated directly in IPv4 or IPv6. This would be directly analogous to existing encapsulation protocol number assignments, such as IPv4-in-IP (decimal 4) and IPv6-in-IP (decimal 41). The relatively scarce assigned Internet Protocol numbers are managed by IANA [265]. Per RFC 5237 [36], the default policy to assign such a Protocol number requires *Standards Action*, i.e., an IETF Standards Track RFC. Alternatively, an *IESG Approval* can suffice, but most likely still requiring a standards document. We expect SCION standardization to make rapid progress starting in 2022. Consequently, this will become a viable option only once SCION starts to undergo a standardization process.

As the SCION packet encapsulation is an AS-internal operation and the encapsulation is removed when transitioning to another AS, the choice of the addressing and encapsulation scheme is an AS-local decision and we can keep support for encapsulation in both IP as well as in UDP/IP. As the SCION-encapsulating IP packets never leave a domain, we may enable encapsulation of SCION in IP without an officially assigned IP protocol number, by using a protocol number set aside for experimentation (decimals 253-254) or recycling any unused legacy protocol number.

13 Deployment and Operation

Laurent Chuat, Matthias Frei, David Hausheer, Samuel Hitz, Jonghoon Kwon, Markus Legner, Nicola Rustignoli, Lars-Christian Schulz, Joel Wanner, François Wirz*

Chapter Contents

13.1 **Global Deployment** **319**
    13.1.1 Stakeholder Incentives 320
    13.1.2 Deployment Considerations 320
    13.1.3 ISP Core Network Deployment 321
    13.1.4 Customer Site Deployment 323
    13.1.5 IXP Deployment Scenarios 325
    13.1.6 SCION and Network Address Translation (NAT) 326
13.2 **End-Host Deployment and Bootstrapping** **327**
    13.2.1 Terminology 327
    13.2.2 Design 327
    13.2.3 Discovery Mechanisms 328
    13.2.4 Bootstrapping Server 331
    13.2.5 Security 331
13.3 **The SCION–IP Gateway (SIG)** **332**
    13.3.1 Overview of the Problem Space 332
    13.3.2 Interoperability between SCION and IP 334
    13.3.3 The SIG Encapsulation Protocol 334
    13.3.4 SIG Discovery 335
13.4 **SIG Coordination Systems** **336**
    13.4.1 The SCION Gateway Routing Protocol (SGRP) 336
    13.4.2 SIAM: A Global SCION–IP Address-Mapping System 337
13.5 **SCION as a Secure Backbone AS (SBAS)** **345**
    13.5.1 Overview of SBAS 346
    13.5.2 Use Cases for SBAS 346
    13.5.3 SBAS in Detail 347
13.6 **Example: Life of a SCION Data Packet** **354**

---

*This chapter reuses content from Hu, Klausmann, Perrig, Reischuk, Shirley, Szalachowski, and Ucan [250], Krähenbühl, Tabaeiaghdaei, Gloor, Kwon, Perrig, Hausheer, and Roos [305], Sridhara, Wirz, de Ruiter, Schutijser, Legner, and Perrig [487].

This chapter presents deployment alternatives and network-operation approaches of the SCION Internet architecture. Deploying a next-generation architecture is a challenging task, as it needs to be integrated with and operate alongside existing infrastructure. In the following, we discuss deployment approaches of SCION in the real world, supporting both SCION-enabled hosts and legacy hosts.

We first present considerations around a global SCION deployment. Then we discuss the stakeholder incentives needed for such a deployment to be successful (cf. §13.1). To this end, different deployment scenarios for ISPs, IXPs, and end-user domains are outlined.

§13.2 discusses deployment considerations for end hosts, in particular end-host bootstrapping.

To enable legacy hosts also to benefit from SCION, the SCION–IP Gateway (SIG) provides an interface between SCION and the legacy IP world (cf. §13.3). Different types of SIG coordination systems have been developed to facilitate a large-scale deployment of the SIG, including SGRP and SIAM (cf. §13.4.1 and §13.4.2). Furthermore, different SIG deployment scenarios are discussed.

The Secure Backbone AS (SBAS) enables a partial SCION deployment to offer secure routing benefits not only to customers of participating ISPs, but also to hosts all across the legacy Internet, as is explained in §13.5. To illustrate the operation of SCION on real networks, the life of a packet is followed in §13.6.

Required Changes for Different Deployment Scenarios. Table 13.1 shows the required changes for the different technologies that drive deployment of SCION. The table considers changes that are needed at the leaf AS (AS which the end host is in), the ISP (the service provider of the leaf AS), the Operating System (OS) of the end host, and the application – considering both the source (src) and the destination (dst).

The table uncovers the large difference in terms of required changes for different technologies, ranging from a full SCION deployment to SBAS. In the short term, SBAS is a promising approach to make the benefits of SCION widely available, which in turn will enable SCION to expand its deployment with increasing use. With the growing availability of SCION at ISPs, the other deployment options will be supported, resulting in a virtuous cycle of increasing adoption.

Table 13.1: Required changes for different deployment scenarios. Elements that require change are marked with a ✗, those not needing change are marked with a ✓.

Scenario	ISP		Leaf AS		OS		App	
	src	dst	src	dst	src	dst	src	dst
Full SCION deployment (§13.1.4a, §13.1.4c)	✗	✗	✗	✗	✗	✗	✗	✗
SCION–IP Gateway (SIG) deployed at local AS (§13.1.4b)	✗	✗	✗	✗	✓	✓	✓	✓
Carrier-Grade SIG deployed at ISP [1] (§13.1.4d)	✗	✗	✓	✗	✓	✓	✓	✓
In-application deployment without OS-support[2] (§12.2.5)	✗	✗	✗	✗	✓	✓	✗	✗
"Happy Eyeballs" standard with SCION support (§12.2.4)	✗	✗	✗	✗	✓	✓	✓	✗
Secure Backbone AS (SBAS) approach (§13.5)	✗	✓	✓	✓	✓	✓	✓	✓

[1] This approach can be used to secure home office connectivity.

[2] This approach takes advantage of SCION if present in local AS, and it makes use of the libcurl library. For more information, see §12.2.3.

13.1 Global Deployment

Since August 2017, SCION has been in production use by a central bank, with two main goals: Test the long-term reliability of SCION, and replace leased lines. Over time, several of their branches have been connected to their data centers over the SCION network. Their positive experiences have fueled adoption by ISPs, as well as by commercial, education, and government entities. Today, eight ISPs offer SCION connections, and several banks and government entities benefit from the BGP-free backbone for production use.

This section describes how SCION is deployed in production networks and used for real-world traffic at ISPs, their customers (leaf ASes) and IXPs. Before we describe the technical deployment details, we first discuss the stakeholder incentives that led to the first production use of SCION in 2017, and then briefly discuss deployment considerations that are needed to achieve the salient SCION properties.

13.1.1 Stakeholder Incentives

An important aspect for the deployment of a new Internet architecture is the incentives driving initial deployment. Similar to the question of "Who bought the first fax machine?", the case for a next-generation architecture is even more challenging, given the plethora of commercial communication offerings.

The initial customer incentive was to test the reliability of SCION and to use a SCION connection to replace a leased line. A leased line–often provisioned via dedicated layer-2 circuit switching or layer-3 MPLS–is a premium connectivity service that provides availability and confidentiality. On the other hand, leased lines often have long lead times (in some cases several months), lack flexibility for short-term changes, and are often expensive to operate. SCION approximates leased-line properties, offering geofencing, path transparency, high reliability thanks to fast failover, and flexibility for changes. Furthermore, as SCION adoption grows and converges towards today's pricing, costs will be reduced compared to leased lines in the long term. For instance, to connect N branches with K data centers, which can be implemented using $N \cdot K$ leased lines, $N + K$ SCION connections are required (and for even larger savings if redundancy is needed). Since SCION can reuse the existing IP or MPLS-based network, the additional capital and operational expenditures to run SCION are marginal, requiring only a few standard servers or VMs.

The long-term incentives for using SCION are to achieve higher performance and quality of communication through the use of multipath and optimized path selection based on application requirements (e.g., latency, bandwidth, jitter, or loss).

Today, eight ISPs offer SCION connections, and several banks and government entities benefit from their BGP-free backbones for production use. This demonstrates that the initial deployment incentives have been sufficient, but additional ones are needed to further drive deployment of native SCION connectivity on endpoints used by applications. Other use cases that benefit from SCION's path awareness and multipath properties include industrial control systems ([273], [272]), bulk file transfers, C02-optimized routing [454], access to cloud environments, communication infrastructure for blockchain systems, and many more. We further discuss such use cases in detail in Chapter 15.

13.1.2 Deployment Considerations

An overlay deployment on top of today's Internet was not desirable, as SCION would inherit the vulnerabilities of its weak underlay. Thus, a challenge was to deploy SCION in parallel to existing networks in an economically viable way, while preserving the security properties. In particular, there should not be any dependence on External BGP for the SCION network to operate, which we refer to as a "BGP-free" deployment.

Since deploying a completely new parallel network infrastructure would be uneconomical and inefficient, intra-AS networks are re-used. However, care

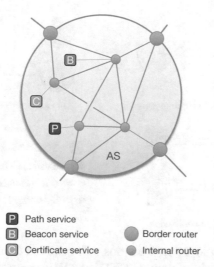

Figure 13.1: Core SCION components needed for an AS deployment. SCION
services can be deployed anywhere within the AS, including on
single or distributed machines.

needs to be taken that traditional IP traffic cannot be used to crowd out SCION
traffic, for instance by causing IP-level congestion.

13.1.3 ISP Core Network Deployment

As Figure 13.1 depicts, a SCION ISP needs to set up border routers and run instances of the control service. The border router and control service instances
are deployed on standard x86 commercial off-the-shelf (COTS) servers, supporting up to 100 Gbps links, while with P4 hardware it is possible to forward
SCION traffic even at terabit speeds ([155]). The ISP internal IP or MPLS-
based network can be re-used to enable the SCION infrastructure to communicate within the AS. If dedicated links are not available, queuing disciplines on
internal switches can provide separation of IP and SCION traffic.

Customer connections and SCION connectivity between the border routers
of neighboring ISPs can be achieved in three different ways. Ideally, SCION-
enabled adjacent ISPs would be connected via a native SCION link (Figure 13.2a). That is, two SCION border routers are directly connected via a
layer-2 cross-connection at a common point-of-presence, achieving connectivity with high reliability, availability, and performance. The native SCION link
is unaffected by BGP failures, achieving a "BGP-free" deployment.

To minimize changes to the current infrastructure, ISPs may also reuse existing cross-connections to carry SCION traffic, e.g., in a *Router-on-a-stick*
deployment model. As shown in Figure 13.2b, the SCION border routers can
be attached to the existing legacy border routers. A SCION border router en-

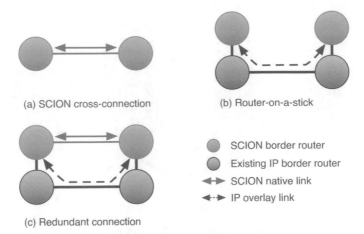

(a) SCION cross-connection   (b) Router-on-a-stick

(c) Redundant connection

Figure 13.2: ISP deployment scenarios. The various deployment models support a mix of native SCION and IP overlay links.

capsulates SCION packets into IP packets and forwards them to a neighboring SCION border router over a short IP connection, which can be "BGP-free" through the setup of static routes. The main advantage of this deployment model is that ISPs can simultaneously use their network infrastructure for the new network architecture. Since the *Router-on-a-stick* model is a short-range direct cross-connection, potential shortcomings of using an IP encapsulation, such as non-optimal routing, BGP hijacking, and slow route convergence, are typically not an issue. Given that an adversary could overload the shared link with IP traffic, it is important to define a queuing discipline on the link to ensure that SCION traffic obtains at least a minimum fraction of the link bandwidth to achieve availability properties.

Finally, Figure 13.2c shows the deployment of a *redundant connection*, combining the aforementioned two deployment models, with one L2 and one L3 link. The two links can be exposed as two separate links with different SCION interface numbers, enabling multipath selection for either of the links. Alternatively, more traditional approaches are also possible (i.e., combining redundant links at a layer 2 into a single logical one using Link Aggregation [260]).

SCION-enabled ISPs should seamlessly communicate with each other even in partial-deployment scenarios; i.e., two SCION-enabled ASes may not be neighbors. Bridging two SCION islands – e.g., by creating an IP tunnel to forward SCION packets through the public Internet – however, introduces BGP vulnerabilities. To this end, Anapaya has established a SCION-transit service [236], a global backbone service for SCION-enabled ISPs. The SCION-transit service provides native SCION connectivity at 100+ data centers located in the largest metropolitan areas across the world. With such distributed points-of-presence, ISPs can readily establish one-hop access to the SCION-transit service, forwarding SCION traffic through the BGP-free network.

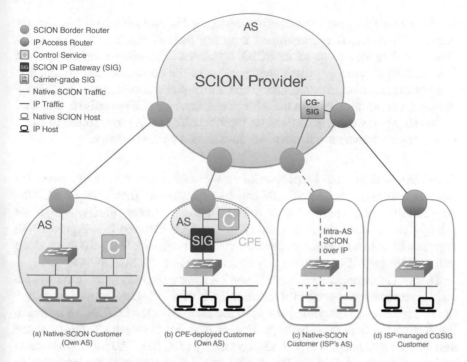

Figure 13.3: Example deployments for end customers. With a SCION–IP Gateway, customer networks are enabled to SCION connections, without requiring any changes to end hosts or applications.

13.1.4 Customer Site Deployment

A customer can use SCION in two different ways: (1) native SCION applications (Cases a and c), and (2) transparent IP-to-SCION conversion (Cases b and d). The benefit of using SCION natively is that the full range of advantages becomes available to applications, at the cost of installing the SCION endpoint stack and making the application SCION-aware. In the short term, approach (2) is preferred, leveraging a *SCION-IP-Gateway (SIG)* that encapsulates regular IP packets into SCION packets with a corresponding SIG at the destination that performs the decapsulation (see, §13.3).

An end customer's network can constitute an independent SCION AS (described below as Cases a and b), or it can leverage its provider's AS (Cases c and d). So far, all deploying enterprises elected to become their own AS. The required cryptographic certificates are issued by the core ASes, and the AS numbers are re-used from today's AS numbers or, if needed, allocated from the larger 48-bit space of SCION AS numbers.

Case a: Native SCION Customer Running Own AS. As shown in Figure 13.3a, native SCION hosts can send SCION traffic directly to a SCION border router (BR) over the existing customer internal routing infrastructure. Native SCION hosts are equipped with the SCION stack components, enabling applications to generate SCION packets. The data-plane component (i.e., *SCION dispatcher*) dispatches packets to the corresponding application and performs packet transmission. The control-plane component (i.e., *SCION daemon*) communicates with the AS's *control service (CS)* to build end-to-end forwarding paths for applications on their behalf. In this case, suited to enterprises, the customer runs its own AS, including a control service.

Case b: SIG-Based Deployment. We understand that many customer hosts may initially not be SCION capable. Customers purchasing a SCION-connection from a provider ISP therefore obtain a *customer-premise equipment (CPE)* that provides the functionality of the SIG, BR, and CS. Figure 13.3b depicts a high-level topology of an end-customer network that SCION-enabled with a CPE; the SIG enables legacy hosts to opt into the SCION network.

The SIG is responsible for encapsulating legacy IP packets in SCION packets, to provide interoperability between SCION and legacy networks. When the SIG receives an outgoing packet, it first determines the SCION AS to which the destination IP address belongs. For the mapping between IP address space and AS, the SIG keeps a SIAM *ASMap* table (§13.4.2.2). The SIG then obtains paths to the remote AS from the control service, encapsulates the packet with a SCION header, and routes it via a BR.

Case c: Native SCION Customer Using ISP's AS. Similarly to case a, Figure 13.3c depicts a native SCION deployment. The customer's end host runs the full end-host SCION stack and communicates with the ISP's SCION router over IP. In this case, the customer is fully part of its ISP's AS. This has the advantage of not requiring the customer to run an AS and its infrastructure. This case is therefore better suited to small domestic customers.

Case d: Carrier-Grade SIG Customer. End customers can also be transparently SCION-enabled with a *carrier-grade SIG (CGSIG)* as depicted in Figure 13.3d, requiring no changes to the customer premises; the CGSIG is operated by the provider ISP, it aggregates upstream traffic towards remote ASes and carries out SCION packet routing on behalf of its customers, while legacy hosts residing in the end-domain networks remain SCION-unaware. The CGSIG-driven SCION service is designed to minimize the impact on existing infrastructure and is suitable for small business and home office users.

Internal Routing of SCION Traffic. To transport SCION packets to an egress BR, the customers do not need to change their internal routing infrastructures; the SCION packets are IP-routed by an IGP, e.g., OSPF or IS-IS. Given that the AS's internal entities are considered to be trustworthy, the IP overlay

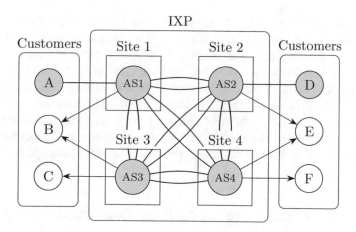

Figure 13.4: IXP deployment. Shaded circles denote Core ASes.

for the first-hop routing does not compromise or degrade any security properties SCION delivers. To exchange SCION packets with the provider network,
the customer-side SCION border routers directly connect to the provider-side
border routers using last-mile connections. It is important to note that the current customer connections to the SCION production network are native SCION
connections, not shared IP / SCION connections, i.e., while IP is used to route
SCION traffic AS internally, those existing SCION customer connections *cannot* forward regular IP packets.

13.1.5 IXP Deployment Scenarios

Internet Exchange Points (IXP) play an important role in today's Internet, as
they let ISPs, content delivery networks (CDNs), and other providers exchange
traffic with each other. We envision two models describing how the role of
IXPs can be reflected in the SCION infrastructure.

"Big Switch" Model. In the big switch model, IXPs would be considered
as a large L2 switch between multiple SCION ASes (i.e., customers of the IXP).
The role of the IXP is then to facilitate bilateral (peering) links among those
ASes. This role is entirely transparent to the SCION control plane. Today,
SwissIX already follows this model by offering a dedicated SCION VLAN
on which it prohibits non-SCION traffic. The automatic interconnection of
SCION ASes over an IXP can be facilitated with a SCION Peering Coordinator
[461].

Enhanced Model: Exposing the ISXP Internal Topology. Figure 13.4
shows an enhanced model in which the internal topology of an IXP is exposed
within the SCION control plane. Here, the IXP operates its own SCION ASes,

whereas each AS represents an IXP site and the links between them represent redundant connections between these sites. This enhanced model enables IXP customers to use SCION's multipath and fast failover capabilities to leverage the IXPs internal links (including backup links) and to optimize paths depending on the application's needs. This model would entirely replace the IXP's traditional interconnection fabric that is mostly based on Ethernet switching or MPLS today. We believe that IXPs have an incentive to expose their rich internal connectivity as the benefits from SCION's multipath capabilities would increase their value for customers and provide them with a competitive advantage.

13.1.6 SCION and Network Address Translation (NAT)

In network address translation (NAT) [488, 489], a middlebox rewrites the IP and transport headers of packets. NAT was originally created to mitigate issues stemming from the limited address space of IPv4 [124, 531]. Today, IPv4 NAT is ubiquitous on consumer routers and often deployed at residential and mobile ISPs in the form of "carrier-grade NAT" [535]. In the most common use case, a LAN using private address space [431] is connected to the Internet through a single gateway router with a single external globally routable IP address. The private network appears as a single node to the public Internet, thus hiding the number of hosts and topology of the internal network, and blocking incoming connection requests by default.

While the above mentioned properties are perceived as benefits, NAT has also been identified as the cause of many networking issues, including the breakage of several applications [240]. IPv6 solves address scarcity, thanks to a substantially increased address space compared to IPv4. The additional benefits discussed above can be achieved by other means [517] such as IPv6's privacy extensions [215] and firewalls.

SCION decouples end-host addressing from inter-domain routing. While the latter is based on ISD-AS tuples (as described in §5.6), end-host addresses do not need to be globally unique—they can be assigned independently by each AS including private address space [431]. Additionally, IPv6 can be used for end-host addresses, in which case all benefits of IPv6 are transferred to SCION. To address privacy issues, additional mechanisms like OTA [325] and APNA [324] can be used in combination with SCION. NAT is therefore not necessary to increase the number of addressable hosts.

Network address translation, by definition, alters the source and/or destination addresses of packets in transit. This generally impedes address-based host authentication systems like DRKey (§3.2.3) and PILA (Chapter 20). Additional systems relying on DRKey, such as EPIC and COLIBRI (§11.3, that provide strong availability guarantees for communication), and SCMP authentication (§4.7.3), are thus adversely affected by NAT. We discuss challenges with NAT and PILA in greater detail in §20.6. We conclude that SCION-level

NAT is not necessary in a SCION Internet. *Not* deploying NAT would avoid its negative impacts on end-to-end connectivity, and disruptions of end-host based secure mechanisms.

As we discuss in §12.1, neighbor SCION routers communicate via a UDP/IP connection. While NAT is neither necessary nor desirable in a SCION Internet, it is, of course, possible to establish inter-AS links on top of a NATed IPv4 connection. This is frequently done in SCIONLAB, e.g. for systems running in public clouds. In these cases, the router's public IP is statically mapped to its private IP. In such a scenario, when both a SCION router and client are deployed behind an IPv4 NAT, SCION Internet-wide communication is possible using the end-host private IP, as long as it is unique within its own AS. The SCION header, however, discloses the end host private IP.

13.2 End-Host Deployment and Bootstrapping

Low entry barriers are crucial to facilitate the broad adoption of SCION. End hosts within an AS that have adopted SCION can use an automated end-host bootstrapping mechanism to connect to the SCION network. All an end user needs to do to benefit from SCION, is to install a software package.

Such a SCION package contains a bootstrapper daemon, in addition to the dispatcher and SCION Daemon (SD), which retrieves hints from the local network using *zero-configuration* (zeroconf) [112] services, downloads the required SCION configuration files from a local discovery service, and starts the SCION Daemon. A zeroconf service is a service provided by the network that requires no network-specific configurations on the clients making use of it.

13.2.1 Terminology

Bootstrapping Server. A bootstrapping server exposes the endpoints required by the bootstrapper. We discuss further details in §13.2.4.

Hint. A hint is a piece of information returned by a zeroconf service deployed in the local network. Depending on the discovery mechanism, a hint can either be used to contact a bootstrapping server (e.g., providing its address) or to further query the local network (e.g., a DNS PTR response).

Discoverer. A discoverer is a client of a zeroconf service. It communicates with the service and provides hints to the bootstrapper.

13.2.2 Design

Bootstrapping Process. On end hosts, an external orchestrator (e.g., the init system on Linux machines, more specifically systemd on Ubuntu) manages the bootstrapper daemon and starts the SCION Daemon once the bootstrapper daemon finishes successfully.

Bootstrapping Steps.   The end host bootstrapper daemon performs the following steps:

1. Probe the local network for hints about a bootstrapping server address using the available discovery mechanisms (i.e., DHCP, DNS, and mDNS).

2. Wait for hints from the discoverers.

3. Once a hint is received, try to download the TRCs and the topology of the AS from the bootstrapping server. While there is no maximum amount of TRCs to be served, the bootstrapping server must provide at least the TRC of the ISD in which the AS is located.

   a) On success, prepare the SD's files and exit successfully; the SD is then automatically started by the orchestrator.

   b) On failure, go back to step 2.

If no hint is received after a certain period, the bootstrapper daemon times out and exits with a non-zero value.

Note that the TRCs retrieval is a transition solution to ease adoption; ideally they are installed on a device out-of-band, before the device gets connected to a network (more details are given in the security considerations on page 331).

13.2.3 Discovery Mechanisms

A bootstrapper can leverage DHCP, DNS or mDNS in order to find the IP address of the bootstrapping server. We describe each case, where we assume that

- the end host is located in the example.com domain; and

- the IP address of the bootstrapping server is 192.168.1.1.

DHCP.   The DHCP mechanism relies on the presence of an existing DHCP server in the network. This mechanism is advantageous in environments where there is a managed DHCP server, but no dedicated DNS infrastructure is operated for the local network.

The DHCP server has to be configured to announce the address of the discovery services using one of the DHCP options. One natural choice is to use the option field with ID 72 "Default WWW server", given that HTTP, the same application-level protocol as used in the WWW, is used to retrieve the configuration files. In our example, we would set the option value to 192.168.1.1.

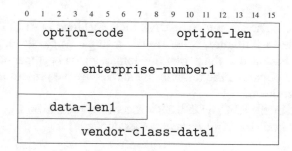

Figure 13.5: DHCP Vendor-Identifying Vendor Option format.

Code	Len		Vendor-specific information			
tc	n	i_1	i_2	...	...	i_n

Figure 13.6: DHCP option field format.

Using an existing DHCP option has the advantage of enabling easy deployment without the need to standardize a separate option.

This mechanism is well suited to small networks, covering scenarios such as consumer residential networks. The drawback is that many DHCP server implementations currently only allow an IP address inside option 72, hence a default port needs to be used.

A more advanced solution is to use a DHCP Vendor-Identifying Vendor Option as defined in RFC 3925 [339]. This allows both an IP address and port for the bootstrapping server to be specified. The main drawback is that it is a bit more complex to configure, and support for RFC 3925 is limited in many DHCP server implementations. The option's format is shown in Figure 13.5. We propose to use as an enterprise number the Private Enterprise Number assigned to Anapaya Systems via IANA, PEN 55324.

The IP address and the port of the bootstrapping server are encoded as a sequence of code/length/value fields as defined for the DHCP standard [14] section 2, "DHCP Option Field Format". An IPv4 address is encoded as a 4 byte sequence with type code 1. A UDP port is encoded as a 2 byte sequence with type code 2. The option fields' general format is shown in Figure 13.6.

DNS. The DNS-based mechanism relies on records on the DNS server used by end hosts. This offers a quick and easy option to configure bootstrapping on multiple networks, regardless of whether they have DHCP configured or not. This approach is well suited for enterprise or ISP networks where DNS is centrally managed. We describe a minimal setups.

DNS SRV.   The advantage of DNS SRV, specified in RFC 2782 [224], is that it is simple to configure, consisting of a single entry. From a conceptual perspective, it has the drawback that the domain name is prepended with information that is not part of the domain, which is why we present an alternative solution further down.

As specified by the RFC, a DNS SRV record redirects to an A or AAAA record, pointing to the bootstrapping server. For example:

- `_sciondiscovery._tcp.example.com`
  `IN SRV 8041 sciondiscovery.example.com`

- `sciondiscovery.example.com IN A 192.168.1.1`

DNS-SD.   As specified in DNS-Based Service Discovery (DNS-SD, RFC 6763 [111]), a list of DNS PTR records points to SRV records, each of which defines an instance of a SCION discovery service. This has the advantage of potentially allowing multiple redundant discovery service instances to be announced for high availability.

Here is an example configuration for a single instance:

- `_sciondiscovery._tcp.example.com`
  `IN PTR instance._sciondiscovery._tcp.example`

- `instance._sciondiscovery._tcp.example.com`
  `IN SRV 8041 sciondiscovery.example.com`

- `sciondiscovery.example.com IN A 192.168.1.1`

DNS-NAPTR.   In this variant, a DNS NATPR record (Naming Authority Pointer DNS resource record [361]) redirects to an A or AAAA record pointing to the bootstrapping server. This is cleaner than a DNS SRV record from a conceptual perspective, since the domain name part is not mixed with the attributes we are defining or requesting, but cleanly defined in the content part of the record.

- `example.com IN NAPTR "A"`
  `"x-sciondiscovery:tcp" "" sciondiscovery.example.com`

- `sciondiscovery.example.com IN A 192.168.1.1`

- `_sciondiscovery._tcp.example.com IN TXT "8041"`

A TXT record is used to convey the port information. Like the DNS-SD approach, multiple NAPTR records for different discovery services can be defined.

This mechanism is very suitable for large-scale networks having control over their DNS domain and infrastructure.

An end-host bootstrapper supporting both DNS-SD and DNS-NAPTR gives network operators more flexibility to choose which mechanism to support depending on their DNS setup.

mDNS. mDNS, a decentralized DNS based on IP multicast [110], is often used in combination with DNS-SD in zeroconf networks. It removes the need for a centralized DNS server, but requires proper support for multicast traffic on the network where it is deployed. It is a lightweight solution that requires just one entity besides the client in the network. If the client does not reside in the same network, the intermediate routers need to be configured to propagate multicast traffic between the subnets in order for the discovery to work.

It is thus not well suited to large, segmented enterprise networks, but can be advantageous in small domestic networks or in a cluster environment where multicast is already configured.

13.2.4 Bootstrapping Server

The bootstrapping server can in principle be a simple web server. It exposes the following endpoints to serve the bootstrapping configuration files:

- `/trcs/isd{isd}-b{base}-s{serial}` and `/trcs/isd{isd}-b{base}-s{serial}/blob`: to retrieve the TRCs needed by the SD, and

- `/topology`: to retrieve the topology of the AS.

In the long term, we envision the TRCs being be installed on devices via an out-of-band mechanism (i.e.,in the factory or at device initialization).

13.2.5 Security

Guaranteeing the authenticity of the retrieved resources is crucial to ensure that a user connection is not hijacked. Bootstrapping resources can be signed, so that the bootstrapper can verify their authenticity.

In the current Internet, the root of trust is based on an oligopoly of CAs. In SCION, this root of trust is represented by one TRC per ISD. Nonetheless, as in the current Internet, a device joining a network for the first time needs to have some pre-shared knowledge to determine what is authentic or not.

While we can consider the discovery of TRCs a temporary solution, the same is not true for the topology—which is at the heart of the automatic bootstrapping. For this, a signing solution based on the cryptographic keys of an AS should be implemented.

The bootstrapper has a configuration option to download an initial TRC from the local AS infrastructure. Only this initial retrieval is allowed to be unauthenticated, under the *trust on first use* (TOFU) principle, and subsequent requests must be authenticated and the user warned if there is a conflict with an existing TRC. Alternatively, a user needs to manually import a trusted TRC onto its system.

13.3 The SCION–IP Gateway (SIG)

Successfully deploying a new Internet architecture requires incremental deployability and the ability to reuse existing hard- and software. Interoperability with the existing Internet is therefore key. We previously discussed how SCION minimizes the need for new infrastructure by reusing existing intradomain networks (in §13.1). Additionally, during a long period of partial deployment of SCION, we expect many applications and end hosts to be SCION unaware, and only use legacy IP for communication.

This section describes the SCION–IP Gateway (SIG) and illustrates how it enables SCION to interoperate with the legacy IP world. In particular, our mechanism enables legacy IP end hosts to benefit from a SCION deployment by transparently obtaining improved security and availability properties.

We first describe the challenges that deploying a new Internet architecture entails: ensuring interoperability with the current Internet through minimally invasive changes, enabling transparent operation, and preventing downgrade attacks to the legacy Internet should the more secure SCION Internet be available for a given destination. We then introduce the SIG and explain our mechanisms for addressing the challenges presented.

13.3.1 Overview of the Problem Space

We first describe the requirements of the SIG and provide an overview of the problems we intended to address with our design.

13.3.1.1 IP-in-SCION Encapsulation

Transporting legacy IP traffic over a SCION network requires encapsulating the IP traffic in SCION packets. The encapsulation protocol should be specifically *non-reliable*, to avoid problems with stacking retransmission timers.[1]

Recall that the maximum payload size of a SCION packet varies depending on the path length and MTU, and other factors (e.g., extensions used). As IP-path MTU discovery only allows an MTU to be decreased (it has no mechanism to increase an MTU again), the IP MTU for an encapsulated connection will decrease over time as paths change, which results in wasted bandwidth.

---

[1]Tunneling a reliable protocol over another reliable protocol can cause retransmission storms (also known as "meltdown") in the event of packet loss [504].

Thus, the encapsulation protocol should insulate the IP traffic from the underlying SCION maximum payload size.

13.3.1.2 Routing and Connectivity

Providing proper interoperability requires legacy IP connectivity to be transparently supported (i.e., communicating legacy hosts should be oblivious to SCION, and their connectivity should not be impacted by SCION's involvement). This means that traffic routing must be fully supported between two legacy IP hosts—one in a legacy (i.e., non-SCION) AS and one in a SCION AS. The same applies to traffic exchanged between two legacy IP hosts that both reside in SCION ASes.

As a consequence, the same routability rules apply regarding public and private (RFC 1918 [431]) IP ranges. Hosts in SCION ASes that wish to be reachable by legacy hosts in other ASes must have public IP addresses. However, network-address translation (NAT) between the legacy host and the entry point to the SCION network (e.g., at a home router) is still possible.

13.3.1.3 Addressing

As legacy hosts (and clients) will not have support for SCION's name resolution service (RHINE), they will still rely on the legacy name resolution service (DNS). The latter does not provide any specific routing information to the legacy host, as the SCION AS is not mentioned in DNS; nor would a legacy host know how to route to a SCION AS in any case. As a consequence, interoperability requires that bare IP addresses are sufficient for legacy addressing of hosts in SCION ASes.

13.3.1.4 Support for Layer-4 Protocols

The legacy Internet heavily uses TCP and UDP, but it also uses many other layer-4 protocols—including SCTP, L2TPv3, IPIP, and ICMP.[2] Any interoperability solution for SCION must be layer-4 agnostic, i.e., it must work for any layer-4 protocol that is in use by legacy traffic.

13.3.1.5 Support for SCION-only ASes

Some SCION ASes may decide to be directly connected to both the legacy IP Internet and the SCION Internet. Other networks may decide they do not want (or need) a direct connection to the IP Internet. Both of these cases should be fully supported.

---

[2]ICMP is considered as a layer-4 protocol in this context.

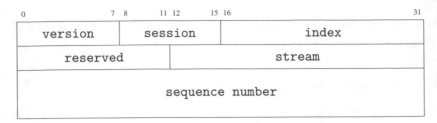

Figure 13.7: Format of the SIG frame header.

13.3.2 Interoperability between SCION and IP

The SCION–IP Gateway (SIG) service is responsible for providing interoperability between SCION and the legacy IP world. Every SCION AS that wants to enable legacy IP connectivity between its legacy hosts and those in other ASes deploys a SIG service. The service is responsible for routing and encapsulating legacy inter-AS traffic. All legacy traffic between SCION ASes is handled by the SIG service, with the sending side encapsulating the traffic, and the receiving side decapsulating it again back into regular IP packets.

All legacy traffic into (or out of) a SCION AS passes through the SIG service by means of legacy IP routing rules. This means that the SIG service must be fast, in order to keep up with the traffic flow. It must also be robust, and able to deal with any packet loss in the encapsulated traffic.

13.3.3 The SIG Encapsulation Protocol

The SIG encapsulation protocol is built on top of UDP/SCION. It converts legacy IP packets into a byte stream to the remote SIG service (see Figure 13.8). The stream contains the original layer-3 (i.e., IP) and above contents of the encapsulated IP packet(s). The IP packets transported via the SIG are encapsulated in SIG frames. There can be multiple IP packets in a single SIG frame. A single IP packet can also be split into multiple SIG frames.

SIG traffic can be sent over multiple SIG sessions. The SIG uses different sessions to transport different classes of traffic (e.g., priority vs. normal or geofenced vs. open.) Within each session there may be multiple streams. Streams are useful to distinguish between traffic sent by different SIGs. For example, if a SIG is restarted, it will create a new stream ID for each session. That way, the peer SIG will know that the new frame with a new stream ID does not carry a trailing part of the unfinished IP packet from a different stream.

Each SIG frame starts with a SIG frame header with the following format:

- The `version` indicates the SIG framing version. We describe version 0, therefore this field must be set to zero.

- The `session` indicates the SIG session to be used.

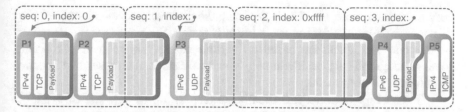

Figure 13.8: Encapsulation of five IP packets.

- The `index` is the byte offset of the first beginning of an IP packet within the payload. If no IP packet starts in the payload, for example, if the frame contains only a trailing part of an IP packet, the field must be set to 0xFFFF.

- The `reserved` is reserved and must be set to zero.

- The `stream`, along with the `session` identifies a unique sequence of SIG frames.

- The `sequence number` field indicates a position of the frame within a stream. Consecutive frames can be used to reassemble IP packets split among multiple frames. The sequence number starts from zero for a given direction of traffic and pair of SIG services, and increases monotonically by one with every SIG frame. It resets whenever the sending SIG service restarts or the value reaches $2^{64} - 1$.

The SIG frame payload may contain multiple IPv4 or IPv6 packets, or parts thereof. No other types of packets can be encapsulated. The packets are placed one directly after another, with no padding. The SIG uses the IPv4/6 *payload length* field to determine the size of the packet. To make the processing easier, it is required that the fixed part of the IP header is in the frame where the IP packet begins. In other words, the initial fragment of an IPv4 packet must be at least 20 bytes long and the one of an IPv6 packet must be at least 40 bytes long.

13.3.4 SIG Discovery

When a SIG service wants to send encapsulated traffic to another AS, it needs to first discover the instances of the SIG service in the remote AS. One option is to rely on service-level anycast (see §4.6) by sending a query to the remote SIG SVC address. The remote instance (resolved by the ingress SCION border router) responds with its own network endpoint where it accepts incoming SIG frame streams.

The other option is to use the discovery service (§4.6.2) of the remote AS. The local SIG queries the remote discovery service for instances of the SIG service in the remote AS. The reply from the discovery service includes a list of network endpoints for the available SIG instances. Note, that the discovery service can curate that list based on various parameters such as availability, load, or geographical considerations. The local SIG then chooses one instance to initiate communication.

SIG discovery is done periodically to keep an up-to-date list of available SIG instances in the remote AS. This allows failover in case the currently used remote SIG instance becomes unreachable.

13.4 SIG Coordination Systems

In order to facilitate the transition from legacy IP-based Internet to SCION, we introduce mechanisms that allow SIGs to coordinate and automatically exchange IP prefix information. We first present SGRP §13.4.1, a system enabling SIGs to mutually discover and announce IP prefixes. Such an approach is well suited to enterprise networks, where SIGs deployed at branches can mutually exchange prefixes for each location.

We then introduce SIAM, a global and scalable SIG coordination mechanism that translates between legacy public IP and SCION (§13.4.2). SIAM transfers authorizations in RPKI [328] to SCION, making it a viable global transition mechanism for public IP networks and ISPs.

Finally, we introduce SBAS, a transition mechanism that aims at offering SCION's benefits to the wider legacy IP Internet, by routing regular IP traffic over SCION (§13.5).

13.4.1 The SCION Gateway Routing Protocol (SGRP)

The SCION Gateway Routing Protocol (SGRP) enables SIGs to discover and announce IP prefixes directly from/to remote SIGs. To that end, each SIG is configured with a prefix policy than contains the IP prefixes announced to remotes and also IP prefixes accepted from remotes (see Listing 13.1). Optionally, SIGs can learn or advertise prefixes using a dynamic routing protocol as BGP.

A SGRP policy rule contains the following fields:

- `ACTION` can be `accept`, `reject`, `advertise`, or `redistribute-bgp`. `accept` and `reject` define IP prefixes that the SIG will accept or reject. `advertise` defines what should be advertised (statically). `redistribute-bgp` defines IP prefixes learned from BGP that should be redistributed.

- `FROM` and `TO` enable scoping of the action to certain source or destination ISD-ASes. '0' indicates a wildcard and means any ISD or AS.

Listing 13.1: Example of an SGRP policy file.

```
# ACTION     FROM  TO   PREFIX-LIST        COMMENT
  reject      0-0   0-0  10.0.0.0/8         # globally reject 10/8
  accept      1-2   0-0  1.2.3.0/24         # accept prefix
  reject      2-5   local                   # reject everything from ISD-AS 2-5
  reject      0-0   0-0  0.0.0.0/0,::/0     # default reject policy
  advertise   0-0   3-6  !1.2.5.0/24        # advertise all local prefixes
                                            # except 1.2.5.0/24 to 3-6
  redistribute-bgp                          # redistribute all prefixes
              0-0   4-7  0.0.0.0/0          # from BGP to 4-7
```

- PREFIX-LIST lists the IP prefixes (in CIDR notation) that the action should be applied to. There is no limitation of what IP prefixes can be specified here, i.e., both public and private IP address space is allowed.

- COMMENT allows the operator to add a comment to a specific rule.

A SIG will periodically fetch the advertised IP prefixes from a remote SIG. The remote SIG uses its SGRP policy to determine which IP prefixes to advertise, while the local SIG matches the advertisements against its own accept and reject rules. IP prefixes that were previously advertised by a remote SIG, but are not included anymore in the most recent advertisement will be removed from the local SIG routing table.

To ensure authenticity of the SGRP advertisements two SIGs can establish an authenticated channel based on TLS and use the SCION AS certificate to authenticate the connection. Furthermore, public IP prefixes can be validated using RPKI. To that end, the advertising SIG includes route origin attestations (ROA) for each prefix it advertises. Clearly, this is not possible for private IP prefixes. For that, operators need to rely on appropriate filtering policies to allow only specific private ranges to be advertised by certain peers.

13.4.2 SIAM: A Global SCION–IP Address-Mapping System

As the SCION network continues to grow and reaches a scale comparable with today's Internet, neither manual configuration of SIGs nor the SGRP are sufficient. To make SIGs a viable transition mechanism in an Internet-scale network with tens of thousands of ASes [59], an automatic configuration system is required.

In this section, we describe such a system to create mappings between legacy IPs and SCION addresses, and transfer the authorizations in the Resource Public Key Infrastructure (RPKI) [328] (see also §1.1.3) to SCION for use with the SIG. This SCION–IP address-mapping system (SIAM) is sufficiently scalable to be used in a global SCION network and can thus support the use of SIGs as a global transition system.

13.4.2.1 Problem Description

SIAM is meant to serve as a global mapping system between IP prefixes and SCION ASes. It should therefore provide the following functionalities to authorized entities. They can:

F1 Publicly register a mapping between their IP prefix and the SCION AS, thus indicating that the IP prefix is reachable through the SCION network.

F2 Publish dynamic mappings at timescales of several hours, i.e., they can be updated and removed.

F3 Any SIG can query SIAM on the mapping to a SCION AS for a particular IP prefix or the absence thereof.

This system should be secure, and only authorized entities should be able to register mappings. Concretely, we require SIAM to

S1 prevent hijacking attacks, i.e., an AS should not be able to add mappings for IP prefixes which it is not authorized to announce;

S2 prevent flooding attacks on SIGs, i.e., a SIG should only be able to create mappings for their own AS;

S3 be resilient to downgrade attacks to the legacy Internet where SCION connections are possible; and

S4 ensure high availability, i.e., the system should not have single points of failure.

Scale. We plan SIAM to work as a *transition mechanism* at Internet scale, designing the mapping system for up to 1000 ISDs with a total of up to 10^6 IP–AS mappings.

Adversary Model. In our design and analysis, we consider the adversary described in §7.2. For S4 we must also assume that packets eventually arrive at the correct destination.

Challenges. The problem and security requirements presented above pose several challenges:

- **Different network structures and address hierarchies:** On the one hand, the main network architecture is SCION, which suggests tying SIAM to SCION's ISD meta structure. On the other hand, the ultimate goal is to map a given legacy address to a SCION AS, which suggests basing the resolution on the IP-prefix hierarchy.

- **Connecting multiple PKIs:** Properties S1 and S2 imply that SIAM must be tied to both the legacy RPKI and the SCION CP-PKI.

- **Consistent and complete mapping:** Properties S3 and S4 imply that every entity querying SIAM gets a complete view of all existing mappings and that the failure of a single system does not disrupt the functionality. SIAM therefore requires a mechanism to prove the *absence* of mappings for a given IP address.

13.4.2.2 SIAM Overview

SIAM enables an entity that controls both a legacy AS and a SCION AS with a SIG (with the same AS number) to publish a mapping from an IP prefix it owns to the SCION AS. Such a mapping (F1) enables the AS to receive IP traffic through the SCION network. To improve scalability, SIAM introduces intermediate *mapping services* (MSes) located in SCION core ASes to mediate between SIGs and the globally distributed *publishing infrastructure* (PI).

13.4.2.3 SIAM Components

The system deploys components that form the global PI, and components local to a SCION ISD that are responsible for submitting and retrieving lists of mappings to/from the PI. An overview of all SIAM components and their interaction is shown in Figure 13.9 on the next page.

Publishing Gossip Node (PGN). PGNs are nodes in the PI that store IP prefix to SCION AS mappings and propagate those in periodic intervals. They accept lists of mappings and respond to queries. PNGs are globally deployed in core ASes and contain all mappings within their ISD. They form a gossip network that achieves eventual consistency. For SIAM, eventual consistency is acceptable as entries can be independently verified based on attached signatures and RPKI, and PGNs provide signed statements of non-existence of missing lists.

Publishing List Node (PLN). To discover other PGNs, both PGNs and MSes (see below) rely on a secondary gossip network of PLNs, which are deployed in all core ASes with either a PGN or an MS. A PGN registers with a PLN by sending its ISD and AS number (IA). PLNs periodically propagate this information to other PLNs. As PLNs are not configured with addresses of nearby PLNs, they rely on SCION's service discovery (see §4.6) to discover other PLNs.

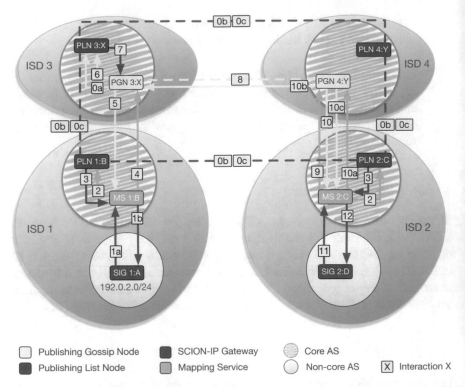

Figure 13.9: Overview of the SIAM system in an example SCION network
with four ISDs. The numbered interactions are described in
§13.4.2.4. Dashed lines represent gossip messages enabling a
bidirectional exchange of the current state.

Mapping Service (MS). To use SIAM, an ISD must deploy an MS (and,
consequently, a PLN) in at least one of its core ASes. The MS accepts mappings from SIGs within the same ISD, validates them based on both the CP-
PKI and RPKI,[3] combines them to form a list, and submits the list to the PI.
Periodically, it pulls lists of all other ISDs from the PI, validates the mappings,
and stores them locally in the form $ipPrefix \rightarrow ia$ ($ia = isd : as$). Based on the
stored data, it responds to queries for IP addresses from SIGs in the same ISD.

SCION–IP Gateway (SIG). A SIG creates mappings from the prefixes that
it is authorized to announce to its ASN, signs them with its SCION CP-PKI
key, and submits them to an MS in its ISD. To be able to correctly encapsulate
IP packets, SIGs query the MS for IP-to-SCION-AS mappings, which they
then store locally.

---

[3]The details of messages and their protection and validation are described in the SIAM paper [487].

13.4.2.4 Messages and Protection

In this section, we discuss the messages exchanged between the SIAM components.

PGN Startup. As discussed in §13.4.2.3, PGNs are global components in SIAM that are deployed in core ASes. [0a] When a PGN is started, it registers itself by sending its ISD-AS to the PLN it was configured with. The PLN that receives the registration checks the CP-PKI signatures on the message to ensure that the register request is from the correct AS.

PLN Gossip. [0b] PLNs periodically send out pings using SCION's anycast feature to SCION core ASes up to h_{pln} hops away. The pings are used for detecting liveness of other PLNs and for announcing presence when a new PLN is added to the gossip network. [0c] PLNs periodically combine all the registrations and propagate them to k_{pln} PLNs chosen from the PLNs discovered using [0b]. The propagation interval and the number of PLNs k_{pln} are configurable.

Register Mapping with SIAM. [1a] A SIG deployed in AS *as* creates $map_{ip,as}$ for a prefix *ip* that it is authorized to originate and sends it to an MS in a core AS of its ISD. Timestamps ensure that consecutive add/remove messages can be ordered. The MS validates the origin of the message using the CP-PKI signature and whether the *as* is authorized to originate *ip*. The MS uses the standard RPKI validation to ascertain the validity of a mapping. [1b] The MS acknowledges the mapping registration/removal by sending back a hash of $map_{ip,as}$ along with the current timestamp. SIGs can use this token as proof against misbehaving MSes if a mapping that was submitted was not added to the PI.

Send Mappings to PI. The MS periodically combines all the mappings in its local store into a list and sends it to the PI. [2] If it does not have a locally configured or cached PGN to which to send the list, the MS sends a request to the PLN it was configured with. [3] The PLN replies to the MS with $list_{pgn_{ia}}$ from its local store. [4] The MS picks a PGN to send the list of mappings to based on its local policy. It forms the corresponding message including a timestamp and sends it to the PGN. The timestamp ts_{ms} defines until when the list is valid.

The PGN verifies the CP-PKI signature on the message to validate its origin, checks that the ISD of the sender and the $list_{isd}$ are the same, and locally stores the signed list. [5] The PGN replies with a signed message containing a hash of the MS's message, $Hash(obj_{ms})$, and the current timestamp. This confirmation token can be used by the MS to prove misbehavior of PGNs.

PGN Gossip. The PGN periodically propagates gossip messages to k_{pgn} PGNs in different ISDs. The gossip messages contain two objects: a list of mappings and a list of signed "empty tokens" used to prevent downgrade attacks, see §13.4.2.5. "Empty tokens" are signed with the CP-PKI.

$\boxed{6}$ To find PGNs to send the list to, the PGN contacts the PLN it was configured with at startup. $\boxed{7}$ The PLN replies with the $list(pgn_{ia})$ from its local store. $\boxed{8}$ The PGN picks k_{pgn} based on its local policy to send the list to. The PGN forms the message containing the mappings and sends it to the selected PGNs.

A PGN that receives the gossip message validates the CP-PKI signatures of each of the objects in the list, checks that they are fresh, and locally stores the mapping list. The PGN iterates over the empty tokens, checks in its local store if a mapping list for the ISD with the empty token exists, and, if not, stores the empty tokens indexed by ISD number. It then replies to the sender with a list of mappings from its local store that are newer than or not present in the received gossip message. This process corresponds to a bidirectional exchange of the current state.

Pull Mapping Lists from PI. $\boxed{9}$ The MS queries the PGN for lists of mappings. To find a PGN, it contacts the PLN it was configured with and picks a PGN based on its local policy (e.g., a PGN in the same ISD or an ISD it trusts). $\boxed{10}$ The PGN that receives the request combines all mapping lists in its local store that are not stale and the signed empty tokens and sends them to the MS.

Validate and Reverse Mapping Lists. The MS that receives the list of mappings from the PGN performs two checks on each of the lists:

- Validates the list's CP-PKI signature;
- Checks that the list is not stale.

Afterwards, the MS validates and then stores each entry $map_{ip,as}$ in the mapping list by checking its CP-PKI signature and using RPKI to check for a valid ROA.

For empty tokens lists, the MS stores the fresh tokens indexed by ISD number. To ensure that it has full information on the network, the MS checks that for each ISD in the network it either received a valid list or has f_{ms} valid empty tokens from different ISDs. The "fault parameter" f_{ms} is configured at MS startup and defines the number of faulty PGN nodes the MS will tolerate.

Fetch Missing Lists. $\boxed{10a}$ If, for a particular ISD, both checks fail, the MS queries the list for the ISD by sending a message to a PGN. The PGN that receives this message checks in its local store if a fresh list for the *isd* is present. If there is no fresh list, it checks in its local store for at least f_{ms} fresh empty tokens for *isd* and signed by different ISDs. $\boxed{10b}$ If f_{ms} valid tokens are not present, the PGN contacts the required number of PGNs in different ISDs for the *isd* list. The PGNs that receive this request check their local store for the *isd* list and either send back the corresponding mapping list or the signed empty token. Through this process the MS can ascertain that it has all the mappings that are published in SIAM.

Request IP-to-AS Mapping. If a SIG receives an IP packet with a destination for which it has no mapping stored locally, it uses SIAM to look up the SCION ISD-AS number corresponding to the IP address. ⑪ The SIG sends a request with the IP address to an MS in any of the core ASes in its ISD. ⑫ The MS looks up its local store for IP prefixes that match the IP address. If there is more than one mapping in the MS local store, the MS performs a longest-prefix match for the IP address and returns the corresponding message to the SIG. If there are no matching prefixes, the MS replies to the SIG with an empty *ia* field.

13.4.2.5 Security Analysis

In the following, we explain how SIAM achieves the properties laid out in §13.4.2.1 and defends against common attack scenarios.

Hijacking and Flooding Attacks. To achieve properties S1 and S2, SIAM uses two trust anchors: RPKI and the SCION CP-PKI. For property S1, the MS uses RPKI's ROAs. This validation is performed when entities in its ISD submit $map_{ip,as}$ and when the MS pulls the mapping lists from the PI. For property S2, the SCION CP-PKI is used to validate the origin AS of $map_{ip,as}$ is *as*. This ensures that ASes can only create mappings for themselves.

Downgrade Attacks. There are two entities that are in a position to perform downgrade attacks violating property S3 (i.e., convince SIGs that no mapping exists even though it does): PGNs and MSes.

A PGN could simply send empty tokens instead of an existing mapping list in ⑧. However, to prevent legitimate PGNs from receiving the correct list, an adversary would need to completely partition the gossip network. This can be made virtually impossible by selecting a default value for k_{pgn} of 8, and considering the fact that these PGNs must be located in different ISDs. Furthermore, a misbehaving PGN can be identified quickly based on the signatures on empty objects. Another possible attack is to provide an empty token to a querying MS in ⑩. The MS defends against such attacks by requiring f_{ms} separate empty objects from PGNs in different ISDs. Furthermore, the MS can explicitly query PGNs in ISDs it trusts.

In SIAM, an MS can attempt to mount downgrade attacks *only* on ASes in its ISD by excluding a $map_{ip,as}$ either in ④ or ⑫. The first case—if an MS does not include a mapping submitted by a SIG in ④—can be detected by the AS that originated $map_{ip,as}$. The integrity of individual mappings returned in $\boxed{12}$ can be verified by any SCION entity by checking CP-PKI signatures and querying the RPKI infrastructure using the same technique that MSes use to validate the mappings, see §13.4.2.4. If an MS excludes an existing mapping in $\boxed{12}$, this can be detected through out-of-bands communication and proven through the signatures on the messages.

DoS Attacks. Availability (property S4) is protected through several mechanisms. Using a distributed PI enables an MS to query any of the PGNs in case some are under a DoS attack. Furthermore, to disrupt the gossip protocol, an attacker would have to partition the gossip network by attacking a large number of PGNs. The long validity periods of most SIAM entries together with local storage at MSes and SIGs ensure that intermittent outages of individual SIAM components do not directly affect SIAM's operation. Amplification attacks exploiting the large size of PGN responses are prevented by verifying source CP-PKI signatures on requests. Finally, additional availability mechanisms proposed for SCION, including source authentication based on DRKey §3.2 and bandwidth guarantees with COLIBRI §10.2.3 enable SIAM components to protect and authenticate traffic to ensure availability for legitimate requests, in particular as MSes only need to accept requests from SIGs within the same ISD.

Sybil Attacks. An adversary could attempt to create many additional PGNs to either partition the gossip network or provide the required number of empty objects to MSes. However, both PGNs and MSes require the PGNs they communicate with to be located in *different* ISDs. PGNs and MSes can prefer ISDs they trust, so that an attacker would need to either control core ASes in multiple existing ISDs or create new ISDs. As creating new ISDs requires a lot of effort in SCION, this would be impractical and highly noticeable.

13.4.2.6 Evaluation

We have chosen parameters k_{pln} and k_{pgn} equal to the Bitcoin network, which has been shown to be well connected for the number of nodes ($\sim 10,000$) that we consider as a maximum for SIAM (assuming 10 core ASes in each of 1000 ISDs) [160]. Our gossip networks (similarly to the Bitcoin network) closely fit the model of "K-out graphs", which are virtually certain to be connected according to theoretical and empirical results [484]. Results from related random regular graphs suggest that, for these values, the network diameter is below 10 [82]. This means that new information reaches all gossip nodes within less than 2 h (even if the nodes are synchronized), which is in line with recommendations for the synchronization interval of RPKI of at most 4–6 h [88].

We have implemented a prototype of SIAM in SCION to demonstrate its feasibility and scalability. Our measurements show that SIAM works well in networks of up to 1000 ISDs with a total of 1 million ASes.

13.4.2.7 Conclusion

Summarizing, SIAM builds a bridge between the IP-based and the SCION Internet by making it possible to transfer authorizations from the RPKI to SCION. This allows SIGs to be configured automatically according to publicly registered mappings of addresses. SIAM prevents hijacking attacks, is resilient to downgrade attacks, and avoids single points of failure.

For future work, we are considering various optimizations and extensions to SIAM including only transmitting hashes of lists in the gossip protocol and fetching full lists on demand; allowing the MS to renew lists that have not changed by pushing only signed hashes with new timestamps for freshness; and extending the PI to accept lists with a *type* tag so the PI can be used by different entities as a general distributed storage mechanism.

13.5 SCION as a Secure Backbone AS (SBAS)

By proposing a more tightly coordinated approach, the Secure Backbone AS (SBAS) enables a partial SCION deployment to offer secure routing benefits not just for customers of participating ISPs, but to legacy Internet hosts.

As a band-aid approach to circumvent the various shortcomings of the BGP routing protocol, products that offer connectivity over a globally deployed private backbone are rapidly gaining traction in today's Internet ecosystem (i.e., AWS Transit Gateway [42], Cloudflare Argo [129], Azure Virtual WAN [367]). However, with such networks, customers seeking higher reliability and security for their Internet connectivity are placing their trust in a single entity.

The inter-domain routing security provided by SCION enables a different approach: to construct a *federated* backbone consisting of a group of entities. For this purpose, we design SBAS, a system that both leverages and drives partial deployment of SCION as a secure backbone that can be used to provide immediate benefits for legacy Internet hosts. Crucially, SBAS requires minimal additions for ISPs that already deploy SCION and is compatible with standard BGP practices.

The SIG coordination mechanisms presented in the previous section (§13.4) allow SCION-deploying ASes to bridge the gap to IP hosts in their networks. However, the benefits of SCION are not carried out into the wider Internet: A service hosted on a SCION endpoint will not offer improved security to customers of ISPs that do not deploy SCION. Using SBAS, the space for use cases is much larger: Even endpoints that are not aware of the system can benefit from it, thanks to the seamless bridge between SCION and BGP provided by SBAS. At a small additional cost, ISPs can therefore deploy SBAS to tap into novel offerings for their customers, such as hijack-resilient server addresses or carbon-optimized Internet connections.

13.5.1 Overview of SBAS

Toward the legacy Internet, SBAS appears on the outside as a regular BGP-speaking AS. On the inside, the network is comprised of many SCION ASes. The different entities participating in SBAS are the following:

- **Customer:** A customer is an entity that resides outside of the backbone and obtains service from SBAS through a contract. This enables the customer to route traffic through the system. SBAS supports both (1) customers that only control single hosts (e.g., server operators or end users), and (2) entities that own entire address ranges and AS numbers.

- **Point of Presence (PoP):** A PoP is a member of SBAS that is located at the edge, i.e., provides connectivity to SBAS customers and interfaces with the legacy Internet. The PoPs form a full-mesh BGP topology over the internal routing protocol of the backbone, which is used to distribute customer announcements to the globally distributed PoPs and to achieve maximum security for traffic to secured prefixes.

- **Regular SCION AS:** ASes that are not located at the edge simply participate in the internal routing and forwarding. This type of member does not need to be aware of the SBAS infrastructure.

Participating customers that are not direct neighbors of SBAS can connect to it via a secure tunnel (e.g., VPN) to one or multiple distributed points of presence (PoPs). A customer can bring its own prefix and announce it through SBAS (e.g., if the customer is an AS with its own address space), or use IP prefixes assigned by SBAS (e.g., if the customer is an individual client without control over their address space).

13.5.2 Use Cases for SBAS

Routing Security. The prevalence of BGP hijacks and outages is a major driver for the adoption of private backbones on the Internet today. By leveraging SCION's secure routing, SBAS addresses this problem effectively and provides improved security for addresses protected by it, as illustrated in Figure 13.10. This enables use cases for, e.g., availability-critical services (which suffer high costs in the event of a BGP outage) and domain validation processes (which are an attractive target for BGP hijacking attacks).

Latency-Optimized Connections. With the path choice provided by SCION, it is possible to optimize packet routes inside SBAS with regard to various metrics. This allows an ISP to sell latency-optimized connections to its customers at a much lower price than is possible with a dedicated backbone network, which is extremely expensive to operate.

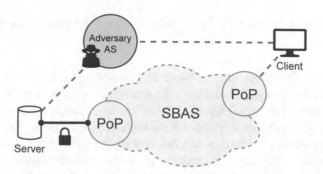

Figure 13.10: A server connects to a nearby SBAS point of presence (PoP) via
a secure tunnel. The system announces the server's address at
other PoPs, enabling clients to reach the server through SBAS.
This protects the connection against an AS-level adversary that
can make BGP announcements in an attempt to disrupt or eavesdrop on the connection.

Green Routing. SCION enables more carbon-efficient approaches to
packet routing, as we describe in Chapter 16. By utilizing these principles for
forwarding traffic inside the backbone, ISPs participating in SBAS can sell
connections to customers that are optimized to reduce their carbon footprint.
The geographically distributed deployment of SBAS also allows for a second
optimization: For legacy destinations, packets have to leave SBAS for the last
part of their delivery. SBAS typically has various options to select a packet
egress location for any given packet, and this choice can be optimized for
carbon efficiency. This enables SBAS to circumvent ASes or countries that
are known to have a particularly large carbon footprint.

13.5.3 SBAS in Detail

13.5.3.1 Design Goals and Challenges

To offer strong deployment incentives, SBAS has been designed to be (1) readily deployable without modifications to existing Internet infrastructure and protocols, (2) readily available for customers who want to use the system, requiring minimum changes in setup, and more importantly, (3) readily beneficial to
the customer even with a partial deployment of the architecture.

Architectural Continuity. Coupling a secure routing infrastructure and the
rest of the Internet requires architectural continuity. That is, the secure backbone must understand BGP's control plane and seamlessly bridge remote BGP
peers while leaving the leveraged secure routing infrastructure and its security
guarantees intact. To this end, the secure backbone must achieve an architec-

tural abstraction of the underlying infrastructure and provide an interface to customers.

End-to-End Security.   In the context of mediating customers' IP endpoints via a secure backbone, the end-to-end communication path can be segmented into an external (insecure) segment, which is comprised of the Internet links between an IP endpoint and the SBAS ingress/egress point, and an internal segment between an arbitrary ingress and egress pair of the secure routing infrastructure.   Therefore, to ensure end-to-end secure routing, the following conditions must hold: (1) Customers must be able to select trusted ingress/egress points and securely exchange packets with hijack resilience; and (2) the secure backbone must deliver the security properties it promised to any pairs of ingress/egress points even in the presence of internal adversaries.

Routing Priority.   To enable customers to route traffic from/to the Internet through a secure backbone, SBAS must disseminate the customers' prefix announcements to all other customers and external entities. Prefixes will then be announced via SBAS and the Internet, resulting in competing announcements. To maximize the ability to route securely, SBAS must be able to convince the entities receiving the announcements to prioritize routing paths through the secure backbone over the insecure Internet paths.

13.5.3.2 Secure Route Redistribution

The internal structure of SBAS can be abstracted to a full-mesh topology between the PoPs, which communicate over SCION. Over these connections, the PoPs redistribute announcements from SBAS customers as well as the Internet, akin to the operation of iBGP in a regular AS. To prevent tampering by non-PoP members, the iBGP sessions run over an encrypted and authenticated connection (such as a VPN tunnel).

SBAS offers a high degree of flexibility to its customers through support for dynamic route redistribution. Contrary to a traditional AS, which is controlled by a single entity, the redistribution scheme to be used in SBAS must support its federated structure and remain secure in the presence of malicious members. In the following, we describe the design and security aspects of the route redistribution mechanism.

The system distinguishes between three categories of addresses:

- **Secure addresses:** This includes prefixes announced by SBAS customers and SBAS-owned address spaces, which are assigned to customers. Secure address spaces are announced publicly at egress points via BGP.

- **Internal addresses:** In order to provide an internal addressing scheme among PoPs, e.g., to set up iBGP sessions between PoP routers, the PoPs

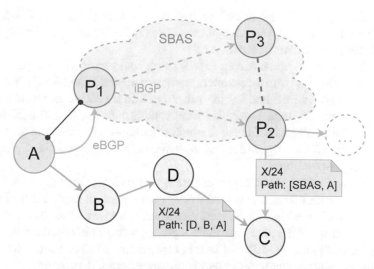

Figure 13.11: The route redistribution process for a prefix X owned by customer AS A. The prefix is being announced in parallel through
SBAS and to neighbors of A. In this scenario, an SBAS-unaware
AS C will receive competing announcements and choose the
path through SBAS that was received from P_2, the closest PoP.

reserve address space for SBAS internal operation. This address space
is not visible outside the SBAS infrastructure.

- **Global addresses:** We use this term to refer to any globally routable
  addresses that do not apply to the above categories.

Federated Bring-Your-Own IP Prefix. Customers that already control
one or multiple IP prefixes can use them directly with SBAS. For this purpose,
the system implements a route redistribution mechanism that enables a customer to route incoming traffic from the Internet through the secure backbone.
The process is depicted in Figure 13.11: The customer (AS A) initiates a BGP
session with the PoP (AS P_1) over a VPN connection. Using this session, A
makes an announcement for its prefix X, which is then redistributed to all other
PoPs over the full-mesh iBGP topology. A remote PoP such as P_2, upon receiving such an announcement over the iBGP session with P_1, sends it out to its
eBGP neighbors, i.e., Internet peers as well as SBAS customers.

Enhanced RPKI-Based Security. The RPKI system provides strong security properties for the first hop of BGP advertisements, but does not protect the
subsequent hops. The design paradigm of SBAS complements this property
well, as it eliminates attack surfaces on the path through its secure backbone.

SBAS leverages RPKI to defend against two distinct threats: (1) customers advertising prefixes that they do not own, and (2) PoPs falsely claiming authorization for a prefix from a customer.

The first threat is prevented in SBAS using route validation at the ingress. Each announcement from a customer must carry a valid ROA, which is verified both at the ingress PoP and by the other PoPs that receive the redistributed announcement. To prevent sophisticated routing attacks, SBAS additionally verifies that the AS path of these announcements does not contain any ASNs other than that of the origin (but still allowing for customer traffic engineering using path pre-pending).

An example of the second type of threat would be P_3 (in Figure 13.11) forwarding the announcement received from P_1 in an attempt to attract traffic to A. To prevent such malicious behavior, a customer can use RPKI to authorize a single or multiple PoPs to re-distribute a given prefix. This approach is similar to path-end validation [130], but in this case, it can be used purely by SBAS members and customers without requiring any external deployment.

SBAS-Only Prefix. Using an SBAS-defined BGP community tag, the customer can instruct the PoPs to only redistribute the announcement internally, i.e., to connected customers. This enables full protection of an address range against hijacking attacks, since secure prefixes are always prioritized by SBAS members and customers.

13.5.3.3 Customer Perspective

Customers sign up for SBAS via an interface by setting up a contract at their local PoP. The connection to SBAS is managed by a client software that receives information about PoPs, including their publicly reachable IP addresses and VPN public keys.

SBAS Connection Setup. A customer can connect to SBAS by connecting to one or multiple PoPs. A customer looking to maximize resilience to BGP attacks should generally prioritize the PoP that is the fewest BGP hops away. The connection to a PoP is usually set up over a VPN tunnel with the PoP's key pair, or, where possible, the customer can set up a link-layer connection directly at the PoP. The latter case has the additional benefit of eliminating the possibility of any BGP attacks on the connection. If the customer uses multiple PoPs, one connection is designated as the primary ingress point, with the others serving as backup alternatives for improved failure resilience. In order to prevent routing loops, the VPN endpoint that is used to connect to the SBAS PoP must be assigned a non-SBAS address. A customer may wish to designate a part of its address range to be routed via SBAS (advertised via SBAS), and separate this from their remaining address space (advertised normally in the legacy Internet).

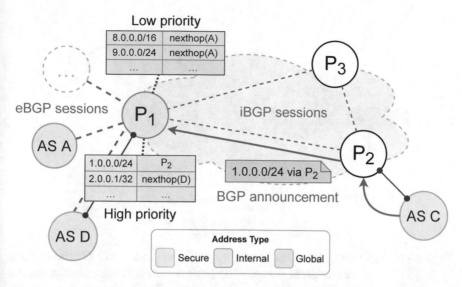

Figure 13.12: Control-plane configuration of P_1. nexthop(X) refers to the address used as the next hop in announcements from an AS X.

Secure Address Assignment. For customers that do not control an address space, SBAS assigns single addresses from a secure SBAS-owned prefix. This option is configured via SBAS client software. Upon assigning such an address to a customer, the PoP announces it to the other PoPs over the existing iBGP sessions. This allows them to route traffic to the appropriate location and keep track of addresses that have already been assigned.

13.5.3.4 Routing at PoPs

As shown in Figure 13.12, each PoP maintains several iBGP sessions to other PoPs over the backbone network, as well as eBGP sessions to customers and Internet peers. From the information received over these sessions, two routing tables are constructed: The first table, which is given the highest priority, maps secure addresses to internal addresses. Each entry may be either a remote customer prefix that is mapped to an internal address representing another PoP (in the example, 1.0.0.0/24 to P_2), or a local customer prefix that can be delivered to the customer's VPN endpoint (2.0.0.1/32 to nexthop(D)). The advertisements for such routes are received over the iBGP session from other PoPs (in the former case) or over the eBGP session from customers (in the latter case).

As a lower-priority table, P_1 maintains an Internet routing table for routes obtained from its Internet peers. These routes will likely also be received via iBGP from other PoPs distributing prefixes they received from their respective neighbors. In this case, route selection can follow standard Internet policies or

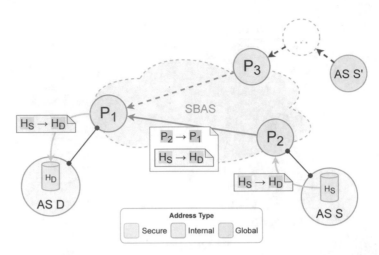

Figure 13.13: Routing logic for incoming traffic to a customer D who owns a
            secure address range. The packet shown is sent from a secure
            address H_S and encapsulated across SBAS using the internal ad-
            dresses for the PoPs P_1 and P_2, before it is delivered to the secure
            address H_D. Packets may also originate from global addresses
            like the AS S'.

custom logic implemented by P_1. While the current implementation of SBAS
relies on iBGP sessions between PoPs, SGRP §13.4.1 could be used as a future
alternative.

In the following, we describe the forwarding process for different scenarios
of source and destination locations, as illustrated in Figure 13.13. By keeping strict priority hierarchy between secure routes and external routes, SBAS
provides resilience to BGP hijacking attacks by design.

Customer-to-Customer.   ($H_S \rightarrow H_D$) In the simplest case, a packet originates from a secure address H_S in a customer AS S and is destined to another
secure address H_D. The packet from H_S is routed through the VPN tunnel to
the ingress PoP P_2. There, P_2 looks up the secure address H_D and finds the
internal address for the egress PoP P_1 associated with it. The original packet
is encapsulated over SCION to P_2, which delivers it across the VPN tunnel to
the destination in AS D.

External Origin.   ($S' \rightarrow H_D$) We now consider a packet that is destined to
the secure address H_D, but originates from a source in AS S' that is unaware of
SBAS. In this case, the data plane operations follow the same sequence: Having received a BGP announcement from P_3 for the secure prefix that contains
H_D, AS S' will forward the packet over SCION to the nearest SBAS PoP. From

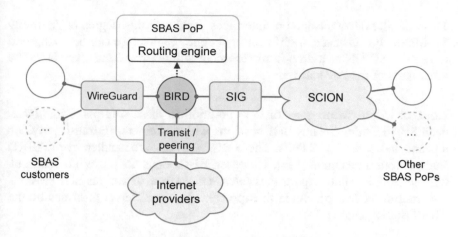

Figure 13.14: The interfaces (shown in filled rectangles) and BGP sessions (solid lines) maintained by an SBAS PoP.

this point, the same logic is applied as in the previous case with a secure source address.

External Destination. $(H_D \rightarrow S')$ For traffic with global destination addresses, the routing decision offers more options through the choice of the egress PoP. Whereas in the previous cases, the traffic was directed to the destination's preferred PoP, the decision to select an egress location is up to the ingress PoP. This decision can be optimized for various metrics depending on the customer's preferences (see §13.5.2).

13.5.3.5 Implementation

We have implemented SBAS in a way that minimizes the need for new software and leverages existing networking components in a synergistic manner. The software automatically configures and runs the various PoP components based on configuration files that describe the setup of the SBAS topology.

Data-Plane Interfaces. Each PoP has three interfaces for different types of destinations, as shown in Figure 13.14:

1. a WireGuard instance to send/receive packets to/from SBAS customers that are connected to that PoP;

2. an Internet interface with IP transit/peering and a BGP routing table; and

3. a SCION–IP Gateway interface, where a (SIG; see §13.3) encapsulates IP packets in SCION packets and sends them over SCION.

This modular decomposition of interfaces enables a high degree of flexibility for SBAS. For instance, a different backbone architecture can be configured to replace SCION as a drop-in replacement without requiring changes to the other parts of the PoP software.

Control-Plane Management. In addition to these data plane interfaces, each SBAS PoP maintains BGP sessions with customers, IP transit providers/peers, and other SBAS PoPs. These BGP sessions are handled by the BIRD Internet routing daemon [146]. However, BIRD does not make the final routing decision; it simply exports the routes learned from its various BGP sessions into routing tables, which are then processed with different priorities by the SBAS routing engine.

Routing Selection. The SBAS routing engine compiles the routes from these BGP sessions to produce the final route table that enforces the security and route preference requirements of SBAS.

13.5.3.6 SBAS Governance Structure

Due to the federated structure of the SBAS PoP operators, a governance structure is needed to coordinate global operation. For instance, the SBAS AS number needs to have an owner, which will also perform registration and redistribution of RPKI ROAs. From a business perspective, the PoP operators will need assurance of a reliable management structure to ensure smooth operation of SBAS credentials. Otherwise, a PoP operator selling a service to their customers could suddenly be faced with an outage due to compromised or expired certificates.

In such a multi-stakeholder setting, the governance structure of a non-profit entity or foundation is well suited, where the PoP operators are all co-owners or members. Once SBAS grows to become a critical Internet infrastructure, its governance can be embedded into the existing structures; for instance, into the regional Internet registries (RIRs), of which the 5 entities are ARIN, RIPE, APNIC, AFRINIC, and LACNIC. Another option would be to embed it in IANA and ICANN, which can then delegate operation to the RIRs, similarly to how IP address spaces and AS numbers are managed.

13.6 Example: Life of a SCION Data Packet

We describe the complete life cycle of a SCION packet: crafted at its source host, passing through a number of routers and middleboxes, and finally reaching its destination host. To this end, we assume that both source and destination are native SCION hosts (i.e., they both run a native SCION network stack). We note that the following description is also valid in the case where a SCION–IP

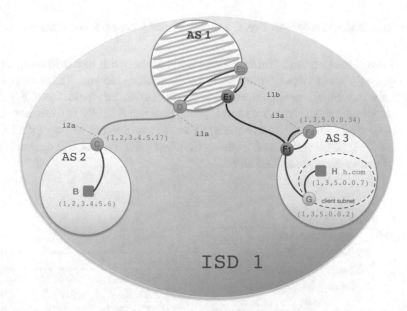

Figure 13.15: Sample topology to illustrate the intra-ISD life cycle of a SCION
packet. AS *1* is a core AS of ISD 1, and AS *2* and AS *3* are non-
core ASes. The red part of the path indicates a traditional IP
connection between AS *1* and AS *3*.

gateway connects non-SCION hosts (as described in §13.3) and sends SCION
packets on behalf of non-SCION hosts.

We start off with an intra-ISD case, i.e., all communication happens *within*
a single ISD. We later extend this simplified example to the inter-ISD case.

Example 1: Intra-ISD Case

Considering the topology depicted in Figure 13.15, we follow a SCION
packet sent from source B to destination H, and we observe how it will be
processed by each router on the path. We show simplified snapshots of the
packet header after each such processing step. The packet header figures below
show the most relevant information of the header, i.e., the SCION path, and IP
encapsulation for local communication.

End host B first queries its local RHINE service (see Chapter 19) for the
SCION address of h.com, which B obtains as (1,3,5.0.0.7). Next, B
queries its local path server for a down-segment to AS *3*, in which destina-
tion host H is located. The local path server (possibly after connecting to a
core path server) returns up to *k* down-segments from the ISD core down to
AS *3* (where the default value of *k* = 5). Figure 13.15 shows only a single path.
Moreover, the local path server returns up to *k* up-segments from AS *2* to the
ISD core. A path segment consists of the interfaces that each AS uses inter-

nally to refer to its inter-AS links. The interfaces have no significance outside the AS.

End host B selects and combines one up-segment with one down-segment, namely $(\bullet,\text{i2a})(\text{i1a},\bullet)$ up and $(\bullet,\text{i1b})(\text{i3a},\bullet)$ down. These two segments are combined to obtain an end-to-end forwarding path from B's AS to the destination AS. In our example, the resulting SCION forwarding path is IF1($\bullet$,i2a)(i1a,$\bullet$) IF2($\bullet$,i1b)(i3a,$\bullet$). It consists of two *info fields*, IF1 and IF2, and a series of *hop fields* that carry the ingress and egress interfaces of each AS, as described in §5.2.

1) B → C

UDP	P_S=35417, P_D=443
SC	SRC=B@(1,2,3.4.5.6) DST=H@(1,3,5.0.0.7) PATH=IF1($\bullet$,i2a)(i1a,$\bullet$) IF2($\bullet$,i1b)(i3a,$\bullet$)
UDP	P_S=30041, P_D=30041
IP	SRC=B@3.4.5.6 DST=C@3.4.5.17
Eth	SRC=B, DST=C

2) C → D

UDP	P_S=35417, P_D=443
SC	SRC=B@(1,2,3.4.5.6) DST=H@(1,3,5.0.0.7) PATH=IF1($\bullet$,i2a)(i1a,$\bullet$) IF2($\bullet$,i1b)(i3a,$\bullet$)
Eth	SRC=C, DST=D

3) D → E2

UDP	P_S=35417, P_D=443
SC	SRC=B@(1,2,3.4.5.6) DST=H@(1,3,5.0.0.7) PATH=IF1($\bullet$,i2a)(i1a,$\bullet$) IF2($\bullet$,i1b)(i3a,$\bullet$)
UDP	P_S=30041, P_D=30041
IP	SRC=D DST=E2
Eth	SRC=D, DST=E2

4) E2 → E1

UDP	P_S=35417, P_D=443
SC	SRC=B@(1,2,3.4.5.6) DST=H@(1,3,5.0.0.7) PATH=IF1($\bullet$,i2a)(i1a,$\bullet$) IF2($\bullet$,i1b)(i3a,$\bullet$)
IP	SRC=E2 DST=F2@5.0.0.34
Eth	SRC=E2, DST=E1

5) E1 → F1

UDP	P_S=35417, P_D=443
SC	SRC=B@(1,2,3.4.5.6) DST=H@(1,3,5.0.0.7) PATH=IF1($\bullet$,i2a)(i1a,$\bullet$) IF2($\bullet$,i1b)(i3a,$\bullet$)
IP	SRC=E2 DST=F2@5.0.0.34
Eth	SRC=E1, DST=F1

6) F1 → F2

UDP	P_S=35417, P_D=443
SC	SRC=B@(1,2,3.4.5.6) DST=H@(1,3,5.0.0.7) PATH=IF1($\bullet$,i2a)(i1a,$\bullet$) IF2($\bullet$,i1b)(i3a,$\bullet$)
IP	SRC=E2 DST=F2@5.0.0.34
Eth	SRC=F1, DST=F2

7)   F2 → F1

UDP	P_S=35417, P_D=443
SC	SRC=B@(1,2,3.4.5.6) DST=H@(1,3,5.0.0.7) PATH=IF1(•,i2a)(i1a,•) IF2(•,i1b)(i3a,•)
UDP	P_S=30041, P_D=30041
IP	SRC=F2@5.0.0.34 DST=H@5.0.0.7
Eth	SRC=F2, DST=F1

8)   F1 → G

UDP	P_S=35417, P_D=443
SC	SRC=B@(1,2,3.4.5.6) DST=H@(1,3,5.0.0.7) PATH=IF1(•,i2a)(i1a,•) IF2(•,i1b)(i3a,•)
UDP	P_S=30041, P_D=30041
IP	SRC=F2@5.0.0.34 DST=H@5.0.0.7
Eth	SRC=F1, DST=G

9)   G → H

UDP	P_S=35417, P_D=443
SC	SRC=B@(1,2,3.4.5.6) DST=H@(1,3,5.0.0.7) PATH=IF1(•,i2a)(i1a,•) IF2(•,i1b)(i3a,•)
UDP	P_S=30041, P_D=30041
IP	SRC=F2@5.0.0.34 DST=H@5.0.0.7
Eth	SRC=G, DST=H

Step-by-Step Explanations

We next explain the packet header modifications at each router, by using the table above. Regarding the notation used in the table, each SRC and DST entry should be read as router (or host) followed by its address, separated by the @ symbol.

1. $\boxed{B \rightarrow C}$ SCION-enabled end host B creates a new SCION packet destined for H with payload P. B learns from the SCION discovery service (see §4.6.2) the mapping from interface fields to IP addresses of the corresponding border routers. For example, the interface i2a (as contained in the combined forwarding path) is mapped to border router C's IP address 3.4.5.17. Based on this information, B knows that it needs to send its packets (for the chosen forwarding path) to border router C, which will then consider the SCION path that B has added to the packet's SCION header. B adds a temporary UDP/IP header for the local delivery to C, utilizing AS 2's internal routing protocol.

   The info field pointer in the SCION header is set to IF1, which is indicated in the packet header figures above by a <u>line</u> below it. The pointer to the current hop field is indicated by a <u>wave</u> below it. Once the information in the path is consumed, the pointers are moved forward.

2. $\boxed{C \rightarrow D}$ Router C inspects the SCION header and considers the info field of the specified SCION path that is pointed at by the current info

field pointer. In this case, it is the first info field `IF1` with its first hop field, which instructs the router to forward the packet on its interface `i2a`. After reading the current hop field, C moves the pointer forward by one position.

Note that, at this point, no IP header is necessary, since the routers C and D are directly connected.

3. $\boxed{\text{D} \rightarrow \text{E2}}$ When receiving the packet, router D checks whether the packet has been received through the ingress interface `i1a` as specified by the current hop field. Otherwise, the packet is dropped by D. The router notices that it has consumed the last hop field of the current path segment, and hence moves the pointer of the current info field to the next info field. There, it starts processing the first hop field, which instructs the router to perform intra-domain routing to transport the packet to the specified egress interface `i1b`, which is on router E2.

4. $\boxed{\text{E2} \rightarrow \text{E1}}$ The dual setup with E1/E2 and F1/F2 models a *router-on-a-stick* configuration (see §13.1.3), in which AS *1* and AS *3* are connected through the BGP-speaking IP routers E1 and F1. This rather conservative setup guarantees that legacy IP traffic is not affected if one of the SCION routers (E2 or F2) fails.

   E2 inspects the current hop field in the SCION header, uses interface `i1b` to forward the packet to F2 (which is part of a static configuration between AS *1* and AS *3*), and moves the current hop-field pointer forward. It adds an IP header to reach F2.

5. $\boxed{\text{E1} \rightarrow \text{F1}}$ Router E1 forwards the IP packet to the IP border router of AS *3*, according to its IP forwarding table.

6. $\boxed{\text{F1} \rightarrow \text{F2}}$ Router F1 performs intra-domain forwarding of the IP packet to F2.

7. $\boxed{\text{F2} \rightarrow \text{F1}}$ SCION router F2 detects the SCION header and realizes that the packet has reached the last hop in its SCION path. Therefore, instead of stepping up the pointers to the current info or hop field, F2 inspects the SCION destination address and extracts the end-host address `5.0.0.7`. It creates a fresh UDP/IP header with this address as destination and with F2 as source. The intra-domain forwarding will first send the IP packet to router F1.

8. $\boxed{\text{F1} \rightarrow \text{G}}$ Router F1 continues the intra-domain forwarding and sends the packet to the next router on the path to H, which in this case is G.

9. $\boxed{\text{G} \rightarrow \text{H}}$ Router G delivers the packet to end host H.

When H sends an answer to the sender, it will flip the source and destination addresses in the SCION header, reverse the SCION path, and set the pointers to

the info and hop fields to the beginning. H sends a response via border router
F2, which has been used for the inbound direction. The address of F2 can be
learned either from the source IP address of the inbound packet or from the
SCION discovery service.

Example 2: Inter-ISD Case

We next discuss the case in which a SCION packet travels from one ISD to
another, as depicted in Figure 13.16.

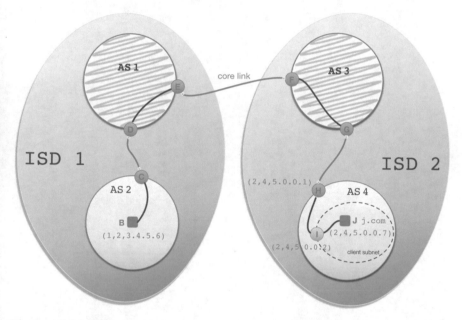

Figure 13.16: Sample topology to show the *inter-ISD* life cycle of a SCION
        packet.

This inter-ISD case is slightly more complex than the previous case inside an
ISD. The increased complexity is not due to the forwarding process (it works
exactly as before with a longer path descriptor), but it comes with a slightly
more complex path resolution process, which we will explain next.

The source end host B requests a path to the destination AS, AS *4* in ISD 2,
from its local path server. The local path server may have a cached path to
the destination AS, or requests one from the core path server located in its ISD
core, AS *1*. In this case, the core path server returns up to *k* down-segments. As
the destination AS resides within a different ISD, the core path server requests
the down-segments from the remote (destination) ISD's core path server and
returns these segments to the local path server, together with core-segments
connecting ISD 1 to ISD 2.

14 SCIONLAB Research Testbed

MATTHIAS FREI, JUAN A. GARCÍA-PARDO, DAVID HAUSHEER, JONGHOON KWON, MARKUS LEGNER, ADRIAN PERRIG, NICOLA RUSTIGNOLI, FRANÇOIS WIRZ*

Chapter Contents

14.1 Architecture **362**

    14.1.1 Infrastructure Operations 363

    14.1.2 Management 365

    14.1.3 User Interface 365

14.2 Research Projects **366**

    14.2.1 Path-Quality Prediction 366

    14.2.2 Monitoring Multipath Communication 366

    14.2.3 Selective Reachability 367

    14.2.4 Trustworthy Internet Infrastructure 367

    14.2.5 Evaluation of SCION over Multiple Fed4FIRE+ Testbeds . . . 367

    14.2.6 A Global-Ring Core Network with KREONET 368

14.3 Comparison to Related Systems **368**

Throughout this book, we have seen how path-aware network architectures enable the creation of new applications and services, as they provide users with advanced communication features such as multipath, efficient key distribution, or geofencing. However, to tap the full potential of these technologies, several open questions must be addressed: For example, which paths and which information should be disseminated to end hosts? What is the API between the network, transport, and application layers? How do the different layers work together to select the best paths with limited overhead? What congestion-control algorithms are suitable when end hosts can switch paths or use multiple paths simultaneously?

To foster research on multipath networking and SCION, we created SCIONLAB, a research and development SCION testbed. The SCIONLAB network is composed of a set of dozens of globally distributed infrastructure ASes managed by the SCIONLAB team, and over 1000 user ASes running on various

---

*This chapter reuses content from Kwon, García-Pardo, Legner, Wirz, Frei, Hausheer, and Perrig [310].

Figure 14.1: Map of SCIONLAB infrastructure ASes.
©OpenStreetMap contributors. OpenStreetMap data is available
under the Open Database License. License details can be found
at https://www.openstreetmap.org/copyright.

systems. The whole network is orchestrated by an open-source coordination
service, which we refer to as the *Coordinator*.

In this chapter, we present an overview of SCIONLAB's architecture and
main components, as well as use cases and research projects powered by
SCIONLAB. We also show what distinguishes SCIONLAB from other testbeds
such as PlanetLab [134] and RIPE Atlas [438].

14.1 Architecture

SCIONLAB is a globally distributed network in which users can easily set up
their own AS, connect and run experiments. Figure 14.2 illustrates how the
network's main components are organized. There are two types of SCION-
LAB ASes: infrastructure ASes (hosted by partner networks and managed by
SCIONLAB administrators) and user ASes, that are created and self-managed
by end users. All SCIONLAB core ASes are infrastructure ASes. User ASes
run on user-provided infrastructure, following the notion of "bring your own
computation" (BYOC). This way, they are free to set up infrastructure that
suits their experiments best.

All SCIONLAB ASes run SCION control services—including a beacon service (BS), a certificate service (CS), a path service (PS), and a COLIBRI service (§10.2.3)—and one or multiple border routers. Some ASes additionally
feature as attachment points that provide connectivity to user ASes. SCION-

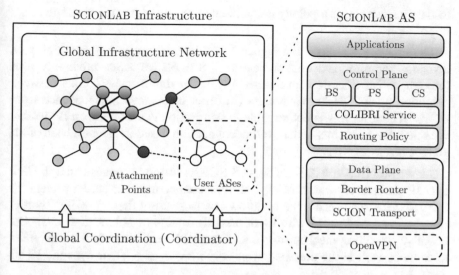

Figure 14.2: Overview of the SCIONLAB architecture. Users obtain ASes and connect to the infrastructure network orchestrated by a global coordination service (the Coordinator).

LAB's topology offers a variety of paths between AS pairs as well as several attachment points.

SCIONLAB operates with full cryptographic support. Each ISD defines its own roots of trust within a TRC. Each AS has a globally unique AS number (ASN), a key pair, and a certificate.

The life cycle of a user AS starts with the user creating an account on the Coordinator hosted at https://www.scionlab.org. The user selects an attachment point, and the Coordinator provides the AS configuration that includes the certificates required for operation. Once a user deploys its own AS, it starts receiving PCBs from the AS of the attachment point, and registers its down-paths in the core path service. From this moment on, the user AS is reachable from other SCIONLAB ASes.

14.1.1 Infrastructure Operations

To guarantee the reliability of the infrastructure, it is essential that ASes and inter-AS links are stable. We achieve this through continuous monitoring and close interaction with the AS host networks. Configuration changes for infrastructure ASes are managed by the SCIONLAB team and orchestrated with Ansible [430].

The infrastructure topology is continually expanding. As of late 2021, the SCIONLAB network consists of a backbone ISD and 8 regional ISDs of more than 50 infrastructure ASes connecting more than 1000 user ASes. The up-to-date infrastructure topology is available at https://www.scionlab.org/topology.

The network topology is designed to achieve high capacity and high path diversity for multipath communication. SCIONLAB ASes preferably peer through native SCION connections, thanks to direct layer-2 links between neighboring SCION border routers (in order to achieve independence from BGP). Whenever this is not possible, a connection is set up via an IP overlay. We select the endpoints of the overlay connection based on latency, bandwidth, and the number of physical hops.

As we describe in §12.1.1, inter-AS SCION traffic is encapsulated in UDP over IP packets, regardless of the link type connecting peer border routers. In SCIONLAB, UDP encapsulation allows us to flexibly deploy SCION border routers, even co-located with traditional IP routers or behind firewalls. Optionally, tunnelling mechanisms (e.g., OpenVPN or GRE) allow users' ASes to connect even in the presence of multiple layers of NAT or dynamic IP addresses.

To achieve additional path diversity, SCIONLAB is augmented by a backbone ISD, deployed on geographically distributed Amazon AWS EC2 instances. Border routers in AWS connect to nearby ASes within their region. The AWS backbone provides low-latency interconnection across regions. As the number of SCIONLAB host networks grows, nodes deployed on global research networks (such as GÉANT, KREONET2, and Internet2) are taking over the role of the global backbone ISD.

14.1.1.1 Attachment Points

Attachment points allow user ASes to connect to the SCIONLAB infrastructure ASes, extending the network topology with the experimenters' own resources. This allows users to experiment with different topologies. Experimenters can grow these resources in their preferred way, while the SCIONLAB infrastructure provides global connectivity. If desired, a user AS can also become an attachment point, enabling a scalable extensibility of the topology.

When a new user AS selects an attachment point, the Coordinator defines a new connection (link) between the border router of the user AS and one border router of the attachment point. The link's configuration includes a unique port number for both sides, and the link type. The attachment point periodically polls the updated configuration from the Coordinator, and downloads it if it has changed. In this way, no human interaction is needed to process these user AS requests. Once the SCIONLAB control-plane packets (i.e., PCBs) are successfully forwarded through the new border router, the new AS becomes a part of the network.

14.1.1.2 Selective Path Properties

It is important for network experimenters to be able to test their ideas under various network conditions. The SCIONLAB infrastructure comprises, therefore, a region where links are purposefully degraded. On these links, bandwidth is constrained to three different rates, latency is increased, and loss is introduced. With the selective path properties, applications can be evaluated under deteriorated but predictable network conditions.

14.1.2 Management

The Coordinator orchestrates the creation and connection of user ASes, by providing AS numbers, cryptographic keys and certificates, and by initializing overlay connections to the attachment points.

As a central authority, the Coordinator issues a unique ISD-AS number to each AS in SCIONLAB. This number follows the scheme described in §2.5: each ISD-AS number is 64 bits long, where the top 16 bits represent the ISD of the attachment point, and the remaining 48 bits indicate the AS. Whenever a user connects to an attachment point, it is assigned the ISD number of the selected attachment point, and an AS number.

For each new user AS, the Coordinator generates a key pair and issues an X.509 certificate [136]. The cryptographic key information is stored in the Coordinator database along with the details of the AS, such as the AS owner, the organization, and the current status including network connectivity.

14.1.3 User Interface

Users can instantiate a user AS through a simple and intuitive web-based interface provided by the Coordinator. Users enter local information (i.e., installation type, border router IP, attachment point), then the Coordinator issues a unique AS number as well as cryptographic keys with certificates and reconfigures the selected attachment point. Depending on the installation type, it provides a Vagrant file, a developer-oriented configuration based on Supervisord or instructions on how to install the Debian packages. In all cases, the AS is automatically instantiated, and the SCIONLAB services, corresponding packages, and configuration are set up. Although SCIONLAB runs on Linux, execution on other systems is supported through a virtual machine (VM).

Most SCIONLAB users rely on an IP overlay to connect their AS to an attachment point. Border routers require bidirectional UDP connectivity between peers. Users behind NAT or with dynamic IP addresses therefore face additional challenges (i.e., configuring port forwarding if behind NAT, or updating IP if using a dynamic IP). To facilitate such cases, users can establish a link to their attachment point over a VPN connection—this also enables traversal of multiple layers of NAT devices.

14.1.3.1 User AS Capabilities

SCIONLAB users receive a full-fledged AS, participating without limitations in the control plane. They can configure the routing policy of the AS. Each AS can also set up native SCION end hosts that can use the local AS infrastructure to communicate with other SCIONLAB hosts. The control plane enables user ASes to interact with the rest of the network and implement their own fine-grained routing decisions. Through beaconing to neighbor ASes and registration of paths in the core path servers, user ASes propagate connectivity information and thus announce their presence. Users can control their path-exploration policies through several parameters or even implement completely new algorithms. In addition, users can constrain the set of available paths to their own AS.

End hosts can also define path policies, by selecting their preferred paths. This allows experimenters to lock end-host traffic to particular paths or avoid certain untrusted or inefficient paths. For example, by defining a path policy with unwanted ASes = {ASes in Backbone ISD}, an AS can avoid all paths that traverse the backbone ISD.

Optionally, end hosts can connect to an existing AS, and communicate with the AS' services to fetch paths and certificate information. This allows experimentation with a native SCION host without having to run a fully fledged AS. The end-host stack is as described in §12.1.

14.2 Research Projects

We now present research projects that leverage the SCIONLAB testbed.

14.2.1 Path-Quality Prediction

In path-aware networking, end hosts need a good understanding of the network topology to choose forwarding paths. Even more, sources must be able to predict the quality of feasible paths to ensure efficient data transmission regardless of dynamic changes of the link condition. Research on path-quality prediction has been conducted on SCIONLAB by changing path conditions (e.g., artificial link failures), and evaluating the performance of the prediction model [298, 543]. Such experiments are difficult to realize on other testbeds.

14.2.2 Monitoring Multipath Communication

With the emergence of multipath communication as facilitated by SCION, monitoring customers' traffic flows across different network paths becomes a challenging task for operators to ensure optimal network utilization, fault tolerance and fairness. Traditional single-path flow monitoring such as sampling-based mechanisms fall short, since packets may be spread across a potentially large number of different paths. To address this challenge, SpeedCam [196] proposes a new approach that aims to achieve scalable and efficient flow monitoring in multipath networks. The approach which has been evaluated with SCIONLAB is based on probabilistic probe selection to significantly reduce the number of required monitoring probes, while enabling an effective gathering of flow information.

14.2.3 Selective Reachability

The unlimited reachability on today's Internet, allowing all entities to send packets to anyone without explicit permission, makes end hosts potentially vulnerable to unwanted traffic. Given the path-awareness offered by SCIONLAB, each entity can express the destination-driven decision about their reachability to the network. They do so by announcing the path availability only to trusted parties, in order to prevent untrusted sources from sending traffic [311].

14.2.4 Trustworthy Internet Infrastructure

2STiC (pronounced "to stick")[420], short for Security, Stability, and Transparency in inter-network Communication, is a joint research program run by Dutch research institutions and network operators. It aims at increasing the trustworthiness of Internet infrastructure through a hands-on approach that leverages existing testbeds and experiments. The program uses SCIONLAB to investigate how multipath routing can mitigate network outages and contributed a SCION P4 implementation [155].

14.2.5 Deployment and Evaluation of SCION over Multiple Fed4FIRE+ Testbeds

In the scope of a Fed4FIRE+ experiment [397], SCION was deployed and evaluated on multiple Fed4FIRE+ testbeds [178], specifically the GEANT Testbed Service (GTS) [226], Virtual Wall [262], the French large-scale testbed Grid5000 [219], and ExoGENI [202]. SCIONLAB facilitates the interconnection between SCION ASes deployed on the different Fed4FIRE+ testbeds as well as with other ASes in the global SCIONLAB network such as SWITCH [497], the Swiss National Research and Education Network, its German counterpart DFN [163], and SIDN [473], the operator of the .nl country-code top-level domain. All those connections were established in an entirely BGP-free manner via native L2 links.

This experiment demonstrated the benefits of path awareness within Fed4FIRE+, showing the diverse path opportunities between the testbeds enabled by SCION, the feasibility of aggregating the bandwidth over multiple paths as well as the possibility to choose low-latency paths and quickly fail over to alternative paths in case of path failures.

14.2.6 Rich Path Diversity: A Global-Ring Core Network with KREONET

Korea Research Environment Open Network (KREONET [291]) is a research and education network operated by the Korea Institute of Science and Technology Information (KISTI). It is a Layer-2 based carrier-grade network composed of a regional and international network: 1) KREONET1 is a regional research network that connects Korean universities, research institutes, and public institutions, providing collaborative research environments for national research activities, and 2) KREONET2 is its international counterpart, connecting to more than 36 research networks operated in the US, China, Russia, Canada, and Europe. It provides connection-oriented network service and is being used to develop high-speed data transmission, next-generation Internet, and availability guarantee technology.

Thanks to a close collaboration since 2017, KREONET is currently being used as one of SCIONLAB's critical core networks. In January 2017, with two nodes deployed in the Seoul and Daejeon data centers, it participated as the first Asian regional ISD. Then, in July 2019, SCIONLAB infrastructure nodes connected with 20G links were deployed to five PoPs in Chicago, Seattle, Amsterdam, Daejeon, and Hong Kong. The nodes form the SCIONLAB core network in the northern hemisphere, connecting Europe-America-Asia. In particular, the layer-2 peering connection to European regional research networks (e.g., GEANT) and the American regional research networks (e.g., Internet2) supports direct connectivity with other SCIONLAB nodes deployed at other research networks, forming a cooperative core network. In 2022, an establishment of a new KREONET PoP in Singapore is planned. By deploying a new SCIONLAB node in the Singapore PoP, we expect to finalize the SCIONLAB global-ring core network (connecting Amsterdam-Chicago-Daejeon-Hong Kong-Singapore-Amsterdam), enriching path diversity.

14.3 Comparison to Related Systems

Two categories of projects are similar to SCIONLAB: projects related to network measurement platforms, and those related to network testbeds. In the category of measurement platforms, RIPE Atlas [438] focuses on diagnosing and troubleshooting real-world network infrastructure [44]. Passive and active measurement systems related to BGP include Route Views [394] and BGP Beacons [352].

Projects in the second category allow researchers to test new networking services. These projects use network testbeds based on dedicated connections, overlays, or emulation. Among the best-known testbeds are PlanetLab [134], EmuLab/Netbed [539], VINI [60], and GENI [65], each of them with a particular focus and catering to a specific class of experiments. Mininet [317] provides the possibility to even run a complete virtual network on a single

Table 14.1: Comparison between SCIONLAB and other network testbeds; the symbols indicate whether the benefit is offered by the respective testbed (●) or not (○).

Capabilities	PlanetLab	Mininet	EmuLab	VINI	GENI	RIPE Atlas	PEERING	SCIONLAB
Global distribution	●	○	○	●	●	●	●	●
Expandable topology	●	●	●	●	●	●	●	●
Run user's code	●	●	●	●	●	○	●	●
Routing control	○	○	○	●	○	○	●	●
Instant participation	○	●	○	○	●	○	●	●
Multipath support	○	○	○	○	○	○	○	●
Path awareness	○	○	○	○	○	○	○	●
Embedded cryptography	○	○	○	○	○	○	○	●

● = offers the benefit; ○ = does not offer the benefit.

machine. All of these testbeds are mainly designed for intra-domain research and do not allow their users to affect the inter-domain routing system. PEERING [458] combines intra-domain routing based on VINI, EmuLab, or Mininet, with some control over the inter-domain routing system.

Compared to all these systems, SCIONLAB provides a number of innovations and new opportunities as summarized in Table 14.1. A first important difference, in particular compared to the measurement platforms, is that SCIONLAB is not intended for research on the current Internet architecture and ecosystem. Instead, SCIONLAB seeks to provide research opportunities on a future Internet architecture providing path-aware networking and security mechanisms. Another difference is that users receive a fully fledged AS, including all cryptographic credentials needed to participate in the global SCIONLAB control plane. There is also the opportunity for user ASes to create additional network connections. Thanks to the "bring your own computation" (BYOC) approach, computation scalability is also available, which further facilitates the operation of the network infrastructure.

15 Use Cases and Applications

David Hausheer, Samuel Hitz, Nicola Rustignoli, François Wirz

Chapter Contents

15.1 Use Cases **372**
   15.1.1 End User: Improved Performance 373
   15.1.2 End User: Guaranteed Communication 373
   15.1.3 End User: Trust and Transparency 374
   15.1.4 Enterprise: Geofencing 374
   15.1.5 Enterprise: Private and Leased Lines Replacement 374
   15.1.6 Enterprise: Multi-Organization Ecosystems 376
   15.1.7 Enterprise: Cloud Connectivity 376
   15.1.8 Critical Infrastructure Operator: High Availability 377
   15.1.9 Critical Infrastructure Operator: Low-Latency Communication 378
   15.1.10 Critical Infrastructure Operator: The Internet of Things (IoT) . 379
   15.1.11 Distributed Infrastructure Operator: Blockchains 381
   15.1.12 Network Operator: 5G Networks 381
   15.1.13 Network Operator: Inter-domain Path Control 382
15.2 Applications **382**
   15.2.1 SCION-Native Applications 383
   15.2.2 High-Speed Multipath File Transfer 383
   15.2.3 Multipath Video Streaming 384
15.3 Case Study: Secure Swiss Finance Network (SSFN) **385**
   15.3.1 SSFN Infrastructure 386
   15.3.2 SSFN Evaluation 387
   15.3.3 SSFN Observations and Outlook 388
15.4 Case Study: SCI-ED, a SCION-Based Research Network . . **389**
   15.4.1 Hercules for Science-DMZ 389
   15.4.2 GeoVITe Proof of Concept 390
   15.4.3 Petabyte Archive: Transmission of Large Science Datasets . . 391

Numerous events have made the inadequacies of today's Internet increasingly apparent. As SCION is able to achieve desirable properties for stakeholders in a wide variety of use cases, there exist many reasons to adopt SCION [1] to overcome those limitations. In this chapter, we highlight some of the most promising use cases and applications that can benefit from SCION. Additionally, we also provide insights into concrete case studies where SCION has been applied successfully.

Those use cases are motivated by several recent trends which are of relevance for a new Internet architecture:

- **Digitalization.** Across the world, the digitalization of government, industry, education, and society in general is rapidly progressing — the COVID-19 pandemic has further accelerated this transition. Digitalization means increased reliance on the Internet for our modern society to function, demanding higher levels of reliability and security.

- **Internet of Things (IoT).** The number of IoT devices is increasing rapidly, creating a huge market for IoT applications. At the same time, weak protection measures for many of those devices and applications raise significant security and privacy concerns, demanding mechanisms to increase IoT security.

- **Cloud computing.** IT infrastructure is shifting from on-premise or local data centers to cloud operators. This comes with two challenges: First, connectivity between users and the cloud becomes critical. Second, there is a trust issue in cloud infrastructure providers operating under a foreign jurisdiction. Moreover, the demand for sovereign and hybrid clouds is emerging.

- **Distributed applications.** Decentralized Internet-based applications such as cryptocurrencies are transforming the financial sector. These applications need a fast and reliable network for their operation.

- **Environmental challenges.** Climate change requires to reduce CO_2 emissions in all sectors including ICT, which consumes an increasingly significant fraction of the world's total electricity. This challenge demands for an increased energy efficiency in the Internet.

15.1 Use Cases

SCION's security and availability properties are valuable for a multitude of applications and stakeholders. We provide an overview of use cases, reflecting each stakeholder's perspective: end users, enterprises, critical and distributed infrastructure operators, and network operators.

---

[1]https://www.scion-architecture.net/pages/reasons/

15.1.1 End User: Improved Performance Thanks to Path Optimization

End users desire high-quality network communication: fast downloads, seamless failover during mobility, and low latency for interactive applications (e.g., voice communication or gaming). The current Internet provides a single path at a time to a given destination. Applications are therefore constrained by the single path's latency and bandwidth. In case of path failures, end users must wait for BGP to converge and react to a failure. Although multipath TCP is becoming available more widely, it only provides only a single path per network interface or IP address. This means that alternative paths with potentially better performance cannot be used. In a native multipath Internet, an application can obtain several paths to the destination and select the ones that best cater to the desired characteristics. Such path optimization will enhance the quality of experience for users: high bandwidth paths for large downloads or streamed video, rapid failover in case of broken links while being mobile, or low-latency paths for immersive gaming.

Multipath positively influences Internet browsing experience as well. Although it is intuitively clear that multiple paths provide improved connectivity over a single path, several previous studies confirm this [109, 335, 525].

15.1.2 End User: Guaranteed Communication for Home Office and Critical Services

An increasing number of employees are working from home and require a reliable access to their company's network. This trend has been accelerated by the Covid pandemic, when home office work became compulsory in some instances. Additionally, more and more critical services such as e-government services, public announcements, or financial services are accessed from home as well. A disruption of home connectivity over a longer period of time would therefore have detrimental consequences. Today's best-effort Internet does not provide any mechanisms to ensure reliable communication with adequate bandwidth and latency guarantees. It is also missing mechanisms for source authentication or rate limiting that can prioritize and protect enterprise and critical traffic.

Several approaches exist to connect end users to SCION as described in §13.1.4. Stronger connectivity guarantees can be provided when an end-user and an enterprise are both connected to SCION ISPs. First, end hosts can obtain end-to-end bandwidth reservations using COLIBRI. Such hosts rely on their ISPs to set up COLIBRI segment reservations and themselves set up end-to-end reservations when needed. This results in end-users having guaranteed bandwidth to communicate to the enterprise. Second, the enterprise can provide hidden paths §8.1 that provide preferential and protected access for their remote employees. In addition, the enterprise can deploy LightningFilter in order to authorize the user's home AS clients and filter out unwanted or attack traffic. Such mechanisms offer strong availability guarantees to end users. When combined with SCION native multipath, they offer all building blocks needed for providing performant and uninterrupted connectivity to home users, even when working on mission-critical services.

15.1.3 End User: Trust and Transparency

Currently prevalent public key infrastructures (PKIs) rely on two models: monopoly and oligopoly. In these models, roots of trust do not scale to a global environment because mutually distrustful entities cannot agree on a single trust root (monopoly model) and because the security of a plethora of trust roots is only as strong as its weakest link (oligopoly model) [51].

RPKI (which is used in BGPsec) and the DNSSEC PKI are examples of the monopoly model, as both rely on a single or small set of keys that serve as roots of trust. SCION's CP-PKI (§3.1) allows each ISD to define its own set of trust roots, along with the policy governing their use. Such approach provides support for global but heterogeneous trust. This model fits well to the nature of the Internet, where its constituents lay in multiple jurisdictions and have divergent interests.

Specifically, compromise outside of an ISD cannot affect operations within the ISD. In addition, geofencing (§15.1.4) gives end hosts the ability to restrict paths where packets travel. These two properties make SCION a good fit for workloads that require a high level of trust in the network infrastructure (i.e., by Military, Government, Healthcare).

15.1.4 Enterprise: Geofencing

In many industries, such as finance, healthcare, or government, the confidentiality of sensitive data must be protected. While network traffic is typically encrypted, studies show that even encrypted traffic can leak information [479].

Thanks to its path-awareness, SCION users can explicitly select ISPs carrying their traffic. This property can be used to set up geofencing rules, where sensitive traffic is only routed on trusted (i.e., sovereign) infrastructure. With such geofencing, we can be certain that makes sure traffic does not leave a certain jurisdiction. With the use of EPIC-SAPV (§10.1.5), the sender can also obtain a cryptographic proof that the packet indeed traversed the desired path, further adding to the strong guarantees offered by SCION.

15.1.5 Enterprise: Private and Leased Lines Replacement

Enterprise WANs traditionally rely on MPLS and leased lines to provide inter-branch and inter-data-center connectivity. These are private connections that are segregated from the Internet. This approach has proven inflexible over time. First, such lines are significantly more expensive than Internet connectivity. Second, they are typically operated by a single ISP, limiting coverage and global reach. Third, they are slow and cumbersome to provision, as the ISP running the private network needs to procure local partners for last mile connectivity. Fourth, those networks typically have a star-like network topology, where the central entity also represents a single point of failure.

Therefore, enterprises are currently transitioning from private lines to SD-WANs and VPNs, that in most cases rely on the regular Internet as an underlay. This means that such deployments are as vulnerable as today's Internet to DoS attacks and hijacks. In addition, recent research [494] shows that VPNs are susceptible themselves to DoS attacks. To increase reliability, VPN overlays are typically established over multiple Internet lines, where traffic is shifted to the best performing line. However, this approach only allows a basic level of path selection, as only the first hop can be selected. This means that SD-WANs/VPNs are unable to provide end-to-end path control, and that they cannot route around failed or congested links in the core of the Internet.

SCION provides a more robust underlay that can be used to augment SD-WAN/VPN solutions. Multipath mechanisms can be used to achieve the same level of availability and reliability as redundant private lines, even in face of DoS attacks. Additionally, multipath can also help to enhance data confidentiality [343]. Moreover, as the global SCION network is composed of federated networks similar to today's Internet, there are no limitations due to a single provider or SD-WAN/VPN vendor.

We consider a company C that aims to guarantee communication between branch offices and a headquarter or data center. These elements are needed:

- Each location and branch office forms a SCION AS.

- Each AS is connected via one or more SCION Internet links, depending on availability requirements. Such links can be provided by the same or different ISPs.

- Access links can be public or remain hidden. SCION hidden paths §8.1 offer an alternative to leased lines. Such paths are not publicly available and cannot be revealed by remote network scans, allowing secure and reliable communication between mutually trusted organizations.

- The service-level agreement (SLA) between C and its ISP includes the availability of COLIBRI §10.2 reservations.

- C's ASes set up and regularly renew COLIBRI reservations so that traffic between branches is forwarded with strong bandwidth guarantees.

- Additionally, C can deploy LightningFilter §9.2 or an existing network-layer security mechanism to filter incoming traffic.

15.1.6 Enterprise: Multi-Organization Ecosystems

In many sectors, reliable communication between multiple organizations is critical and strong availability guarantees are necessary. As an example, in power grids, all operators and power stations must exchange real-time data, so that energy production exactly meets demand. Financial institutions need to exchange real-time transaction information with low latency. Traditionally, cross-organization connectivity is built with site-to-site VPNs over the Internet or dedicated lines. These approaches, however, suffer from either of the Internet's unreliability or the limitations of private lines.

SCION's core goal is to achieve comparable availability properties—such as traffic isolation, minimal bandwidth guarantees even under active attacks, and resilience against re-routing attacks—in a more robust, more flexible, and economical network. The additional capital and operational expenditures to run SCION are marginal because it can reuse the existing IP or MPLS-based network, requiring only a few additional standard servers or VMs. With SCION, high security and reliability can be achieved in a public network where infrastructure is managed by different entities (similar to today's Internet).

Leased-line properties in a multi-organization ecosystem can be achieved using SCION hidden paths:

- Each organization has an AS and obtains native SCION access links from two independent ISPs. This ensures that the failure of any single ISP cannot disrupt communication.[2]

- All organizations and ISPs set up and enable COLIBRI.

- Each organization sets up COLIBRI reservations to other members that require a high-availability communication link, and periodically renews them such that valid reservations are available at all times.

- Additionally, members can deploy LightningFilter or an existing network-layer security mechanism to filter incoming traffic.

Payment processing networks are an example of a multi-organization ecosystem. In §15.3 we present in detail a case study on the SCION deployment for the Swiss Interbank Clearing (SIC) system, processing payments for hundreds of billions Swiss Francs every day [47, 379].

15.1.7 Enterprise: Cloud Connectivity

With enterprise workloads moving from on-premise data centers to public clouds, connectivity between users and the public cloud infrastructure becomes critical. However, by default public clouds rely on plain Internet connectivity.

---

[2]Availability even during the failure of a single ISP is guaranteed under the assumption that every tier-2 ISP is a customer of at least two different tier-1 ISPs.

This is often not sufficiently reliable nor performant enough for enterprise applications. In order to remediate the shortcomings of plain Internet connectivity and improve reliability, many public cloud providers offer dedicated private connectivity between their data center and enterprises. Some examples are Azure Express Route, AWS Direct Connect, or Google Direct Interconnect.

Cloud providers can adopt SCION to enable their customers to reliably and efficiently connect to cloud data centers, without the need for dedicated solutions. This comes with the advantage that customers could leverage the same SCION connections for reliable connectivity to multiple cloud sites, and to their partners.

15.1.8 Critical Infrastructure Operator: High Availability

Highly available communication is important in many contexts, in particular for critical infrastructures such as financial networks and industrial control systems used for power distribution. Internet outages have been known to wreak havoc on day-to-day operations, for example preventing ATM withdrawals or payment terminal operations [505]. SCION's control-plane isolation through ISDs, its stable data plane, its built-in DoS defense mechanisms, and its multipath operation all contribute to higher availability.

Business continuity is currently highly dependent on communication. We can witness the increasing inter-connectedness required for business operations when network outages cause a disruption of a surprising number of operations. For instance, when Telecom Malaysia wrongly announced 179,000 IP prefixes to Level3, it caused global outages for 2 hours, even affecting ATM operations in Sweden [505].

Here are a few examples of sectors where availability is crucial:

- Financial services require highly available communication networks, for instance for the distribution of stock market data, real-time market trading, or transaction processing. While critical communication is often sent over leased lines, it is not economical to pervasively use leased lines between all communicating parties. In this setting, SCION can offer high availability across a public network infrastructure.

- Critical command-and-control infrastructures—such as air-traffic, power-grid, rail, or power-plant control systems—require very high communication availability. Communication disruptions can lead to outages with significant cost for industry and danger for society.

- Governments require high communication availability especially during crisis situations. Examples of critical communication include connectivity to embassies, military, and blue-light organizations.

Conversations with some critical infrastructure operations revealed that operators today still operate on legacy segmented private networks (i.e., MPLS, leased lines) that are completely segregated from the Internet. While enterprises shift from expensive private lines to Internet-based solutions (SD-WAN, SaaS applications), such operators are hesitant to change, due to security and reliability concerns. SCION bridges this gap, allowing critical infrastructure to run reliably over the Internet. Thanks to its built-in DoS defense mechanisms as discussed in §11.2, SCION is able to provide high availability even if the network is under attack. The DoS defense mechanisms ensure that paths can be discovered, traffic reaches its intended destination, and unwanted traffic can be filtered at very high rates.

15.1.9 Critical Infrastructure Operator: Low-Latency Communication and LEO Satellite Networks

Low-Earth-Orbit Satellite Networks. Several "NewSpace" companies have launched the first wave of thousands of planned satellites for providing global broadband Internet service. The resulting low-Earth-orbit (LEO) satellite networks (LSNs) will not only bridge the digital divide by providing Internet connectivity to remote areas, but they also promise much lower latency than terrestrial fiber for long-distance routes [73]. These networks have been designed and deployed as "eyeball" ISPs: Customers directly connect to the satellite network using proprietary antennas; the satellites then relay traffic to ground stations; and finally ground stations provide connectivity to the Internet. The internal operation of the LSN is therefore a hidden *black box* to external networks and users.

In principle, great benefits can be achieved by more closely integrating LSNs into the Internet fabric [208, 296]. First, ISPs and end hosts could directly benefit from the low-latency nature of satellite networks—even without owning a terminal directly—and could optimize for the lowest-latency end-to-end paths (as opposed to just the first hop). Second, this integration would rekindle competition not only in the last-mile market, but in the transit and IXP markets as well, possibly creating new service models such as "AS-peering in Space".

However, satellite networks are challenging to integrate into today's Internet for a variety of reasons. Foremost, single-path communication cannot route all traffic to these specialized low-latency networks because of their much higher cost of operation but also due to the limited bandwidth available on those networks. Ideally, only critical traffic is forwarded to the LSN. Further, LSNs provide inherently variable connectivity due to satellite motion and atmospheric phenomena, which today's Internet is ill-suited to accommodate. The connection and disconnection events caused, e.g., by rain fade would generate unsustainable amounts of route churn—resulting in permanent routing instability—if directly exposed to BGP.

Microwave Networks. A similar breed of low-latency, low-bandwidth networks is gaining widespread deployment thanks to the demand for fast and long-range communications from the financial trading industry. These networks are composed of chains of microwave towers linking trading hubs, as for example between Zurich–Frankfurt–London or Chicago–New York, and drastically reduce communication latency compared to the regular Internet or leased lines.

These networks offer comparable opportunities and challenges to LSNs. Bhattacherjee et al. [74] explore the promise of using microwave networks for more than just trading, but also highlight the difficulty of (i) mitigating atmospheric losses and (ii) directly integrating these networks in the user-facing Internet.

Low-Latency Networks in SCION. A path-aware and multipath Internet infrastructure is then ideal to integrate such specialized networks into the Internet fabric.

First, in SCION paths can be selected by the end hosts to match the requirements of the application at hand. Thus, only critical and latency-sensitive traffic can be transmitted using the higher-cost, higher-performance networks (LSN or microwave). The rest of the traffic can be forwarded through the inexpensive and high-bandwidth fiber path. Second, beacons can be modified to disseminate a compressed representation of the weather forecasts above ground stations. These forecasts can then be used by end hosts to infer the real-time status of the ground-to-satellite or tower-to-tower microwave links [208]. This drastically reduces the amount of global routing announcements and allows for further path optimization. Finally, COLIBRI (§10.2) can be used to efficiently allocate and bill the constrained bandwidth in these networks, increasing profitability for network operators while enabling new use-cases at the same time.

15.1.10 Critical Infrastructure Operator: The Internet of Things (IoT)

With the emergence of low-power wide-area communication technologies, such as LoRa WAN and 5G additions, new IoT devices are emerging and are being connected to the Internet. In the industrial sector, the long-predicted IoT revolution is now taking place, fueled by artificial-intelligence (AI) and machine-learning (ML) technologies that enable the optimization of supply chains and manufacturing. We increasingly rely on automated data gathering— for real-time applications or to make our systems learn and adapt to our needs. Examples include driverless cars, smart homes, factories, or general sensors, e.g., to detect and quickly report wildfires. Many new applications will be connected via the cellular network due to their mobile nature or remote location.

Connecting many IoT devices comes with a series of challenges. The large number of deployed devices and lax security in many of them make them a prime target for cyber attacks. As IoT vendors are under high cost pressure, numerous systems have serious security vulnerabilities. Examples include systems with easy-to-crack default passwords that do not force the user to change them on initial configuration, and zero-day vulnerabilities that can be exploited by hackers. On the user's end, many people do not regularly patch their systems. Barely noticeable IoT devices (e.g., in the context of a smart home) will most likely receive even less care. The problem is exacerbated by the remote nature of some devices such as sensors, and the lack of convenient user interfaces, constituting another hurdle for keeping our systems up to date and secure. With this in mind, an interesting question is what properties are desirable for networks handling large-scale IoT deployments—and how SCION could be used in the protocol stack to make them a reality.

SCION offers many features that can support IoT applications, especially in terms of security, privacy, and availability. IoT applications can utilize the path-aware networking properties of SCION to realize geofencing for their traffic, and achieve high availability with fast failover and multipath networking. Furthermore, with SCION's Dynamically Recreatable Key (DRKey) approach (§3.2), IoT applications can benefit from a fast and secure authentication mechanism, which helps them to protect against unautorized access to IoT devices. Additionally, source authentication also helps to lower the attack surface for DDoS attacks by blocking malicious unauthenticated traffic using Lightning-Filter (§9.2). Furthermore, service-level objectives can be guaranteed, even during a DDoS attack, thanks to COLIBRI (§10.2), SCION's global quality-of-service (QoS) system.

S3MP [272] is a secure SCION-based platform for smart-metering applications. It demonstrates how SCION can be utilized in a typical smart-metering network to achieve security and fault tolerance. Leveraging SCION, S3MP achieves geofencing, resilience against DDoS and man-in-the-middle attacks, and fast failure recovery. Through a prototype implementation, S3MP shows that existing libraries can be extended to support SCION. It also shows that applications can utilize SCION without any modification using the SCION–IP gateway (§13.3).

LINC [273] is an industrial network gateway designed to provide low cost and secure interconnectivity for industrial control systems (ICS). Traditionally, ICSes are interconnected over dedicated private networks such as leased lines or MPLS backbones due to their high security and availability requirements. This incurs high cost and management overhead to establish and maintain these networks. LINC makes it possible to provide connectivity for ICSes over a SCION-enabled public network without compromising security while providing high availability through its various failover modes. LINC supports both fast failover and redundant multipath communication for the highest reliability in the communication channels. Additionally, LINC can be programmed to select the optimal path for an application based on path metrics. All these features are made possible by the path-aware networking of SCION, which also enables LINC to support geofencing.

15.1.11 Distributed Infrastructure Operator: Blockchains

Public blockchain systems rely on decentralized networks, where each participating node is generally operated by an independent entity. Because blockchains rely on a consensus algorithm to process transactions, these systems require a functioning communication infrastructure for their operation. While such systems are designed with fault tolerance in mind, a minimum number of nodes need to be online to reach consensus. In addition, several algorithms rely on the availability of a leader node in order to make progress. Such nodes are therefore particularly vulnerable to DoS attacks, which have the potential of slowing down, if not halting, the consensus mechanism.

The decentralized nature of the consensus nodes complicates usage of a SD-WAN solution or DDoS prevention systems, therefore current blockchains use standard Internet-based communication. This approach suffers from Internet outages, routing/DoS attacks. For example, Bitcoin has been demonstrated to be prone to blockages and double spending in case of routing attacks to its nodes [32].

SCION is well suited to provide reliable connectivity to nodes, as it offers strong communication guarantees even in a heterogeneous provider environment, while at the same time optimizing node to node communication (i.e., to reduce latency). Additional DoS resilience can also be obtained thanks to LightningFilter (§9.2), providing high performance filtering by only allowing authenticated traffic between nodes.

15.1.12 Network Operator: 5G Networks

New technologies in cellular and wireless communications are rapidly developed and deployed. New standards for connecting billions of devices–called massive machine-type communication (mMTC), massive Internet of Things (mIoT), or low-power wide-area networks (LPWAN)–are emerging and are integrated in 5G and 6G deployments.

Today's 5G networks strive for performance in terms of low latency and high reliability on the access network, but they need a QoS infrastructure to achieve predictable performance in the 5G backbone across several providers as well. SCION/COLIBRI can provide the QoS fabric across heterogeneous providers.

Another challenge is network slicing in the 5G backbone, enabling resource separation and access control. Network slicing enables to create application-specific "zones" spanning the network towards the participating entities. The goal is to ensure isolation between different applications to protect against network-wide threats outside the zone. SCION already provides control-plane isolation between ISDs. Additionally, recent research on SCION-based network virtualization (MONDRIAN [312]) is being considered as a basis for the zoning mechanism.

MONDRIAN is a new network zoning architecture that secures inter-zone communication—which operates on layer 3, supporting heterogeneous layer 2 architectures—while ensuring scalable cryptographic-key management and flexible security policy enforcement. MONDRIAN flattens the current hierarchically-complex network zone topology into a collection of horizontal zones connected to a Zone Translation Point (TP), a unified security gateway interconnecting zones over the insecure public Internet, thus simplifying inter-domain network virtualization. With MONDRIAN, complex application-specific zone initializations and migrations are simplified, while ensuring source authentication, zone transition authorization, and illegitimate access filtering at the ingress point of zones.

15.1.13 Network Operator: Inter-domain Path Control

In the legacy Internet, only rudimentary forms of inter-domain path control and traffic engineering are possible for an AS. For outgoing traffic, one can at best control the next ISP, but only when an AS is multi-homed. For incoming traffic, a limited amount of path control is available to operators, thanks to techniques as BGP MED, AS path prepending, selective announcements. As we discuss in §6.2.2.1, such techniques are often filtered by upstream ASes, reducing their effectiveness .

In intra-domain networks, software-defined networking (SDN) has revolutionized path control; for example, Google has achieved higher network utilization with their B4 system [270]. Analogous to B4's intra-domain path control, SCION makes inter-domain path control available through path registration. An AS can select which down-segments are announced to the path servers. Hidden paths (§8.1) can be used, which are only communicated to senders who are selected to use them. In addition, ongoing research is exploring wether consent routing scheme could provide a mechanism for traffic senders and receivers to mutually agree on a network path. In conclusion, while SCION gives much path control to the sender, who can select which end-to-end path the packet will follow, network operators retain the ability to select which down-paths to use for incoming traffic.

In addition, as we discuss in Chapter 16, SCION enables CO_2-based traffic optimizations, so that traffic can be steered on lower carbon intensity paths.

15.2 Applications

After introducing SCION use-cases, we now look at some end host applications that benefit from SCION, before finally looking at specific deployments and case studies.

To simplify the use of SCION's native multipath capability for application developers, research is currently ongoing on expanding existing network sockets into multipath sockets, providing built-in bandwidth aggregation and failover [200, 571]. Such multipath extensions are designed to support different kinds of applications, such as file transfers or video conferencing. For instance, PARTS [197] is a current project on transport protocols using SCION's multipath capability, providing efficient and reliable data transfers optimized for file transfer applications. More details on some of these ongoing efforts can be found in §12.2.

15.2.1 SCION-Native Applications

To facilitate the creation and development of new path-aware applications, we have created a small set of native SCION applications [572] that use SCION libraries to perform basic functions. These applications are useful not only as examples for the use of SCION libraries, but also as current tools for debugging and performance analysis. All of these applications and tools are kept up to date with the current version of SCIONLAB, and are also meant to be run and tested using SCIONLAB, as well as local deployments.

Some of these applications are:

bat Similar to `curl`, it interacts with HTTP servers.

bwtester Measures the bandwidth of a path. It comprises a `bwtestserver` and `bwtestclient`.

netcat Similar to the well-known `netcat`, this application sends data using a specified path. It can also receive data, in which case it prints it to `stdout`. It works with both UDP and QUIC as transports.

skip Browser integration for SCION. It uses a proxy auto-config file to forward all requests with a SCION destination to a proxy server running as a native binary on `localhost`.

web-gateway A SCION web server that proxies web content from the TCP/IP web to the SCION web.

15.2.2 High-Speed Multipath File Transfer

SCION's path-awareness feature promises performance improvements for applications by avoiding congested paths and being able to aggregate bandwidth over multiple paths. Therefore, a number of ongoing projects aim to achieve high bandwidth for file transfers by leveraging multipath over SCION.

Using Linux kernel features such as AF_XDP [191], Hercules [386, 570] provides fast point-to-point file transfers over multiple paths at a bandwidth of up to 100 Gbit/s on a single host. AF_XDP allows Hercules to bypass large parts of the Linux network stack, enabling significant performance improvements compared to userspace applications.

Supporting even multipoint-to-multipoint transfers, Bittorrent over SCION [198, 300] adds multipath support to the Bittorrent protocol. Based on the initial SCION Swarm work [445], it enables connections between peers over multiple paths. This way Bittorrent over SCION is able to aggregate the available download and upload bandwidth to achieve high-speed transfers. Bittorrent over SCION runs as a userspace application, but benefits from several optimizations in the SCION libraries to achieve file transfers in the Gbit/s range. Additionally, the application relies on a replacement of the SCION dispatcher based on eBPF as described in §12.1.1.

15.2.3 Multipath Video Streaming

Video streaming is also a promising candidate to benefit from multipath communication. SCION Video Setup [199] is an ongoing project that focuses on multipath video streaming over SCION, in order to increase the bandwidth and reduce the latency for a better streaming experience.

Video calls are particularly sensitive to changing network conditions and SCION presents the opportunity to leverage path-awareness to optimize video call Quality of Experience (QoE). The SCION-WebRTC [204, 205] project combines SCION and WebRTC in an iOS/macOS application and performs video call-specific path selection at the application layer. Latency and packet loss metrics are measured in real-time to discover the live conditions of numerous available paths. SCION-WebRTC uses these QoS metrics to select paths that have favorable network conditions and are likely to yield a high call QoE. Since the app performs path selection at the application layer, WebRTC metrics like video/audio freezes, video resolution and video frame rate can also be considered in the path selection process. These metrics more closely reflect the call QoE than network QoS metrics do, and SCION-WebRTC continuously monitors them during a call to decide whether to switch to a different path for outgoing call traffic. This path selection strategy has shown to substantially increase the quality of video calls in case of sudden bandwidth restrictions.

SCION-WebRTC also makes use of redundant transmission as an additional method to increase the reliability of video calls. Call traffic, especially the critical and low-bitrate audio track, can be sent redundantly over disjoint paths. The redundancy is able to mask adverse events such as packet loss or high packet delay variation as long as it does not affect all redundant packets. Experiments have also shown the efficacy of this technique [204].

15.3 Case Study: Secure Swiss Finance Network (SSFN)

Payment information between Swiss banks is exchanged via the Swiss Inter-bank Clearing (SIC) system. In 2018, it processed payments for 156 billion Swiss francs per day [47, 379]. According to the Swiss National Bank, "an operational disruption or indeed a temporary failure of the SIC system would greatly impair cashless payment transactions in Swiss francs" [379].

The Secure Swiss Finance Network (SSFN) therefore started as a vision initiated by the Swiss National Bank (SNB). The goal was to build a next-generation communication fabric for the financial sector in Switzerland fit for the needs of the 21st century. The SSFN as a network infrastructure should have the following properties:

- Highly available and resilient to support system-critical applications like the Swiss Interbank Clearing (SIC) real-time gross settlement (RTGS) system.

- Highly secure to thwart cyber-attacks at the network level such as traffic hijacking or DDoS attacks.

- Flexible to support any-to-any communication to support new use cases such as open banking.

- Strict governance and control of the entities that can join the SSFN, both as a user or a service provider.

SCION was identified as the ideal network technology to meet the re-quirements of the SSFN. Through its native multipath capability and the ability to seamlessly include multiple network service providers in a fed-eration, SCION leads to exceptional availability and resilience of the net-work infrastructure and end-to-end connectivity. SCION's control-plane PKI and path-authorization property prevent any form of traffic hijacking (see Chapter 7) while COLIBRI (§10.2) and hidden-path groups based on EPIC (Sections 8.1 and 10.1) can be used to thwart DDoS and remote exploitation attacks.

A SCION network enables Internet-like communication, i.e., in principle any AS can communicate with every other AS thus enable the flexibility of any-to-any communication without requiring star-shaped network topologies, which are suboptimal due to the single point of failure at the central node. Finally, the concept of the SCION ISD provides an ideal mechanism to imple-ment strict governance and access control (through the issuance of AS certifi-cates) independently of any network service provider.

In 2019, the concepts of an SSFN based on SCION technology have been worked out and throughout 2020 and the first half of 2021 a pilot has been con-ducted to evaluate the feasibility and effectiveness of the SCION-based SSFN

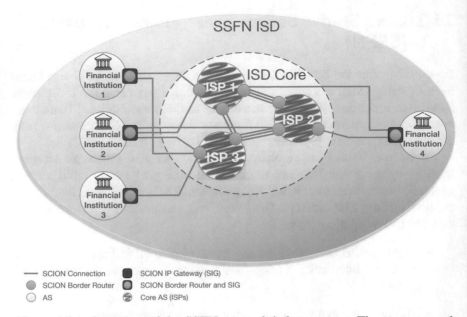

Figure 15.1: Overview of the SSFN network infrastructure. The core network
is provided by three independent service providers. Participants
connect to the SSFN via one or more of the service providers
and can communicate with every other participant on the network.
This flexibility is crucial to support future use cases such as open
banking.

that culminated in the official announcement of the SSFN in July 2021 [132].
In the following, we describe the SSFN infrastructure in further detail and
present the results of the technical evaluation that was performed as part of the
pilot phase of the SSFN.

15.3.1 SSFN Infrastructure

The SSFN is implemented as a SCION ISD, where the core is formed by a
federation of three network service providers (core ASes) deploying SCION
routers at the borders of their network. The SCION routers are then interconnected to form the redundant core network that powers the SSFN. Members of
the SSFN—i.e., financial institutions—are directly connected to one or more
of these core ASes. They are themselves SCION ASes that additionally deploy SCION–IP gateways (SIGs). These SIGs provide the entry points for
members to use the SSFN as a communication platform—they transparently
enable traditional IP-based communication to use a SCION network ("IP-in-
SCION encapsulation") (see §13.3). Secondly, SIGs are in control of which
network path to choose when sending data through the SSFN. This path control

combined with the availability of many different paths (native multipathing) is what enables the exceptional reliability of SCION-based networking.

While network service providers supply network infrastructure, the governance of the SSFN ISD lies with the SNB and SIX (the main financial infrastructure and service provider in Switzerland) as well as one of the network providers. Those three are the voting ASes of the SSFN and thus in charge of creating and maintaining the TRC. Furthermore, SIX acts as a certificate authority (issuing AS) of the SSFN and thus has control over who can join the SSFN ISD through the issuance of AS certificates. As such the SSFN makes great use of the separation of concern enabled by SCION's CP-PKI (see §3.1): the network service providers act as infrastructure providers while the governance of the communication platform lies with the governing entities, i.e., the SNB and SIX plus one network provider.

15.3.2 SSFN Evaluation

To evaluate the resilience of the SSFN, two financial institutions were each connected to two network providers at two different sites. At each site, a SIG is deployed acting as the entry point to the SSFN core network. Each SIG offers a SCION tunneling interface towards every other SIG on the remote destination. IP prefix information needed to route traditional IP traffic between the two financial institutes via the SSFN is learned locally through an interior routing protocol and redistributed between SIGs using the SCION Gateway Routing Protocol (SGRP) (see §13.4.1) ensuring dynamic and seamless integration of the SIGs in the existing network infrastructure of the financial institutes.

Finally, two test instances of the SIC application have been set up to evaluate the impact of the various test scenarios on one of the most system-critical applications of the Swiss financial market. The SIC application is a classical client–server architecture, where the client and server establish a secure session over which clearing information is exchanged.

To demonstrate the reliability and resilience of the SSFN, a series of tests have been performed and their results evaluated. The tests involved various failure scenarios including

- link failure in the access network,
- link failure in the core network,
- complete SCION router failure in the core network,
- complete SIG failure (site failover), and
- complete failure of an entire network provider.

As a baseline, two continuous and active SIC sessions were maintained between a pilot bank and the central SIC test system and the effect of each failure scenario was observed on those active sessions. Furthermore, failover time on the network level was measured. In a final step, the results of each failure scenario in the SSFN were compared against the results obtained using the existing financial network infrastructure (where possible).

Table 15.1: Test results and comparison to existing infrastructure. "Gateway failure" refers to a failure of the SIG in SSFN or a failure of the access router in the existing network. Several tests could not be performed in the existing network as it could have led to a widespread outage; those are marked by "—"

Scenario	SSFN		Existing Network	
	Failover Time	Session upheld?	Failover Time	Session upheld?
Link failure (access)	< 1 s	✓	> 3 min	✗
Link failure (core)	< 1 s	✓	—	✗*
Core router failure	< 1 s	✓	—	✗*
Gateway failure	< 5 s	✓	> 3 min	✗
Provider failure	< 1 s	✓	—	✗*

*The outcome is estimated, as this test could not be performed in the existing network.

Table 15.1 shows the results of the various test scenarios. It clearly demonstrates the exceptional reliability and resilience of the SSFN infrastructure. Throughout the entirety of the testing, no single active SIC session needed to be re-established and clearing information was flowing uninterrupted—despite all the failure cases that were evaluated. Looking at the effective failover times (i.e., the time needed to find and switch over to a working network path), the SSFN shows a sub-second failover time in most cases. Only when an entire site failed, the site-to-site failover took in the order of 5 seconds, but even that did not lead to an interruption on the application level. Compared to the results of the existing network infrastructure, SCION shows a clear improvement in terms of reliability. With legacy infrastructure, in every single scenario a SIC session had to be re-established. The reason for that is not only insufficient failover time but also that network endpoints change because of the failover requiring the establishment of a new application session.

15.3.3 SSFN Observations and Outlook

The Secure Swiss Finance Network is the next-generation communication infrastructure for the Swiss financial market and as such part of critical infrastructure. It has been designed to meet high requirements in terms of availability, resilience, security, flexibility, and governance. The new network infrastructure relies on SCION as the underlying technology to achieve its goals.

After two years of conception, implementation, and evaluation the SCION-based SSFN became reality in 2021. Extensive evaluation has shown the exceptional resilience of the infrastructure against a wide range of failure conditions. Furthermore, the inherent security mechanisms prevent many attacks on the network fabric while also enabling strict governance and access control. As such, the SSFN is a perfect case study to show the potential and applicability of SCION to build critical communication infrastructures.

15.4 Case Study: SCI-ED, a SCION-Based National Academic and Research Network

The SCION for the ETH Domain (SCI-ED) project[3] made SCION available to Swiss higher education and research institutions. Specifically, the project connected ETH Zürich (ETHZ), EPF Lausanne (EPFL), Paul Scherrer Institute (PSI), Swiss National Supercomputing Centre (CSCS), the Swiss Federal Institute for Forest, Snow and Landscape Research (WSL), the Swiss Federal Laboratories for Material Science and Technology (EMPA), the Swiss Federal Institute of Aquatic Science and Technology (EAWAG), and the Singapore-ETH Centre (SEC). During the two-year project (2019–2021), a number of use cases corresponding to the specific needs of those partner institutions were explored. Thanks to a collaboration with SWITCH (the Swiss NREN [281]) and Anapaya Systems, SCI-ED also allowed to build up operational experience in building and connecting a production SCION network. A sample AS deployment in SCI-ED is depicted in Figure 15.2.

The focus of the project was on the following core use cases:

- High-performance data transmission (see §15.4.1, §15.4.2, §15.4.3),
- Secure communication of sensitive information,
- High availability of critical infrastructure, and
- Network platform enabling research in computer networks, security and applications.

15.4.1 Hercules for Science-DMZ

Science DMZ [150] has been proposed as a set of network design patterns that address specific needs of scientific computing, with focus on performance. SCION offers several mechanisms that focus on high-throughput communication. For example, the host-to-host throughput achievable through the high-speed multipath file transfers using Hercules (cf. §15.2.2) is depicted in Figure 15.3.

In a use case, Hercules enabled a 2.5x speedup versus wget when used to download the large-scale data sets at PSI using SCION. This speedup was achieved based on a dataset of 5455 files of 4 MB and 1771 files of 26 MB with a total size of 69.8 GB.

---

[3]https://scied.scion-architecture.net

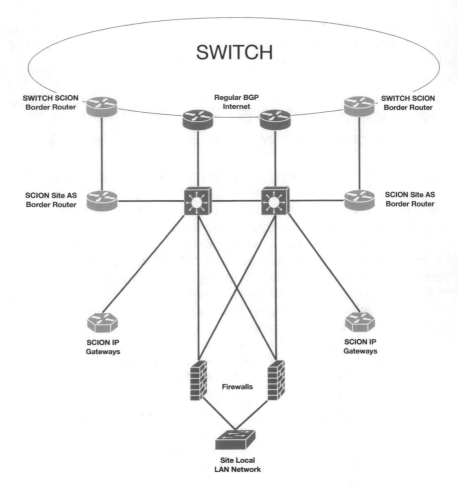

Figure 15.2: Local topology of a sample AS deployment in SCI-ED with redundant uplinks to the transit AS and redundant IP gateways for
legacy IP traffic.

15.4.2 GeoVITe Proof of Concept

GeoVITe is a user-friendly geodata access service. The goal of this proof-of-
concept [4] was to improve the large-scale geodata transmission from WSL to
ETH, benefiting from the additional bandwidth provided by SCI-ED and fast
file transfer using Hercules, while integrating with existing SMB file servers.
In the past, due to the sheer volume of data, the transmission process was
manual. This involved copying the data to removable storage, which was distributed to the recipient via courier service, and then copied again to a storage

---

[4]https://scied.scion-architecture.net/geovite-poc/

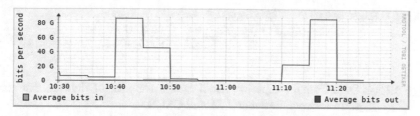

Figure 15.3: A Hercules transfer on a single machine over the SWITCH network is able to fill a 100Gbps network link over several minutes (time is shown in the x-axis). The figure shows two file transfers, whose total peak sustained throughput fills the link capacity.

cluster for processing. This resulted in an end-to-end transfer time in the range of hours to days.

With the network transfer speeds achieved during this use case, data transmission takes no longer than the travel time of the courier service between the locations, with the additional benefit of providing a higher degree of automation.

SCION connectivity made it possible to use efficient network transfers, where previously the process involved handling physical disks, arranging courier pickups, and manually transferring the data from the removable media to the server storage.

15.4.3 Petabyte Archive: Transmission of Large Photon Science Datasets from PSI to CSCS

PSI maintains two 100 Gbps direct links to CSCS, over which datasets from the Photon Science facilities are regularly transmitted to be archived in a Petabyte Archive System [5]. As of 2021, these transmissions add up to around 4 Petabytes per year, but are expected to reach up to 70 Petabytes per year in the future. This amounts to 17.8 Gbps of network transfer sustained, or 71.2 Gbps one fourth of the time. Without control over which paths such large data volumes are transferred, even a high-speed 100 Gbps link can easily become congested.

To address this use case, both PSI and CSCS were linked up to the SCION network as part of the SCI-ED project [6]. To transfer the large volumes of data, a solution for high-speed bulk file transfer for large datasets was implemented, which works in a pipelined fashion as depicted in Figure 15.4.

The solution is based on Hercules, which splits large files into 32GB blocks for the transmission and verification. The files can be arbitrarily large (petabytes). With Hercules it is possible to transfer these blocks at up to 100

---

[5]https://www.psi.ch/en/photon-science-data-services/data-catalog-and-archive

[6]https://scied.scion-architecture.net/photo-science-datasets-psi-cscs/

Figure 15.4: Data transfer pipeline for the bulk transfer of datasets for long term archival between PSI and CSCS.

Gbps as shown earlier in Figure 15.3. Several parameters can be tweaked, such as block size, concurrent transfers, checksum algorithm, etc. The current implementation loads the block to transmit in memory through the HugePages filesystem. This allows Hercules to run as fast as possible, with options such as preload disabled. The receiving side runs Hercules writing on a HugeTable filesystem as well, and then performs a checksum and copies the block to its appropriate place in the final filesystem. A final checksum can be performed on the final file once all blocks have been reassembled, for further confidence before archiving. These steps conform a pipeline, and several can be run concurrently, limited by the parallelism of each step (at the moment Hercules is not run concurrently in a given machine).

The current bottleneck resides at the loading process (around 5 Gbps on a single machine), but this can be optimized with faster disks or through parallelization across several machines. Thanks to SCION, these file transfers can be easily balanced among the two paths between PSI and CSCS in order to avoid congestion.

16 Green Networking with SCION

SIMON SCHERRER, SEYEDALI TABAEIAGHDAEI*

Chapter Contents

16.1 Direct Power Savings with SCION **394**

    16.1.1 Power Analysis for Core Routers 394

    16.1.2 Power Analysis for WDM Devices 396

    16.1.3 Power Analysis for Metro and Edge Routers 397

    16.1.4 Power Analysis for Access Networks 398

    16.1.5 Power Analysis for Communication Networks 398

16.2 SCION Enables Green Inter-domain Routing **399**

    16.2.1 Green Routing Design 400

    16.2.2 Green Routing Evaluation 402

16.3 Incentives for ISPs to Use Renewable Energy Resources . . **404**

    16.3.1 Models for Traffic Demand and AS Reaction 405

    16.3.2 Preliminary Results 405

Networking infrastructure (excluding data centers) is responsible for 270 TW h/yr or 1.1% of worldwide electricity consumption (in 2020) [28], which is expected to experience a ten-fold increase by 2030 due to the steady growth of Internet traffic volume [28, 29]. As electricity production emits considerable amounts of greenhouse gasses (475 gCO_2/kWh [258]), Internet traffic growth presents a serious concern regarding climate change.

In this section, we argue that SCION is a promising weapon to combat climate change as it not only reduces the power consumption of the Internet by more than 9%—despite its high security, high efficiency, and its packet-carried forwarding state—but also provides end hosts with an infrastructure to send their traffic to the greenest data center over the greenest paths, incentivizing Internet service providers and data centers to acquire their electricity from greener sources. This increased demand for green electricity pushes the electric power industry to deploy such resources at a faster pace, simplifying the use of green electricity for other sectors.

Therefore, SCION fights climate change on three different fronts:

---

*This chapter reuses content from Tabaeiaghdaei [500].

© The Author(s), under exclusive license to Springer Nature Switzerland AG 2022
L. Chuat et al., *The Complete Guide to SCION*, Information Security
and Cryptography, https://doi.org/10.1007/978-3-031-05288-0_16

Table 16.1: A Pareto analysis breakdown of the core router power by Hinton et al. [232].

Router component	Share of total power consumption (%)
Power supply and cooling	33
Forwarding engine	32
Switching fabric	15
Control plane	10
Input/output	6
Buffers	4

1. SCION directly reduces power consumption of the Internet;

2. SCION provides information about the carbon footprint of different Internet paths and thus enables CO_2-aware inter-domain routing; and

3. SCION provides incentives for ISPs to use green energy resources by providing carbon-footprint information of Internet paths to end hosts.

We discuss these three ways in the following sections.

16.1 Direct Power Savings with SCION

The first source of power and emission reductions enabled by SCION is the elimination of a particularly power-intensive router component: the ternary content-addressable memory (TCAM) required for longest-prefix matching. The resulting power saving is more than the power overheads of cryptographic operations and communication overhead due to longer packet headers. We describe how SCION reduces communication networks' power by analyzing its effect on the power consumption of core routers, metro and edge routers, access networks, and wavelength division multiplexing (WDM) devices.

16.1.1 Power Analysis for Core Routers

Table 16.1 shows the power breakdown for an IP core router. According to this table, the forwarding engine, which makes forwarding decisions, consumes 32% of the router's power. Almost all this power is consumed by the table lookup process [131] using TCAM chips, which enables longest prefix matching over large IP forwarding tables at line speed. TCAMs are among the most power-hungry memory technologies, at least 75 times more power-intense than other technologies (e.g., SRAM) [106]. They are the main culprit for the forwarding engine's significant contribution to routers' consumption. As the global IP prefix space size increases (to almost one million entries in

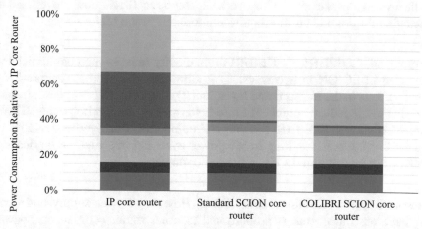

Figure 16.1: Cumulative power consumption by core router components in IP and SCION routers with and without COLIBRI, relative to IP core router total power.

2021 [59]), IP routers need larger TCAMs to store ever-larger forwarding tables. With the expected decline in the power-efficiency improvement rate of network devices in the future [8, 28, 29], larger forwarding tables will continue to increase routers' power usage.

SCION Savings. SCION routers do not perform longest-prefix matching to forward packets, so they do not need TCAMs. Instead, they look up the exact match of the current hop field's egress interface in interface lookup tables. These tables have as many entries as the number of inter-domain interfaces in an AS (thousands for large or hundreds for small ASes). With memories at least 75 times more power-efficient than TCAM (e.g., SRAM) [106] and at least two orders of magnitude smaller tables, the table lookup power in SCION routers is 7500 times less than IP routers, resulting in a 31% power decrease in a router.

SCION Overheads. Verifying hop fields' MAC using an FPGA AES-CMAC module [482] introduces a 0.5% power overhead to a large router at its peak throughput. We calculate this number by estimating the number of such AES modules a Cisco CRS 16-slot [119] router needs to reach its peak throughput. Moreover, including a path as long as the Internet's average AS-path in packet headers introduces 18.4% communication overhead relative to the average packet size. This overhead results in 18.4% power overhead in switching fabrics and buffers, equivalent to 3.5% of the router power.

Impact of COLIBRI.   COLIBRI dramatically reduces the communication overhead of SCION in two ways: 1) It eliminates the need for acknowledgment packets in the transmission layer, and 2) it reduces SCION packet header size due to the reduced number of info fields and the shorter hop fields. Therefore, SCION with COLIBRI has only 5% communication overhead relative to the current Internet, resulting in 5% power overhead in switching fabrics and buffers, equivalent to only 1% of the router power.

Power Supply, Cooling, and Total Power.   Power supply and cooling have an impact that is proportional to other router components' power. If we exclude power supply and cooling from a router, our estimations indicate that the net power savings of SCION are 44% and 40% (with and without COLIBRI, respectively). As savings for power supply and cooling are proportional, there are similar savings in total router power consumption. Figure 16.1 illustrates cumulative router components' power in IP, standard SCION, and SCION with COLIBRI, and Figure 16.2 illustrates SCION's power savings and overheads in a router.

Global Impact.   Baliga et al. [46] model the global power consumption by core routers as a function of users' population and their average access rate. Since network devices' power efficiency has been improving by around 10% per year, the proposed model takes these improvements into account so that the model would be applicable to future devices as well.

In 2021, there were 4.3 and 1.2 billion mobile and fixed broadband Internet users [151, 266] with average access bandwidth of 54 Mbps [395] and 105 Mbps [395], respectively.

According to Baliga et al.'s model, core routers consume 1.3 W and 2.07 W per user at 50 Mbps and 100 Mbps access rates, respectively. Therefore, core routers consume 8.1 GW globally, of which SCION can save 3.5 GW with and 3.2 GW without COLIBRI.

16.1.2 Power Analysis for WDM Devices

Impact on a Single WDM Device.   Assuming wavelength division multiplexing (WDM) power is proportional to traffic rate, SCION's communication overhead proportionally increases WDM power, similarly to the power utilization of the switching fabric and the buffer of a core router, yielding 5% and 18.4% power increase with and without COLIBRI, respectively. However, since WDM devices do not look up tables, eliminating forwarding tables does not affect their consumption.

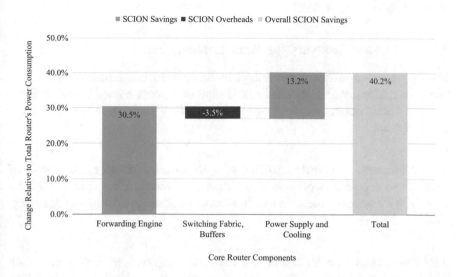

Figure 16.2: Core router power savings and overheads by SCION with COLIBRI.

Global Impact. Using the model proposed by Baliga et al. [46] for the WDM power per user per access rate, similar to the one for core routers (see Section 16.1.1), we estimate that SCION increases the global WDM power consumption by 0.1 GW with and 0.35 GW without COLIBRI.

16.1.3 Power Analysis for Metro and Edge Routers

Impact on a Single Metro or Edge Router. SCION does not modify the forwarding engine power of metro/edge routers because 1) these routers do not store the global forwarding table, so packet-carried forwarding state does not reduce their power significantly, and 2) SCION does not modify intra-domain forwarding.

However, SCION's communication overhead increases their switching fabric and buffer power by 5% with and 18.4% without COLIBRI, similar to core routers. According to the model proposed by Vishwanath et al. [519], the metro/edge router power is dominated by the switching fabric power, resulting in an increase in router power of 5% with and 18% without COLIBRI.

Global Impact. Using the same methodology we used in Section 16.1.1 and the model proposed by Baliga et al. [46], we estimate that SCION increases the global power consumption by metro and edge routers by 0.2 GW with and 0.74 GW without COLIBRI.

16.1.4 Power Analysis for Access Networks

For the same reasons we mentioned in Section 16.1.3, eliminating longest-prefix matching by SCION does not change the access network power. However, SCION's communication overhead can change access networks' power consumption.

Wired Access. According to Baliga et al. [46], the power consumption of most wired access networks does not change as access rate increases. Therefore, SCION's communication overhead does not change their power consumption.

Mobile Access. In mobile access networks, however, SCION's communication overhead increases the power proportionally. Pihkola et al. [415] estimated the power intensity of mobile access networks to be 0.3 kW h/GB in 2017 and predicted that it would diminish to 0.1 kW h/GB in 2020. Given this power intensity and the global mobile traffic (55 EB per month in 2021 [491]), the 5% and 18.4% communication overheads of SCION result in global power overheads of 0.38 GW with and 1.41 GW without COLIBRI.

However, we think that this overhead can be lightened through path negotiation between mobile devices, and base stations where the base station fills the packet headers with the negotiated paths. Therefore, the header size of the packet sent from/to the end host to/from the base station would be close to the IP header size, resulting in a negligible overhead.

16.1.5 Power Analysis for Communication Networks

By adding the overall power savings and overhead that SCION introduces to different parts of communication networks, we conclude that SCION decreases the networking power by 2.88 GW with, and 0.73 GW without COLIBRI—see also Figure 16.3, which illustrates the power savings and overhead that SCION with COLIBRI introduces to communication networks.

The 2.88 GW power saving of SCION corresponds to almost 9.4% and 1.3% of power consumption by the networking and the whole ICT sectors in 2020, respectively [28].

Assuming TCAM's relative contribution to the core router power remains the same, SCION can save 9.3 GW in 2030 according to the prediction made by Andrae [28]. However, with growing IPv6 adoption and an expected slowdown in network devices' power-efficiency improvement rate, it is expected that TCAM chips will have an even more significant share of the core router

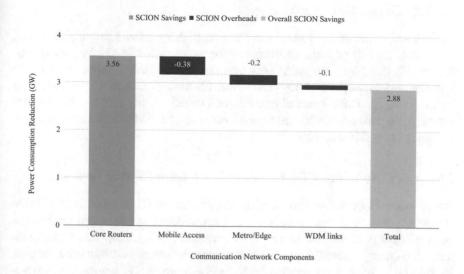

Figure 16.3: Networking power savings and overheads by SCION with COLIBRI.

power [8, 28, 29], increasing the power savings by SCION beyond the estimated 9.3 GW in 2030.

16.2 SCION Enables Green Inter-domain Routing

The second way SCION can reduce the CO_2 emission of the Internet is to disseminate information about the CO_2 emission of Internet paths using PCBs. ASes can use this information to discover "green" paths—the ones emitting less CO_2 than others—and use them to send traffic. This can reduce the Internet's CO_2 emissions substantially as the CO_2 emissions of different paths can vary strongly: Different geographic regions and countries have different energy resource mixes, and network operators can buy or produce electricity from various sources.

Although forwarding packets over greener paths has been proposed for intra-domain routing (inside one AS) [201, 437, 463, 567] or between neighboring ASes [380], no CO_2-aware inter-domain routing and forwarding approach has been proposed so far. This is mainly because BGP, which forwards packets only on a single path, does not provide the opportunity to optimize paths based on multiple criteria such as CO_2 emissions. In contrast, SCION as a multi-path Internet architecture enables path optimization for different criteria, including CO_2 emission, see §4.4. We use this feature of SCION to optimize inter-domain paths for CO_2 emission, and present our estimate for the environmental impact of such a design.

16.2.1 Green Routing Design

To forward traffic on the lowest-emitting path, ASes should provide their end hosts with the sets of paths emitting the lowest amount of CO_2 for forwarding traffic. To discover such paths between all ASes, ASes (1) need to estimate their intra-domain paths' CO_2 emissions, (2) disseminate this information to other ASes, (3) make beaconing decisions based on this information to construct green path segments, and (4) end hosts use this information to construct the greenest end-to-end paths.

16.2.1.1 Calculating CO_2 Emissions of Internet Paths by ASes

We propose a model that allows ASes to calculate the CO_2 emissions of Internet paths. In general, researchers use either top-down or bottom-up models for network energy consumption and CO_2 emission [140]. A top-down model divides the energy a network or a part of it consumes over a certain time interval by the traffic volume it carries over the same interval. By contrast, a bottom-up model uses the energy consumption of every single device in a network. We propose one top-down and one bottom-up model to estimate inter-domain paths' CO_2 emissions. In both models, we calculate the CO_2 emission per bit of data (*CEPB*) on an inter-domain path as the sum of *CEPB*s of all its intra-domain path segments.

Top-Down Model. In the top-down model, ASes estimate the *CEPB* of an intra-domain path between a pair of ingress and egress interfaces using the CO_2 emission of the whole AS over a certain time interval—which can be derived using their electricity consumption over that interval and their electricity contracts—and the distance between the ingress and the egress interfaces. This is justified since the longer the distance between a pair of interfaces is, the more energy is consumed to transmit packets between them, and thus more CO_2 is emitted.

We model this relation between the distance and the energy consumption for forwarding a packet on the path using a constant function for short distances and a linear function for long distances. The constant for short distances is equal to the average CO_2 emission of the AS per bit of data it has carried over the last interval. The slope of the linear function for long distances is equal to the average CO_2 emission per bit of data the AS has carried over the past interval per kilometer between all its interface pairs.

Bottom-Up Model. The bottom-up model is more precise but requires more detailed knowledge about an AS's network. Each AS estimates the *CEPB* of its AS entry included in an inter-domain path as the sum of the expected *CEPB*s of all network devices on the intra-domain path between the ingress and the egress interface of the AS. The expected *CEPB* of each networking device is calculated as the product of the carbon intensity of its electricity resources and its expected per-bit energy consumption. We break down each network device's energy consumption into per-bit, per-packet (for operations performed per packet such as table lookup), and idle energy consumption [519]. We model the *expected* per-bit energy consumption as the sum of per-bit energy consumption, per-packet energy consumption divided by average packet size, and the idle power consumption divided by traffic load on the device.

The last term is of great importance as the idle power is responsible for the majority of network devices' power consumption [233], and including it ensures that the model estimates the actual CO_2 emission of a path, not just the carbon intensity of its energy resources. Since each AS has fine-grained information about its devices, their locations, their electricity resources, and their traffic load, they can compute the expected *CEPB* of all intra-domain paths that connect all their border routers.

As *CEPB* values vary over time, ASes need to re-calculate them dynamically and include the updated values in the PCBs they disseminate.

16.2.1.2 Dissemination of CO_2 Emission Information

To distribute the per-AS CO_2 emission information to other ASes, we use the *static info extension*, which is presented in §8.3. An AS computes the intra-domain *CEPB* between any pair of external interfaces that are part of policy-compliant paths (child↔child, child↔peer, child↔parent, child↔core, core↔core) and includes them in the extension. Notably, this includes the energy consumption of the border routers of both interfaces. We neglect the consumption of inter-domain links, as they are usually short links within the same location and have therefore a negligible impact.

The *CEPB* of the inter-domain path to a destination AS would then be equal to the sum of *CEPB* information encoded in all AS entries of a PCB. The way the data is included in PCBs ensures that all information necessary for all possible segment combinations is available. As the metadata extension is protected through the AS's signature, the AS can be held accountable for the information it provides. In addition, the information can be certified by a trusted third party.

16.2.1.3 Constructing Green Paths

Optimizing forwarding paths for CO_2 emissions requires two steps: (1) a green beaconing algorithm, which allows ASes to construct segments that cause low emissions, and (2) green path combination and selection at end hosts.

Green Beaconing. To construct green sets of core-, up-, and down-segments, the beaconing policy at each AS should take into account the *CEPB* information encoded in PCBs as well as the *CEPB* of the local AS's intra-domain paths between its interfaces. Therefore, we introduce the green beaconing policy such that it examines all possible paths from a destination AS to each of its neighboring ASes before disseminating PCBs, and only disseminates the lowest-emitting ones—the ones with the lowest amount of emissions among all others. This green beaconing policy constructs all possible paths from every destination AS to every neighboring AS by taking all PCBs originating from that AS and appending all egress interfaces connecting to neighboring ASes. The algorithm then computes the *CEPB* of all these prospective paths and sorts them based on their *CEPBs* and disseminates the lowest-emitting ones. This beaconing policy ensures that each core AS receives the set of lowest-emitting possible paths to any other core AS in the inter-ISD network, and each leaf AS receives the set of lowest-emitting possible paths to core ASes of its ISD. Note that this green beaconing algorithm can run in parallel with other beaconing policies that optimize paths for other metrics such as AS-path length, bandwidth, or latency. Thus, an AS discovers the best paths with respect to multiple different metrics.

Green Path Combination.   An end host can combine the green path segments constructed in the beaconing process and the emission information stored in the extension to construct green end-to-end paths. This can be achieved by modifying the path-construction algorithm described in §5.5.4. Instead of simply using the segment length as a metric, the cost of segments in the graph construction can be set to the corresponding emissions or any combination of metrics such as latency or bandwidth. Then, searching for minimal-cost paths automatically yields low-emission paths.

16.2.2 Green Routing Evaluation

To evaluate the impact of green inter-domain routing in SCION, we simulate the green beaconing policy using a simulator [501] we have developed based on the ns-3 network simulator [393]. To compare the results with BGP, we also simulate BGP using the SimBGP simulator [481]. We run our simulations on an inter-domain topology of 2000 core ASes extracted from the CAIDA AS-Geo-Rel dataset [101], which contains the locations of the border routers of the largest ASes in the Internet.

To estimate the *CEPB* of intra-domain paths between ingress and egress interfaces of each AS hop, we use our bottom-up model with several assumptions and approximations to compensate for unavailable data. We look up the number and locations of routers between all the border routers of each AS using the CAIDA Internet Topology Data Kit (ITDK) dataset [103]. Then, we assume that all routers on these paths have the same power intensity as the CISCO CRS-3 router (10 W/Gbit [229], and based on the country where each router was located we estimate the carbon intensity of its electricity resources [172, 263]. Using the power intensity and the carbon intensity of

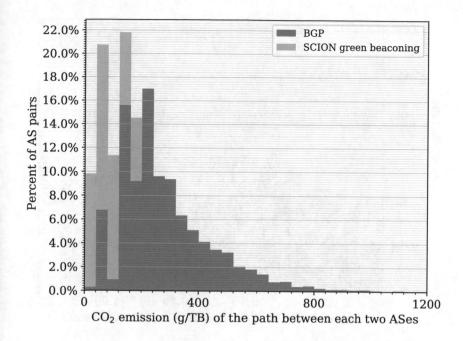

Figure 16.4: Distribution of the per-terabyte CO_2 emission of the paths found
        by BGP and the SCION green beaconing policy between each
        pair of ASes in a core topology containing 2000 ASes, including
        all Tier-1 ASes and the highest-degree Tier-2 ASes.

routers, we calculate their *CEPB*. Using a similar method, we also estimate
the *CEPB* of optical devices between routers. By summing up the *CEPB* of all
devices between each pair of border routers, we estimate the *CEPB* of these
intra-domain paths.

By comparing the *CEPB* of the inter-domain paths constructed by green
beaconing with the paths constructed by BGP, we find that the green beaconing policy in SCION reduces the *CEPB* of paths for 82 percent of source–
destination AS pairs. An exciting result is that the reduction is more than
50 percent for almost 50 percent of source–destination pairs, indicating the
tremendous potential of this approach. By modeling the traffic between these
ASes, our very conservative estimate finds that this could have saved 210,000
tons of CO_2 emission in 2021. Figure 16.4 shows the distribution of the per-
terabyte CO_2 emission of the paths found by BGP, and the SCION green beaconing policy between every two ASes. Figure 16.5 shows the distribution of
the CO_2 emission of the paths found by the SCION green beaconing policy
relative to the paths found by BGP between the same pair of ASes.

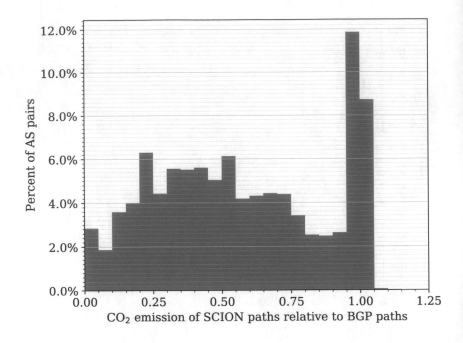

Figure 16.5: Distribution of the CO_2 emission of the paths found by the SCION
green beaconing policy relative to the paths found by BGP in a
core topology containing 2000 ASes including all Tier-1 ASes
and the highest-degree Tier-2 ASes. Note that a few paths constructed by the green beaconing policy show more emission than
BGP paths. This is because the green beaconing algorithm running at each AS optimizes the CO_2 emission of paths from each
origin AS to its neighboring ASes, but not to further downstream
ASes (e.g., its neighbor's neighbors). In a few cases, these optimized paths by upstream ASes lead to suboptimal paths in further
downstream ASes.

16.3 Incentives for ISPs to Use Renewable Energy Resources

Providing end hosts with the green routes to every destination and giving them
the power to select the path to send their traffic on, environmentally consci-
entious users can configure their devices to select greener paths even if they
are slightly more expensive or have lower quality than other paths. This is not
limited to private end users but also includes large content providers, which
are responsible for a large share of Internet traffic.

Thus, some amount of traffic shifts to greener paths, and ASes on polluting
paths lose some portion of traffic they used to transit. We hypothesize that

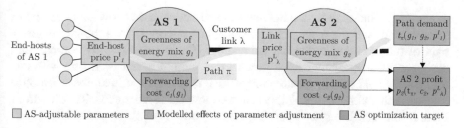

Figure 16.6: Visualization of economic model.

this shift of traffic could set in motion a virtuous cycle through the following mechanism: As the business model of ISPs has typically low profit margins given high static and low variable costs, a small decrease in the amount of traffic they transit can significantly drop their profit, or even make their business unprofitable [515]. This incentivizes these ASes to make their electricity resources greener, or to make their devices and traffic engineering mechanisms more energy-efficient so they can win back their lost traffic. This creates a virtuous cycle between different ISPs to attract traffic by becoming more energy efficient and greener.

16.3.1 Models for Traffic Demand and AS Reaction

In order to test this virtuous-cycle hypothesis, we extend our beaconing simulator with an economic model that attempts to reflect competitive dynamics (cf. Figure 16.6). These dynamics consist of a feedback loop connecting two elements, namely (1) the response of traffic demand to path characteristics, and (2) the business calculation of ASes striving to optimize profit.

To model the demand response, we build on an established model [68] for predicting the demand distribution onto competing goods, or in our context, paths. This demand model takes into account both the quality (of which emissions are one aspect) and price of all paths. To quantitatively anchor this demand model in reality, we rely on an empirical estimate of the increase in average willingness-to-pay for socially responsible products.

Based on the traffic distribution created by the demand model, the ASes can derive their transit traffic as well as the resulting revenue, cost, and profit. Each AS can thus optimize its profit by adjusting its prices (charged to customer ASes and end hosts) and its energy mix, thereby affecting the quality of transiting paths, the pattern of its transit traffic, and the resulting money flows.

16.3.2 Preliminary Results

Interestingly, choosing a greener energy mix generally increases both the revenue of an AS (because the AS thereby increases the attractiveness of all transiting paths), and the AS cost (because greener energy is modeled as more expensive). The relationship between the attractiveness of low-emission paths to customers and the price premium of green energy is thus the key determinant of the emission reduction achieved by the competitive dynamics. Our early results suggest that for a green price premium that is in line with current energy-cost estimates, the competitive dynamics outlined above produce a network that almost completely relies on low-emission electricity. Therefore, these results confirm the plausibility of the virtuous-cycle hypothesis.

17 Cryptography

MARKUS LEGNER, ADRIAN PERRIG, MARC WYSS*

Chapter Contents

17.1 How Cryptography Is Used in SCION **408**

   17.1.1 Authentication as the Main Security Goal 408

   17.1.2 Short Validity Periods 409

   17.1.3 Local Algorithms 409

17.2 Cryptographic Primitives **409**

   17.2.1 Symmetric Primitives 409

   17.2.2 Asymmetric Primitives 410

17.3 Local Cryptographic Primitives **410**

   17.3.1 Key Generation 411

   17.3.2 Internal Key Derivation 411

   17.3.3 Deriving First-Level DRKeys 411

   17.3.4 Hop-Field Authenticators 411

   17.3.5 Hash Functions for Filtering 412

17.4 Global Cryptographic Primitives **412**

   17.4.1 Control Plane: PKIs and Signatures 412

   17.4.2 Data Plane: MACs, PRFs, and Hash Functions 413

17.5 Post-Quantum Cryptography **415**

With security as a central objective, SCION naturally makes extensive use of cryptography. Different SCION components require different types of cryptographic primitives: symmetric and asymmetric. For example, while the control plane mainly relies on asymmetric cryptography in the form of signatures to enable any entity to verify the authenticity of the signed data and achieve non-repudiation, only symmetric primitives such as message authentication codes (MACs) are sufficiently efficient for data-plane packet processing.

Unfortunately, there is no fundamental security proof for most symmetric cryptography used today. Similarly, existing security proofs for asymmetric primitives rely on unproven assumptions about the complexity of certain problems; even worse, these assumptions are already known to be violated as soon

---

*This chapter reuses content from Perrig and Szalachowski [410].

© The Author(s), under exclusive license to Springer Nature Switzerland AG 2022

L. Chuat et al., *The Complete Guide to SCION*, Information Security and Cryptography, https://doi.org/10.1007/978-3-031-05288-0_17

as quantum computers reach a certain scale. In general, it is unpredictable which algorithms will be made obsolete by advances in cryptanalysis and increased computing power. Therefore, a crucial feature that must be considered for every protocol that uses cryptography is *cryptographic agility* [246], which refers to the ability to deploy new algorithms in case previously used algorithms are no longer deemed secure.

In this chapter, we first perform a risk assessment and discuss SCION features that facilitate cryptographic agility (§17.1). Then, we describe which cryptographic primitives are used in SCION, which requirements they have, and how cryptographic agility is achieved for them (Sections 17.2–17.4). Finally, we discuss the threat of quantum computers for cryptography and how it affects SCION (§17.5). In Sections 17.3 and 17.4, we also provide examples of cryptographic algorithms for various primitives based on standards, recommendations, best practices, and performance evaluations available at the time of writing [207, 396, 469, 478].

17.1 How Cryptography Is Used in SCION

Several features and properties of the SCION architecture limit the risk stemming from the potential future discovery of vulnerabilities in cryptographic algorithms and facilitate cryptographic agility.

17.1.1 Authentication as the Main Security Goal

SCION almost exclusively uses cryptography for the purpose of authentication, not encryption (with a few exceptions such as the encryption of COLIBRI authenticators or hidden paths). This has several benefits with respect to the risks associated with cryptographic vulnerabilities and agility:

- Breaking a cryptographic algorithm used for authentication does not (directly) leak any secret data and thus does not affect the secrecy of previously protected data. One of the main goals of authentication in SCION is to guarantee availability, not secrecy. As opposed to secrecy, the availability of a communication path cannot be compromised at a later date by an attacker who has dedicated computing resources to break a cryptographic algorithm over a long period of time.

- If efficiency is not the prime goal of a certain protocol, multiple cryptographic algorithms can be used simultaneously to authenticate data. For example, an AS could have multiple different keys in the CP-PKI and protect PCBs with multiple signatures using different signature algorithms. As long as one algorithm is secure, the AS certificates and control-plane messages remain protected.

17.1.2 Short Validity Periods

With the exception of a subset of keys and certificates in the various PKIs (§3.1 and Chapters 18 and 19), all cryptographic material and cryptography-protected data in SCION is short-lived—with a maximum validity period of around one day. This is in stark contrast to a large portion of traffic encrypted through TLS or VPNs, which should still remain secret long after it has been sent.

An additional advantage of short validity periods is that any attack would need to be highly efficient to be able to recover secret keys or forge authenticators in this short time period. It is highly unlikely that such an attack becomes suddenly available for algorithms that are believed to be secure in the long term.

17.1.3 Local Algorithms

Many applications of cryptography do not require coordination among multiple ASes, as they are only used internally by one AS (e.g., the MAC algorithm used to protect hop fields). This substantially simplifies the deployment of new algorithms because the infrastructure components of a single AS are typically fully controlled by a single company. We further discuss these local algorithms in §17.3.

17.2 Cryptographic Primitives

SCION uses both symmetric and asymmetric cryptography. We summarize the different types of primitives that are used in the following sections. The applications of these primitives and their requirements will be described in further detail in Sections 17.3 and 17.4.

17.2.1 Symmetric Primitives

SCION uses symmetric cryptography to generate and derive keys and enable high-speed authentication of data:

- **Pseudorandom number generator (PRNG):** ASes need PRNGs to generate symmetric and asymmetric keys. As a PRNG's output is critical for the security of AS operations, it must be selected carefully. The most common PRNGs are provided by underlying operating systems, and may be software or hardware components.

- **Internal key-derivation function:** Key-derivation functions are used by ASes to generate temporary symmetric keys used in production (e.g., for keys for hop-field generation).

- **Pseudorandom function (PRF):** PRFs are mainly used for deriving keys in the DRKey hierarchy.

- **Message authentication codes (MACs):** MACs are used as hop-field authenticators, to compute hop-validation fields in EPIC and COLIBRI, and for packet authentication in the SCION Packet Authenticator Option (SPAO) or COLIBRI control messages.

- **Cryptographic hash function:** Cryptographic hash functions are used in SCION as part of digital-signature schemes and optionally in the SPAO together with a MAC. Furthermore, they are used in the probabilistic data structures of filtering mechanisms.

17.2.2 Asymmetric Primitives

The only direct use of asymmetric cryptography in SCION is **digital signatures**, which are used for two main purposes:

- In public-key infrastructures, signatures are used to certify the binding between a key and a name or identifier. Notably, signatures are used at all levels of the CP-PKI (§3.1), from the TRC to AS certificates; in the F-PKI end-entity PKI (Chapter 18); and in RHINE (Chapter 19) to authenticate naming data.

- Signatures are used to authenticate information contained in control-plane messages, which prevents other ASes from forging or modifying this information and enables non-repudiation (i.e., a signed message can be attributed to the entity that owns the corresponding private key).

Additionally, SCION relies on higher-layer security protocols like TLS (e.g., for the first-level DRKey exchange), where asymmetric primitives are additionally used for key establishment. These systems are orthogonal to SCION and provide their own mechanisms for the coordination of algorithms and cryptographic agility.

17.3 Local Cryptographic Primitives

Some cryptographic operations in SCION are performed locally, i.e., only within an AS's infrastructure. This is ideal for algorithm agility, as the algorithms used by an AS neither have to be negotiated with other parties, nor even known by other ASes or end hosts. ASes can thus change their internally used PRNG, MAC, hash, and key-derivation algorithms at any time, without coordination with the rest of the network. It is only required that the AS's infrastructure elements use a consistent algorithm.

17.3.1 Key Generation

Each AS must generate its own keys. At least one symmetric key (master key) and one asymmetric key pair are required. To generate them, a strong PRNG with reliable true randomness must be used. The open-source SCION implementation relies on the cryptographically secure random number generator implemented in the `crypto/rand` package[1] from the Go standard library.

17.3.2 Internal Key Derivation

All symmetric keys used within an AS (e.g., keys used for hop-field authentication or as secret values in DRKey) are derived from the master key, which is shared among relevant infrastructure elements. ASes can freely select their key-derivation algorithm, as this decision influences only locally used keys.

As these are only applied periodically, they do not need to be highly efficient. Consequently, standard key-derivation functions like the Password-Based Key Derivation Function 2 (PBKDF2) [372] can be used with multiple iterations to defend against brute-force attacks.

17.3.3 Deriving First-Level DRKeys

First-level DRKeys are only derived by trusted entities within the fast AS (e.g., border routers, the certificate service, DNS resolvers). This derivation should be more efficient than the derivation of AS-internal keys, as it is performed frequently by these entities. Therefore, a single application of an efficient pseudorandom function (PRF) should be used. One algorithm that fulfills these requirements is an AES-based CMAC [483], as AES is commonly used, and CMAC requires only two AES executions for the computation of authentication tags (provided that sub-keys are cached).

17.3.4 Hop-Field Authenticators

Hop-field authenticators (in standard SCION, EPIC, and COLIBRI) are created through a MAC algorithm. However, these authenticators are never computed or verified by entities outside the AS that computed the authenticators; therefore, the AS does not have to even inform other parties which algorithm it is currently using. After hop-field authenticators are created, they are either directly included in the path headers (SCION path type) or are used to compute hop-validation fields (EPIC and COLIBRI). In all cases, they only need to be recomputed by border routers of the AS who created them.

The MAC scheme used for hop-field authentication/verification is one of the most important cryptographic primitives in SCION. This function is executed by border routers for every single packet. Hence, besides security, it must

---

[1]https://pkg.go.dev/crypto/rand

provide outstanding performance, and it must be possible to truncate the output. AES-CMAC fulfills all these requirements; in particular, as a PRF, its output can be truncated [483].

17.3.5 Hash Functions for Filtering

All filtering systems (replay suppression, LOFT, LightningFilter) use hash functions internally. These do not need to be known by other entities; on the contrary, adding a salt or secret value in the input keeps hash functions secure against various poisoning attacks. One candidate algorithm, which also fulfills the stringent efficiency requirements, is SipHash [39].

17.4 Global Cryptographic Primitives

While some algorithms can be chosen by each AS independently, there are several primitives that require global coordination. SCION elements that require consensus on cryptographic algorithms follow best practice for cryptographic algorithm agility [246]. In general, SCION defines a set of mandatory-to-implement cryptographic algorithms, and every allowed algorithm has a unique identifier assigned. In the following, we describe the different algorithms that an AS must support, their requirements, and their coordination.

17.4.1 Control Plane: PKIs and Signatures

The control plane relies on several global cryptographic primitives:

- **TRC creation and verification:** TRCs are authenticated through digital signatures. Each entity within a TRC is specified via its certificate, which in turn specifies the digital signature algorithm used. The list of used algorithms must be specified globally, as TRCs are verified by remote end hosts.

- **Beaconing:** Similarly to TRCs, PCBs are protected with digital signatures. Also, AS-level certificates are used to identify the deployed algorithm. Note that each AS can use a different algorithm (as long as it is supported by other parties) to authenticate its AS entry. A PCB is only forwarded if every AS that received it can verify all signatures. However, to support algorithm agility, SCION permits ASes to add multiple signatures to PCBs. This can result in PCBs (and path segments consequently) with different security properties.

- **Path-segment registration and lookup:** Every path segment is authenticated by a single AS, and similarly the algorithm used to verify path segments is inferred from the AS certificate.

- **End-entity authentication:** End-entity authentication relies on the TLS protocol, which provides algorithm agility and defines its own algorithm suites [165].

17.4.1.1 Requirements

The requirements for signature schemes used in SCION are (1) security (as certificates, PCBs, and TRCs are protected with them), (2) efficiency (as signature creation and verification are executed frequently), and (3) short signature and public key. At the time of writing, these requirements are fulfilled, for example, by the Edwards-curve Digital Signature Algorithm in its instantiation called Ed25519 [66]. This signature scheme provides outstanding efficiency and signatures created are relatively short (64 bytes).

Additionally, digital signatures rely on a cryptographic hash function, for which there are no special requirements besides the standard security properties. As a result, any of the standard hash-function families (e.g., SHA-2 [391] or SHA-3 [419]) can be used.

17.4.1.2 Coordination

The authentication of TRCs, PCBs, and path segments is based on AS-level certificates (§3.1). Every certificate contains the `algorithm` field, which specifies the digital-signature algorithm used. An identifier of the digital-signature algorithm implicitly determines the size of the private and public keys used, and the size of signatures. The certificates are short-lived, so an AS that wishes to change its digital-signature scheme can do it with the next certificate reissuance (up to a few days).

17.4.1.3 Composition of Signatures

By using multiple asymmetric keys simultaneously (with corresponding certificates for all of them), an AS can create multiple signatures with different algorithms on a single message (e.g., on a PCB). If one signature algorithm is later discovered to be insecure, the message is still protected through the second signature. This also facilitates the transition to new signature algorithms.

17.4.2 Data Plane: MACs, PRFs, and Hash Functions

The SCION data plane relies exclusively on symmetric cryptography, as asymmetric cryptography is not efficient enough.

- **DRKey:** The DRKey system requires an efficient PRF for the second- and third-level key derivation.

- **Packet authentication:** The SPAO is used to authenticate data packets and SCMP packets. It requires a MAC scheme and for some applications a cryptographic hash function.

- **Authentication in COLIBRI control packets:** All COLIBRI control messages are authenticated with MACs for each on-path AS.

- **Hop validation fields (HVFs):** In both EPIC and COLIBRI, HVFs require a MAC algorithm. This algorithm has additional requirements as we describe below.

- **Authenticated encryption for COLIBRI:** Hop authenticators for COLIBRI end-to-end reservations are sent in encrypted form to the source AS. This requires an authenticated encryption mechanism.

17.4.2.1 Requirements

For all use cases, efficiency is the prime objective besides security. As a result, algorithms for which hardware acceleration is available are generally preferred. For example, the AES block cipher can be used to construct a MAC scheme, a PRF, and achieve authenticated encryption [223].

It should also be possible to truncate the MACs used in the HVFs of EPIC and COLIBRI (similarly to how hop-field authenticators are computed, as described in §17.3.4). Furthermore, the MAC algorithm used in EPIC-SAPV must not leak any input. Both requirements can be achieved by using a MAC scheme that also fulfills the requirements of a PRF [528].

Unfortunately, AES-CMAC is a suboptimal choice for short inputs when keys change frequently—which is the case for EPIC, COLIBRI, and DRKey— due to the overhead of sub-key generation. In this case, an AES-based CBC-MAC is preferred, which is secure for fixed-length messages or when defining the input length in the first input block [63].

17.4.2.2 CASA: Cryptographic Agility for SCION ASes

In contrast to protocols like TLS or IPsec, the applications of symmetric cryptography in SCION cannot rely on negotiation of cryptographic algorithms as the first packet sent by an end host should already be authenticated. On the other hand, defining a global algorithm to be used with EPIC, COLIBRI, and DRKey is undesirable from an agility perspective, and border routers and other infrastructure components may be severely restricted in what algorithms they support, as they rely on hardware acceleration to achieve sufficient throughput.

SCION thus relies on CASA, a more sophisticated approach where each AS can specify the cryptographic algorithms it expects for these systems in (signed) PCB extensions (§4.1.2). In CASA, every AS chooses exactly one algorithm per algorithm category. Those categories are globally defined and can either be general (PRF, MAC, hash function) or protocol-specific (e.g., PRF for DRKey, MAC for EPIC HVFs, MAC for COLIBRI HVFs).

The idea behind the general categories is to cover a large set of protocols that only need algorithm agreement between different entities. On the other hand, the protocol-specific categories serve the needs of protocols that have additional requirements like high performance. If an algorithm for a protocol-specific category is announced, it has priority over the one specified in the general category.

Note the asymmetry in the CASA design: There is in general one side with higher and one with lower efficiency requirements. The side that requires higher efficiency chooses the algorithm to be used, as it is the one that is primarily affected by the algorithm's performance. This mirrors the design of DRKey, where one side must locally fetch the key (slow), and the other side can recompute the key on the fly (fast).

CASA only introduces a small overhead and does not affect the layout of data-plane packets in any way; that is, no additional fields are necessary. The border routers of the ASes only need to implement the small number of algorithms promoted by their AS, which minimizes their overhead and prevents performance degradation. Announcing the expected algorithm in PCBs ensures that this information is directly available in path segments and does not need to be fetched separately (from every on-path AS) by the sender.

CASA could be further extended to not only support per-AS but also per-border-router algorithm specifications, which further increases the flexibility and improves the granularity of systems at which cryptographic algorithms can be deployed.

17.5 Post-Quantum Cryptography

Quantum computing, for a long time an exclusively academic research area, has matured significantly in recent years with practical applications on the horizon. An important milestone for the field of quantum computation is *quantum supremacy*, which describes a state where quantum computers are able to solve problems that are infeasible for classical computers in a reasonable time frame [80, 418]. This milestone was reportedly reached in 2019, albeit for an artificial problem without practical application [37]. Even though these claims were promptly called into question [406], it is clear that quantum supremacy, even if not reached yet, will likely be reached in the near future.

Unfortunately, these advances in quantum computing also threaten many cryptographic primitives that are in use and considered secure today. While most symmetric primitives such as block ciphers and hash functions are believed (though not known) to remain secure even in the face of quantum computers (as long as the keys and hash-function outputs are sufficiently long), virtually all asymmetric cryptography used today is vulnerable to quantum computers [355].

This has triggered extensive research into post-quantum cryptography [67]—cryptographic algorithms that are secure even under the assumption that large-scale quantum computers exist—and related standardization efforts by NIST [108]. While it is still unclear which of the various existing proposals for post-quantum primitives are actually secure, it is likely that a secure system will be developed.

In the context of SCION, we mainly need to discuss two questions:

1. Can asymmetric primitives be upgraded with reasonable effort to post-quantum cryptography?

2. Can an attacker cause damage by recording data and later breaking cryptographic primitives when quantum computers become powerful enough ("capture now, exploit later")?

Both questions have already been answered in earlier sections: As multiple signature algorithms can be used simultaneously, it is possible to gradually deploy new signature algorithms. Also, not all ASes need to transition simultaneously. Furthermore, as protected data is only valid (and thus useful) for relatively short time periods and no secret data can be leaked (Sections 17.1.1 and 17.1.2), the "capture now, exploit later" approach is ineffective in SCION as long as all authenticity aspects involved are short-lived.

A special situation arises when an adversary breaks the signature scheme used for the CP-PKI. In this case, the chain of TRCs can no longer be trusted, and an immediate trust reset becomes necessary to restore trust. Such a reset should be used to switch to a post-quantum cryptography scheme, if available. Efficient mechanisms to update the keys are in place, so that the TRC update will not last longer than one day. For more detailed information on TRC updates and trust resets, see §3.1.1 and §3.1.6.

In summary, we conclude that the threat of quantum computation for SCION is manageable as long as secure and sufficiently efficient post-quantum cryptography becomes available at the same time as powerful quantum computers.

Part V

Additional Security Systems

18 F-PKI: A Flexible End-Entity Public-Key Infrastructure

Laurent Chuat, Cyrill Krähenbühl, Prateek Mittal, Adrian Perrig*

Chapter Contents

18.1 Trust Model **421**

    18.1.1 Adversary Model 422

18.2 Overview of F-PKI **423**

18.3 Policies **424**

    18.3.1 Validation Policies (Trust Levels) 424

    18.3.2 Domain Policies 425

18.4 Verifiable Data Structures **426**

    18.4.1 Comparison to CT Logs 426

    18.4.2 Hierarchical Naming 427

    18.4.3 Entry Format 427

    18.4.4 Revocations 428

    18.4.5 Proving Consistency over Time 428

18.5 Selection of Map Servers **428**

18.6 Proof Delivery **428**

    18.6.1 Stapling 428

    18.6.2 Fetching via DNS 429

    18.6.3 Alternative Delivery Methods 430

18.7 Certificate Validation **430**

In contrast to the control-plane PKI (presented in §3.1), whose main purpose is to authenticate ASes, an end-entity PKI enables the authentication of end systems such as web servers. It also facilitates the design of RHINE, a next-generation secure and reliable Internet naming system (Chapter 19). The most widely used end-entity PKI is the HTTPS public-key infrastructure (or Web PKI, for short). Unfortunately, today's Web PKI suffers from a weakest-link problem: Any certification authority (CA) can issue certificates for any domain

---

*This chapter reuses content from Chuat, Krähenbühl, Mittal, and Perrig [116].

at will, and several events in the past few years have proven that this constitutes a real threat.

In March 2011, news broke that Comodo—a security firm operating a certification authority—had been hacked. The intrusion resulted in the unwarranted issuance of 9 certificates for several high-profile domain names [133]. A few months later, DigiNotar suffered a similar attack [242]. These events led Google to create the Certificate Transparency (CT) framework [318]. About 8 years later, CT is in the final stages of its deployment [490]. Transparency greatly facilitates the *detection* of illegitimate certificates, but there remain the questions of how to *react* after misbehavior is observed and how to *prevent* misbehavior altogether. Unfortunately, simply revoking the certificates of vulnerable CAs would have serious consequences: All the certificates issued by these CAs would become invalid, rendering countless websites unavailable. An ideal public-key infrastructure would prevent a vulnerable or misbehaving CA from jeopardizing the security of the entire system in the first place, and it would give users and browser vendors an option to demote CAs without completely distrusting them.

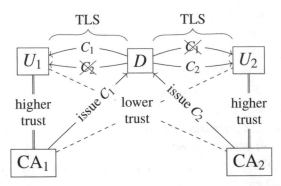

Figure 18.1: A dashed line indicates a lower trust and a double line indicates a higher trust. An arrow indicates certificate issuance or a certificate sent during a TLS handshake.

Our central observation is that trust is highly heterogeneous across the world. A PKI that supports trust flexibility would provide a foundation for domain owners and users/browsers to express trust preferences, penalize misbehaving or vulnerable CAs, and reward CAs that implement strong security measures. The core challenges in enabling different levels of trust in CAs are to achieve a meaningful overall system behavior with concrete security properties, ensure global verifiability of all certificates, and prevent downgrade attacks to lower security levels. An example to illustrate this is shown in Figure 18.1. In this example, user U_1 trusts CA_1 more than CA_2 for issuing certificates for domain D because CA_1 supports multi-perspective domain validation [1], while user U_2 trusts CA_2 more than CA_1 because CA_2 is an American CA and D's top-level domain is .us. In this example, U_1 should be able to express higher trust in CA_1 than in CA_2, while retaining the ability to use certificates issued by CA_2.

To address this issue, F-PKI introduces trust flexibility in two dimensions: Each domain can set a domain policy and each verifier can set (or choose) a validation policy using trust levels for CAs. Domain owners can specify policies in their certificates to restrict the set of valid certificates for their domain. Clients are then presented with a comprehensive set of certificates for each domain they visit and can make informed decisions based on their validation policy. F-PKI allows these clients to express a preference for certain CAs. Our new notion of trust is ternary and name-dependent: Every authority may be either untrusted, trusted, or highly trusted, for each domain name. In our example, user U_1 would treat CA_1 as highly trusted and CA_2 as trusted for domain D. U_1 would then reject C_2 if it conflicts with C_1. We envision that browser vendors would initially dictate trust levels, but users could modify these default trust levels according to their preferences. Once trust levels are defined, a user should only accept a certificate if it comes with evidence that no highly trusted CA has issued a conflicting certificate.

Domain owners are given the option to define what constitutes a "conflicting" certificate for their domain, as F-PKI offers the ability to specify policies through certificate extensions. Users then receive and consider all policies signed by highly trusted CAs. For example, domain owners can specify which CAs are authorized to issue certificates for their domain name. Giving domain owners the ability to define policies through certificate extensions would be futile, however, as an attacker who is able to obtain a bogus certificate could simply hide those policies. Therefore, we introduce a verifiable log server that can provide users with a view of all certificates and revocation messages relevant for any given domain name. This new log server, which we call map server, is meant to complement existing Certificate Transparency servers.

18.1 Trust Model

In this section, we introduce a more flexible trust model for the Web PKI. Our primary goal is to prevent a CA from attacking a domain if a certificate has already been issued by a CA that the **relying party*** trusts more for the domain name in question. To achieve this goal, we extend the trust model of today's Web PKI in two ways. First, we introduce a new trust level: Each relying party may consider some CAs more trusted than others, for all names or a subset thereof. In other words, trust in F-PKI is *ternary* and *name-dependent*. Second, we introduce new domain policies, which are set by the respective domain owner and are certified by a CA alongside the certificate. The relying party considers all domain policies certified by highly trusted CAs. As opposed to X.509 name constraints [138], all CAs can still issue certificates for any domain as long as they don't interfere with the certificates issued by highly trusted CAs (highly trusted from the perspective of the verifier). The three trust classes of our model are the following:

- **Untrusted:** As in today's trust model, only a public key that is part of a valid CA certificate can be used to verify certificate signatures. Other public keys are untrusted.

- **Standard trust (non-highly trusted):** This corresponds to the current notion of trust in a CA. Any CA in this trust class can keep issuing certificates, as in today's Web PKI, as long as it does not violate a policy defined by a highly trusted CA.

- **Priority trust (highly trusted):** This is the new trust level we introduce. Some CAs may be highly trusted for a set of names. Relying parties only consider the policies defined by CAs that they highly trust. This is defined by the following function:

  $f(N)$: set of authorities highly trusted for name N.

This model could be extended to support more trust classes. However, the number of trust classes should be kept low (for usability reasons), while still allowing relying parties to distinguish between reputable, neutral, and untrustworthy CAs.

18.1.1 Adversary Model

We assume that the attacker's capabilities are constrained as in the Dolev–Yao model [168]. Cryptographic primitives are unbreakable but their operation is known by the adversary, so are all public keys, but private keys are only known by their respective owners. Moreover, the attacker can obtain any message passing through the network, initiate a communication with any other entity, and become the receiver of any transmission. The objective of this attacker is to obtain a certificate for a victim's domain, and then perform an impersonation attack using that certificate and the corresponding private key. The adversary's goal is to remain undetected as long as possible if the attack succeeds. F-PKI is designed to completely prevent impersonation attacks under a set of assumptions, but even if the attacker's capabilities go beyond these assumptions, attacks can be detected. For this reason, we use two adversary models: one for prevention and one for detection. Also, because F-PKI supports trust heterogeneity, our adversary model can only be defined from the perspective of one client establishing a connection to a web server with a specific domain name. Let N be the domain name in question, $f(N)$ the set of CAs highly trusted by the client for that name, $g(N)$ the set of non-highly trusted CAs for N, and M the set of map servers that the client uses.

- **Adversary Model 1 (Prevention):** The attacker may compromise all CAs in $g(N)$ and a number of map servers such that (1) a subset of map servers $M_1 \subseteq M$ is honest, and (2) the map servers in M_1 collectively support all CAs in $f(N)$.

- **Adversary Model 2 (Detection):** The attacker may compromise all map servers in M and all CAs in $f(N)$ and $g(N)$.

18.2 Overview of F-PKI

We seek to accomplish two main goals. First, browser vendors and users can define a validation policy, i.e., label CAs as highly trusted, trusted, and untrusted for each domain. Second, to clearly identify these conflicts, domain owners must be able to define domain policies. No attacker should then be able to hide or downgrade these policies. Therefore, it is necessary to provide clients with a comprehensive view of all certificates, policies, and revocation messages relevant to the domain they are contacting.

We introduce an entity called map server, which provides a comprehensive view of certificates for its supported set of CAs. The goal of the map server is to aggregate certificate-related data in a verifiable manner and provide a meaningful interface to both clients and domain owners. Map servers use a sparse Merkle hash tree to effectively produce proofs of presence or absence. The data provided by map servers complements the traditional certificate validation procedure. The user gains a higher degree of assurance that the binding between a public key and a name is authentic by checking that there exists no conflicting certificate for the domain in question.

Figure 18.2 illustrates how the different entities interact. At a high level, the steps needed to establish a secure HTTPS connection within F-PKI are the following:

1. The domain owner requests a certificate for *www.example.com* from a certification authority. The certificate signing request sent to the CA may contain additional domain policies we introduce in §18.3.2.

2. The map server maintains a complete set of certificates (by periodically fetching certificates from CT logs, or acting as a special CT log).

3. The CA returns the certificate to the domain owner. The certificate should contain the parameters and policies the domain owner specified in Step 1.

4. The domain owner configures the web server so that it uses the newly obtained certificate.

5. The browser connects to the web server via HTTPS.

6. As part of the TLS handshake, the browser receives the certificate, and possibly stapled data periodically fetched by the web server from the map server (see below).

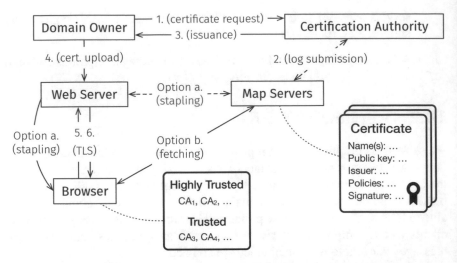

Figure 18.2: Overview of communication flows before and during the establishment of an HTTPS connection. A map server can either act as a special CT log or aggregate data from existing logs. The dashed lines indicate asynchronous communications.

Two main options allow the client to obtain a comprehensive set of certificates for *www.example.com* and the corresponding inclusion proof from a map server:

a. The web server fetches the certificate set and inclusion proof periodically or on-demand and provides it to the browser by stapling it to the TLS handshake.

b. The browser fetches the certificate set and inclusion proof from the map server, directly or via DNS (we describe the DNS technique in more detail in §18.6).

18.3 Policies

In this section, we describe two types of policies: the validation policies in which clients dictate which CAs are more trusted than others for a given name, and the policies that domain owners may define through X.509v3 extensions to opt-in and benefit from the stronger certificate validation that we propose.

18.3.1 Validation Policies (Trust Levels)

Clients can specify which CAs they highly trust. We envision that such policies would be defined by browser vendors, but users should be free to modify their default policies. Validation policies govern the behavior of $f(N)$ (see §18.1), which we defined as the set of authorities that are highly trusted for name N. Below are some examples of policies browser vendors and/or users may want to define:

- **CA-Based Policies:** Some CAs may be more trusted than others, regardless of the domain name. This judgment may be based on past events (such as security incidents), the validation methods that the CA employs, a reputation for following best practices, and/or geopolitical factors. As a concrete example of a reason to trust a CA over others, Let's Encrypt now validates domains from multiple vantage points or "perspectives" [1] to mitigate routing attacks on BGP [75, 76].

- **TLD-Based Policies:** CAs operating in a given country or region (as defined in their certificate) can be designated as more trusted for certain top-level domains (TLDs). For example, a policy may state that American CAs are more trusted for domain names ending in ".us" and ".gov" [302], while Chinese CAs are more trusted for ".cn".

- **Enterprise Policies:** Employees might be required to comply with policies for domains owned by their company. Online banking or trading platforms, as security-critical applications, may also provide their customers with policies they should enforce when connecting to their websites, specifying which CAs are highly trusted.

We also envision that such policies could be downloaded by users in the form of a "trust package" provided by organizations such as the Electronic Frontier Foundation, the CA/Browser Forum, the Mozilla Foundation, or ICANN.

18.3.2 Domain Policies

Domain owners can also define policies through X.509v3 extensions. There are two reasons to define policies as certificate extensions: security and backward compatibility. Domain policies allow domain owners to provide stronger validation of their certificates, for example, prohibiting wildcard certificates or listing authorized issuers. Relying parties receive all relevant certificates through map servers so an attacker with a fraudulent certificate cannot hide any policy. Policies also allow domain owners to opt in, which is necessary because F-PKI mandates a stronger certificate validation procedure that could potentially break the validation of existing certificates. Specifically, a certificate may contain the following policies:

- ISSUERS: A set of public keys (CAs) that may be used to verify signatures on this domain's certificates. If this extension is not present, then all the CAs the client trusts are authorized to issue certificates.

- SUBDOMAINS: A set of subdomain names for which certificates can be issued. Ranges of subdomains can be covered with wildcards (e.g., *.sub.example.com). If this extension is not present, then all subdomain names are authorized.

- WILDCARD_FORBIDDEN: Prohibits the use of wildcard certificates.

- MAX_LIFETIME: The maximum lifetime of a certificate.

Each attribute is marked as either *inherited* or *non-inherited*. An inherited attribute will be passed on to subdomains. Non-defined, non-inherited attributes default to the browser's policy.

Multiple Certificates for the Same Domain.   Although we presented F-PKI as an alternative to multi-signed certificates (to increase resilience against CA compromise), the two approaches can be combined. A domain owner can obtain certificates from several CAs, as long as the certificates respect the domain policies. Ideally, policies and other parameters defined in the certificates would be identical, but this is impossible to guarantee if the certificates are issued independently. Therefore, clients must consider the strictest of all policies.

18.4 Verifiable Data Structures

We introduce an entity called map server that provides a mapping from domain names to a set of certificates and revocations for this domain and parent domains. Each map server supports a set of CAs and keeps track of certificates issued by these CAs by leveraging the existing CT infrastructure and log data. The efficient audit of map servers is enabled by sparse Merkle hash trees, which we extend with nested trees to support hierarchical naming (similar to DNS).

18.4.1 Comparison to CT Logs

Even with Certificate Transparency, it is not possible to query a log server to directly verify that no certificate has been illegitimately issued for a given domain name. Instead, domain owners must rely on monitors, which keep entire copies of several logs. Only a few monitors exist at the moment. Li et al. [334] recently reported that none of the active third-party monitors they found could guarantee to return a complete set of certificates. The critical interface missing from CT log servers is thus one for fetching all valid (i.e., unexpired) certificates for a given domain name.

Table 18.1: The map entry for *example.com*.

Content	Type	Domain(s)
Certificates	list<X.509 certificate>	*example.com*
Revocations	list<revocation message>	*example.com*
Certificates	list<X.509 certificate>	**.example.com*
Revocations	list<revocation message>	**.example.com*
Tree Root	cryptographic hash	[subdomains]

18.4.2 Hierarchical Naming

Map servers distinguish between effective second-level domains (e2LD) [439] (i.e., domains where the parent domain is a public suffix [192]) and descendants of e2LDs (hereinafter subdomains). The remaining domains—parent domains of e2LDs and domains without a valid TLD—are invalid and the map server rejects certificates for such domains (e.g., *ac.jp* or *test.invalid*). e2LD entries are stored in a single sparse MHT. An advantage of such a hierarchy over a label-based hierarchy such as DNSSEC is that certificates for e2LDs can be stored at a lower depth (e.g., example.blogspot.co.uk has depth 0 instead of 3 in DNSSEC), which reduces the proof size.

The entries of subdomains are stored in nested sparse MHTs located below the parent domain's entry. The data structure for subdomains is the same as for the e2LDs, except that the key used for the index calculation only includes the name of the subdomain (without the parent domain). Since the Merkle proof of a subdomain contains all parent-domain entries, policies issued by parent domain owners are included in the subdomain proof. A parent domain owner can thus restrict CAs from issuing certificates for its subdomains or allow certificates only for certain subdomains. The reason map servers reject entries of public suffixes is that a certificate for a public suffix would enforce policies specified in the certificate for all e2LD domains using this public suffix. The map server is also pruned periodically to remove expired certificates.

18.4.3 Entry Format

The map server creates an entry for each domain with at least one certificate (valid or revoked) or one active subdomain. The entry consists of the certificates and revocation messages of both the domain (*example.com*) and the corresponding wildcard (**.example.com*), and the root of the subdomain MHT, as shown in Table 18.1. Each certificate might contain policies in the form of X.509v3 extensions.

18.4.4 Revocations

An end-entity certificate can be revoked either by a CA in the certification path (using the private key that corresponds to the CA certificate) or by the domain owner (using the private key that corresponds to the certificate itself). A revocation message for certificate C, revoked using a private key k, has the form $R_C = \text{Sig}_k(H(C), \text{revoke})$. Such a certificate revocation message can either be pushed to map servers by the domain owner or sent to the issuing CA which forwards the message to map servers. Both the revoked certificates and the revocation messages will be stored at the map server for transparency.

18.4.5 Proving Consistency over Time

In addition to the sparse e2LD MHT, the map server maintains a consistency MHT with chronologically ordered signed map heads (SMH) coming from the sparse e2LD MHT. The consistency tree can prove that the modifications between log revisions are correct, and prove that the log did not include non-existing certificates or exclude existing certificates. In other words, the signed map head represents the map at a given point in time, while the signed consistency head represents the entire history of the map.

18.5 Selection of Map Servers

Each client must find a minimal set of map servers such that each highly trusted CA is supported by a quorum of servers in this set. We envision that map servers would support most (if not all) major CAs and that the quorum would typically be equal to 1 (especially during the initial deployment of F-PKI). Therefore, clients only need proofs from a small set of map servers.

Users may either use a set that is pre-configured on their machine or manually pick a set. The set of required map servers may differ for each TLS connection, as it depends on validation policies.

18.6 Proof Delivery

A key element in preventing impersonation attacks and detect fraudulent certificates is to provide clients with proofs of presence/absence from map servers when (or even before) the TLS connection is established. In this section, we describe several options for this.

18.6.1 Stapling

The first approach to delivering proofs to clients is to have the web server embed them into a TLS extension, similarly to how revocation messages can be delivered with OCSP stapling. This technique does not cause privacy issues but requires that servers be updated to fetch proofs related to their own name periodically from a set of map servers.

18.6.2 Fetching via DNS

In case the web server does embed relevant proofs from a map server in its response, the client should fetch such proofs on its own. Instead of directly contacting map servers, the client can fetch log data through DNS [319, 498] or RHINE (Chapter 19). For example, to request a proof from a map server *mapserver1.net* for the domain name *www.example.com*, the client can send a recursive DNS query for *www.example.com.mapserver1.net* to its DNS resolver. The DNS resolver finds the IP address of the map server and forwards the client's query to that address. Acting as a name server, the map server replies with one or more TXT records[1] containing the entry of *www.example.com*. Finally, the resolver returns the records to the client. The advantages of using DNS for the purpose of delivering proofs from map servers to clients are numerous:

- **Decentralization:** The decentralized and hierarchical nature of DNS falls in line with our objective of avoiding reliance on a global authority.

- **Caching:** Records can be cached at several levels, for a configurable amount of time (TTL), to minimize latency.

- **Privacy:** Although DNS is not in itself a privacy-preserving system, the resolver is always aware of the domain names the client is trying to reach, so the privacy implications of asking a public DNS resolver for extra information about those names are limited.

- **Timeliness:** DNS resolution occurs before the TLS connection is established (to obtain the web server's IP address), which minimizes perceived page-load time increase.

DNS already supports the delivery of certificate-related data (e.g., through CAA, CERT, and TLSA records). DANE [170], in particular, binds certificates to domain names directly in the name system with the objective of solving the issues caused by rogue CAs. Unfortunately, DANE relies upon DNSSEC for protecting name-to-certificate bindings, and is not widely supported by browsers at the moment; moreover, as discussed in Chapter 19, DNSSEC has many security, reliability, and performance drawbacks that must be addressed in fundamental ways. In contrast, F-PKI does not require an authenticated name system such as DNSSEC, because the cryptographic objects we transport through DNS (merely as a channel for distribution) are authenticated independently, i.e., users know the public keys of their trusted map servers.

One concern over using DNS for transporting cryptographic data is that the size of messages sent over UDP is limited by the DNS standard. However,

RFC 6891 [149] describes extension mechanisms to enable using UDP for messages with sizes beyond the limits of traditional DNS. The RFC suggests that requestors try initially selecting a maximum payload size of 4096 bytes, which is sufficient to contain a cryptographic proof (such as the ones produced by Trillian) in many cases. If necessary, falling back to TCP enables the transport of larger packet sizes, at the price of increased latency.

18.6.3 Alternative Delivery Methods

Directly fetching a proof about a specific domain from a map server has privacy implications, but there might be situations where the user is willing to send requests directly to a map server. In particular, if the user relies on a public DNS server (such as CloudFlare's 1.1.1.1 or Google's 8.8.8.8), and if one of the companies operating the name server also operates map servers, then the privacy implications of fetching proofs directly are limited (assuming a secure channel is established between the client and the map server). In principle, any entity can act as forwarder and cache for F-PKI proofs. A potential candidate is the user's local SCION certificate service, which would improve privacy by hiding the identity of the requesting user from the map server. Another approach would be to use a middlebox (instead of updating the web server) to staple proofs by having the middlebox detect new connections and append relevant data to the TLS handshake (which is in plain text) [321, 499].

18.7 Certificate Validation

The certificate provided by the web server during the TLS handshake is validated by the client using the normal validation procedure and additional data provided by the map server(s), which may include a list of other certificates for the same domain, a list of certificates for parent domains, and a list of revocations for both. The client will then check revocations, resolve the domain's policy (considering only policies defined by highly trusted CAs), and verify that the certificate respects the resolved domain policy.

The validation algorithm starts by verifying that the legacy X.509 validation procedure succeeds and that the certificate is not revoked. Then, invalid and revoked certificates are removed from the list of additional certificates. Certificates issued by non-highly trusted CAs are also removed from the list. The algorithm then iterates through these certificates and resolves the domain's policy, considering the strictest of all policies in the received certificate and the filtered list of additional certificates. Finally, the algorithm checks that all domain policies are respected.

19 RHINE: Secure and Reliable Internet Naming Service

HUAYI DUAN, ADRIAN PERRIG

Chapter Contents

19.1 **Background** **433**
   19.1.1 Threats 433
   19.1.2 Data Authenticity 434
   19.1.3 End-User Privacy 435
   19.1.4 Service Availability 435
   19.1.5 Robustness 436
19.2 **Why a Fresh Start?** **437**
   19.2.1 Weak Practical Security and Performance 437
   19.2.2 Fragile Architecture and Trust Model 438
   19.2.3 Inefficient Secure Communication 439
   19.2.4 The Call for a Fresh Start 440
19.3 **Overview of RHINE** **440**
   19.3.1 Innovations 441
   19.3.2 System Architecture 442
19.4 **Authentication** **444**
   19.4.1 Secure Delegation 444
   19.4.2 Designing for Robustness 446
   19.4.3 Cyber Sovereignty, Split View, and Transparency 451
19.5 **Data Model** **452**
   19.5.1 Assertions 453
   19.5.2 Assertion Contexts 454
   19.5.3 Redirections 454
   19.5.4 Queries 455
19.6 **Secure Name Resolution** **455**
   19.6.1 Secure Communication 456
   19.6.2 Achieving Higher Availability 457
19.7 **Deployment** **457**
   19.7.1 RHINE for the Internet 457
   19.7.2 Name Resolution for SCION 458

Translating human-readable names to network addresses (among other information), a naming service is indispensable in making digital communication practically usable. Almost every connection over the Internet starts with a name resolution query to the Domain Name System (DNS). Likewise, for the setup of SCION connections, proper name resolution should precede path lookup: Before an endpoint can ask for paths to another endpoint, it must obtain a SCION address corresponding to the destination name.

Despite its proven performance and scalability, DNS was not created with security in mind—numerous vulnerabilities have been discovered over the past decades. According to a recent report, almost 80% of organizations globally suffered DNS attacks with an average cost reaching $950,000 [190].

The corruption of DNS data constitutes the first and foremost security concern. The standardized solution to this problem—DNS Security Extensions (DNSSEC)—promises to ensure the authenticity of naming data and is being slowly deployed around the world. The increasing privacy risks of using DNS have also spurred active development of countermeasures such as DNS over TLS (DoT) and DNS over HTTPS (DoH) in recent years. Furthermore, the frequent Internet-wide outages caused by disrupted DNS under cyber attacks (e.g., the DDoS attack on Dyn in 2016 [3]), software bugs (e.g., the failure of Akamai Edge DNS in Jul 2021 [122]), or operational errors (e.g., the historic outage of Facebook in Oct 2021 [354]) are constantly troubling users and require fundamental mitigation measures. Evidently, the Internet is in pressing need of a more secure and reliable naming infrastructure.

We investigate the fundamentals of such an infrastructure through a modern lens, articulating the gaps between the status quo and the ideal system. We then introduce RHINE (Robust and High-performance Internet Naming for End-to-end security), a *secure-by-design* naming system that provides a set of desired security, reliability, and performance properties beyond what the DNS security infrastructure offers today. RHINE addresses the inherent fragility of DNSSEC and is robust against both operational risks and malicious actors throughout the system. It also enables efficient end-to-end authentication of naming data, among other salient features. While RHINE is primarily designed for SCION, its architectural compatibility with the existing DNS ecosystems allows it to fit into (with minor modifications) and benefit the current Internet at large.

RHINE builds partly on a previous system called RAINS [508], but it has departed radically from its predecessor and is still under active development with possible future enhancements. This chapter provides a high-level view of the rationale and design of RHINE.

19.1 Background

We start by introducing basic concepts and terms related to name resolution, following the conventions of DNS. In a hierarchical naming system, the global namespace is organized into *zones*, each of which is administrated by an *authority*. Delegating a portion of a zone creates another (sub)zone, thereby forming a tree structure beginning from the root zone. A zone maintains mappings from its names to a range of information including but not limited to network addresses, email services, arbitrary texts, etc. The data of a zone is typically served by a set of distributed *authoritative name servers*, which in many real-world cases are not under the zone authority's direct control [282]. For ease of presentation, we sometimes use the term "zone" to also mean its authority.

The resolution of a name, typically initiated by a client but mostly undertaken by a recursive *resolver*, involves iteratively querying corresponding authoritative name servers to find the matching naming data; most servers on the resolution path return referrals to the resolver, eventually directing it to the server authoritative for the queried name. Caching naming data at resolvers speeds up name resolution and saves network resources, allowing the entire naming system to scale. In practice, other types of intermediary servers are also likely involved in the resolution process, e.g., a forwarder that simply relays client queries to a recursive resolver. We abstract them away here and consider only one logical resolver sitting between the client and authoritative name servers. This will simplify but not weaken our security analysis.

DNS has no security built in: All messages in name resolution are transmitted in the clear and thus are subject to tampering, forgery, and eavesdropping; name servers hosting unprotected naming data are inviting targets for attackers; the naming service itself is frequently abused for attacks such as reflection amplification; and new vulnerabilities are still discovered regularly. Several security-enhancing solutions for DNS have been proposed over the years. Some of them have overlapping but non-identical security goals, which are sometimes confused by the general public and even experts. It is thus imperative to first develop an end-to-end view of the threat landscape, the desired security properties for a modern naming system, and the design space thereof.

19.1.1 Threats

We consider a simplified DNS-style name resolution architecture, which consists of four types of entities as depicted in Figure 19.1: client, resolver, authoritative name server, and zone authority. The solid arrows represent logical communication links between entities for name resolution, and the dotted arrows indicate out-of-band channels for administration and management purposes (e.g., zone data publication).

Before discussing desired security properties in detail, we give an overview of potential threats. Our focus is on threats *to* the naming system, in particular

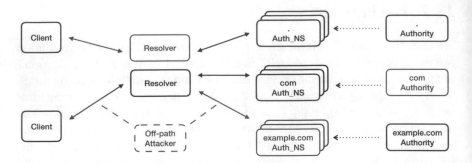

Figure 19.1: The typical architecture of a hierarchical naming system, with components where common threats arise marked in red.

those faced by the clients, as protecting their best interest is the primary goal of a secure naming service. There are also threats *from* the naming system, such as abusing it for covert command-and-control channels; these go beyond the scope of our discussion.[1] In short, the main threats to the naming system are the following:

- Network nodes on the communication paths can access, modify, fabricate, or drop messages exchanged during the name resolution process.

- Off-path adversaries may intervene in the normal data flow, altering the final or intermediate resolution result. They are best demonstrated by the infamous Kaminsky attack and its variants [351].

- Authoritative name servers and resolvers may also deviate from their expected behavior due to domain hijacking [520], malware infection [147], business incentives [533], or regulatory requirements [405].

- A misbehaving zone authority poses probably the most daunting security challenge because it has ultimate control over its own zone and implicitly all subzones as well.

19.1.2 Data Authenticity

The most fundamental property of a secure naming system is origin authentication and data integrity. More specifically, the name mappings received by a client should come with an assurance that they match exactly what is published by the genuine authority of the target zone. The lack of such a guarantee puts clients in a perilous position, as they may be obliviously directed to an attacker-controlled destination.

---

[1]Nevertheless, the common threat of abusing DNS for reflection amplification attacks is prevented by our secure resolution protocol.

DNSSEC was proposed to address this very problem. In brief, it allows a zone to sign its DNS records and have the signing key endorsed by the parent zone, thus creating an authentication chain all the way up to the root zone. A signed record can be verified by following the chain terminated at a trust anchor (which in most cases is the root zone key) configured by the validating entity. This eliminates threats of data tampering and forgery that originate outside zone authorities.

In general, end-to-end authenticity is achievable only by having zone authorities themselves secure the zone data with publicly verifiable authenticators. One may attempt a similar assurance by establishing hop-by-hop secure (in particular, authenticated) channels between entities involved in the name resolution process. However, channel security alone does not imply end-to-end security: While such a solution can foil attacks launched from the communication network, it fails to address risks arising from authoritative servers, resolvers, and any intermediary servers pertinent to name resolution. Clients must place trust in all these entities, instead of just the zone authority in the ideal case, to be assured of data authenticity. We thus strive for an end-to-end guarantee, where intermediaries do not need to be trusted from an authenticity perspective.

19.1.3 End-User Privacy

The privacy implications of DNS have received a surge of attention in the past few years. Name-resolution messages transmitted in the clear directly leak the websites or applications accessed by clients; even if subsequent connections are secured with TLS, the leakage from DNS is already information-rich enough to build accurate profiles of end users.

All privacy-enhancing solutions follow the same principle of establishing encrypted channels between entities (mostly between clients and resolvers). Examples include DoT and DoH as well as DNSCurve, which makes use of customized encryption methods. One caveat to these solutions is that they protect data confidentiality against eavesdropping on the communication links but not the entities themselves; countermeasures to further address this concern are also emerging, e.g., Oblivious DNS (ODNS) [459], and they typically work by using proxy servers to break the linkage of client identities to decrypted queries.

Regardless of the exact design, encrypting name-resolution messages is essential to preserving the privacy of end users and is complementary to naming data authentication.

19.1.4 Service Availability

Another more subtle security aspect relates to availability. DNS is designed to be available by redundancy: Each zone is typically replicated to multiple distributed authoritative name servers. This simple mechanism is effective in balancing workload and mitigating accidental events (e.g., system failures), but it is insufficient to mitigate threats from attackers that can actively intervene in the resolution process.

DNS servers are frequent targets for DoS attacks. Any data authentication mechanism will further amplify the effectiveness of such attacks because of the inflated bandwidth and computational costs. Common mitigation techniques include increasing the level of redundancy, blacklisting, and rate limiting. Yet, customized and optimized solutions are necessary to fulfill the global availability requirement of an Internet-scale naming service.

The interweaving of availability and integrity is a less obvious and often overlooked issue. Referrals are not authenticated as authoritative naming data by design, and this has profound implications [532]. Through manipulating referrals, an attacker can divert a resolver to non-responsive authoritative servers, or malicious ones that misguide the resolver further, hence wasting resources and rendering the naming service unusable to clients. These attacks are stealthier and more difficult to detect than direct DDoS attacks; they also require potentially fewer resources to launch. The design challenges lie in developing cost-efficient authenticators for referrals without compromising the overall resolution efficiency.

19.1.5 Robustness

In a robust system, the failure or compromise of a single component should not propagate to others and cause a cascading effect. In the context of a hierarchical naming system, this means that a corrupted zone should not prevent the correct resolution of other zones and in particular its subzones. DNS delivers a basic robustness guarantee through the common practice of hosting each zone at disparate operators: The existence of at least one correctly functioning authoritative server for a zone (and recursively all superordinate zones) guarantees the existence of valid resolution paths to all the subzones.

However, such a property does not hold in adversarial settings where zone authorities themselves turn against other honest entities in the naming system, as a result of security incidents, business competition, or regulatory pressure, etc. A corrupted zone could, for example, have its delegation records modified and redirect clients to bogus subzones. The hijacking of TLDs, as evidenced by the infamous Sea Turtle campaign [7], can lead to adversarial control of a large portion of the namespace and put entire countries' domains at stake.

DNSSEC-style authentication only alleviates the problem up to key compromise but does not completely solve it. An attacker controlling the secret signing key of a zone can create secure delegations that eclipse and invalidate the genuine subzones. The situation may be even worse than having an unsecured system, because users gain a false sense of security and this complicates the detection of wrongdoing.

Furthermore, DNSSEC-style authentication makes DNS more delicate in real operation (see more discussion in §19.2.2). The infrastructure's sheer complexity has led to pervasive mismanagement [117] and numerous outages [257]. Such a deficiency is intrinsic in that the chain of trust is established entirely within the DNS hierarchy. Conceptually, we can only overcome the limitation by re-designing the security architecture from the ground up, likely introducing trusted parties outside of DNS. This aspect has rarely been investigated before, and a large design space remains to be explored.

19.2 Why a Fresh Start?

Given the maturity of the DNS ecosystem, it might seem more judicious to use DNS with mandatory security and privacy extensions (e.g., DNSSEC and DoT/DoH) than to build a new naming system from the ground up. However, several reasons (that we explore in this section) lead us to believe that a clean-slate design is more appropriate than ad-hoc fixes. To justify our approach, this section discusses existing security solutions and their drawbacks.

19.2.1 Weak Practical Security and Performance

Although DNSSEC was intended to provide end-to-end data authenticity, end users rarely enjoy this assurance in reality. They almost always rely on security-aware resolvers for the validation of DNS records. Standard stub resolvers (e.g., `systemd-resolved` and `Dnsmasq`) come with buggy implementations of DNSSEC features and disable validation by default [26]. Indeed, even the official specification [33] carries a connotation that a validating stub resolver is optional and secondary to a validating recursive resolver. The actual security benefits reaped by end users thus depend on their trust relationships with the resolvers. For instance, a resolver located within an enterprise network may be considered more trustworthy than an ISP-operated one, as the latter may implement DNS surveillance and censorship [30]. In any case, non-validating clients run the risk of accepting data that may already be tampered with at the resolvers or during the last-mile communication [428].

The recent rise of privacy-enhancing name resolution is driving users to increasingly use open resolvers running in centralized and arguably more opaque environments (e.g., 1.1.1.1 by Cloudflare and 8.8.8.8 by Google), thereby further complicating the trust management issue [243].

Besides the operational complexity of DNSSEC, its significant performance impact also discourages client-side validation. The security-related DNS records are much larger than the regular ones and require much more resources to process. It is estimated that, with failure cases taken into account, DNSSEC-enabled zones should prepare themselves to handle 10 times the query load and 100 times the response load of their unsigned counterparts, and the overall resolution latency may soar from tens of milliseconds to seconds [255].

A validating client must retrieve and process the entire signature chain up to the trust anchor, with extra costs imposed even surpassing that of establishing the subsequent TLS connection. Such performance characteristics call into question the scalability of DNS when client validation becomes prevalent. The situation will get even worse when DNSSEC is adapted to use post-quantum cryptography that requires larger keying materials and more computational power [376].

Another byproduct of high performance overhead is the greatly increased risk of DoS attack. Akamai reported over 400 reflection and amplification attacks exploiting DNSSEC-signed domains just between Q4 2015 and Q1 2016 [12]. A recent study further shows that an attacker can potentially reach an amplification factor over 400 when the victim authoritative server enables DNSSEC [9].

19.2.2 Fragile Architecture and Trust Model

The key hierarchy of DNSSEC parallels the DNS hierarchy itself, converting the plain delegations created by NS records into their secure counterparts. Such seamless integration is beneficial in that it turns DNSSEC into a self-contained PKI with trust ultimately anchored in the well-established global root. On the flip side, the resulting security architecture has proven to be unanticipatedly fragile and has caused recurring problems to practitioners. We discuss the fragility of DNSSEC from two aspects: the (1) operational environment, and the (2) adversarial settings.

Fragile in operation. In order to establish a chain of trust, DNSSEC requires each zone (except the root) to synchronize its keying materials (i.e., public keys or their digests) with the parent zone. This imposes stringent consistency requirements on the entire DNS hierarchy. Any inconsistency (e.g., missing keys and signatures, mismatching keys, algorithms and parameters [257]) will cause validation and hence resolution failure; worse yet, the failure of a zone terminates the validation chain and blocks all the subzones. Such intricacy is also the culprit of many implementation bugs that can be exploited for cache poisoning, denial of service, and information leakage [257].

It is thus not surprising that DNSSEC outages happen frequently at all levels of the hierarchy (including the root and many cases of TLDs) and among all sorts of entities: Internet governance bodies (e.g., ICANN, RIPE, and APNIC); official authorities (e.g., NIST, NASA, and numerous .gov sites); large network and security service providers (e.g., Verisign, Dyn, and Google DNS); universities and news outlets; and many others [257]. DNSSEC problems are often difficult to locate due to the existence of many possible causes. It is reported that the outages last for 8 days on average and some peculiar ones even lasted for several years, leaving large portions of the Internet unreachable.

Somewhat ironically, to mitigate prevalent validation failures and keep resolver operators motivated for the deployment of DNSSEC, the community introduced Negative Trust Anchor (NTA) as a transitional solution that can temporarily *disable* validation for misconfigured domains [540]. Such a band-aid solution corroborates again the widespread mismanagement of DNSSEC.

Fragile against attacks. The top-down construction of DNSSEC's chain of trust dictates that a zone's signing keys presented to validators are ultimately determined by its parent. The parent zone has unlimited control over all its child zones through the signing keys; while the same principle applies to plain DNS, the cryptographic enforcement of DNSSEC makes the hierarchy more rigid. The closer to the root, the more power and impact a zone has to the global namespace. The single root key embodies an effective kill switch—whoever controls the key can potentially paralyze the entire Internet [441].

As with any security systems, proper management of secret keys is vital to DNSSEC. Unfortunately, empirical studies show that DNS authorities are not as prudent and reliable as expected: A significant portion of signed zones (including TLDs) use weak keys and algorithms [472]; in many cases, a single signing key pair is shared by multiple zones, and some large DNS service providers even reuse the same key across hundreds of thousands of zones [117].

Once the keys are compromised by adversaries, not only the zones themselves but all their subzones are affected. Notably, governments are notorious for compelling network operators (e.g., country-code TLDs and registrars) to collaborate as demanded or even stealthily sponsoring attacks [234]. In contrast to Web PKI, DNSSEC does not provide end users with the flexibility to configure trust preferences and the agility to switch to alternative authorities in case of security incidents—the trustworthiness of naming data always rests on DNS authorities alone.

The risks of DNSSEC's fragile trust model only recently gained attention from a technical perspective. An IETF draft describes a monitoring mechanism to mitigate the threat of a delegation-only parent zone (especially the root and TLDs) overriding secure delegations to child zones [548]. However, such incremental designs cannot fundamentally address the fragility of DNSSEC rooted in its very architecture.

19.2.3 Inefficient Secure Communication

Compared to the intricate DNSSEC, secure channel solutions are simpler to design and deploy. They have so far mostly focused on the last-mile communication between a client and a resolver. However, securing the interaction between resolver and authoritative name servers is of equal importance. The operation, policy, and security considerations here are different, and according to the IETF DPRIVE Working Group, relevant standardization work is still in a nascent stage. One crucial aspect is surely to keep the costs of secure communication to a reasonable amount such that the operators of resolvers and authoritative servers have sufficient motivation to adopt encryption [238].

Most privacy-enhancing solutions, including DoT, DoH, and DNS over QUIC (DoQ), build on off-the-shelf secure communication protocols. They all incur extra RTTs for session establishment as well as non-trivial CPU and memory resources. Whether it is sensible to adapt them to the iterative resolution process is still an open question.

Following a different approach, DNSCurve leverages DNS to distribute keys for secure channels, and the use of elliptic-curve cryptography makes it more efficient than aforementioned solutions. It also aims to protect the communication between resolvers and authoritative servers. Despite these features, piggybacking on DNS with unprotected transport means that DNSCurve can be abused for reflective amplification attacks.

Due to their performance penalty, all existing secure channel designs are prone to resource exhaustion attacks. It remains worthwhile to explore more efficient solutions that can survive large-scale DDoS attacks.

19.2.4 The Call for a Fresh Start

To sum up, existing solutions offer unsatisfactory security and performance guarantees even when they are fully deployed. In fact, even after years of promotion, DNSSEC has not yet reached a satisfactory adoption rate—only 25% of all DNS responses worldwide are validated as of mid-2021 and many zones are not yet signed [314]. Many interoperability issues exist in partial deployment and they are not easily resolvable [230]. Also, the complexity of DNS and DNSSEC makes it extremely difficult to introduce technical improvements. All these factors contribute to increasing arguments for revisiting DNS security and developing a modern secure-by-design naming system from scratch [257].

19.3 Overview of RHINE

RHINE represents our systematic effort to address the security, reliability, and performance issues discussed so far. A clean-slate approach enables us to overcome the litany of legacy issues in existing protocols and to innovate in radical ways. RHINE works with the same global namespace as DNS, which has a well-established ecosystem and business models that are difficult to change. An important design principle for us is thus to maintain as much operational compatibility with today's DNS entities as possible, and meanwhile leave enough room for future improvements. This section summarizes RHINE's major innovations and overall system architecture.

19.3.1 Innovations

19.3.1.1 New Security Architecture

DNSSEC embodies a simple and natural idea to authenticate naming data: establishing and aligning the chain of trust with the DNS hierarchy itself. All nodes on the chain are treated equally. However, a closer look reveals that the chain is composed of two distinct parts: the authentication of naming data and the authentication of authority delegation. Our key insight is that the two types of objects can be separated and handled in different ways: Clients only need to validate the authenticity of the final answer from the corresponding zone, whereas the heavier task of authenticating the delegation chain can be offloaded to other trusted parties and completed via out-of-band channels. This gives birth to a new security architecture with appealing properties.

Specifically, we introduce a dedicated RHINE certificate (RCert) to certify zone authority. RCerts can be issued by CAs through an end-entity public key infrastructure (EE-PKI), e.g., F-PKI (Chapter 18); but unlike regular TLS certificates, RCerts are solely used to sign zone data and require stricter polices for their issuance. Anyone with the RCert of a zone can directly validate its naming data, without going through a potentially lengthy validation chain. Compared with DNSSEC, this design promises much lower resolution costs and facilitates client-side validation, which becomes no harder than establishing a TLS connection. Furthermore, the introduction of external trusted parties opens up opportunities to design a robust system where no single entity can overpower others.

On a conceptual level, the new architecture flattens the fragile hierarchical trust model of DNSSEC and reduces the attack surface. It also draws a clear line between the functionality of distribution and the security of a naming system, enabling the two aspects to evolve independently. Improvements to the EE-PKI, such as public logging systems, can be readily incorporated into the naming system; inversely, changing the name resolution architecture will not downgrade the desired end-to-end authenticity property. The modularity also implies that RHINE itself does not serve as a PKI as DNSSEC, the abuse of which has been the subject of controversy for years.

19.3.1.2 Modernized Data Model

While the data model used by DNS has proved its efficacy for decades, it suffers from vulnerabilities and inefficiency. For instance, the relative time to live (TTL) used by DNS records can be maliciously extended at caching resolvers, which may lead to domain hijacking [271]; glue records are not mandatory and the lack thereof can create not only circular dependencies but also DoS attack vectors [9].

To this end, we introduce a security-centric data model that both incorporates best DNS operation practices and accommodates RHINE's new design.

At its basic level, the model distinguishes between two types of data in a zone: authoritative naming data and non-authoritative redirection information. The former is always signed by the zone authority for end-to-end authenticity, whereas the latter is protected with secure channels during the resolution process. This distinction is implicit in DNS; we make it explicit to facilitate modular design for data authenticity and service availability.

The new data model also comes with several desirable features that improve the versatility and transparency of the global naming service. For example, RHINE integrates the functionality of WHOIS by treating registrars and registrants as a type of named information. Moreover, it introduces the concept of name context, which allow alternative roots, split DNS, captive portals, and other real-world scenarios where naming inconsistency may exist, to be expressed explicitly and transparently to clients.

19.3.1.3 Secure Resolution with High Efficiency and Availability

Securing communication links involved in name resolution is vital to end-user privacy and naming-service availability. Given the performance-critical nature of an Internet naming service, the primary design goal here is to keep security overhead as small as possible. All solutions to date have suboptimal performance because they rely on asymmetric cryptography to establish shared keys between entities. Persistent connections are usually needed to amortize the high setup costs across DNS transactions [84], but this requires name servers to maintain excessive states.

Fortunately, the DRKey system (§3.2) already provides a stateless and highly efficient mechanism for key establishment based on symmetric cryptography. It allows each pair of RHINE entities during name resolution to set up secure channels with marginal overheads. As another salient feature, DRKey enables high-speed authentication of the first packet and its source address, thereby effectively thwarting reflection and amplification attacks. To ensure the naming service is still available to legitimate users amid DDoS attacks, we can further incorporate the DRKey-powered traffic filtering and bandwidth reservation systems. These techniques together will guarantee unprecedented name-resolution availability, whereas the deployment costs will be much smaller than those of a naïve availability-through-redundancy approach.

Secure channels cannot prevent the misbehavior of a rogue authoritative server. To obtain a high assurance in this unlikely but not impossible case, RHINE further allows resolvers to optionally validate a redirection message through its signed counterpart from the authoritative zone.

19.3.2 System Architecture

RHINE resembles DNS for the system architecture (see Figure 19.1). A RHINE server provides transient or permanent storage for *assertions* about names, and a lookup function that finds assertions for a given *query* about a name, either by searching local storage or by referring the querier to another server. Thus, RHINE servers can be configured to provide authoritative zone hosting service on behalf of zone authorities, recursive query service on behalf of clients, or any intermediary service (e.g., forwarding and caching) in the existing DNS ecosystem. The architecture compatibility makes RHINE a drop-in replacement of DNS and facilitates incremental deployment.

A zone authority is a logical entity responsible for the setup, update, and publication of the zone. An authority normally performs administrative tasks at a primary authoritative name server. For security reasons, it is preferable to sign the zone data offline with an air-gapped machine (sometimes called a "hidden primary" server unreachable from the Internet). Each zone should be published to a set of RHINE servers that are ideally managed by disjoint operators. The centralization of a naming service is detrimental to the Internet's robustness and stability, as evidenced by the massive outages caused by small bugs at large cloud-based DNS service providers [122, 354, 363]. For each RHINE server, an anycast constellation can be set up to improve latency and service uptime. The interaction between all entities is specified through two complementing sets of protocols:

- *Online* protocols are run between clients and RHINE servers for secure name resolution. The process follows DNS: Upon receiving a client query, a resolver iteratively contacts authoritative servers to get the matched assertion if no relevant data is found in its local cache. It is possible to use other mechanisms for matching queries and assertions. Thanks to the efficient authentication architecture of RHINE, assertions can be flexibly served from arbitrary infrastructure without breaking the end-to-end authenticity.

- *Offline* protocols are run by authorities to publish zone data, and more importantly, to delegate zone authority and issue RCerts. The issuance of RCerts differs from that of regular TLS certificates in that it requires the validation of correct zone delegation. Intuitively, this will establish implicit authentication chains for zone delegations in an offline manner, in contrast to what is explicitly constructed as part of zone data by a DNSSEC-style authentication architecture. All management operations of RCerts (e.g., update, renewal, and revocation) are also performed through offline protocols.

Once a zone sets up a proper RCert, it can simply use the corresponding secret key to sign naming data, without the need to synchronize any information with the parent zone. Such simplicity eliminates many mismanagement issues of DNSSEC by design [117], minimizing the chance of validation failures and system outages caused by operational mistakes [257].

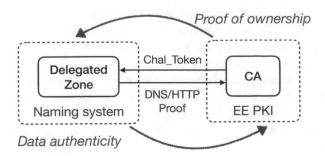

Figure 19.2: A circular dependency caused by adapting ACME-style domain
validation to RHINE for certificate issuance.

19.4 Authentication

RHINE shifts the heavy lifting of authentication away from the online name-resolution process to an offline stage where zone authorities should get themselves certified by the EE-PKI. The certification process revolves around the authentication of delegated authority and takes place only among zone authorities and CAs at infrequent intervals. This new security framework comes with both exciting opportunities and unique challenges. At the most basic level, it enables an augmented trust model such that, instead of the unconditional trust placed on the naming infrastructure (as in DNSSEC), a robust system of *checks and balances* can be achieved between naming authorities and CAs.

A large design space exists for the authority delegation and certification process. This section lays out the key considerations and builds up the authentication architecture progressively. Although RHINE takes into account the protection against malicious zone authorities and CAs, we make an important assumption that the two types of entities do not collude, since otherwise any security guarantees would be trivially broken. Also, unless otherwise specified, a delegation can apply to any level of the naming hierarchy, with the real-world relation between the delegator and delegatee abstracted away.

19.4.1 Secure Delegation

The Problem. When a delegation is created, the child zone should request an RCert from the EE-PKI. The issuing CA must verify that the requesting entity indeed controls the claimed zone. This process is analogous to the domain validation (DV) required for certificate issuance in the Web PKI. The standard approaches, as specified by the CA/Browser Forum [90] and the ACME protocol [48], let the CA securely send a random token to the requester, which must publish the token in the claimed domain via a certain channel to prove the ownership. An insecure channel can lead to the mis-issuance of certificates and domain impersonation; unfortunately, all practical validation methods hinge on

DNS and thus are subject to the manipulation of naming data [462]. DNSSEC becomes a requirement to fundamentally mitigate such threats in today's Internet [86, 462].

Can we directly apply these validation approaches for RCerts? The answer is unfortunately no, as otherwise a circular dependency will be created (as illustrated in Figure 19.2): a CA from the EE-PKI depends on a secure naming system for *proof of ownership* to issue certificates, but the requesting zone relies on a certificate to ensure *data authenticity* in the first place. One plausible solution to break the circle is using DNSSEC to authenticate the initial data for zone authority validation. However, this will bring the entire system's security guarantee back to the fragile DNSSEC—which we strive to eschew from the onset.

Secure Delegation Solution. We address the dilemma with the observation that, unlike a TLS certificate whose issuance is independent for each domain, an RCert is issued as a result of delegation and thus the parent of the requesting zone should naturally engage in the certification process. Assume the parent zone is already properly set up, it can then provide a *proof of delegation* attesting the authority of the child. Only upon receiving a request with such a valid proof should a CA issue the RCert for the child. If the parent's authority is also established through an RCert (probably from another CA) and so forth, we will have an entire authenticated chain of delegation all the way up to the root zone (Figure 19.3), analogous to the case of DNSSEC. As will be discussed later (§19.4.2.1), the root zone needs special handling given its unique position in the hierarchy.

To facilitate ACME-style automation for practical use, we specify the communication between the three parties involved in the delegation and certification process through a well-defined *secure delegation protocol*. A legitimate delegation agreement between the parent and child zone must be established as a precondition but not part of the protocol execution. This should follow existing practices, e.g., domain registration at registrars, or an organization internally assigning a subzone to an independent branch. After that, the three parties will communicate over authenticated channels for zone authority validation and RCert issuance. Note that the establishment of these channels will not cause further circular dependency. Also, we refrain from using RHINE to deliver the protocol messages (as in the case of one configuration of ACME, which uses DNS TXT records to publish the token), for the sake of a clean separation of the online and offline aspects of RHINE and hence low operational complexity. The three-way protocol serves as the basis for our subsequent design refinements.

RCert Management. The lifespan of an RCert should be aligned with that of the legitimate delegation. A delegated zone should be able to make changes to the RCert and get it re-issued by the CA as long as the delegation is valid. It

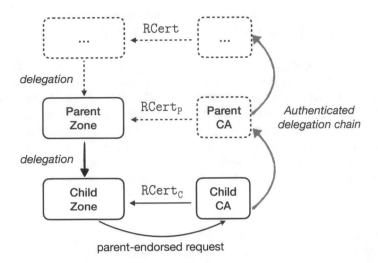

Figure 19.3: Illustration of the implicit authenticated delegation chain established in RHINE.

can also renew the RCert before expiration if the delegation is to be extended
upon agreement with the parent zone, e.g., extending the registration period
of a domain at some registrar. The revocation of RCerts follows existing practices used for regular TLS certificates; it is possible to have short-lived RCerts
with automatic renewal [523], obviating the need for and drawbacks of revocation [506]. All these management procedures will be specified with protocols
that allow automation, and operated in an offline manner without disrupting
the online name resolution process.

19.4.2 Designing for Robustness

Using RCerts instead of DNSSEC-style key chains to authenticate naming data
makes RHINE much more robust against operational errors, since no strict consistency requirement for the keying materials is imposed on a zone and its parent. We also aspire to protect the system against different Internet authorities
that may misbehave in adversarial environments, in response to the alarming
incidents of CAs and TLDs over the years [234, 242, 244]. The overarching
goal is to ensure that the entities counteract each other and no single one of
them can pose a kill switch or undue influence to the global namespace.

19.4.2.1 Mitigating Malicious CA

Augmentation with Public Logging. Unauthorized certificate issuance is a common problem in PKIs. A corrupted CA can issue illegitimate RCerts to take over the control of arbitrary zones. An effective and practically deployed solution is to submit all issued certificates to public and auditable logs, i.e., Certificate Transparency (CT). This mainly functions as a deterrence mechanism, since any mis-issued certificates will be publicly detectable. In other words, the protection provided by public logs is not preventive— between the acceptance of a certificate and the point when it is ready for audit, there is a time window during which attacks can occur [320]. Moreover, the detection of malicious RCerts can be more complicated than regular TLS certificates, as the former involves not only a single domain but an entire delegation chain. Nonetheless, we do recognize the logging infrastructure as a good starting point for a comprehensive design. As has already been widely enforced by browsers, in RHINE, an RCert is valid only if it is included in some public certificate logs.

Unlike the use case of Web PKI where CAs are the only authorities, here the parent of a zone has definite control of the child yet to be delegated, and therefore the RCert's validity should be contingent upon the parent. Yet, embedding the approval from the parent (e.g., a signature) explicitly in the RCert will roll the design back to DNSSEC-style authentication. Our workaround is to leverage the public logging infrastructure for safeguarding—the logs should only admit RCerts accompanied by valid approvals from the corresponding delegators. Note that this add-on security checking function is non-invasive, and it can be readily integrated as a rejection criterion for logging requests [162]. With this design, even if a CA can issue illicit RCerts on its own, the certificates should not be accepted by log-compliant clients.

Hybrid Authentication Architecture. Our authentication architecture based on EE-PKI changes the traditional trust model of DNSSEC. Zones are authenticated by their RCerts and clients rely on the PKI for the naming service's security. Here arises a vital question: How to accommodate the root zone—should it be certified with an RCert? The following reasons suggest that this should not be the case.

- Unlike registrable domains, the root zone itself represents an Internet-wide authority governing the global namespace. It is not sensible to certify the root as an end entity. Moreover, such a radical change of security infrastructure at the root level will surely raise tremendous concerns and resistance from the existing DNS ecosystem. These arguments also apply to TLDs.

- The root and TLDs in principle contain only delegation records that will not terminate the resolution of most queries. Thus, they do not need

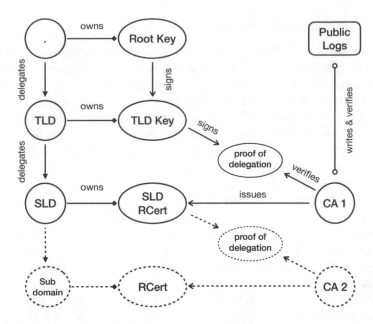

Figure 19.4: RHINE's hybrid authentication architecture, where all zones below TLDs are authenticated by RCerts.

RCerts to guarantee the end-to-end authenticity for the practical usage
of the naming service.

- Considering the issuance of the root RCert: Since the root has no parent,
  its governance body must define a trust anchor in some form to facilitate the secure delegation (or rather, bootstrapping) protocol (§19.4.1).
  Otherwise, a malicious CA can issue root RCerts at will and take over
  the entire namespace. However, maintaining the trust anchor at the root
  level renders the root RCert redundant and unnecessary.

In light of these concerns, we introduce a hybrid authentication architecture: While second-level domains (SLDs) and all subzones use RCerts, the
root zone and TLDs continue to use a two-layer key hierarchy as in DNSSEC
(Figure 19.4). This provides the following benefits:

- RHINE is made compatible with perhaps the most immutable part of
  the DNS ecosystem: No change is imposed on the root zone; the TLDs
  only need to run an ACME-style protocol to facilitate the secure delegation of SLDs as "offline" operation, replacing the error-prone online
  maintenance of DNSSEC keying materials for SLDs.

- From a security perspective, the hybrid architecture allows the root and
  TLDs to retain their innate authority in the naming hierarchy (the restric-

tion of their power will be discussed in §19.4.2.2). A CA can never issue valid RCerts for a zone without the parent's consent.

- Clients continue to enjoy the desired security and performance benefits from the EE-PKI for practically meaningful names. Even if clients query signed names directly from the root or TLDs, the costs are acceptable because the authentication chain consists of at most two levels.

19.4.2.2 Confining Zone Compromise

Recall that DNSSEC suffers from an intrinsic weakness: the compromise of a zone implies the control of all its descendants (see §19.2.2). The authentication architecture described so far suffers from the same problem. In particular, a compromised zone can manipulate its subzones by (1) directly publishing assertions for names belonging to the subzones, and (2) claiming alternative RCerts for the subzones despite the existence of legitimate ones. Concurrent to our work, this overlooked yet potentially devastating risk has gained attention only recently in the DNS community [548]. Interpreted another way, DNS zones other than the root have never enjoyed actual *authority independence*! Fortunately, RHINE opens up opportunities to address this issue in practical and efficient ways.

Enforcing Authority Independence.   To address the first malicious case, we limit the scope of an RCert, allowing its use to sign assertions about names one level below the identified zone and the zone apex. For example, an RCert for zone example.com can only be used for names abc.example.com but not xyz.abc.example.com. If a zone hosts authoritative data for names further down the namespace hierarchy, it must possess corresponding RCerts, i.e., (recursively) delegating to itself. A corrupted zone cannot affect its descendants unless it manages to obtain the RCerts for them.

The second case is addressed based on the public logging infrastructure already in use. We introduce an option flag to an RCert indicating whether the identified zone is supposed to have independent authority, which is decided by and agreed between the delegator and the delegatee upon delegation creation (§19.4.1). The processing of authority independence should be incorporated into the secure delegation protocol: A CA will issue an RCert for an independent zone only after verifying that there exists no valid RCert of the zone in the public logs. Note that such absence proofs are already efficiently supported by F-PKI (Chapter 18). With this design, once a zone has been legitimately delegated and is not yet expired, there is no way for its parent to eclipse it with another simultaneously valid RCert[2].

---

[2]To avoid being locked up due to the loss of the private key, a zone can generate a pre-signed revocation request (e.g., similar to the revocation certificate in GPG) in advance. This allows the zone to revoke the old RCert and request a new one immediately after the old private key becomes inaccessible, without waiting until the old RCert expires.

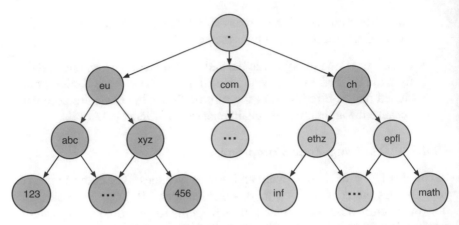

Figure 19.5: Illustration of authority independence, where non-independent
zones ("abc" and "xyz") are vulnerable to their corrupted parent
("eu") whereas independent zones ("ethz" and "epfl") are not.

Authority independence can be made configurable but not mandatory
to zones, since the property may be unnecessary in some use cases, e.g.,
when both parent and child zones are managed by the same entity. For non-
independent zones, it is possible to have multiple valid RCerts simultaneously.
We impose constraints to maintain security: (1) The root and TLDs are assumed to be independent, and all SLDs should be made independent by default
during registration unless explicitly configured by their owners, which is in
line with the current domain registration model; (2) delegation is only allowed
between two consecutive levels in the hierarchy; and (3) delegation from a
non-independent zone to an independent subzone is prohibited and otherwise
allowed. The last two conditions ensure that a malicious naming authority
cannot manipulate the parts of its namespace that are already legitimately delegated, and the conditions should be checked by CAs in the secure delegation
protocol.

The above mechanism can protect all zones below TLDs from the potential
risks of their corrupted antecedents. When the root zone is corrupted, it can
manipulate only the keys of TLDs but not the RCerts of SLDs, as in practice
SLDs should always be created with authority independence. With authority
independence enforced (Figure 19.5 depicts an illustration), a malicious zone
can compromise at most the availability but not the integrity of subzones, a
significant improvement over DNS and its security infrastructure today [441,
548].

Further Enhancements. The solution described so far is not yet perfect.
The apex of a zone (i.e., the zone's name itself) is still subject to manipulation
by the parent zone. We outline several directions for enhancements.

- For SLDs or registrable domains (w.r.t. the Public Suffix List [192]) in general, the risk can be reduced by having a validator always reject assertions about registrable domains returned from their parents.

- To eliminate the risk altogether, a validator can consult the public logs by itself to check the correctness of RCerts. For instance, if it receives an RCert of zone `example.com` from zone `example.com` for an assertion about `abc.example.com`, but the public logs record a valid RCert of zone `abc.example.com`, then it can conclude that the parent zone `example.com` must be cheating. As suggested by F-PKI [116], verifiable inclusion or absence proofs of the certificates can be published to corresponding zones and delivered through the naming system itself.

- An alternative and probably more efficient approach is to embed delegation information in space-efficient approximate membership query (AMQ) data structures, e.g., variants of the Bloom filter, in RCerts as extensions. This allows a validator to verify that a subzone (`abc.example.com`) does not exist for an assertion in question (for the potential apex domain `abc.example.com`) when validating it with the RCert (of zone `example.com`). Note that an AMQ data structure provides affirmative answers to non-membership queries and thus the validator is always assured that a subzone has not been delegated yet.

These methods have different trade-offs. A closer examination is left for future work.

19.4.3 Cyber Sovereignty, Split View, and Transparency

While DNS is intended to provide a consistent view of naming data to the entire Internet, this principle is violated in various ways in reality. The global DNS namespace has a default root managed by ICANN, but the attempts of different entities to operate alternative roots have never ceased over the years, especially when it comes to the sovereignty of cyberspace [448]. Each such root defines its own set of TLDs that may be identical to, based on, or completely distinct from those in the ICANN root, effectively creating an alternative namespace serving its target user groups. Debates over this subject have gone far beyond technology and are still ongoing. Despite that, we believe it is meaningful to provide technical means to accommodate such inconsistency and reduce its impact on end users as much as possible.

Split-view DNS is another common practice where servers selectively return naming data to clients depending on their network location or other information. It has found wide application in CDN, captive portal, internal usage of DNS, etc. Compared to alternative roots, these deployment scenarios do not affect the global namespace and are acquiesced by the broad community.

To address these inconsistency issues, we introduce the concept of *context* to RHINE. Contexts specify information about the global namespace identified

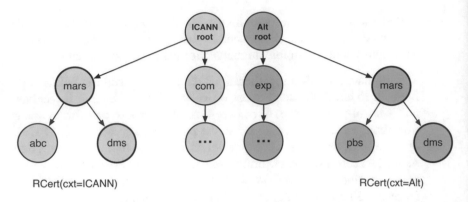

Figure 19.6: Illustration of certificate contexts accommodating namespace defined by alternative roots.

by some root or local use of the naming system. Each RCert must be associated with one and only one context of either global or local type. An assertion is valid only if signed with an RCert with a matching context (see §19.5.2).

The global RCert contexts of a zone and its parent must match, with the consistency checked by CAs during the secure delegation process. For the issuance of RCerts for SLDs, CAs should check signatures from the TLDs up to the corresponding root keys. The security checks based on public logs apply to each root separately, allowing all roots to operate in parallel without interfering with each other. A zone wishing to operate in multiple global namespaces must obtain multiple RCerts with respective contexts (as illustrated in Figure 19.6). The issuance of RCerts with local contexts is less strict because they are detached from any global namespace; an entity can easily request them from a CA without going through the security checks.

With the effective use of context, RHINE makes naming data inconsistency *transparent* to everyone on the Internet. Regardless of social or political issues, governments can implement their virtual boundaries at lower costs without building new infrastructures from scratch [448]. End users can make more informed decisions on what they (dis)trust.

19.5 Data Model

This section gives an overview of the RHINE data model used for secure name resolution. A zone contains two main types of data: An *assertion* is a signed mapping between a name and a property of that name; a *redirection* is an unsigned mapping between a name and an authoritative server that is hosting the zone containing the name. In addition, a *query* data type is used in the resolution process to express interest in information about names. Except for the root and TLDs (see §19.4.2.1), each zone is associated with at least one RCert for assertion signing.

19.5.1 Assertions

An assertion consists of the following elements:

- **Context:** The context in which the assertion is valid.
- **Subject:** The non-qualified name about which the assertion is made. A non-qualified name is a local, not necessarily globally unambiguous identifier (e.g., 'foo'), which—in combination with the zone (e.g., 'example.com')—yields the fully qualified (i.e., unambiguous) name (e.g., 'foo.example.com'). The domain name separator is '.' and it separates subject and zone. For a domain apex, a special identifier '@' is put in this field.
- **Zone:** The zone (e.g., 'example.com') in which the assertion is made.
- **Object:** The data associated with the name of the given type.
- **Type:** The type of information (aka property) about the subject contained in the assertion. Each assertion is about a single type of data. The supported types include but are not limited to the following:
    - **Address:** One or more addresses associated with the name, given an address family (analogous to the A and AAAA DNS record type).
    - **Authority:** Metadata about the zone authority, including the name and address of the RHINE server(s) authoritative for the zone (analogous to the SOA DNS record type).
    - **Alias:** One or more aliases of the name (analogous but not identical to the CNAME and the PTR DNS record types).
    - **Registrar:** Information about the registrar responsible for the appearance of a delegation within the zone.
    - **Registrant:** Information about the registrant of the zone within a TLD (analogous to the WHOIS service).
- **Issued:** Time at which the assertion was issued.
- **Expires:** Time after which the assertion is no longer valid.
- **Signature:** A signature generated by the authority to authenticate the assertion. This signature covers all elements within the assertion except the signatures themselves.

Timestamps are always expressed in terms of UTC (i.e., absolute instead of relative time). Since the signature protects the timestamps as well, it is necessary to sign new assertions before the old ones expire. At a single point in time, it is possible to have multiple active valid assertions with overlapping validity times for a given ⟨subject, zone, context, type⟩ tuple. The union of the object values of all of these assertions is considered to be the set of valid values at that point in time.

19.5.2 Assertion Contexts

All assertions are held to be valid within an explicit context. Contexts are used to provide explicit inconsistency, while allowing assertions themselves to be globally valid regardless of the query to which they are given in reply. Explicit inconsistency is the simultaneous validity of multiple sets of assertions for a single subject name at a given point in time. Explicit inconsistency is implemented by using the context to identify the authority whose RCert should be used to verify the validity of an assertion. If an assertion is from a root zone or TLDs, it should be verified with the corresponding keys. In what follows, we assume assertions are not from the root zone or TLDs for ease of presentation. In RHINE, two types of context are defined.

- A global context identifies a namespace defined by a root (§19.4.3). It defaults to the ICANN root. Assertions in the global context are signed by the authority for the subject zone with the corresponding RCert. For example, assertions about the name 'inf.ethz.ch.' in the default context are only valid if signed with an RCert for zone 'ethz.ch' (or zone 'inf.ethz.ch' if it exists) with the default context.

- A local context consists of an authority-part and a context-part separated by a marker 'cx-'. The authority-part identifies the authority whose RCert is used to sign the assertion; the context-part provides additional contextual information. Authorities have complete control over how their local namespaces are arranged and over the names within those contexts, independent of any global namespace.

Example: Captive Portal. A captive portal resolver could respond to a query for a website (e.g., 'www.google.ch') with an assertion whose context identifies the network location and ISP:

'sihlquai.zurich.ch.cx-starbucks.access.some-isp.net'.

This informs the client of the existence of this particular captive portal and that the result for 'www.google.ch' is not globally valid. The client can also validate that the assertion is signed by some local authority with an RCert with local context sihlquai.zurich.ch.

19.5.3 Redirections

Redirections correspond to NS and glue records in DNS. Unlike its predecessor, a redirection fuses both the name and the address of an authoritative server into an inseparable whole. This not only prevents the common DNS resolution failures caused by missing glue records but also reduces DoS attack surface [9, 375], as no extra lookup is required to resolve names of RHINE servers themselves. Note that the combination does not make the zone management any harder: When a downstream RHINE server changes its address, the corresponding redirection in an upstream RHINE server should be updated accordingly—such synchronization is necessary in DNS as well.

19.5.4 Queries

A query is a request for a set of assertions supporting a conclusion about a given subject-object mapping. It consists of the following information elements:

- **Context:** The context in which responses will be accepted. A query may also name a special *any* context, signifying a willingness to receive information about names in any context available at the query server.

- **Subject:** The name about which the query is made; in contrast to assertions, the subject name here is fully qualified.

- **Types:** A list of the types of information about the subject that the query requests.

- **Valid-until:** A client-generated timestamp for the query after which it expires and should not be answered.

- **Token:** A client-generated token for the query, which can be used in the response to refer to the query.

- **Options:** A set of options through which a client may specify trade-offs.

A response to a query contains a set of assertions or redirections bound to the token supplied by the client in the query. A response from the final authoritative server should include the RCert to enable validation of assertions.

19.6 Secure Name Resolution

The basic name-resolution flow of RHINE is similar to that of DNS. A client sends queries to a pre-configured query server and expects responses. A query server may send queries to other RHINE servers to express interest in information about a given name. An authoritative server returns assertions or redirections in response to queries.

In a SCION-based deployment, each AS can operate a set of anycast query servers for its customers. Each ISD specifies the global roots to use within its realm. A strict policy, for instance, may require AS-operated RHINE servers to reject queries for any root other than specified by the ISD. Despite this, a host is always able to validate received assertions with the RCerts on its own and decide whether to accept the assertions or not.

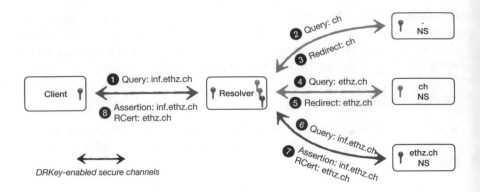

Figure 19.7: Illustration of RHINE secure resolution.

Since the end-to-end authenticity guarantee of RHINE does not depend on
the organization of the servers, networks can use different inter-server topologies with varying performance trade-offs. For instance, query and intermediary
servers can be organized as hierarchical caches. Queries that cannot be served
out of an AS-operated query server's cache are delegated to the service anycast
query server of one or more upstream ASes. Query and intermediary servers
may also recurse to the authoritative server for the zone, according to their configuration. Alternatively, intermediary servers may connect to each other via
a distributed hash table, and instead of a request-response resolution, authoritative servers may proactively provide assertions to intermediary and query
servers via a push-based mechanism.

19.6.1 Secure Communication

RHINE protects the entire resolution path with high-speed secure channels.
Each pair of communicating nodes on the path uses the DRKey system to
derive a shared key (a third-level key in particular). For comprehensive protection, the key is used to encrypt and authenticate RHINE messages at the
packet level, with network addresses also authenticated to thwart packet spoofing. The authentication of client addresses prevents reflection attacks, and the
authentication of RHINE server addresses assures the querier that it is not maliciously diverted.

Key derivation in DRKey is inherently asymmetric in terms of resource requirements (see details in §3.2). To avoid creating DoS attack vectors against
the name-resolution infrastructure, the client of a connection here should always be the slow side and the server be the fast side. Specifically, an end host
should contact the certificate service run by the source AS to fetch a third-level
key, whereas a client-facing query server is entrusted with the zeroth-level or
first-level key of the destination AS to efficiently derive third-level keys on the
fly. Similarly, when a query server contacts an authoritative server, the former

should fetch any third-level key from its AS-hosted certificate service while the latter can compute the key locally.

Encrypting RHINE packets foils network eavesdropping, but the initial query server or intermediate servers may still pose privacy risks to end users. This can be addressed by adopting emerging solutions (e.g., introducing additional proxy servers as in Oblivious DNS [459] and Oblivious DoH [474]), or SCION-enabled anonymous communication protocols (e.g., HORNET [105], PHI [104], and TARANET [107]) to RHINE. We leave this for future work.

19.6.2 Achieving Higher Availability

Recall that the subtle threat of corrupted authoritative servers returning incorrect redirections cannot be addressed with authenticated channels. One possible solution is to have each zone authority also sign its redirections, but this blurs the boundary of zone authority and complicates operation.

A better approach is for a resolver to optionally ask for an assertion of type **Authority** (§19.5), and verify if it matches the redirection that leads itself to the answering server. This enables the detection of any bogus redirections and preempts DoS attacks that create long fake redirection chains. Even with this security add-on, the resolution process of RHINE is still less expensive than that of DNSSEC, and the extra costs are never imposed on clients.

As a further step to ensure service availability when the naming infrastructure is under a DDoS attack, we can integrate LightningFilter (see §9.2) and COLIBRI (see §10.2) into RHINE. SCION-aware ASes may already deploy these systems at different operational levels, yet specialized deployment options could be configured for RHINE. For example, an AS can set up a LightningFilter instance dedicated to filtering RHINE traffic, with traffic profiles adjusted to suit the processing capacity of RHINE servers. This guarantees that even if a DDoS attack is launched from multiple ASes, (limited) naming service is still available to hosts in uncorrupted ASes.

19.7 Deployment

We expect the deployment of RHINE to be an incremental process and that it will co-exist with DNS for an extended period of time, in a similar way to the transition from IPv4 to IPv6. This section discusses the deployment issues in both current Internet and SCION.

19.7.1 RHINE for the Internet

Architecture-wise, RHINE is fully compatible with DNS (despite its flexibility in arranging different network topology for name resolution, see §19.6). It also bridges the DNS ecosystem and EE-PKI in an unusual way, allowing the latter to augment the security but not overthrow the authority of the former.

From an operational perspective, RHINE will come with a set of software equivalents of DNS and modern EE-PKI.

- A full-fledged RHINE server that can be configured as a forwarder, resolver, or authoritative nameserver, performing operations specified by RHINE protocols.

- A client-side RHINE library exposes the same resolution APIs as standard DNS stub resolvers (e.g., `systemd-resolved`) to userspace applications, and handles mandatory validation of assertions.

- A CLI tool `rdig` is also provided as a replacement of the familiar `dig`.

- A RHINE manager consolidating offline functions, including secure delegation, zone signing and publication, etc.

- A set of ACME-style tools for automated issuance and management of RCerts.

- An extension to the public logging infrastructure to implement additional security checks for RCerts.

DNS entities can migrate to RHINE by substituting legacy software with their RHINE counterparts. Zone authorities should additionally communicate with the EE-PKI and manage RCerts with the new toolset. A remarkable exception is that the operation of the root zone, which is complicated and requires extreme prudence, remains unaffected. This significantly reduces the friction of adopting and deploying RHINE.

It may be useful to introduce a Happy Eyeballs [541] approach to facilitate the co-operation of RHINE and DNS. This allows clients to send queries to both systems and prefer authenticated answers from RHINE. One caveat here is that potential downgrade attacks may disrupt the delivery of RHINE messages and force clients to always fall back to the untrusted DNS. This risk can be mitigated by deploying COLIBRI to RHINE for guaranteed low-latency transmission of name queries and responses.

19.7.2 Name Resolution for SCION

While there are multiple ways for legacy Internet hosts to benefit from SCION (see Chapter 13), the resolution of native SCION addresses is necessary for many applications. We describe a transition solution that enables the lookup of SCION addresses via DNS and the integration between RHINE and SCION.

DNS Support for SCION. Given the versatility of DNS, adding the support is technically simple. For backwards compatibility and low overheads, we should make a sensible decision on which type of DNS resource record (RR) to use for SCION-related information. Introducing a new RR type may require a tedious standardization process and impose non-trivial changes to DNS software. On the other hand, most standardized RR types are designed for specific purposes, and overloading them for SCION may interfere with the intended services in unexpected ways. We identify the TXT RR as the most suitable type, as it is designed to hold arbitrary textual data that allows application-specific parsing.

The TXT RR represents a straightforward textual encoding of the SCION address, which contains ISD and AS identifiers as well as the IP address of the host, e.g., `scion=1-ff00:0:101,192.0.2.1`. Including the network address in the TXT record may appear redundant as it already exists in A/AAAA RRs for the same name. Yet, such an encoding allows the resolution of a complete SCION address via a single DNS query and avoids unnecessary complexity, e.g., when a name is mapped to multiple mixed IP and SCION addresses.

No change to existing DNS software is required to support the resolution of SCION addresses. Zone authorities just need to add these additional TXT RRs in their zone files. On the client side, a DNS query with the type TXT must be made and the answer be parsed accordingly. The special query and parsing functionality can be implemented by individual applications (e.g., a SCION-enabled browser §12.2.5) or wrapped in a reusable API library (e.g., snet §12.1.3).

Transition to RHINE. The versatile toolbox of SCION provides an array of methods to use RHINE for secure name resolution. In addition to the above application-level and API-level integration, a stub resolver can be integrated into the SCION daemon (§12.1.2), which handles control-plane functionality including path lookup that should come after name lookup by design.

20 PILA: Pervasive Internet-Wide Low-Latency Authentication

CYRILL KRÄHENBÜHL, ADRIAN PERRIG*

Chapter Contents

20.1 Trust-Amplification Model 463
20.2 Overview of PILA 464
20.3 ASes as Opportunistically Trusted Entities 464
20.4 Authentication Based on End-Host Addresses 465
    20.4.1 End-Host Certificates 465
    20.4.2 Additional Local Identifiers 465
    20.4.3 Anycast 466
20.5 Certificate Service 466
20.6 NAT Devices 467
20.7 Session Resumption 467
20.8 Downgrade Prevention 468
    20.8.1 Signature-Based Approach 468
    20.8.2 Log-Based Approach 468

PILA provides end-to-end encryption and authentication, increasing the minimum level of security of *all* communication. The core idea in PILA is to reduce the attack surface of trust-on-first-use (TOFU) key exchanges from all on-path ASes to only the source and destination ASes and to hold these ASes accountable enabling detection of misbehavior. Based purely on AS local network addresses, PILA provides a base security level for pervasive encryption. User or service-based key establishment provides stronger properties, but in many settings, such as to secure IoT communication, PILA provides a strong default level of security.

In SCION, DRKey (§3.2) can be used to derive symmetric keys (DRKeys) between two entities and efficiently authenticate data-plane and control-plane messages. To achieve scalability for global symmetric-key distribution, the certificate service of an AS derives end-host keys and could use these keys

---

*This chapter reuses content from Krähenbühl, Legner, Bitterli, and Perrig [304].

itself to produce message authentication codes (MACs). This is generally un-problematic for the use cases of DRKey, such as network-layer source authenti-cation, providing authenticity of SCMP messages, and packet filtering. These use cases focus on preventing spoofing attacks to achieve network connectivity and availability; furthermore, a misbehaving AS forging a MAC for a message does not cause damage to other ASes (only to the AS itself), since the misbe-having AS can only forge MACs for messages originating from or received by the AS itself.

However, if two end hosts want to securely communicate, for example using TLS, simply encrypting and authenticating the messages using the (symmetric) DRKeys does not give the end hosts the desired end-to-end security properties: AS's certificate servers can always generate the same DRKeys and decrypt the messages or forge and authenticate new messages. Even if the DRKeys are only used by the end hosts to create a MAC over an anonymous Diffie–Hellman key exchange, an AS can intercept the key-exchange message, forge its own key-exchange message including the MAC, and thus break the end-to-end security through such a man-in-the-middle attack.

Such attacks by source and destination ASes—either on data traffic or the key exchange—can be prevented if the two end hosts authenticate each other through asymmetric cryptography. To achieve this, each end host should have its own private key and the certificate service of its AS should issue certificates containing the end hosts' respective public keys.

Pervasive Internet-Wide Low-Latency Authentication (PILA) [304] is an au-thentication system designed to provide ubiquitous end-to-end authentication in SCION by issuing such certificates to end hosts. PILA is not intended to re-place existing strong authentication protocols, such as server certificates based on the end-entity PKI (Chapter 18), but to provide an improved minimum level of security if strong authentication protocols are not available.

Existing anonymous Diffie–Hellman approaches inhibit off-path attackers and thus reduce the set of possible attackers to only (active) on-path attack-ers. PILA (and DRKey) provide strictly stronger security properties by further restricting the set of possible attackers to only the ASes of the respective end hosts. PILA furthermore disincentivizes these ASes from misbehaving through cryptographic proof of any misbehavior. In summary, PILA is an authentica-tion protocol providing a strong minimum security guarantee for all end-to-end encryption in SCION.

We first introduce the trust-amplification model, which is a general model to increase the security of an authentication system, and then describe PILA, an instantiation of trust amplification based on the control-plane PKI (CP-PKI) and opportunistically trusted ASes providing end-host authentication.

20.1 Trust-Amplification Model

Our authentication system builds on a trust-amplification model, which is a certificate-based authentication model relying on three key principles: *crude authentication*, *accountability*, and *leverage*. Trust amplification provides a generic model to increase the security of a certificate-based authentication system indirectly by deterring misbehavior of involved certificate-issuing entities. The meaning of misbehavior depends on the actual system used and typically means equivocating by issuing conflicting certificates. Trust amplification guarantees correct authentication if the certificate-issuing entities selected in *crude authentication* consist of curious-but-cautious (CuBC) attackers, which only launch coward attacks (i.e., attacks that cannot be detected).

The trust-amplification model makes several assumptions. First, we assume that there exists a single trust anchor, agreed on by the communicating entities, which provides keys and certificates to ASes; this is provided through the CP-PKI (§3.1). Second, we assume that participating ASes are able to authenticate their end hosts. Third, end hosts must have access to an authentic version of the TRC of their ISD, which represents their trust anchor. Fourth, rough time synchronization with a precision on the order of a minute is essential for a certificate-based system with certificate lifetimes of several hours.

Crude Authentication. The first step is to significantly reduce the number of entities that can issue certificates for the communicating end hosts to reduce the attack surface. Such a reduction is only meaningful if the certificate-issuing entities are not omnipotent, i.e., cannot issue certificates for arbitrary end hosts. Ideally, the entities manage disjoint sets of end host identifiers which reduces the certificate-issuing entities for an end host to a single entity. For the SCION CP-PKI, only the local AS (i.e., the local certificate service) can issue certificates for an end host; thus achieving the optimal reduction to a single certificate-issuing entity. In the trust-amplification model, the relying end host establishes a trust relation to a certificate-issuing entity of the end host that is authenticated, based on the following two principles.

Accountability. In order to increase trust in a certificate-issuing entity, which might initially be untrusted, certificate-issuing entities are held accountable for their actions. In the trust-amplification model, this property is achieved by generating irrefutable evidence that proves the misbehavior of a certificate-issuing entity. Important properties are resilience to slander (cannot forge false evidence) and framing (cannot manipulate an entity to produce false evidence itself), such that evidence is necessarily a result of improper behavior of a certificate-issuing entity.

Leverage. As a third principle, misbehavior must be disincentivized. After detecting misbehavior of a certificate-issuing entity M, other entities must have some form of leverage over M. For end hosts that are issued certificates by M, leverage could be economic detriments through loss of customers or legal recourse. For other end hosts, leverage could be a global or local trust rating of certificate-issuing entities based on collected evidence of misbehavior. Such ratings might be used to determine trust policies in case F-PKI (Chapter 18) is used.

Trust amplification is similar to the approach used in certificate transparency (CT) [318] since misbehavior is deterred by providing cryptographic proof thereof. However, with trust amplification, the power of each certificate-issuing entity is restricted to a subset of identifiers (i.e., only the end hosts within the AS), in stark contrast to the omnipotent highly-trusted certificate authorities in the Web PKI with CT.

20.2 Overview of PILA

PILA provides authentication based on the end-host address[1] as an extension to existing protocols, such as TLS or SSH. PILA reduces the attack surface to the end hosts' ASes, and produces proof of misbehaving ASes that create illegitimate certificates—e.g., to perform man-in-the-middle attacks on their end hosts. The underlying protocol—which PILA extends to provide authentication for—must have (or must be extended to have) the property that an entity can authenticate itself using an X.509 certificate.

A PILA workflow where an initiator (relying party) A authenticates a responder B, works as follows. First, B's AS uses its private key from the CP-PKI (for which it has an AS certificate) to issue a short-lived certificate to B over B's public key and end-host address. B uses this certificate to, for example, authenticate an SSH or TLS handshake. A verifies the authenticity of the handshake or signed reply using the certificate chain starting at the TRC of B's ISD. A also keeps track of the used CP AS certificates and end-host certificates locally or adds them to an append-only log to retain the irrefutable proof of misbehavior, which can be detected through an out-of-band channel or an external auditor.

20.3 ASes as Opportunistically Trusted Entities

In PILA, trust anchors (TRCs) are axiomatically trusted, but end hosts interact with few ASes, which are only *opportunistically* trusted. ASes are not omnipotent, as they are identified by a unique AS number and implicitly by an ISD number corresponding to the TRC used as the trust anchor. PILA uses ASes to bootstrap connection establishment, and then increases the trust placed into these ASes through trust amplification.

Each entity in the Internet is part of at least one AS, which is under the control of a single administrative entity. This facilitates providing a common

---

[1] An end-host address consists of the ISD, AS number, and local end-host address.

service that authenticates end hosts (e.g., using a challenge–response protocol or pre-installed keys and certificates) and issues certificates. Another advantage is the typically close relationship between an end host and its AS, which allows for a stronger *leverage* in case of misbehavior. Since it is infeasible for an end host to authenticate each AS by itself (there are ∼75,000 active ASes in 2021 [59]), the CP-PKI is used as a trust anchor to authenticate ASes.

Using ASes as opportunistically trusted entities promotes an incremental deployment model of PILA since there is the immediate benefit for end hosts to set up secure connections.

20.4 Authentication Based on End-Host Addresses

PILA authenticates end hosts based on their local end-host addresses. The benefit compared to name-based authentication, like domain names, is that all devices participating in Internet-wide communication have an end-host address and can thus make use of PILA. If an AS uses IP addresses as local end-host addresses, end-host certificates can be represented as X.509 resource certificates [253] in order to be compatible with existing PKI technologies. X.509 resource certificates add several extensions, most notably certificate policies [253] and IP-address and AS-number resources [349], which authorize subdomains to use these resources. End-host certificates are issued by the AS of the end host and contain a single local end-host address or a set of local end-host addresses delegated to this end host.

The relying end host constructs the chain of trust based on a given TRC and the CP CA and CP AS certificates and verifies it as explained in §3.1.7.2. In combination with the short-lived end-host certificate, the relying end host can authenticate its communication.

20.4.1 End-Host Certificates

An end host requests its certificate ($CERT_E$) from the local AS certificate service. An end-host certificate binds the public key of an end host to a local end-host address unique within the AS. End-host certificates are typically short-lived on the order of hours to allow changing address assignments without the necessity for revocation. In scenarios where a more dynamic address allocation is desirable, certificates can be issued with lifetimes on the order of minutes if the increase in certificate issuance overhead is acceptable.

20.4.2 Additional Local Identifiers

In addition to the local end-host address, end-host certificates might contain other (AS-)local identifiers, e.g., a username valid within the AS or a port range for which this certificate is valid, see §20.6. In order to enable seamless transitions between short-lived certificates, an AS issues multiple certificates with overlapping validity times as long as the public key and all identifiers are identical. An AS might prefer not to add a local identifier to a certificate to protect itself against framing attacks if it cannot verify the correctness of the end host's claim of the identifier.

20.4.3 Anycast

Anycast can be used to run a service from multiple locations in the network associated with a single SCION service address. In the case that at least two service instances are located in the same AS, the service operator has two options: (1) All service instances can use the same end-host certificate and share a private key; or (2) the service operator requests a single end-host certificate for this AS and issues separate certificates for each service instance. The second approach reduces the impact of a private-key compromise but requires the additional intermediate certificate to be sent during connection establishment.

20.5 Certificate Service

All PILA-related functionality within an AS is handled by its certificate service. The certificate service generates and distributes short-lived end-host certificates ($CERT_E$) and, for that purpose, must be able to authenticate end hosts within the AS. It also provides the CP AS certificates to end hosts that are necessary for the verification of other end hosts' certificates and signatures.

In PILA, issuing end-host certificates is an automatic process similar to the Automatic Certificate Management Environment (ACME) [48]. End hosts are authenticated either with a challenge–response protocol, which requires setting up an HTTP server on the client as in ACME [471], or based on a signature of the certificate signing request (CSR) by the end host.

While the challenge–response protocol is automatic and does not require a pre-existing trust relation between the certificate service and the end host, the signature-based authentication allows managing certificates from entities other than the end host. Additionally, ASes allow relying end hosts to retrieve their local CP AS certificate including the certificate chain. The remote CP AS certificate including the certificate chain can be fetched from the local certificate service or the remote end host. Finally, the end hosts can request proofs from the remote AS whether a given end-host address supports a specific PILA protocol. Section 20.8 provides a detailed explanation of the different downgrade prevention mechanisms of PILA. These are the corresponding calls an end host can make:

- getEPCert(local end-host address, [local identifier], public key) either returns a short-lived certificate $CERT_E$ or an error message.

- getASCert(ISD and AS number) returns the CP AS certificate for the given ISD and AS number including its certificate chain. The certificate service fetches the certificate either from its cache or from a core AS of the specified ISD that originates a down-segment to the specified AS. Such an eligible core AS can be found through the regular path lookup process (§4.5).

- getProof(local end-host address, protocol(s)) returns a signed statement as to whether the end host at this local address has been issued an end-host certificate for the given protocol(s).

20.6 NAT Devices

As we discuss in §13.1.6, network address translation (NAT) is neither necessary nor used in a SCION Internet. If NAT *was* performed between the source and destination, it would pose obstacles for PILA: Since PILA relies on the fact that ASes can identify end hosts by their local end-host addresses to distribute per-end-host certificates, a simple mapping of an end-host address to a (single) certificate is impossible. The second issue is that due to the NAT device, both end hosts have a different view of the opposite end host's address, which breaks address-based authentication. To authenticate end hosts behind NAT devices, end hosts must be able to use the public address of their NAT device as an identifier unique within the AS. While SCION hosts do not use NAT at all, in our research paper [304] we present a system that allows PILA to work with NAT devices.

20.7 Session Resumption

If the underlying protocol supports session resumption, end hosts can combine the session resumption with a PILA handshake and derive the keying material of the new session based on both sources. TLS 1.3 [433], for example, supports combining pre-shared key and certificate-based authentication to increase the security of a session [247]. The derived keying material is authentic if either the pre-shared key derived from previous keying material or the keying material produced by the PILA handshake is authentic and no secret values were leaked. Since PILA reduces the attack surface to the end hosts' ASes, authenticated session resumption over different ASes increases the number of ASes that an attacker has to compromise in order to launch a successful undetected man-in-the-middle attack.

20.8 Downgrade Prevention

Whenever an initiator communicates with an unknown responder, an attacker might perform a downgrade attack to reduce the security to a less secure protocol (e.g., a TOFU protocol such as TCPCrypt). In such an attack, an attacker attempts to convince the initiator, i.e., the relying end host, that either the responder's AS does not support PILA or that the responder does not support a specific PILA-supported protocol.

20.8.1 Signature-Based Approach

In this approach, the relying end host fetches the necessary proofs for AS and end-host downgrade prevention from the certificate service of the local or remote AS, respectively. A fresh proof is generated and signed for each individual request from an end host.

AS downgrade is prevented by locally keeping a regularly updated list at each AS containing all PILA-enabled ASes. End hosts then request certificates for a specific AS from their local certificate service, which responds with a signed list of AS certificates.

An AS that supports PILA must provide proof that a service at a given end-host address does not support a specific PILA-supported protocol to assure the relying end host that its communication is not being downgraded. An end host sends a request including the end-host address, the PILA-supported protocol, and the current time as a timestamp. The certificate service replies with a signed proof that contains the hash of the request and a (possibly empty) list of certificate entries valid at the requested time. A certificate entry consists of the hash of the certificate and its validity period. The end host then verifies the signature and that the returned list is empty before falling back to a non-PILA protocol.

20.8.2 Log-Based Approach

While the signature-based approaches for both the AS and end-host downgrade prevention method work well and are easy to implement, they have a large computational overhead due to the signature operation necessary to create each proof. A more elaborate approach that scales better to a large number of requests is organizing AS and end-host certificates in public append-only logs as in certificate transparency. The AS certificate log must provide a globally consistent view of all AS certificates, while the end-host certificate log can also be implemented as a separate log per AS.

Each log is accompanied by a verifiable log-backed map [174], which provides a verifiable key-value store that can efficiently derive proofs of presence for a specific key-value mapping and proofs of absence for non-existing keys.

The log and the log-backed map only require one signature operation per maximum merge delay (MMD) regardless of the number of requests. The log-backed maps allow end hosts to fetch an AS certificate for an ISD and AS number and a list of certificate entries from an $\langle$ ISD and AS number, end-host address, protocol $\rangle$ tuple.

Part VI

Formal Verification

21 Motivation for Formal Verification

DAVID BASIN AND PETER MÜLLER

Chapter Contents

21.1 Local and Global Properties 474
21.2 Quantitative Properties 475
21.3 Adversarial Environments 475
21.4 Design-Level and Code-Level Verification 476

The construction of distributed systems is extremely challenging—especially for systems that operate in adversarial environments. SCION is no exception. A SCION network consists of a variety of different components such as routers, beacon servers, path servers, and edge devices. Each of them must satisfy complex requirements in order to ensure the correctness and security of the overall Internet architecture, even in the presence of malicious actors.

State-of-the-art engineering techniques such as design reviews, code reviews, and elaborate testing are integral parts of quality assurance, but they cannot guarantee the correctness and security of real-world distributed systems. Many errors and vulnerabilities are subtle and extremely difficult to uncover during reviews, and testing can cover only a tiny fraction of the possible system behaviors.

We therefore complement traditional engineering techniques with formal verification in order to achieve strong guarantees about the correctness and security of the SCION architecture and its implementation. The goal of verification is to construct a mathematical proof that a system satisfies its specification for all possible executions. Such proofs require a formalization of the system, typically either a mathematical model of the design or the actual implementation, as well as a formal specification that expresses the system's intended behavior precisely and unambiguously. The process of formal verification consists of developing these formalizations together with the actual proofs. The proofs are checked by tools to ensure that they do not contain flaws, such as missed corner cases.

In contrast to testing, formal verification does not execute the system and thus neither imposes runtime overheads nor depends on any concrete inputs or interactions with the environment. To illustrate the difference, consider a

© The Author(s), under exclusive license to Springer Nature Switzerland AG 2022
L. Chuat et al., *The Complete Guide to SCION*, Information Security
and Cryptography, https://doi.org/10.1007/978-3-031-05288-0_21

procedure isPrime(x) that should return true if and only if its parameter is a prime number. Assume that isPrime is implemented incorrectly and simply returns whether its argument is an odd number. Even though this implementation is obviously wrong, test code might not reveal the problem: Testing the procedure on inputs 3, 4, 5, 6, 7, 8, and 13 would always yield the expected result. In contrast, verification would attempt to prove that *for all inputs* x, x is odd implies x is prime. This proof would fail, indicating that the procedure implementation does not satisfy its specification. In this toy example, the proof considers all possible inputs x; for the verification of SCION, it must also cover all possible interactions between all network components, all thread schedules within each component, and all possible actions of an adversary. These aspects make the proof much more complex, but also increase the advantage that verification has over testing.

Formal verification has been applied successfully to a variety of non-trivial systems, including distributed systems. Lighthouse verification projects include the seL4 operating system [292], the CompCert compiler [332], and the Everest HTTPS implementation [71]. Even though these and other projects have advanced the state of the art, verifying an entire network architecture like SCION poses several challenges, which we summarize in the following. The subsequent chapters explain our solutions.

21.1 Local and Global Properties

To verify SCION, one must reason about both *local* and *global* properties. A local property might specify the input-output behavior of a router. For example, when a router receives a well-formed packet on an ingress interface, and cryptographic checks succeed, then the packet is transformed and sent out on the appropriate egress interface. Required local properties also include mundane requirements such as the absence of crashes or concurrency errors.

Many local properties are *trace properties*, that is, properties of a single execution of a system component. In security, when cryptography is used and components manipulate secrets, it may also be necessary to locally establish *hyperproperties* of components. A hyperproperty is a property of multiple (typically two) executions. For example, information flow control is typically formulated as a hyperproperty: A router does not allow undesired information flows like leaking secret keys during computations. Information flow security can be expressed by stating that two different runs of the system that differ only on values of the secret key have equivalent (non-secret) outputs; in other words, the observable output does not depend on the secret key.

Ultimately, we are also interested in establishing global properties of systems, namely that all of the routing components, working together, achieve the intended system-wide behavior. For example, for SCION we can show that packets forwarded in the data plane actually follow previously authorized paths. Making such statements precise, and proving them, requires reasoning about the overall system, namely the constellation of all routers and their inter-connection topologies.

21.2 Quantitative Properties

The above properties are all *qualitative*. In practice, when reasoning about networks, we are often also interested in *quantitative* properties such as performance. For example, in the conventional Internet, one might want to establish an upper bound on the time taken for BGP convergence. SCION operates differently, and different quantitative properties are relevant. For example, SCION supports inter-domain bandwidth reservation, and one might wish to show that some quantity of bandwidth can always be reserved and that this quantity is subsequently available, independent of adversarial reservations and attempts to flood the network. This requires different formal methods than those used to specify and establish qualitative properties as typically one must specify and reason about the *probability* of a certain actions, not just their *possibility*.

21.3 Adversarial Environments

SCION is designed to work in an adversarial network environment where communication may be overheard, modified, and spoofed. To make matters worse, different infrastructure components themselves may be corrupted by the adversary. The challenge here is to be precise about the adversary's capabilities and to provide guarantees under the strongest possible adversaries.

Reasoning in an adversarial environment is hard, as has been demonstrated by over three decades of research on cryptographic protocols [58]. These protocols are small distributed programs that use cryptography to achieve cryptographic objectives—for example, authenticate entities or exchange a session key. However, despite their apparent simplicity, many such protocols fail to work as intended, i.e., they are insecure. A classic example is the authentication part of the Needham–Schroeder Public Key protocol [384]: Even though this subprotocol consists of just two participants and three messages, it was believed to be secure for 17 years, at which point a relatively simple man-in-the-middle attack was discovered [344].

Intuitively, these security vulnerabilities are extremely difficult to detect because even small protocols, consisting of only a few message exchanges, give rise to very large numbers of possible executions, which are interleavings between the operations and messages of the involved parties. This complexity is aggravated by the presence of an active adversary, which may perform arbitrary sequences of operations according to the chosen attacker model. For instance, an adversary may analyze all network traffic and construct infinitely many different messages from the messages that have been previously sent, and the resulting messages can be injected into the network at any time. Moreover, the adversary can dynamically compromise components or, in SCION, even entire autonomous systems. Therefore, the potential presence of an adversary greatly increases the number of possible protocol executions that must be considered during formal verification to ensure that the SCION Internet works securely, even under attack.

21.4 Design-Level and Code-Level Verification

Problems can arise both during *design* and *implementation*. We emphasize here that even a design that works perfectly on paper may fail to be secure when implemented and deployed. The implementation may crash (and thereby enable denial-of-service attacks), use cryptographic primitives incorrectly (this topic will not be our central focus), fail to correctly reflect the design, or contain backdoors leaking secrets. Hence verification is essential at both levels: the design level, where one establishes, in particular, global, system-wide properties, and the implementation level, where one establishes properties of the actual code running on the network components. It is particularly critical that these different kinds of proofs are integrated: The design proofs ultimately rest on the assumption that the implemented systems' behaviors conform to that given by the design. Code-level proofs must establish that this is actually the case.

In the following chapters, we will address these challenges. Our goal is to give a high-level overview of our approach. For details about the techniques and tools we use, we refer the reader to the specialized research literature.

We will show in Chapter 22 how we make system-wide models at a high level of abstraction, and refine them down to low-level models describing how network components operate concurrently. We also explain how we verify both qualitative and quantitative properties for these models. We present a methodology for code verification, both of trace properties and hyperproperties in Chapter 23, where we also present a methodology for linking design-level verification with code verification. Taken together, our proofs give us extremely strong guarantees about SCION's security, functional behavior, and performance. The verification of SCION is ongoing; we summarize the current state of our effort in Chapter 24.

22 Design-Level Verification

DAVID BASIN, TOBIAS KLENZE, SI LIU, CHRISTOPH SPRENGER

Chapter Contents

22.1 **Overview of Design-Level Verification** **478**

22.2 **Background on Event Systems and Refinement** **482**

    22.2.1 Event Systems 483

    22.2.2 Properties of Event Systems 484

    22.2.3 Refinement 485

22.3 **Example: Authentication Protocol** **488**

    22.3.1 Authentication Protocol 488

    22.3.2 Approach Followed 488

    22.3.3 The Abstract Protocol Model 489

    22.3.4 The Concrete Protocol Model (Dolev–Yao) 491

22.4 **Verification of the SCION Data Plane** **494**

    22.4.1 Verification Scope 496

    22.4.2 Abstract Model 496

    22.4.3 Concrete Model 501

    22.4.4 Refinement 506

    22.4.5 Instantiation with SCION Protocol 509

    22.4.6 Conclusion 510

22.5 **Quantitative Verification of the N-Tube Algorithm** **510**

    22.5.1 Maude and Statistical Model Checking 511

    22.5.2 Formalizing N-Tube in Maude 512

    22.5.3 Statistical Analysis 513

    22.5.4 Conclusion 517

In this chapter we present our verification of SCION at the protocol level. We describe two projects that are completed: the verification of SCION's data plane forwarding protocol, in §22.4, and the verification of COLIBRI's bandwidth reservation algorithm N-Tube, in §22.5. Before we turn to these concrete verification efforts, we begin in §22.1 with a general overview of design-level verification and our methodology. In §22.2, we introduce protocol verification by refinement, which is the technique that we use to verify the data plane. To

© The Author(s), under exclusive license to Springer Nature Switzerland AG 2022
L. Chuat et al., *The Complete Guide to SCION*, Information Security
and Cryptography, https://doi.org/10.1007/978-3-031-05288-0_22

illustrate our refinement approach, we give an example of verifying a simple authentication protocol in §22.3. The section on the verification of N-Tube is self-contained and does not depend on the other sections beyond the overview given in §22.1.

22.1 Overview of Design-Level Verification

As observed in Chapter 21, the design of correct distributed systems is challenging and calls for formal methods. This has generated a slow but steady industry trend to incorporate formal methods into the design of critical systems; for notable examples see [43, 62, 388] and for examples of the use of formal methods in the standardization process see [70, 143].

Verification Approaches. Verification is important for correctness and security, but it is also challenging! Most realistic verification problems are undecidable as models may include system aspects such as unbounded data structures, complex control flows (e.g., exceptions), clocks, concurrency, or nondeterminism (e.g., input from the environment). Even when decidable, such aspects result in a combinatorial explosion of the system's state space.

Despite these problems, a large variety of practical formal modeling and verification methods and techniques have been proposed by researchers. On the methodological side, one can distinguish between the *post-hoc verification* of completed designs and stepwise, *correct-by-construction* system design using refinement. On the tool side, the most common verification tools are theorem provers and model checkers. Theorem provers may be fully automated (e.g., for first-order logic) or require user interaction (e.g., for higher-order logic); they may also be general purpose, or specialized for some theory or problem domain. Model checkers (e.g., for temporal logics) are mostly fully automated. As can be expected, interactive approaches require more user expertise than fully automated ones. Fully automated tools have different strategies to cope with undecidability or the state explosion problem. For example, a tool may fail to terminate in some cases (but still strive for termination in most practically relevant cases), sacrifice completeness, or use techniques to reduce the size of the state space such as abstraction or state-space reduction.

Verification approaches for the quantitative analysis of systems include probabilistic and statistical model checking, where systems are modeled as a form of probabilistic automata and properties expressed, for example, in a probabilistic variant of temporal logic. Probabilistic model checking performs an exhaustive analysis of the state space and therefore suffers from the same state-space explosion problems as its possibilistic counterpart. In contrast, statistical model checking (SMC) does not try to be exhaustive. Instead, it samples and verifies system executions until the user-specified confidence level and error margin are reached. SMC is therefore applicable to larger and more complex systems than exhaustive model checking techniques.

In our design-level verification of SCION components, we mainly employ stepwise refinement and statistical model checking. We will give a high-level overview of stepwise refinement in the rest of this section. In §22.2, we will formally introduce our system models and refinement. In §22.3, we will apply refinement in a case study to develop a security protocol. In §22.4, we use stepwise refinement to derive SCION data-plane protocols. Finally, in §22.5, we will use SMC for the analysis of the N-Tube bandwidth allocation algorithm. For a more detailed overview on verification approaches, focused on formal methods for security, see [56].

Stepwise Refinement. In our work, we follow a correct-by-construction approach using stepwise refinement [2, 348]. This approach is now well-established and has proven its usefulness in large, practical verification projects. These include the seL4 secure operating system [292], the CompCert C compiler [332], the Paris metro line 14 [62], and the Roissy airport shuttle [43]. The main benefit of stepwise refinement is to enable abstraction and to reduce the complexity of proofs by proving the desired system properties at the most appropriate level of abstraction. Moreover, this abstraction and the associated proofs provide valuable insights as to why systems are correct and what correctness depends upon.

We give a short overview of how development by stepwise refinement works. Our focus is on distributed systems and we can distinguish several phases of the development process.

I. System Requirements and Environment Assumptions. A development starts by (informally) defining the requirements that the system should fulfill and the assumptions one makes about the system's environment. The subsequent steps will successively model and formalize these requirements and assumptions.

For example, the system requirements for a distributed leader election protocol might be specified as follows. The nodes running the protocol must elect at most one leader among the participating nodes (this is a safety property, expressing that "nothing bad ever happens") and the nodes must eventually elect a leader (a liveness property, expressing that "eventually something good will happen"). The assumptions about the leader election algorithm's environment may include assumptions about the network's reliability and a failure model in case the algorithm is supposed to tolerate certain faults. As a second example, a security protocol might be required to establish a fresh session key between two parties and operate in an environment with a Dolev-Yao (network) attacker.

II. Initial High-Level Abstract Model. The formal part of the development begins with a very high-level model of the system to be constructed. This model captures some central aspects of the system, but it also abstracts away many details and is therefore usually not directly implementable. Often this model gives a global, centralized view on a distributed system and, in an adversarial setting, it involves no attacker or only a weak one. This abstraction makes it easy to prove some of the system requirements relevant for functional correctness or security.

For example, a leader election algorithm could elect a leader among n nodes in a single step by picking the one with the maximum identifier. Alternatively, in a security protocol, the protocol's roles might directly access each other's local memory, thereby excluding the possibility for an attacker to interfere with their communication.

III. Iterative Model Refinement to a Protocol Model. In this phase, one incrementally introduces details to incorporate further system requirements and environment assumptions. System requirements are formalized as trace properties (such as invariants) and proved at the highest possible level of abstraction, which simplifies their proofs. This results in a series of models, each of which is proved to refine the previous one. As refinement implies trace inclusion, trace properties are preserved from any model in the sequence down to the most concrete model. This phase ends with a model that incorporates all of the system requirements and environment assumptions. For distributed systems, we call this the *protocol model*.

For example, for a distributed system like a leader election protocol, this model will be distributed, where each node has its local state and communicates with the other nodes over a communication medium with certain properties stated in the environment assumptions. Likewise, for a security protocol, this will result in a model where each role has a local store and communicates with the other roles by exchanging messages over an adversarial network, which is identified with a Dolev-Yao attacker.

IV. Further Refinement towards an Implementation. While the protocol model incorporates all system requirements and environment assumptions, it might still be too abstract for a direct implementation. It may therefore require further refinement to be implementable. This may also include a decomposition step, which decomposes a monolithic model into different system components and a separate environment model.

For example, a security protocol model might use abstract terms instead of concrete bytestrings for representing messages, and the interfaces with the environment may still omit details such as network addresses. Moreover, some data may be modeled abstractly such as using sets, which must be refined to an implementation for example by lists or balanced trees.

V. System Implementation. In this phase, each component model is implemented in a programming language. This might be achieved by hand-translation or automatically using a code generator. Code written by hand can be verified against a component's specification using a code verification tool. The code verifier or code generator ideally comes with a soundness result proving that the code implementing a component refines its model. The overall soundness guarantee is intended to be that the entire implemented system satisfies the formalized and proven requirements, provided that the environment model faithfully captures (i.e., is refined by) the real environment.

Our Tool Set for Refinement. There exist a variety of refinement methods, techniques, and tools. We use a refinement method that borrows elements from Event-B [4], embedded in the proof assistant Isabelle/HOL [390]. Various alternatives exist, most prominently Event-B itself, which is implemented in the Rodin tool set [5], and the temporal logic of action (TLA+) with its TLC model checker [316] and TLAPS proof tool [141].

There are several reasons for our choice. First, we wanted to have the expressiveness offered by higher-order logic to formulate models. Isabelle/HOL provides support for specifying inductive datatypes, recursive (and co-recursive) functions, and inductively defined predicates and it automatically generates induction principles for each of these objects. The other tools do not offer this flexibility. Second, although it is an interactive prover, Isabelle/HOL offers strong support for automated theorem proving using its simplifier, classical reasoner, and the (foundational!) integration of external automated theorem provers. Third, Isabelle/HOL follows a so-called foundational approach, meaning that all proofs are certified by a small logical kernel. This kernel is well-tested and thus offers strong soundness guarantees. Finally, by embedding the refinement framework ourselves into higher-order logic we can easily adapt it to our needs and integrate it with tools for code verification.

Our Isabelle/HOL refinement framework covers the development of mathematical models of systems rather than their actual implementations. This corresponds to Phases I–IV above. These phases employ abstraction to simplify the models and their correctness proofs. However, as already mentioned in Chapter 21, a verified design leaves open the possibility of an incorrect implementation, which either fails to follow the design, or introduces additional details that lead to subtle bugs. We will address the correctness of the implementation in Chapter 23. In particular, in §23.4, we will explain our Igloo [486] methodology, also embedded in Isabelle/HOL, to correctly and flexibly link the design models with implementations written in realistic programming languages.

Table 22.1: Summary of notation and definitions.

$\mathbb{N}, \mathbb{B}$	natural numbers, Booleans
$A \times B, A^*$	cartesian product, finite sequences
$\mathbb{P}(A), A_\perp$	powerset, option set (i.e., sum of A and $\{\perp\}$)
$A \rightarrow B$	total function
$A \rightharpoonup B$	partial function
$dom(f), ran(f)$	function domain and range
$(\!|x \in A, y \in B|\!)$	set of records
$(\!|x = a, y = b|\!)$	concrete record
$x(r), r(\!|x := v|\!)$	record field x access, and update
$f(x := v)$	function update
$\langle\rangle, x \# xs, \langle a, b, c\rangle$	empty, cons, concrete sequence
$xs \leqslant ys, x \in xs$	sequence prefix, sequence membership
$hd(xs), tl(xs)$	list head and tail if cons, else $\perp$
$xs \cdot ys, rev(xs)$	concatenation of sequences, sequence reversal

22.2 Background on Event Systems and Refinement

We present next some basic theoretical background in modeling and refinement. We formally model protocols using event systems and we use invariants to express properties of these models. Stepwise refinement allows us to develop protocol models in several steps, starting from simple abstract models and gradually adding details in successive models. We illustrate these concepts by applying them to a simple file transfer protocol adapted from [4]. In §22.3, we will further illustrate refinement by showing how to develop security protocols. However, this additional example is not required to understand the remainder of this chapter. In §22.4, we will use refinement to model and prove security properties about the SCION dataplane.

We have formalized all definitions and results in this section and Sections 22.3 and 22.4 in the interactive theorem prover Isabelle/HOL. Despite our use of Isabelle/HOL, we largely use standard mathematical notation (summarized in Table 22.1) and deliberately blur the distinction between types and sets.

22.2.1 Event Systems

We first define our notion of a system model. We will then give an example and compare a model's components to similar notions from programming.

Event systems are labeled transition systems, whose labels we call *events*. Formally, an event system $\mathscr{E} = (S, s^0, E, \{\xrightarrow{e}\}_{e \in E})$ consists of a set of states S, an initial state $s^0 \in S$, a set of events E, and a transition relation $\xrightarrow{e} \subseteq S \times S$ for each event e. As usual, we write $s \xrightarrow{e} s'$ for $(s, s') \in \xrightarrow{e}$. We assume that all models $\mathscr{E}$ in our developments include a special stuttering event $\mathsf{skip} \in E$, defined by $s \xrightarrow{\mathsf{skip}} s$.

In our models, we often use parameterized events and states that are structured as records. We use the notation

$$e(\bar{x}) : \ g(\bar{x}, \bar{v}) \ \rhd \ \bar{w} := u(\bar{x}, \bar{v})$$

to specify the transition relation associated with such events. Here, $\bar{x}$ are the event's parameters (the bar representing a vector), $\bar{v}$ are the state record's fields, $g(\bar{x}, \bar{v})$ is a *guard* predicate defining when the event $e(\bar{x})$ may execute, $\bar{w} \subseteq \bar{v}$ are the updated fields, and u is an *update* function. This notation denotes the transition relation defined by $s \xrightarrow{e(\bar{x})} s'$ iff $g(\bar{x}, \bar{v}(s))$ holds, $\bar{w}(s') = u(\bar{x}, \bar{v}(s))$ and the values of the state fields that are not updated remain unchanged. For example, the event $dec(a) : z > a \rhd z := z - a$ decreases z by the parameter a, provided that the guard $z > a$ holds.

Example 1 (Abstract file transfer model). *Consider an abstract file transfer protocol specified as the guarded event system $\mathscr{E}_f$. Its state space is $S_f = (\!|\, f, g \in file, done \in \mathbb{B} \,|\!)$ with $file = I \to D$, where I is a finite index set and D is a set of data blocks. The initial state s^0 is given by $(\!|\, f = f^0, g = g^0, done = \mathsf{false} \,|\!)$, where f^0 and g^0 are arbitrary constant files. The set of events is given by $E_f = \{\mathsf{skip}, xfer\}$ and contains, besides skip, the event xfer defined by*

$$xfer : \ done = \mathsf{false} \ \rhd \ (g, done) := (f, \mathsf{true}).$$

Thus, this event transfers the file f in one shot to g and sets the variable done to true.

This example clearly shows some differences between states and state transitions in modeling and in programming. Models need not be directly implementable as programs. This gives us a great freedom to represent a model's state variables using abstract mathematical objects. For example, we may use sets to represent collections of (unordered) data items or, as in the example above, functions from an arbitrary index set to data blocks to represent files. Moreover, a model's events, which are associated with state transitions, can be very coarse-grained (e.g., transfer an entire file in one shot) and non-deterministic (e.g., transfer data blocks in an arbitrary order, see Example 3).

We can use stepwise refinement to make states more concrete (data refinement) and events more fine-grained (atomicity refinement). For example, a set of data items may eventually be refined into a balanced tree and a one-shot file transfer into a blockwise transfer (see Example 3).

22.2.2 Properties of Event Systems

We are interested two types of properties of event systems, expressed as sets of traces or sets of states, which we define next.

Trace Properties. We extend transition relations to finite sequences of events $\tau = e_1 \cdots e_n$, called *traces*, by defining, for all $n \geqslant 0$, $s \xrightarrow{\tau} s'$ whenever there are states $s_1, \ldots, s_{n-1}$ such that $s \xrightarrow{e_1} s_1 \xrightarrow{e_2} \cdots \xrightarrow{e_{n-1}} s_{n-1} \xrightarrow{e_n} s'$ (note that $s \xrightarrow{\diamondsuit} s$ for $n = 0$). The set of traces starting in a state s is defined by $traces(\mathscr{E}, s) = \{\tau \mid \exists s'. s \xrightarrow{\tau} s'\}$. A *trace property* ϕ is a subset of E^*. It is a *trace invariant* of $\mathscr{E}$, written $\mathscr{E} \models \phi$, if $traces(\mathscr{E}, s^0) \subseteq \phi$.

State Properties. The set of states reachable from a state s is defined by $reach(\mathscr{E}, s) = \{s' \mid \exists \tau. s \xrightarrow{\tau} s'\}$. A *state property* P is a subset of S. It is an *state invariant* of $\mathscr{E}$, written $\mathscr{E} \models P$, if $reach(\mathscr{E}, s^0) \subseteq P$.

Example 2 (File transfer properties). *The desired correctness property of a complete file transfer can be expressed as the state property*

$$P(s) \equiv done(s) = \mathsf{true} \longrightarrow f(s) = g(s),$$

where we use $\equiv$ the definitional equality to avoid confusion with the equality $=$ in the formula. Note that the property P explicitly refers to values of state record fields in a given state s (e.g., $done(s)$), while specifications of parametrized events leave states implicit, but have a precise semantics as a transition relation between pairs of states, see §22.2.1. The property P above cannot be directly stated as a trace property, since xfer has no parameters. A standard way around this is to add an observer *event $obs(f', g')$: $done = \mathsf{true} \wedge f' = f \wedge g' = g \triangleright -$ to the event system, where "$-$" denotes the identity update. Then we can express the correctness property as*

$$\phi(\tau) \equiv \forall f, g. \, obs(f, g) \in \tau \longrightarrow f = g.$$

These properties are trivially satisfied by the event system $\mathscr{E}_f$. Termination can be expressed as the trace property $\psi(\tau) \equiv xfer \in \tau$ and will be further discussed below.

Trace properties are more abstract than state properties, as they are formulated over sequences of (observed) events and need not be aware of the structure of a system's state space.

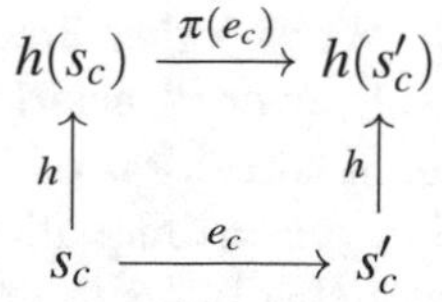

$$h(s_c) \xrightarrow{\pi(e_c)} h(s_c')$$

$$h \uparrow \qquad \qquad h \uparrow$$

$$s_c \xrightarrow{e_c} s_c'$$

Figure 22.1: Refinement of events.

Safety and Liveness Properties. Another important classification of properties distinguishes safety and liveness properties [17]. Informally speaking, *safety* properties state that "nothing bad ever happens", while *liveness* properties express that "something good eventually happens" (in a finite number of steps). Examples of safety properties are deadlock freedom, mutual exclusion, and the correctness properties expressed in Example 2. Termination and the eventual response to a request are typical liveness properties.

Violations of safety properties can be detected after a finite number of steps by pointing to a "bad" event or state. In contrast, violations of liveness properties can never be detected in finite time as their satisfaction can always be deferred to a later time. Hence, liveness properties require modeling infinite system behaviors (e.g., infinite traces). Moreover, in general, reasoning about liveness properties requires fairness assumptions (which abstract event schedulers) and ranking functions to measure progress towards "good" events. For instance, the termination property $\psi(\tau)$ from Example 2 does not hold without fairness assumptions: it is violated by all traces consisting only of skip events. Reasoning about liveness properties is substantially more complex than reasoning about safety properties. We therefore currently only support reasoning about safety properties in our Isabelle/HOL modeling and refinement framework. Note that state invariants as defined above (i.e., as sets of states rather than sequences of states) are safety properties.

22.2.3 Refinement

Given an event system $\mathscr{E}_a = (S_a, s_a^0, E_a, \{\xrightarrow{e}_a\}_{e \in E_a})$ representing an abstract model of a system and an event system $\mathscr{E}_c = (S_c, s_c^0, E_c, \{\xrightarrow{e}_c\}_{e \in E_c})$ representing a more concrete model, we say that $\mathscr{E}_c$ *refines* $\mathscr{E}_a$ if there are refinement mappings $h \colon S_c \to S_a$ on states and $\pi \colon E_c \to E_a$ on events such that

1. $h(s_c^0) = s_a^0$, and

2. $s_c \xrightarrow{e_c}_c s_c'$ implies $h(s_c) \xrightarrow{\pi(e_c)}_a h(s_c')$, for all $s_c, s_c' \in S_c$ and $e_c \in E$.

This is essentially a version of functional forward simulation [348]. The first point states that h maps the concrete initial state to the abstract one. The second point, illustrated in Figure 22.1, expresses that each concrete event maps to an abstract event under h and π. We say that the concrete event e_c *refines* the abstract event $\pi(e_c)$. As a result, every transition of the concrete model can be simulated by a corresponding transition of the abstract model. One can understand the abstract model as prescribing allowed behavior and refinement as showing that the concrete model adheres to this behavior.

For the structured event systems that we consider, condition (2) can be decomposed into two simpler conditions, called *guard strengthening* and *update correspondence*. Guard strengthening requires that the concrete guard implies the corresponding abstract guard, which ensures that the abstract event $\pi(e_c)$ is executable whenever the concrete event e_c is. Update correspondence means that state abstraction using h and state updates commute, i.e., $s'_a = u_a(h(s_c), \bar{x})$ equals $h(s'_c) = h(u_c(s_c, \bar{x}))$, where $\bar{x}$ are the events' parameters.

A refinement implies the inclusion of sets of reachable states and traces (modulo h and π). Namely, $h(reach(\mathcal{E}_c)) \subseteq reach(\mathcal{E}_a)$ and $\pi(traces(\mathcal{E}_c)) \subseteq traces(\mathcal{E}_a)$. Here, we extend π to traces by event-wise application and to trace properties by element-wise application. As a consequence of these set inclusions, refinement preserves both state invariants and trace invariants from the abstract model to the concrete model.

Property preservation: For a state property P and a trace property ϕ,

- $\mathcal{E}_a \models P$ implies $\mathcal{E}_c \models h^{-1}(P)$, where $h^{-1}(P) = \{s \in S_c \mid h(s) \in P\}$, and

- $\mathcal{E}_a \models \phi$ implies $\mathcal{E}_c \models \pi^{-1}(\phi)$, where $\pi^{-1}(\phi) = \{\tau \in E_c^* \mid \pi(\tau) \in \phi\}$.

From a methodological perspective, this allows us to prove (state or trace) invariants on abstract models, knowing that they are inherited by all subsequent refinements. Invariant proofs on abstract models are often much simpler than the corresponding direct proofs on more detailed, concrete models.

Example 3 (Refined file transfer model). *We define a "protocol" implementing the abstract file transfer model $\mathcal{E}_f$ by the concrete model $\mathcal{E}_p$, where $S_p = S_f + (\!\mid b \in I \rightharpoonup D \!\mid)$ extends the state record S_f with a buffer field b, modeled as a partial function from I to D. The buffer b is initially empty. The set of events is defined by $E_p = E_f \cup \{blk(i) \mid i \in I\}$. The protocol non-deterministically transfers blocks from the file f into the buffer b using the events $blk(i)$. The buffer is assigned to g using the event xfer, once all blocks have been transferred to b. The transition relations of these events are defined as follows:*

$$blk(i): \quad i \in I \setminus dom(b) \qquad \rhd \quad b := b(i \mapsto f(i))$$
$$xfer: \quad done = \mathsf{false} \wedge dom(b) = I \quad \rhd \quad (g, done) := (b, \mathsf{true}).$$

Example 4 (Refinement proof for file transfer). *Let us now establish that the concrete model $\mathcal{E}_p$ refines our abstract specification $\mathcal{E}_f$. We therefore define*

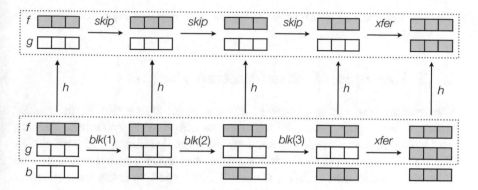

Figure 22.2: Refinement of file transfer protocol steps.

the refinement mappings h on states and π on events as follows. The state function h simply removes the buffer field b from the concrete state and the function π maps xfer and skip to themselves and $blk(i)$ to skip, for all $i \in I$. Figure 22.2 illustrates this refinement on an example protocol execution where a three-block file is transferred. The individual block transfer events $blk(i)$ have no corresponding effect on the abstract state as they operate on the newly introduced buffer b. Therefore, we use skip to simulate them. The dotted red boxes highlight in both models the identical evolution of the state variables f and g, which h extracts from the concrete state.

We then prove the refinement conditions (1) and (2) above. The former is trivial. For condition (2), we must show that $blk(i)$ and skip both refine skip and the concrete xfer refines its abstract counterpart. We decompose these proofs into their guard strengthening and update correspondence parts. The guard strengthening parts are trivial since, in the former two cases, skip's guard is true and, in the latter case, the refinement adds a conjunct to xfer's guard.

The update correspondence for $blk(i)$ and skip (the identity update) requires showing $h(s(\!|\, b := (b(s))(i \mapsto f(i))\,|\!)) = h(s)$, for $i \in I$. This holds trivially since $blk(i)$ only modifies the state field b, which h removes. The update correspondence for xfer requires that

$$h(s(\!|\, g := b(s), done := \mathsf{true}\,|\!)) = h(s)(\!|\, g := f(h(s)), done := \mathsf{true}\,|\!),$$

or equivalently, $b(s) = f(s)$. To prove this, we need additional information about the relation between b and f, which we express and prove as the invariant $J_p = \{s \in S_p \mid \forall i \in dom(b(s)). (b(s))(i) = (f(s))(i)\}$ of $\mathcal{E}_p$. Knowing that $s \in J_p$ allows us to complete the update correspondence proof for xfer.

In further refinements, one could develop a more realistic implementation. For example, one could eliminate non-determinism and model a communica-

tion medium such as an unreliable channel that requires explicit acknowledgment messages.

22.3 Example: Authentication Protocol

Whereas the last section illustrated refinement on a very simple file transfer protocol, we explain here how we bring security into the picture to develop protocols that use cryptography and how we formalize their security properties. We use stepwise refinement to derive a two-party authentication protocol. This protocol, standardized as ISO/IEC 9798-3, is intended for entity authentication. Namely, after running the protocol, a party A should have authenticated her counterpart B, i.e., A knows that she is speaking with B.

This section provides a more realistic example of how refinement is used for verification. However, the protocol discussed here is not directly related to SCION. Readers who are interested primarily in how we apply our techniques to the verification of SCION may skip ahead to Sections 22.4 and 22.5, which are self-contained and do not presume knowledge of this example. However, we will come back to the authentication protocol presented here in §23.4, where we use it as a running example to show how we link protocol verification and code verification.

22.3.1 Authentication Protocol

In standard informal Alice&Bob notation, the authentication protocol reads as follows.

$$\text{M1.} \quad A \to B \quad : \quad A, B, N_A$$
$$\text{M2.} \quad B \to A \quad : \quad [N_B, N_A, A]_{\text{pri}(B)}$$

Here, N_A and N_B are the nonces generated by the initiator A and the responder B respectively, comma denotes the concatenation of messages (by tupling), and $[M]_{\text{pri}(B)}$ denotes the digital signature of the message M with B's private key. First, the initiator generates a fresh nonce and sends it to the responder together with the names A and B. Afterwards, the responder generates his own nonce, signs it together with the initiator's nonce and name, and sends the signed message to the initiator. The nonces (which are pseudorandomly generated numbers, intended to be unguessable) provide replay protection. The protocol's authentication goal is that the initiator agrees with the responder on their names and the two nonces.

22.3.2 Approach Followed

When developing security protocols by refinement, we start with simple models that abstract from cryptographic operations and often do not involve an attacker or at most a very weak one. This abstraction simplifies security proofs.

We then use one or more refinement steps to introduce stronger, more realistic attackers. Accordingly, we must protect the protocol's payloads against interference by these attackers using security mechanisms such as cryptographic operations or channels with security properties, such as authentic or secure channels.

To derive our authentication protocol, we roughly follow the refinement strategy for security protocols proposed by Sprenger and Basin [485], which encompasses four abstraction levels. For brevity, we will only use the following two of these here.

Abstract protocol model where the different protocol roles read from each other's local stores instead of exchanging messages. There is no explicit attacker in this model and the attacker is only represented implicitly by assuming that some agents have been compromised; and

Cryptographic protocol model where the protocol roles exchange messages that are cryptographically secured to protect against an active attacker who controls the network and can eavesdrop and modify messages.

We prove the desired authentication property on the abstract model and that the cryptographic protocol model refines the abstract one. The refinement implies that the concrete model inherits the proven authentication property (see §22.2.3).

In §23.4, we will continue this case study to obtain a formally verified implementation of this protocol with end-to-end soundness guarantees. Namely, the implementation inherits the global security properties established in the present section.

22.3.3 The Abstract Protocol Model

We start our development with an abstract protocol model. In this model, we introduce the local states of the different protocol runs, each run instantiating a role of the protocol. Instead of exchanging messages, the protocol runs read the desired information directly from other runs' local states. Moreover, in this model, there is no active adversary. We only assume that a subset of the protocol agents is compromised. This greatly simplifies reasoning about abstract protocols.

We start by introducing agents, nonces, and runs. Let $\mathscr{A}$ be a set of agents, which is partitioned into a set of honest agents $\mathscr{A}_H$ and a set of compromised agents $\mathscr{A}_C$. We consider protocols with an initiator role and a responder role, denoted respectively by Init and Resp. We also assume a countably infinite set $\mathscr{R}$ of run identifiers, which we use to uniquely identify protocol runs. We construct the set $\mathscr{F}$ of fresh values, whose elements we denote by $\#[role, rid, k]$,

$resp\_read(r_B, B, A, N_B, N_A)$:
$\quad runs_{\mathsf{Resp}}(r_B) = \langle B \rangle \wedge N_B = \#[\mathsf{Resp}, r_B, 0]$
$\rhd \; runs_{\mathsf{Resp}} := runs_{\mathsf{Resp}}(r_B := \langle B, A, N_A \rangle)$

$init\_read\_auth(r_A, A, B, N_A, N_B)$:
$\quad runs_{\mathsf{Init}}(r_A) = \langle A, B \rangle \wedge N_A = \#[\mathsf{Init}, r_A, 0] \wedge$
$\quad (B \in \mathscr{A}_H \longrightarrow (\exists r_B.\; runs_{\mathsf{Resp}}(r_B) = \langle B, A, N_A \rangle \wedge N_B = \#[\mathsf{Resp}, r_B, 0]))$
$\rhd \; runs_{\mathsf{Init}} := runs_{\mathsf{Init}}(r_A := \langle A, B, N_B \rangle)$

Figure 22.3: The two main events of the abstract protocol model.

where $role \in \{\mathsf{Init}, \mathsf{Resp}\}$ is a role, $rid \in \mathscr{R}$ is a run identifier, and $k \in \mathbb{N}$ distinguishes nonces of the same run. Fresh values can be used as nonces or (symmetric) keys.

State. The state of our abstract model consists of two functions, one for each role. These functions map run identifiers to the runs' local stores, which consist of a list of the agent names and nonces that a run has learned during its execution.

$$store = (\mathscr{A} \cup \mathscr{F})^*$$
$$S_{\mathrm{a}} = (\![\, runs_{\mathsf{Init}} : \mathscr{R} \to store, \; runs_{\mathsf{Resp}} : \mathscr{R} \to store \,]\!)$$

For example, $s.runs_{\mathsf{Init}}(r) = \langle A, B, \#[\mathsf{Resp}, r', 0] \rangle$ means that, in state s, the initiator run r has stored agent names A and B (typically the run's participants) and the nonce $\#[\mathsf{Resp}, r', 0]$ generated by the responder run r'. In the initial state, $s.runs_{\mathsf{Init}}(r) = \langle \rangle$ and $s.runs_{\mathsf{Resp}}(r) = \langle \rangle$ for all $r \in \mathscr{R}$, which models that the runs have not yet been created or activated. Note that since our set of run identifiers is countably infinite, we model an unbounded number of runs of each role.

Events. Our abstract model has four events, two of which simply start a run by assigning it an owner and, in case of initiators, also a peer agent. In Figure 22.3, we display the other two, more interesting events. In the first event, $resp\_read(r_B, B, N_B, A, N_A)$, the responder B adds an agent name A and a nonce N_A to its store. Note that these values are completely unconstrained by any guards. This means that they are non-deterministically chosen and model their *unauthenticated* origin. In the second event, $init\_read\_auth(r_A, A, N_A, B, N_B)$, the initiator A reads the nonce N_B from some responder run r_B that knows its own nonce N_A, provided that the responder B is honest. This models an *authenticated* access of the initiator run r_A to A, B, N_A, and N_B in the responder

$$inv1\_iagree(\tau) = \forall r_A, A, B, N_A, N_B.$$
$$init\_read\_auth(r_A, A, B, N_A, N_B) \in \tau \wedge B \in \mathscr{A}_H \longrightarrow$$
$$(\exists r_B.\ resp\_read(r_B, B, A, N_B, N_A) \in \tau)$$
$$\wedge (\forall r'_A.\ init\_read\_auth(r'_A, A, B, N_A, N_B) \in \tau \longrightarrow r'_A = r_A)$$

Figure 22.4: Injective agreement.

run. In case B is compromised, no authenticity can be expected and the nonce N_B is therefore arbitrary. This *authentication guard* will enable us to prove the protocol's authentication property.

Authentication Property.   For this protocol model, we formulate and prove as a trace invariant that the initiator injectively agrees [345] with the responder on both nonces. This property expresses that there is a one-to-one correspondence between initiator and responder runs of the protocol. This means that an attacker cannot replay message M2 from previous protocol runs (which would, for instance, be possible if the nonce N_A were omitted from the protocol).

We formalize injective agreement as a trace property (Figure 22.4), which expresses that whenever a run r_A by initiator A with an honest responder B has finished by executing $init\_read\_auth(r_A, A, B, N_A, N_B)$, then there is a responder run r_B that has executed the event $resp\_read(r_B, B, A, N_B, N_A)$ and hence agrees with r_A on the protocol participants and on the nonces N_A and N_B. Moreover, no replay is possible, that is, no other run r'_A has executed an event $init\_read\_auth(r'_A, A, B, N_A, N_B)$ with the same agent and nonce parameters.

22.3.4 The Concrete Protocol Model (Dolev–Yao)

In the concrete protocol model, the roles exchange cryptographic messages, modeled as elements of an inductive datatype in Isabelle/HOL. Concretely, the initiator A sends a nonce N_A to the responder B and the responder replies with the signed message $[N_B, N_A, A]_{\mathsf{pri}(B)}$, which includes the nonce N_A and his own nonce N_B. We also introduce a standard Dolev–Yao attacker [167] who completely controls the network, but cannot break cryptography. Namely, the attacker is identified with the network and therefore all messages are transmitted via the attacker.

Messages.   We define the set $\mathbb{T}$ of cryptographic message terms as follows.

$$\mathbb{T} ::= \mathscr{A} \mid \mathscr{F} \mid \mathsf{pub}(\mathscr{A}) \mid \mathsf{pri}(\mathscr{A}) \mid \langle \mathbb{T}, \mathbb{T} \rangle \mid [\mathbb{T}]_{\mathsf{pri}(\mathscr{A})}$$

The atomic message terms are agent names $A, B \in \mathscr{A}$ and fresh values $N, K \in \mathscr{F}$. The terms $\mathsf{pub}(A)$ and $\mathsf{pri}(A)$ respectively denote agent A's public and private keys. The term $\langle t_1, t_2 \rangle$ denotes the pair of the terms t_1 and t_2 and $[t]_{\mathrm{pri}(A)}$ denotes the signature of term t with agent A's private key $\mathrm{pri}(A)$. We could easily extend this set (and the Dolev–Yao attacker defined below) with other cryptographic operations such as hashing and symmetric and asymmetric encryption, but this setup is sufficient for our present purposes.

State.   The concrete state space S_c extends the abstract state record with an additional field IK, which is a set of message terms that models the attacker's knowledge:

$$S_c = S_a + (\!| \, IK \in \mathbb{P}(\mathbb{T}) \, |\!).$$

As the Dolev–Yao attacker is identified with the network, the set IK also represents the network state. The attacker's initial knowledge consists of all agent names, all agents' public keys, and the private keys of compromised agents.

$$IK_0 = \mathscr{A} \cup \{\mathsf{pub}(A) \mid A \in \mathscr{A}\} \cup \{\mathsf{pri}(A) \mid A \in \mathscr{A}_C\}.$$

Note that although this set does not include nonces, the attacker can generate and learn nonces by having compromised agents run the protocol.

Events.   The concrete model has seven events. For each role, there is a run creation event, a message sending event, and message receiving event. Additionally, there is an attacker event. The send events, *init_send* and *resp_send*, each construct a protocol message M and send it by adding it to the attacker's knowledge IK. These two events only modify the state variable IK, which has been added in the refinement, and they leave the two *runs* variables unchanged. We therefore have these events refine the abstract model's skip event. In the events *resp_recv* and *init_recv*, the responder and the initiator receive the respective protocol message M, which is modeled by a guard $M \in IK$, and they update their store with any new information received. Note that received messages are not removed from the attacker's knowledge, which enables their later replay. These events respectively refine the events *resp_read* and *init_read_auth* of the abstract model.

The attacker's capabilities are defined by a closure operator $DY(H)$, denoting the set of messages they can derive from a set of observed messages H using both message composition (such as pairing and signing) and decomposition (such as signature verification and projections). This operator is defined inductively by the derivation rules in Figure 22.6. The first rule states that $DY(H)$ contains all messages in H. The second and third rule enable the attacker to form pairs and apply projections. The fourth rule allows the attacker to construct a signature of a message t using A's private key, provided they know both of these elements. The final rule enables the attacker to verify a signature $[t]_{\mathrm{pri}(A)}$ and learn the signed message t, provided they know the corresponding public verification key $\mathsf{pub}(A)$. This models signatures with message extraction. The attacker event updates the attacker's knowledge to $DY(IK)$ and refines skip since it only modifies IK.

$$init\_send(r_A, A, B, N_A):$$
$$runs_{\mathsf{Init}}(r_A) = \langle A, B \rangle \wedge N_A = \#[\mathsf{Init}, r_A, 0]$$
$$\rhd\ IK := IK \cup \{\langle A, B, N_A \rangle\}$$

$$resp\_recv(r_B, B, A, N_B, N_A):$$
$$runs_{\mathsf{Resp}}(r_B) = \langle B \rangle \wedge N_B = \#[\mathsf{Resp}, r_B, 0]$$
$$\langle A, B, N_A \rangle \in IK$$
$$\rhd\ runs_{\mathsf{Resp}} := runs_{\mathsf{Resp}}(r_B := \langle B, A, N_A \rangle)$$

$$resp\_send(r_B, B, A, N_B, N_A):$$
$$runs_{\mathsf{Init}}(r_B) = \langle B, A, N_A \rangle \wedge N_B = \#[\mathsf{Resp}, r_B, 0]$$
$$\rhd\ IK := IK \cup \{[N_B, N_A, A]_{\mathsf{pri}(B)}\}$$

$$init\_recv(r_A, A, B, N_A, N_B):$$
$$runs_{\mathsf{Init}}(r_A) = \langle A, B \rangle \wedge N_A = \#[\mathsf{Init}, r_A, 0] \wedge$$
$$[N_B, N_A, A]_{\mathsf{pri}(B)} \in IK$$
$$\rhd\ runs_{\mathsf{Init}} := runs_{\mathsf{Init}}(r_A := \langle A, B, N_B \rangle)$$

$$attacker:$$
$$\rhd\ IK := DY(IK)$$

Figure 22.5: The main events of the concrete protocol model.

$$\frac{t \in H}{t \in DY(H)} \qquad \frac{t_1 \in DY(H) \quad t_2 \in DY(H)}{\langle t_1, t_2 \rangle \in DY(H)} \qquad \frac{\langle t_1, t_2 \rangle \in DY(H)}{t_i \in DY(H)}\ i \in \{1, 2\}$$

$$\frac{t \in DY(H) \quad \mathsf{pri}(A) \in DY(H)}{[t]_{\mathsf{pri}(A)} \in DY(H)} \qquad \frac{[t]_{\mathsf{pri}(A)} \in DY(H) \quad \mathsf{pub}(A) \in DY(H)}{t \in DY(H)}$$

Figure 22.6: Dolev–Yao closure.

Refinement. To prove that the concrete model refines the abstract one, we define the refinement mapping $h: S_c \to S_a$ on states to simply remove IK from the concrete state record. Moreover, we define the refinement mapping $\pi: E_c \to E_a$ on events such that it maps the concrete run creation events to their (identical) abstract counterparts, the send events and the attacker event to skip, and the receive events as follows:

$$\pi(\mathit{init\_recv}) = \mathit{init\_read\_auth},$$
$$\pi(\mathit{resp\_recv}) = \mathit{resp\_read}.$$

With these definitions, proving refinement consists of establishing the relation $h(s_c^0) = s_a^0$ between the initial states, which trivially holds, and that for any concrete transition $s \xrightarrow{e} s'$ there is an abstract transition $h(s) \xrightarrow{\pi(e)} h(s')$. We decompose the latter proof obligation into guard strengthening and update correspondence proofs for each event e. The interesting case is guard strengthening for the event $\mathit{resp\_recv}$, which states that receiving a signed message $[N_B, N_A, A]_{\mathsf{pri}(B)}$ implies the authentication guard of the corresponding abstract event $\mathit{resp\_read}$. This is equivalent to

$$[N_B, N_A, A]_{\mathsf{pri}(B)} \in IK \wedge B \in \mathscr{A}_H \longrightarrow$$
$$(\exists r_B. \; \mathit{runs}_{\mathsf{Resp}}(r_B) = \langle B, A, N_A \rangle \wedge N_B = \#[\mathsf{Resp}, r_B, 0])).$$

This statement connects the presence of a signed message in the intruder knowledge to the existence of a responder run in a state related to the signed message. It is not directly provable as it stands, but we can prove a slightly strengthened form as an invariant, which we can in turn use to prove the required guard strengthening. This invariant can be understood as explaining the meaning of the signed message.

With the refinement established, we can conclude that the concrete model also satisfies injective agreement. Note that if the concrete protocol model was insecure, we would not be able to establish a refinement with the abstract model. This would manifest in an unprovable proof goal for the refinement of at least one event, which we would then analyze to determine the source of the problem and fix the model.

22.4 Verification of the SCION Data Plane

In this section, we formally verify the path authorization (property P6, page 159) and weak detectability (property P4, page 159) properties of the SCION data plane protocol. We have focused on SCION's data plane first since it is critical to the security of the overall architecture and since insights gained from its formal analysis could be difficult to act upon after the widespread deployment of hardware-based SCION routers has begun. This is in contrast to the control plane, which we expect to be implemented in software, and hence more amenable to updates.

A standard approach in security protocol verification is to used automated tools such as security protocol model checkers like Tamarin [366] and ProVerif [77]. Unfortunately, verifying path authorization poses challenges that these tools are ill-equipped to handle. In particular, they do not

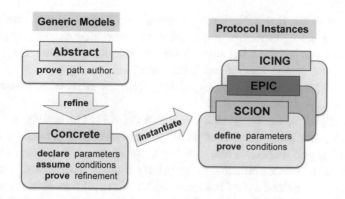

Figure 22.7: Overview of our data plane verification framework, where refinement and instantiation preserve properties.

support proving protocols' security given arbitrary sets of authorized paths of unbounded size. We instead use Isabelle/HOL [390], which supports the modeling and verification of data plane protocols in their full generality. Additionally, we benefit from the theorem prover's strong soundness guarantees.

We follow the refinement approach presented in §22.2. We begin by defining the scope of our verification in §22.4.1. In §22.4.2 we present an abstract model of a data plane protocol without an adversary, where the security properties hold trivially. This model has environment parameters, for instance, to fix the network topology and the set of authorized paths. The concrete model in §22.4.3 introduces both an attacker and cryptographic checks to defend against the attacker. As we show in §22.4.4, the concrete model refines the abstract model, and hence inherits its security properties.

We parametrize our formal models in order to not only capture the behaviors of one specific protocol, but the behaviors of an entire class of path-aware protocols. This class includes SCION, as well as an earlier version of SCION (presented in the first edition of this book) and EPIC Levels 0–2 (cf. §10.1). To formalize and verify this protocol class, we parametrize the concrete model not only over the environment, but we also introduce protocol parameters that abstract the security mechanisms used by protocols (Figure 22.7). The refinement proof of the concrete model requires a set of assumptions on the environment parameters and a set of conditions on the protocol parameters. For each protocol verified, an instance model must be provided, where the protocol parameters are defined and the associated conditions are proven, thereby showing that the model inherits the security properties. We do this in §22.4.5 for the SCION data plane protocol, which uses the chained MACs presented in §5.3.3 (page 99). For brevity, we do not discuss instantiations with other protocols in this book.

We simplify our formalization's presentation for ease of understanding. For instance, our actual formalization extends the presented models by providing features that are required for some protocol instances. Since our models are protocol-agnostic and header fields can serve different purposes in different protocol instances, we sometimes deviate from the SCION nomenclature in naming the generic header fields.

This verification work was first published in [297]. The extensions required for verifying the latest version of SCION, presented in this book, are part of an upcoming publication. Here, we focus on the most important aspects of our verification framework, and only briefly discuss the required extensions when introducing the SCION instance model. Our full formalized framework consists of over 10,000 lines of definitions and proofs in Isabelle/HOL.

22.4.1 Verification Scope

Our work applies to a wide range of future Internet architectures that are path-aware and use cryptographic authenticators. We verify two security properties: *path authorization* (cf. §2.4) and *detectability* (cf. §7.1). They primarily protect ASes against malicious senders. However, our attacker model (§22.4.3.3) also includes colluding on-path adversaries.

As paths are authorized for each segment individually, we only verify the authorization of single segments. When we speak of paths, we therefore mean single-segment paths. We also only verify the security of up-segments. The verification of down- and core-segments is ongoing work.

Intra-AS forwarding is not covered by our verification, since each AS exercises control over its own network, and global coordination is not required for intra-AS security. We also do not verify the control plane here, as its properties are independent of those of the data plane. For instance, path authorization is independent of the property that a path authorized by the control plane is in accordance with the routing policies of all on-path ASes.

22.4.2 Abstract Model

We define an event system that abstractly models the data plane of a path-aware network architecture. This model includes neither cryptography nor an attacker. We prove that it satisfies path authorization and weak detectability.

To distinguish the definitions of this abstract model from those of the concrete model that refines it (§22.4.3), we use the subscripts 'a' and 'c', respectively.

22.4.2.1 Environment Parameters

We model the Internet as a multigraph, where nodes represent ASes and edges represent the network links between them. More precisely, a *network topology* is a triple $(\mathcal{N}, \mathcal{I}, target)$, where $\mathcal{N}$ is a set of nodes, $\mathcal{I}$ is a set of interfaces, and *target* is an environment parameter of our model with the type

$$target : \mathcal{N} \times \mathcal{I} \rightharpoonup \mathcal{N} \times \mathcal{I}. \tag{22.1}$$

This formalizes that *target* is a partial bijective function that models links between ASes. The bijectivity of the *target* mapping models unicast communication (in contrast to, e.g., broadcast). We say that an interface i is *valid* for a node A, if $(A, i) \in dom(target)$, whereby $target(A, i) = (B, j)$ denotes the node B and interface j at the other end of the link. Our definition thus allows for multiple links between a given pair of nodes, with possibly different routing policies.

We often reason directly about paths in the network, rather than the network topology. These paths are defined in terms of both nodes and their interfaces. We define a *path* to be a finite sequence of *hop information fields* from the set

$$HI = (\!(id \in \mathcal{N}, \, prev \in \mathcal{I}_\perp, \, next \in \mathcal{I}_\perp)\!). \tag{22.2}$$

Each hop information field contains the local routing information of a node, i.e., its node identifier and the interfaces that identify the links to the previous and the next AS on the path. Both interfaces are defined as option types, indicated by the subscript $\perp$. When there is no previous or next hop, we assign $\perp$ to the respective interface. The hop fields that will be introduced in the concrete model below augment the hop information fields with a cryptographic hop validation field. Since our abstract model does not contain an adversary, such authenticators are not required here.

Our model's second environment parameter is

$$auth_a \subseteq HI^*, \tag{22.3}$$

the set of *authorized paths* along which packets are allowed to travel. Packets can also traverse just a part of an authorized path. To account for such partial paths, we define $auth_a^{\rightleftharpoons}$, the *fragment closure* of $auth_a$, as the set of paths *his* such that there exist a $his' \in auth_a$ and paths $his_1, his_2 \in HI^*$ such that $his' = his_1 \cdot his \cdot his_2$.

Our third parameter is the set of *compromised* nodes (also called *attacker* nodes)

$$\mathcal{N}_{attr} \subseteq \mathcal{N}. \tag{22.4}$$

All other nodes are called *honest*. This environment parameter only becomes relevant after introducing the adversary in the concrete model (§22.4.3.3), where the attacker has access to the keys of compromised nodes. We nevertheless introduce it here, since using the same environment parameters in all of our models simplifies our presentation.

The environment assumptions (ASM) expressed over these parameters are introduced for the refinement of the abstract to the concrete model (§22.4.4.1).

22.4.2.2 State

We model packet forwarding from a node's internal network to an inter-node link, and vice-versa, using two types of asynchronous channels: *internal channels* (one per node) and *external channels* (two per interface-node pair, one in each direction). We represent these channels as sets of packets PKT_a, defined below. We define the state as

$$S_a = (\!| int \in \mathcal{N} \to \mathbb{P}(\text{PKT}_a),$$
$$ext \in \mathcal{N} \times \mathcal{I} \times \mathcal{N} \times \mathcal{I} \to \mathbb{P}(\text{PKT}_a) |\!).$$

The initial state s_a^0 is the state in which all channels are empty. We overload the set inclusion operator to apply to states: A packet m is in a state s, written $m \in s$, iff $m \in ran(int(s)) \cup ran(ext(s))$. For a valid interface i of A with $target(A, i) = (B, j)$, we define $ext^{send}(A, i) = ext(A, i, B, j)$ and $ext^{recv}(A, i) = ext(B, j, A, i)$.

In the following definition of packets, we abstract from the payload and only model the packet-carried forwarding state:

$$\text{PKT}_a = (\!| past \in \text{HI}^*,\ fut \in \text{HI}^*,\ hist \in \text{HI}^* |\!).$$

A packet consists of the desired future path *fut*, and the (presumed) traversed path *past*, stored in reverse direction. The full path is $rev(past(m)) \cdot fut(m)$. While this splitting of the path simplifies our proofs, the forwarding path could equivalently be defined as a single sequence with a moving pointer indicating the current position on the path. We call a packet m *authorized*, if $fut(m) \in auth_a^{\rightarrow}$. Additionally, each packet records a path *hist*, also in the reverse direction. This path represents the packet's actual trajectory and is used to express security properties. This can be seen as an auxiliary *history variable* [2], meaning that it is not part of the protocol, but serves to specify and prove properties of protocol executions.

22.4.2.3 Events

The events of the abstract model are given on the left-hand side of Figures 22.8 and 22.9. We leave the state implicit when we refer to channels. We write $int(A) \mathrel{+}= m$ to denote the state update where the packet m is added to the channel $int(A)$, i.e., $int(A) := int(A) \cup \{m\}$. Otherwise the state remains unmodified and, in particular, $int(B) := int(B)$ for all $B \neq A$. We use a similar notation for *ext* channels.

The life cycle of a packet is captured by the following events: **dispatch-int**$_a$ creates a new packet containing an authorized future path in the internal channel of a node. The packet is transferred with alternating **send**$_a$ and **recv**$_a$ events between internal and external channels, according to the forwarding path contained in the packet. Finally, the packet is delivered to the end host with an event **deliver**$_a$. The events **dispatch-int**$_a$ and **deliver**$_a$ model the interaction

dispatch-int$_a$ (A, m) :

   $fut(m) \in auth_a^{\rightleftarrows} \wedge$

   $hist(m) = \langle \rangle$

  $\triangleright$ $int(A) \mathrel{+}= m.$

dispatch-ext$_a$ (A, i, m) :

   $fut(m) \in auth_a^{\rightleftarrows} \wedge$

   $hist(m) = \langle \rangle \wedge$

   $(A, i) \in dom(target)$

  $\triangleright$ $ext^{\mathrm{send}}(A, i) \mathrel{+}= m.$

dispatch-int$_c$ (A, m) :

   $m \in DY(ik) \wedge$

   $hist(m) = \langle \rangle$

  $\triangleright$ $int(A) \mathrel{+}= m.$

dispatch-ext$_c$ (A, i, m) :

   $m \in DY(ik) \wedge$

   $hist(m) = \langle \rangle$

   $(A, i) \in dom(target)$

  $\triangleright$ $ext^{\mathrm{send}}(A, i) \mathrel{+}= m.$

Figure 22.8: Dispatching events of the abstract (left) and concrete (right) model, with differences highlighted.

with end hosts, whereas **send$_a$** and **recv$_a$** represent the border routers' packet forwarding actions. The additional **dispatch-ext$_a$** event creates and sends a packet directly to an *ext* channel. This event is not required for normal data plane operations, but serves to introduce a malicious sender at an inter-AS link in the refinement.

We now describe these events in more detail. The events **dispatch-int$_a$** and **dispatch-ext$_a$** create an authorized packet by setting its future path to (a fragment of) an authorized path and inserting it into an internal or external channel. The history is set to the empty sequence in both events, and the past path can be set arbitrarily to allow the refinement into attacker events, where the attacker may disguise the packet's origin. The **send$_a$** and **recv$_a$** events both use the current hop information field, i.e., the hop information field at the head of the future path, to determine where the packet should be forwarded. Hence, they require a non-empty future path. The **recv$_a$** event transfers a packet from the external channel at (A, i) to A's internal channel. The **send$_a$** event takes a packet m from the internal channel and places the transformed packet $fwd_a(m)$ on the external channel at (A, i). The partial function $fwd_a : \mathrm{PKT}_a \rightharpoonup \mathrm{PKT}_a$ moves the current hop information field of m into the past path and adds it to the history. It is defined for m with $fut(m) \neq \langle \rangle$ by

$$fwd_a(m) = (\!|past = hd(fut(m)) \# past(m), fut = tl(fut(m)),$$
$$hist = hd(fut(m)) \# hist(m)|\!).$$

$\mathbf{send}_a(A, m, hi, i):$

  $hi = hd(fut(m)) \wedge hi \neq \bot \wedge$

  $A = id(hi) \wedge i = next(hi) \wedge$

  $m \in int(A) \wedge$

  $(A, i) \in dom(target)$

  $\rhd\ ext^{send}(A, i) +\!\!= fwd_a(m).$

$\mathbf{send}_c(A, m, hf, i):$

  $hf = hd(fut(m)) \wedge hf \neq \bot \wedge$

  $A = id(hf) \wedge i = next(hf) \wedge$

  $m \in int(A) \wedge$

  $(A, i) \in dom(target) \wedge$

  $\psi(hf, hd(tl(fut(m))), tok(m))$

  $\rhd\ ext^{send}(A, i) +\!\!= fwd_c(m).$

$\mathbf{recv}_a(A, m, hi, i):$

  $hi = hd(fut(m)) \wedge hi \neq \bot \wedge$

  $A = id(hi) \wedge$

  $m \in ext^{recv}(A, i) \wedge$

  $(A, i) \in dom(target)$

  $\rhd\ int(A) +\!\!= m.$

$\mathbf{recv}_c(A, m, hf, i):$

  $hf = hd(fut(m)) \wedge hf \neq \bot \wedge$

  $A = id(hf) \wedge$

  $m \in ext^{recv}(A, i) \wedge$

  $(A, i) \in dom(target) \wedge$

  $i = prev(hi)\ \wedge$

  $\psi(hf, hd(tl(fut(m))), tok(m))$

  $\rhd\ int(A) +\!\!= m.$

$\mathbf{deliver}_a(A, m, hi):$

  $fut(m) = \langle hi \rangle \wedge A = id(hi) \wedge$

  $m \in int(A)$

  $\rhd\ int(A) +\!\!= fwd_a(m).$

$\mathbf{deliver}_c(A, m, hf):$

  $fut(m) = \langle hf \rangle \wedge A = id(hf) \wedge$

  $m \in int(A) \wedge$

  $\psi(hf, \bot, tok(m))\ \wedge$

  $\rhd\ int(A) +\!\!= fwd_c(m).$

Figure 22.9: Non-dispatching abstract (left) and concrete (right) events.

We define the functions head $hd : HI_{\bot}^{*} \rightarrow HI_{\bot}$ and tail $tl : HI_{\bot}^{*} \rightarrow HI_{\bot}^{*}$ by $hd(x\#xs) = x$ and $tl(x\#xs) = xs$ and by mapping $\langle \rangle$ and $\bot$ to $\bot$ in both functions.

The $\mathbf{deliver}_a$ event models delivering a packet m containing a single hop information field in its future path to an end host. However, we do not explicitly model end hosts and their state. Hence, we simply add the packet $fwd_a(m)$ to the internal channel of the AS and thereby push the last hop information field into the *past* and *hist* paths.

22.4.2.4 Properties

Path authorization states that packets can only traverse the network along authorized paths. This ensures that the data plane enforces the control plane's routing policies. Formally, for all packets m in a state s, $rev(hist(m)) \in auth_{\mathrm{a}}^{\rightleftarrows}$. Recall that the order of nodes is reversed in *hist*. We strengthen this to an inductive invariant by adding the future path:

$$\forall m \in s.\ rev(hist(m)) \cdot fut(m) \in auth_{\mathrm{a}}^{\rightleftarrows}. \qquad (22.5)$$

The proof in this abstract model is straightforward. New packets are required to have an authorized future path and an empty history. For existing packets, $rev(hist(m)) \cdot fut(m)$ remains invariant during their forwarding. The *past* path is irrelevant for path authorization.

We furthermore formalize *weak detectability* and show that it is an invariant: all traversed hops are recorded on (i.e., are a prefix of) the past path:

$$\forall m \in s.\ hist(m) \leqslant past(m). \qquad (22.6)$$

This property is independent of $auth_{\mathrm{a}}$ and follows directly from the events' definitions. Our presentation will focus on path authorization, as it is the data plane's central security property.

22.4.3 Concrete Model

We refine the abstract forwarding protocol into a concrete model. In this model, packets contain *hop fields*, each of which has the fields contained in the hop information field of the abstract model and a new (generic) cryptographic *hop validation field*. The latter secures the authorized paths against a Dolev–Yao attacker (§22.4.3.3). We present the concrete model's events in §22.4.3.4 and the refinement in §22.4.4.

The concrete model retains the environment parameters of the abstract model (§22.4.2.1), and adds two *protocol parameters*, which we introduce below. One of them is the cryptographic check that ASes apply to their hop validation fields. This allows us to abstract from the concrete cryptographic mechanism used.

22.4.3.1 Cryptographic Terms, Hop Fields, Packets, and States

As in §22.3, we define an algebra $\mathbb{T}$ of cryptographic terms. Since we use different constructors, we re-define the message algebra here.

$$\mathbb{T} ::= \mathcal{N} \mid \mathcal{I}_{\perp} \mid \mathbb{N} \mid K_{\mathcal{N}} \mid \langle \mathbb{T}, \mathbb{T}, \ldots, \mathbb{T} \rangle \mid \mathsf{H}(\mathbb{T}).$$

Terms consist of node identifiers, interfaces, natural numbers (e.g., for timestamps), symmetric keys (one per node), as well as finite sequences, and cryptographic hashes of terms. We define message authentication codes (MACs) using hashing by $\mathsf{MAC}_k(m) = \mathsf{H}(\langle k, m \rangle)$. Our framework also supports encryption and signatures, which we do not use here.

Hop fields (HF), used in the concrete model, extend the hop information fields (HI), used in the abstract model, with a *hop validation field* (*HVF*). This is a cryptographic authenticator that authenticates the hop information:

$$\mathrm{HF} = (\!| \, id \in \mathscr{N}, prev \in \mathscr{I}_\perp, next \in \mathscr{I}_\perp, HVF \in \mathbb{T} \, |\!). \qquad (22.7)$$

In the concrete model, *path* refers to a sequence of HFs. We define the function *abstr-hf*: HF → HI projecting concrete hop fields to abstract hop information fields by dropping *HVF* and we lift it element-wise to paths. To keep our notation compact, we write $\overline{hf_A}$ and $\overline{hfs}$ to denote the application of *abstr-hf* to hop fields and sequences of hop fields.

We next define concrete packets as follows:

$$\mathrm{PKT_c} = (\!| \, tok \in \mathbb{T}, past \in \mathrm{HF}^*, fut \in \mathrm{HF}^*, hist \in \mathrm{HI}^* \, |\!).$$

The past and future paths are sequences of concrete hop fields, while the history remains a sequence of HI fields. Concrete packets contain an additional *packet token field* (*tok*), which in SCION is used for the segment identifier β.

The concrete state space S_c has the same record structure as the abstract S_a, but the channels now carry concrete packets:

$$S_c = (\!| \, int \in \mathscr{N} \to \mathbb{P}(\mathrm{PKT_c}),$$
$$ext \in \mathscr{N} \times \mathscr{I} \times \mathscr{N} \times \mathscr{I} \to \mathbb{P}(\mathrm{PKT_c}) \, |\!).$$

The initial state s_c^0 is defined similarly to s_a^0 as the empty channel state.

Auxiliary Functions. We define the overloaded function *terms* for hop fields, paths, and packets as follows:

$$terms(hf) = \{HVF(hf)\}$$
$$terms(hfs) = \bigcup_{hf \in hfs} terms(hf)$$
$$terms(pkt) = \{tok(pkt)\} \cup terms(past(pkt)) \cup terms(fut(pkt)).$$

For a set T of terms and for a hop field, path, or packet x, we write $x \in T$ for $terms(x) \subseteq T$.

We define a function *hi-term* : HI → $\mathbb{T}$ that maps hop information fields to terms as

$$hi\text{-}term(hi) = \langle id(hi), prev(hi), next(hi) \rangle.$$

We leave the conversion of HI fields to terms via this function implicit below.

22.4.3.2 Protocol Parameters and Authorized Paths

We define two protocol parameters. The first is a *cryptographic validation check*

$$\psi : \mathrm{HF} \times \mathbb{T} \to \mathbb{B}, \tag{22.8}$$

which each border router performs to check the validity of its hop field. This parameter abstracts the cryptographic structure of the hop validation field, which is only determined in concrete protocol instances. Here, $\psi(hf_A, u)$ holds if the hop field hf_A is valid given the packet's *tok* field u.

We also define a function $\Psi : \mathrm{HF}^* \times \mathbb{T} \to \mathrm{HF}^*$, which we apply to the future path of a packet to obtain the longest prefix of *hfs* such that for every hop field hf_A on the path and the *tok* field u, $\psi(hf_A, u)$ holds:

$$\Psi(hf_A \# hfs, u) = hf_A \# \Psi(hfs, u) \qquad \text{if } \psi(hf_A, u)$$
$$\Psi(hfs, u) = \langle\rangle \qquad\qquad\qquad \text{otherwise.}$$

We use this function in the mapping of future paths from the concrete model to the abstract model (§22.4.4.3) to truncate the path at the first invalid hop field. This does not reduce the system's possible behavior, since forwarding is performed by honest agents that do not forward packets along invalid hop fields. We call a path *hfs* or a packet *pkt* with *fut*(pkt) = *hfs* cryptographically *valid (for u)* if $\Psi(hfs, u) = hfs$.

We define the set of concrete authorized paths, $auth_c \in \mathbb{P}(\mathrm{HF}^*)$, as the set of paths *hfs* that are cryptographically valid and whose projection to HI* is authorized for some *tok* field u:

$$auth_c = \{hfs \mid \exists u. \ \Psi(hfs, u) = hfs \land \overline{hfs} \in auth_a\}. \tag{22.9}$$

Similar to the abstract model, a concrete packet m is authorized if *fut*(m) is a fragment of an authorized path, i.e., $fut(m) \subset auth_c^{\leftrightarrow}$, and if it is valid for *tok*(m).

To achieve path authorization, protocols use the *HVF* to protect the future (abstract) path. The second protocol parameter is

$$extract : \mathbb{T} \to \mathrm{HI}^*, \tag{22.10}$$

which is intended to extract this path from a given *HVF*. This function is only required in proofs and not in the definition of the event system. Hence it may use features that would be infeasible to implement in the actual system, such as inverting hashes and MACs.

We lift *extract* to hop fields by $extract(hf) = extract(HVF(hf))$ and to paths by defining $extract(\langle\rangle) = \langle\rangle$ and $extract(hf \# hfs) = extract(hf)$. In §22.4.4.2, we will define a consistency condition (to be discharged by each instance model) that implies that *extract* coincides with $\overline{\cdot}$ on those paths that are both cryptographically valid and derivable by the attacker.

$$\frac{t \in H}{t \in DY(H)} \qquad \frac{\langle t_1, \ldots, t_n \rangle \in DY(H)}{t_i \in DY(H)} \; 1 \leqslant i \leqslant n$$

$$\frac{t \in H}{t \in DY(H)} \qquad \frac{t \in \mathcal{N} \cup \mathcal{I}_\perp \cup \mathbb{N}}{t \in DY(H)}$$

$$\frac{t \in DY(H)}{\mathsf{H}(t) \in DY(H)} \qquad \frac{t_1 \in DY(H) \; \cdots \; t_n \in DY(H)}{\langle t_1, \ldots, t_n \rangle \in DY(H)}$$

Figure 22.10: Rules for Dolev–Yao message derivation.

22.4.3.3 Attacker Model

We model a Dolev–Yao adversary who can eavesdrop on and inject new packets in all *int* and *ext* channels, but has access only to the keys of compromised nodes. We first define the attacker's message derivation capabilities, which are used in the attacker events introduced below.

We model the attacker's knowledge as a set of terms and their message derivation capabilities as a closure operator *DY* on sets of terms, defined in Figure 22.10. We have rules in *DY* for the atomic constructors as well for the sequences and hash constructors. Sequences can be composed and decomposed freely, whereas hashes can only be composed. In other words, the irreversibility of cryptographic hashes is modeled by the lack of a rule allowing the deduction of t from $\mathsf{H}(t)$.

We define the intruder knowledge in a state $s \in S_c$ as the Dolev–Yao closure (*DY*) of the set of terms $ik(s)$, defined by

$$ik_0 = \bigcup \{terms(x) \mid x \in auth_c\} \cup \{K_i \mid i \in \mathcal{N}_{attr}\}, \qquad (22.11)$$

$$ik(s) = ik_0 \cup \bigcup_{m \in s} terms(m). \qquad (22.12)$$

The set $ik(s)$ is the union of the initial intruder knowledge ik_0 and all terms in the packets of state s. The set ik_0 consists of authorized paths (i.e., the *HVF* of their hop fields) and compromised nodes' keys. In this model, we include public information in the derivation rules, rather than in the initial intruder knowledge.

22.4.3.4 Events

Each event of the abstract model is refined into a similar event of the concrete model (Figures 22.8 and 22.9, right, where differences between the models are highlighted in yellow). In the events' guards we omit the state and just write *ik*. The concrete model retains the packet life-cycle of the abstract model (§22.4.2.3). The **dispatch-int$_c$** and **dispatch-ext$_c$** events can send arbitrary attacker-derivable packets, instead of only authorized packets as in the abstract model. To defend against the attacker, we introduce interface and cryptographic checks in **send$_c$**, **recv$_c$**, and **deliver$_c$**. We now discuss the concrete model's events in more detail.

Attacker Events. The two events **dispatch-int$_c$** and **dispatch-ext$_c$** model that the attacker is active: She has the capability to send a packet on any AS' internal or external channel, regardless of whether the AS is honest or compromised. In both events, the packet m created by the attacker may contain arbitrary past and future paths. However, its hop validation fields and packet token field must be derivable from the intruder knowledge, i.e., $terms(m) \subseteq DY(ik(s))$. Note that the event **dispatch-int$_c$** still covers honest senders, as the attacker knows all authorized paths.

Similar to their abstract counterparts, both events set the history *hist* to $\langle\rangle$. The motivation for this is to exclude attacks where the attacker modifies a packet's forwarding path en-route, since these attacks are unavoidable in the presence of a sufficiently strong on-path adversary. For example, consider Figure 5.3 (page 98) and suppose that the attacker has access to E's external channels. Then E may receive a packet arriving on the right path from H, exchange its forwarding path by the left path, and forward the modified packet to C. This would (trivially) violate path authorization. By resetting the history, we effectively consider all packets sent by the attacker as new ones. An additional reason for this modeling choice is that an on-path attacker can not only re-route packets, but also modify their contents arbitrarily. This makes it generally impossible to correlate packets sent by the attacker with those the attacker has previously received. Consequently, path authorization must hold separately for the packets before and after the replacement of the forwarding path by the attacker.

Note that in the **dispatch-ext$_c$** event, the attacker's hop information field is not recorded in the history. This is because the attacker could modify her own hop field in arbitrary ways, and even omit it entirely. The attacker node is still identifiable in the history via the *target* function because the interface identifier *prev* of the next hop points to the link between it and the packet's source AS.

Honest Events. The honest events are **send$_c$**, **recv$_c$**, and **deliver$_c$**. To secure the protocol against the attacker introduced in this model, border routers now perform two validation checks. First, upon receiving a packet from another node, **recv$_c$** includes the guard $i = prev(hf)$ to check that the interface i over which the packet is received matches the interface *prev* of the packet's current hop field *hf*. In the SCION protocol, this corresponds to Item 1 of the checks performed by the ingress border router (page 118). Second, all honest events check the cryptographic *hop validation field* that is added to hop fields in this refinement using the check $\psi(hf, u)$, where hf and u are the packet's current hop field and *tok* field, respectively. This check ensures that the hop field (and indeed the whole or partial path) is authorized. In SCION, this corresponds to Item 6 of the checks performed by the ingress border router. Additionally, the same check is performed by the egress border router. In the event system that we present here, we do not include the update of the packet token field, which is required to model Item 5 of the SCION ingress border router, where the segment identifier is updated. We introduce instead an update mechanism as an extension of our framework in §22.4.5.

The events **send$_c$** and **deliver$_c$** use the function *fwd$_c$* to forward a packet:

$$fwd_c(m) = (\!|tok = tok(m), past = hd(fut(m)) \# past(m), fut = tl(fut(m)),$$
$$hist = \overline{hd(fut(m))} \# hist(m) |\!).$$

This function is defined similarly to *fwd$_a$*, but the *tok* field is not modified and the hop field being moved from the future to the past path is converted from HF to HI using $\overline{\cdot}$ before it is added to *hist(m)*.

Constant Intruder Knowledge. In reachable states s, all packets $m \in s$ are derivable from the static intruder knowledge ik_0. Hence, the attacker does not learn any new messages during the protocol execution. Under the Dolev–Yao closure, we can hence drop the state-dependent part $\bigcup_{m \in s} terms(m)$ of the intruder knowledge and use $DY(ik_0)$ instead of $DY(ik(s))$. We show the following lemma as an invariant.

Lemma 1. *For all reachable states s, $DY(ik(s)) = DY(ik_0)$.*

22.4.4 Refinement

We prove the following refinement theorem for the abstract event system $\mathscr{E}_a = (S_a, s_a^0, E_a, \{\xrightarrow{e}_a\}_{e \in E_a})$ and the concrete event system $\mathscr{E}_c = (S_c, s_c^0, E_c, \{\xrightarrow{e}_c\}_{e \in E_c})$, where for $i \in \{a, c\}$, S_i and s_i^0 are as defined in the previous sections, and Figures 22.8 and 22.9 define both E_i, and $\xrightarrow{e}_i$ for all $e \in E_i$. Below, we define the assumptions (ASM) and sketch the conditions (COND), as well as the refinement mappings h and π.

Theorem 1. *$\mathscr{E}_c$ refines $\mathscr{E}_a$ under the refinement mappings $h : S_c \to S_a$ on states and $\pi : E_c \to E_a$ on events, assuming ASM and COND.*

The refinement proof rests on several global *assumptions* (ASM) about the control plane and on a set of *conditions* (COND) on the authentication mechanism used. To establish that a concrete protocol satisfies the path authorization and detectability properties, it suffices to define the protocol's authentication mechanism by instantiating the protocol parameters and discharge the associated conditions.

In this subsection, we first introduce the assumptions and conditions. Afterwards we define the state and event refinement mappings.

22.4.4.1 Control Plane Assumptions

Our proofs require environment assumptions about the authorized paths $auth_a$ constructed by the control plane. There are two types of assumptions. First, there are two assumptions about the correct functioning of the control plane, which is independent of the data plane. Second, we must make assumptions about the control plane attacker's behavior. These two assumption types provide upper and lower bounds on the set $auth_a$, respectively.

Correctness Assumptions. The first control plane assumption is that authorized paths are *terminated*: The first hop information field's *prev* is $\bot$ and the last hop information field's *next* is $\bot$, except for when the respective hop information field belongs to the attacker. Second, we assume that authorized paths are *interface-valid*: Interfaces of adjacent hop information fields on a path point to the same link, except when both hop fields belong to attacker nodes. This exception accounts for unavoidable out-of-band communication by adversaries, so-called *wormholes* [249].

To formalize interface validity, we introduce the interface validity predicate

$$\phi : \mathrm{HI} \times \mathrm{HI}_\bot \to \mathbb{B}.$$

In the following, we let hi_A (respectively hi_B) denote a hop information field for which $id(hi_A) = A$ (respectively $id(hi_B) = B$). We also use this shorthand for hop fields hf_A. The parameters of ϕ are the current hop information field hi_B and a preceding hop information field hi_A. If there is no previous hop field, no interface must be checked.

$$\phi(hi_B, \bot) = \text{true}$$
$$\phi(hi_B, hi_A) = (target(A, next(hi_A)) = (B, prev(hi_B)))$$
$$\vee\, (A \in \mathcal{N}_{attr} \wedge B \in \mathcal{N}_{attr})$$

We define a function $\Phi : \mathrm{HI}^* \times \mathrm{HI}_\bot \to \mathrm{HI}^*$ as follows. For a sequence of hop information fields *his* and the initial previous hop information field hi_{prev}, $\Phi(his, hi_{prev})$ returns the longest prefix of *his* such that for all fields hi_B on *his* and their respective predecessor hi_A on *his*, $\phi(hi_B, hi_A)$ holds. For the first hop information field on *his*, ϕ must hold with hi_{prev} as the predecessor.

Formally, we define

$$\Phi(\langle\rangle, hi_{prev}) = \langle\rangle$$
$$\Phi(hi \# his, hi_{prev}) = hi \# \Phi(his, hi) \qquad \text{if } \phi(hi, hi_{prev})$$
$$\Phi(hi \# his, hi_{prev}) = \langle\rangle \qquad \text{otherwise.}$$

We write $\Phi(his)$ as a shorthand for $\Phi(his, \bot)$. ASM 1 and ASM 2 formalize the correctness of the control plane.

ASM 1: Terminated
> A hop information field hi with $id(hi) \notin \mathcal{N}_{attr}$ on $hi \# his \in auth_a$ (resp.
> on $his \cdot \langle hi \rangle \in auth_a$) has $prev(hi) = \bot$ (resp. $next(hi) = \bot$).

ASM 2: Interfaces valid
> All paths $his \in auth_a$ are interface-valid, namely $\Phi(his) = his$.

Control-Plane Attacker Assumption. We have modeled a very strong
attacker, who can compromise both end hosts sending packets and ASes (including on-path ASes). As a consequence, we must make assumptions on
the attacker's behavior in the control plane. Concretely, we assume that if all
honest on-path ASes consent to the authorization of a path (and create corresponding hop validation fields), then the compromised on-path ASes consent
as well. This is necessary because in the data plane the attacker can craft a
corresponding forwarding path using the honest nodes' hop validation fields.
It is also acceptable to regard such paths as being authorized, since all honest
agents consented to their authorization. In the extreme case where all ASes are
compromised, all paths must be assumed to be authorized as well.

Formally, we define a number of closure operations and assume that the set
of authorized paths is closed under these operations. These assumptions are
only required given our very strong attacker model where the end host attacker
can collude with on-path ASes. When all compromised ASes are off-path, then
control plane attacker assumptions are not needed. This is in stark contrast to
BGP, which does not achieve security even when all adversaries are off-path.

22.4.4.2 Conditions on Authentication Mechanisms

Our framework defines five conditions that relate the protocol parameters ψ
and *extract* introduced in Equations (22.8) and (22.10) with each other and
with the environment parameters $\mathcal{N}_{attr}$ and $auth_a$ (via $auth_c$). These conditions
are used in the refinement proof. We will need to prove these conditions for any
instance of the concrete model. Here, we show two representative conditions
that are implied by the ones used in our framework.

COND 1 requires that the attacker cannot derive valid hop fields for honest
nodes that are not already contained in $auth_c$.

COND 1: Valid attacker-derivable hop fields are authorized:
> If $hf \in DY(ik_0)$, $\psi(hf, u)$, and $id(hf) \notin \mathcal{N}_{attr}$ then $\exists hfs \in auth_c . \ hf \in hfs$.

COND 2 states that a cryptographically valid path hfs is equal to $extract(hfs)$,
which extracts the subsequent path from the first hop field in hfs. The equivalence holds modulo the projection to abstract paths, $\overline{\cdot \cdot}$.

COND 2: Extract is path for valid paths:
> If $\Psi(hfs) = hfs$ then $extract(hfs) = \overline{hfs}$.

22.4.4.3 Refinement Mappings

We define the refinement mapping $h: S_c \to S_a$ on states as the element-wise mapping of the *int* and *ext* channels under a function to_a that maps concrete packets to abstract packets. This function projects hop fields to hop information fields in the past and future paths, and additionally applies the Ψ and Φ functions to truncate the future path once it becomes invalid. Because of the interface and cryptographic checks that we introduce in the concrete model, no forwarding occurs on invalid hop fields, and they *may* be safely truncated. Since the abstract model does not have such checks, the abstraction function to_a *must* truncate them in order to establish a refinement relation.

The refinement mapping $\pi: E_c \to E_a$ maps each event on the right side of Figures 22.8 and 22.9 to the corresponding event on the left side, where packet and hop field parameters are transformed using to_a and $\overline{\cdot}$, e.g., $\mathbf{send}_c(A, m, hf, i)$ is mapped to $\mathbf{send}_a(A, to_a(m), \overline{hf}, i)$.

22.4.5 Instantiation with SCION Protocol

We now turn to the instantiation of our generic model with the SCION data plane protocol. To ease understanding, we omitted some features in our simplified presentation that are required to accurately model the SCION protocol. An upcoming publication will provide more depth and technical details.

The SCION path authorization mechanism was presented in §5.3. The HF authenticator σ presented in that chapter corresponds to our model's *HVF*, and the segment identifier β corresponds to the packet token field *tok*. The *HVF* of each hop field is a MAC over the local routing information *hi* and the *tok* field. It is computed as:

$$hvf_A = \mathsf{MAC}_{K_A}(\langle hi_A, tok \rangle). \tag{22.13}$$

We instantiate our parameter ψ with this equation.

The *tok* field has two features that we omitted in our simplified presentation: First, it has a parametrized type *'TOK*, which allows us to use a formalization of exclusive-or (XOR). Second, this field is updated en-route using a function $upd\text{-}tok : \mathit{'TOK} \times \mathrm{HF} \to \mathit{'TOK}$, which is a parameter to the concrete model. When an AS receives a packet from an inter-AS channel, it updates its *tok* field using the *upd-tok* function, which takes the current *tok* field and the current hop field. We instantiate it for SCION as

$$upd\text{-}tok(u, hf) = u \oplus HVF(hf), \tag{22.14}$$

where $\oplus$ is XOR. The control plane creates forwarding paths such that for each hop on the path, the *tok* value embedded in its *HVF* is the XOR of the *HVFs* of all previous hops in the beaconing direction starting with a random initialization value *RND*. Recall that in this model, we only consider up-segments, which traverse the network in the *opposite* direction. Hence, during forwarding along authorized segments, the updates of the *tok* field by routers successively remove (by the cancellation property of XOR) the *HVFs* that were added during beaconing until finally, only *RND* remains on *tok*. Since the randomness of *RND* is not essential to achieve security, we set it to 0 in our models.

The *tok* field combines the *HVFs* of all subsequent hops in the forwarding direction with XOR, thereby including the upstream path. We instantiate the *extract* parameter with a function that extracts the *hi*-level path out of the *tok* value contained in a MAC. This is challenging, since the MACs contained recursively in *tok* are combined using XOR and are hence unordered. Once we have the definition of *extract*, we prove the static conditions and thus show that our SCION model is an instance of the parametrized concrete model. Since properties are preserved by refinement and parameter instantiation, SCION satisfies the path authorization and detectability properties proven in the abstract model.

22.4.6 Conclusion

We have successfully verified the path authorization property for packets with up-segments, showing the security of the mechanism that SCION uses to protect the authorization of paths. Our work to this date is focused on the data plane, and hence assumes the correctness and security of the control plane. In particular, we do not verify in this model that the set of authorized paths produced by the control plane reflects the routing policies of each AS (*beacon authorization* property P5, page 159).

We are currently working on generalizing our models and proofs for packets with combinations of up-, core-, and down-segments and we are planning to link our design-level verification with the ongoing verification of the router code. We will discuss this and other plans in Chapter 24 .

22.5 Quantitative Verification of the N-Tube Bandwidth Reservation Algorithm

Networking systems are remarkably hard to get right. Securing system dynamics in adversarial environments further increases this difficulty. When designing such systems, logical correctness and security are necessary, but not sufficient, as *quantitative properties* such as system throughput and latency are equally important. Many algorithms, even if logically correct, become unusable if their quantitative guarantees are poor. This necessitates the formal verification of any proposed algorithm to validate its intended properties and thereby establish not only qualitative correctness and security guarantees but also quantitative guarantees about the system's dynamics.

For design-level verification, many of the best-known automated analysis tools, particularly for quantitative properties, are based on finite automata [87,

309]. The main difficulty is that distributed and networked systems like N-Tube have intrinsic features that are quite hard or impossible to represent in such models. For instance, AS attributes may contain unbounded data structures. Moreover, asynchronous message passing and dynamic AS creation may increase the number of both messages and ASes in an unbounded manner.

Overview. In this section, we show how to leverage a formal approach to design and quantitatively analyze networking systems. We apply this approach to N-Tube, the Neighbor-based, Tube-fair bandwidth reservation algorithm introduced in §10.2.5, focusing on the quantitative assessment of its crucial properties such as stability and fairness. More specifically, we formalize the N-Tube algorithm as an executable probabilistic model and analyze its properties (G1–G5) claimed in §10.2.5 using simulation and statistical model checking [465, 563]. We use *rewriting logic* [127] as the modeling formalism and the Maude's simulation engine and the PVeStA tool [18] for quantitative analysis, which we overview next. Our experimental results provide quantitative guarantees of N-Tube's properties and assess its resistance to attacks in various malicious scenarios.

22.5.1 Maude and Statistical Model Checking

Rewriting logic and the Maude ecosystem [127] have been very successful in formalizing and analyzing high-level designs of a wide range of distributed and networked systems [79, 340, 341, 514, 524, 529]. The Maude ecosystem supports high-performance, machine-checkable, automated formal analysis, including simulation, reachability analysis, linear temporal logic model checking, and statistical model checking [127, 465].

Networking systems like N-Tube can be naturally expressed as *rewrite theories* in rewriting logic and executed in Maude [127]. A rewrite theory consists of an equational theory, specifying the system's data types, and a collection of *labeled conditional rewrite rules* of the form

```
crl [l] :   t => t′ if cond,
```

where l is a label. Such a rule specifies a transition from a system state, represented by the term t, to a new state $t′$, provided the condition *cond* holds. Rewrite theories may be enriched with information on probabilities (and even time) to model probabilistic distributed systems [10] that can be subjected to quantitative analysis.

Statistical model checking (SMC) [465, 563] is a formal approach to analyzing probabilistic systems with respect to temporal logic properties. Compared to conventional simulations or emulations, SMC can verify a property specified, e.g., in a *stochastic temporal logic*, up to a *statistical confidence level*, by running Monte-Carlo simulations of the system model. The expected value $\bar{v}$ of a property query belongs to the interval $[\bar{v} - \frac{\beta}{2}, \bar{v} + \frac{\beta}{2}]$ with $(1 - \alpha)$ statistical confidence, where the parameters α (statistical confidence) and β (error margin) determine when an SMC analysis stops performing simulations [465]. In particular, the performance estimations and predictions from the Maude-based SMC analyzes with the PVeStA tool [18] have shown good correspondence with the implementation-based evaluations under realistic deployments [79, 342].

22.5.2 Formalizing N-Tube in Maude

Many statistical model checkers assume that the system is *purely probabilistic* (no unquantified non-determinism, meaning that all transitions are associated with probabilities). Hence, we formally specify N-Tube as a *timed*, *purely probabilistic* rewrite theory by following the methodology in [10]. In particular, each message is assigned a delay sampled from a *continuous probability distribution*, which determines the firing of the rewrite rule receiving the message.

The system state of our N-Tube model consists of a *multiset* of *objects*, including a *scheduler* object maintaining the global clock, and *messages*. The global clock models well-synchronized local clocks at each AS (achieved using the global time synchronization protocol introduced in §8.2). An object of class C is represented as a term $<o : C \mid att_1 : val_1, \ldots, att_n : val_n>$, with o the object's identifier, and val_1 to val_n the current values of attributes att_1 to att_n. An incoming message of the form $\{t, \ msg\}$ is ready to be consumed at the global time t, while an outgoing message of the form $[t + d, \ msg]$ will be delivered d time units after t where the message delay d is sampled from some continuous probability distribution (e.g., the lognormal distribution). Each message *msg* has the form to o from o' : *mp*, with o, o', and *mp* the message receiver, sender, and payload, respectively. The scheduler object is specified to advance the global time and to deliver outgoing messages at the specified times.

N-Tube's dynamic behavior is axiomatized by specifying its transition patterns as rewrite rules. The following example of a (conditional) rewrite rule [cmp] specifies a message processing event in N-Tube:

```
1    crl [cmp] :
2        {T,  to O from O' : res(M)}
3        < G : Table | links : LS, ATTS' >
4        < O : As | resMap : RM, ATTS >
5    =>
6        < G : Table | links : LS, ATTS' >
7        < O : As | resMap : save(M,O,RM,LS,AVL,IDL), ATTS >
8        [T + lognormal(μ,σ),
9            to next(O,M) from O : compute(M,AVL,IDL)]
10       if (atSrt(M) or onPth(M))    /\ pathCheck(M)
11           /\ resMsgCheck(M,T) /\ resMapCheck(M,RM)
```

```
12              /\ AVL := avail(LS,O,RM,T,M)
13              /\ IDL := ideal(LS,O,RM,T,M) .
```

Upon receiving a reservation message `res(M)` at global time T (line 2), the AS O updates its local reservation map accordingly (using the `save` function, line 7), and forwards the modified message (determined using the `compute` function) to `next` hop in the path (line 9). The network topology and all `links`' capacities are stored in a global "table" G (lines 3 and 6). The message delay is probabilistically sampled from the lognormal distribution [64], parametric on the mean μ and standard deviation σ, each time this rule applies (line 8). The function `avail` computes the available amount of bandwidth on the link at the corresponding egress interface of AS O (line 12). The function `ideal` returns how the adjusted capacity of the corresponding egress link is shared in a so-called bounded tube fair manner among all existing reservations with the same egress link (line 13). The predicates `atSrt` and `onPth` check whether the message is at the source AS or on the path segment, respectively (line 10). The predicates `pathCheck`, `resMsgCheck`, and `resMapCheck` ensure that the message is well-formed and compatible with existing reservations in AS O's reservation map that corresponds to the message (lines 10 and 11). The variables `ATTS` and `ATTS'` refer to the remaining attributes, which do not affect the next state.

22.5.3 Statistical Analysis

With our simulation and SMC analyzes, we investigate the following two questions about N-Tube, concerning qualitative and quantitative aspects, respectively:

- Are the statistical verification results consistent with our claims (G1–G5) in §10.2.5?

- How does N-Tube actually perform in worst- and average-case malicious scenarios with respect to stability and fairness? In particular, how does it resist increasing attack power in terms of total malicious demands, number of attackers, and number of reservation requests per attacker?

22.5.3.1 Benchmark

To statistically analyze N-Tube's properties we implement three *parametric* generators: a topology generator (TG), a path generator (PG), and a workload generator (WG). We use these to probabilistically generate a different initial state for each simulation in an SMC analysis. Specifically, TG generates *scale-free* Internet topologies with strongly connected ASes, which is also characteristic of the realistic AS-level Internet graph in the widely used benchmark

Table 22.2: Generators Parameters.

Parameters	Default Value
# benign ASes	90
# malicious ASes	10
# sources	20
# intermediate ASes	75
# destinations	5
adjusted capacity δ	0.8
minimum acceptable bandwidth	[0,50]
demanded bandwidth	[50,250]
expiration time	20
reservation frequency	10
renewal frequency	2
deletion frequency	50
snapshot frequency	5
reservations per source	5
# paths per source-dest pair	5
segment length	5
link capacity	[100,300]
message delay	lognormal: $\mu = 0.0$, $\sigma = 1.0$

CAIDA for Internet data analysis [92]. Each link between nodes is assigned a bandwidth probabilistically sampled from an interval. PG then explores the generated graph, and collects paths from *sources* to *destinations*. WG provides the generated sources, including adversaries, with reservations, renewals, and deletions on a probabilistic basis, where each of these three types of requests is parametric in the algorithm-specific parameters (such as demanded bandwidth, expiration time, etc.). Table 22.2 lists all 18 parameters used in the statistical analysis of N-Tube, together with their default values, which the three generators use to generate the topologies, paths, and workloads.

22.5.3.2 Experimental Setup

To mimic the real-world network environment, we use the lognormal distribution for probabilistically sampling message delays [64]. We employed a cluster of 50 d430 Emulab machines [539], each with two 2.4 GHz 64-bit 8-Core E5-2630 processors, to parallelize SMC with the PVeStA tool [18]. We set the statistical confidence level to 95% (i.e., $\alpha = 0.05$) and the size parameter β to 0.01 for all our experiments.

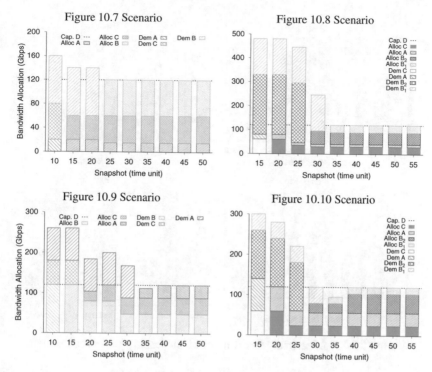

Figure 22.11: Measuring stability and fairness. Time units are defined as logical clock ticks in our probabilistic model.

22.5.3.3 Analysis Results

We have subjected the probabilistic Maude model of N-Tube to the above generators and PVeStA, and performed three sets of experiments according to our
experimental goal with 100 ASes by default. Each simulation or SMC analysis
took up to three hours (in the worst case) to terminate.

Experiment 1: Verifying Properties. In all our SMC analyzes, the probabilities of satisfying N-Tube's global properties (G1–G5) are 100%. This provides a strong validation of our claims for the properties via machine-checked
analysis.

Experiment 2: Stability and Fairness. We report the simulation results
for the scenarios in §10.2.5. Figure 22.11 depicts the bandwidth reservations at
interface D and their state, demanded or allocated, as a function of (simulation)
times where we take "snapshots" of the system state. The allocated bandwidths
adapt over time to self-renewals and other demands. For the scenarios in Figures 10.8 and 10.10, we individually measure the allocated bandwidth for each
of the two demands (B_1 and B_2) through interface B. In all scenarios, the al-

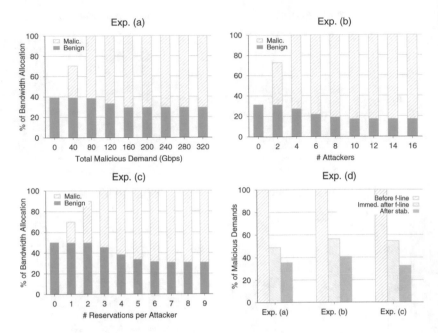

Figure 22.12: Measuring the impact of increasing malicious power.

locations in the entire network *converge* and *stabilize*; the total allocations are always *bounded* by D's adjusted capacity (dashed line), and distributed *proportionally* to the demands after stabilization as expected (§10.2.5). Hence, from the quantitative perspective, these results further demonstrate stability and fairness, in particular for the *worst-case scenarios* (Figures 10.8–10.10) where attacks are mounted directly on an honest path.

Experiment 3: Impact of Increasing Malicious Power. To analyze the influence on bandwidth allocation of increasing malicious power, we randomly positioned the attackers in the network (not necessarily neighbors of the targeted path), and picked a relatively small number of destinations (5 out of 100 ASes) for the reservations so that all demands, including malicious ones, *converge to these destinations*. We then randomly selected one destination and one of its egress interfaces, and reported the associated aggregate allocations for the benign sources and attackers, respectively.

Experiments (a)–(c) in Figure 22.12 show the allocation percentage as a function of attacker capability, represented by *total demanded bandwidth*, *number of attackers*, and *number of issued reservations per attacker*, respectively. With increasing malicious power, the attackers tend to occupy more bandwidth until the entire allocation *stabilizes* (starting from 160 Gbps, 10 attackers, and 7 reservations per attacker, respectively); thereafter their demands are adjusted, and thus limited by the *links' capacities* and *scaling factors*. These results further provide quantitative assessments of N-Tube with varying malicious powers by exploring the large parameter space.

We also measured the adversaries' allocated bandwidth reduction by N-Tube's "frontline defense". A *frontline defender* is the first honest AS on an attacker reservation's path that can mitigate the impact of the attack by limiting the adversary demands. We divided the timeline into three phases: (i) before malicious demands reach the frontline defender; (ii) immediately after those demands pass through the frontline; and (iii) after the network stabilizes.

Experiment (d) in Figure 22.12 reports for Experiments (a)–(c) the percentage of original malicious demands that was allocated to the adversaries in each phase. Phase (iii)'s computation is based on the minimum stabilized "point" (e.g., 160 Gbps in Experiment (a)): The higher the stabilized point is that we consider, the more reduction there will be. As demonstrated in Experiment (d), N-Tube's frontline defense plays an important role in limiting the adversarial demands, e.g., in Experiment (a) roughly 50% of the malicious demands can be reduced, which constitute almost 80% of the total reduction.

22.5.4 Conclusion

Qualitative properties about network security protocols should be complemented by quantitative ones, for example, attesting to the system's performance even under attack. We have shown how this can be done for parts of SCION. In particular, we have developed a formal probabilistic model of N-Tube in Maude and performed a quantitative assessment of its resistance to attacks by statistically exploring the large parameter space and varying malicious scenarios.

The obvious next step is to build an efficient N-Tube implementation, as well as large-scale deployment by, e.g., proceeding along the lines proposed in [412, Chapter 10]. Preliminary results from an N-Tube prototype implementation, realized as part of the COLIBRI QoS infrastructure [440], have demonstrated N-Tube's deployability and scalability.

23 Code-Level Verification

Linard Arquint, Peter Müller, Wytse Oortwijn, João Pereira, Felix Wolf

Chapter Contents

23.1 Why Code-Level Verification? 520
23.2 Introduction to Program Verification 522
    23.2.1 Proving a Program Correct 523
    23.2.2 Functions and Function Calls 524
    23.2.3 Loops and Recursion 526
    23.2.4 Reasoning about Heap Manipulations 527
23.3 Verification of Go Programs 533
    23.3.1 Specifying Go Programs with Gobra 534
    23.3.2 Verifying Concurrent Programs with Gobra 538
    23.3.3 Interfaces in Gobra 544
23.4 Verification of Protocol Implementations 547
    23.4.1 Specification and Verification of I/O Behavior 547
    23.4.2 From Protocol Models to I/O Specifications 549
23.5 Secure Information Flow 555
    23.5.1 Non-interference 556
    23.5.2 Self-Composition 558
    23.5.3 Modular Product Programs 559
    23.5.4 Advanced Topics 560

The verification effort described so far ensures that the SCION protocol guarantees the intended correctness and security properties. These *design-level* guarantees are essential, but by themselves do not guarantee that a SCION network actually works as intended. Errors in the protocol's *implementation* could compromise its availability, correctness, or security—for instance, by causing a component to crash, skip essential checks, or leak confidential information. To guarantee the absence of errors in the SCION implementation, we employ *code-level* verification. Design-level and code-level verification complement each other and target different abstractions of the overall system. In particular, design-level verification provides specifications for many of the properties that need to hold for the implementation to be correct, for instance, the intended

I/O behavior of each component. Other properties, such as memory safety, data-race freedom, and the absence of backdoors are specified directly on the code level.

In this chapter, we will motivate code-level verification for SCION (Section 23.1), provide a gentle introduction to program verification (Section 23.2), and give an overview of the verification technique and tool used for the verification of the SCION implementation (Section 23.3). We will then focus on two crucial aspects of our approach, namely how we formally connect design and code-level verification (Section 23.4), and how we plan to verify that the SCION implementation does not leak secret data such as cryptographic keys to potential adversaries (Section 23.5). We neither assume nor require prior knowledge or experience with program verification and instead give a high-level overview of our verification approach. For the technical details, we refer the reader to `https://gobra.ethz.ch` and `https://www.pm.inf.ethz.ch/research/verifiedscion.html`.

23.1 Why Code-Level Verification?

SCION has been implemented with great care and state-of-the-art quality assurance including thorough code reviews and comprehensive testing. However, even though reviews and testing are helpful, they cannot guarantee the absence of subtle errors in the implementation. Studies have shown that delivered, industrial software contains at least 10 bugs on average in every 1000 lines of code [357]. Any bug could potentially compromise the availability, correctness, and security of the SCION implementation.

While flaws in the protocol design would be caught during the protocol-level verification, there are numerous sources of errors that show up only on the implementation level. These include memory errors (e.g., null-pointer dereferencing and out-of-bounds accesses), deviations from the intended functional behavior (e.g., omitting checks that are required by the protocol), concurrency errors (e.g., data races and deadlock), and security vulnerabilities (e.g., leaking confidential data). The purpose of code-level verification is to prove the absence of such errors. It guarantees the following properties *for all executions of the code*, that is, for all inputs, thread schedules, and interactions with the environment:

- **Memory safety:** The implementation does not abort due to memory errors or uncaptured panics. A recent study shows that memory errors cause 70% of all security vulnerabilities in Microsoft software [503]. Moreover, memory errors, no matter whether they occur sporadically or are triggered by an attacker, reduce the availability of the network. Code-level verification eradicates these errors.

- **Functional correctness:** The code correctly implements the SCION protocol, such that all properties that have been verified on the protocol level carry over to the implementation. This property links design-level and code-level verification, providing end-to-end guarantees for the entire system.

- **Security:** The implementation does not leak confidential information such as cryptographic keys. In particular, attackers cannot obtain secret information by interacting with a SCION component and observing its input/output (I/O) behavior.

Violations of these properties are extremely hard to find—even with the most sophisticated testing approaches—for several reasons.

First, many SCION components are concurrent, that is, use multiple execution threads to increase performance. The occurrence of concurrency errors such as data races, deadlocks, or atomicity violations depends on the interleaving of different threads, which is highly non-deterministic. Consequently, testing does often not reveal concurrency errors, and errors that *do* occur during tests or in production are extremely hard to reproduce and debug. For that reason concurrency errors are sometimes called "Heisenbugs." Code-level verification reliably detects errors in concurrent code by verifying the code for all possible thread schedules.

Second, certain errors in a SCION component might be triggered only by very specific inputs such as a specific sequence of packets. Whereas shallow bugs such as errors in packet parsing are often detected by fuzzing (i.e., exposing a component to a large number of randomized inputs), deeper flaws in the logic that depend on complex inputs are very difficult to find during testing. Code-level verification detects both shallow and deep bugs.

Finally, some essential security properties, such as not leaking secret information, are hyperproperties. They relate two or more executions of a program whereby executing the program multiple times with different inputs does not allow an attacker to learn secret information by comparing the outputs. Standard test cases consider only one program execution at a time and, thus, cannot check hyperproperties; repeated executions are not meaningful for concurrent programs because it is unclear whether any differences indicate leaking of secrets or are caused by different thread schedules. Code-level verification is able to check hyperproperties, as we will explain later in this chapter.

In summary, our code-level verification effort goes far beyond testing by providing mathematical proofs that certain problematic behaviors do not exist in the SCION implementation. These proofs are checked by tools and cover all possible executions of a program, for all inputs, thread schedules, and interactions with the environment.

Nevertheless, testing is still necessary to complement verification. Testing is much cheaper and, thus, useful to detect errors and to increase confidence in the implementation before we embark on the verification effort. Testing is also useful to validate the specifications on which verification is based. Finally, our code-level verification of SCION focuses on the SCION implementation

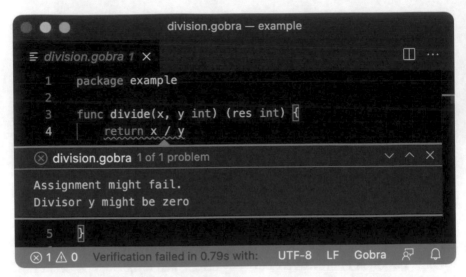

Figure 23.1: Verification failure because the parameter y is possibly zero. The
            screenshot shows the VSCode integration of the Go verifier Go-
            bra.

itself, but does not cover libraries (such as cryptographic libraries) and the entire software and hardware stack underneath, including the operating system.
Moreover, we perform verification on the level of source code and rely on the
compiler to be correct. We also trust the verification tool itself. For all these
reasons, verification eliminates the most common sources of errors, but provides no absolute guarantees. We will further reduce the trust assumptions in
future work, for instance, by verifying libraries and by generating certificates
that can be checked by an independent tool, such as the Isabelle prover that we
use for protocol verification [402].

23.2 Introduction to Program Verification

This section starts by explaining how mathematical proofs are obtained for simple programs consisting only of local variables. The approach is then extended
to cover more interesting control-flow constructs such as function calls, loops,
and recursion. Finally, we introduce the concept of permissions, which enables
reasoning about heap-manipulating and concurrent programs. We will use Go
syntax in our examples, but the concepts developed in this section can be applied to a wide range of programming languages, including Go, Python, Rust,
and Java. Verification techniques for Go-specific features will be explained in
§23.3.

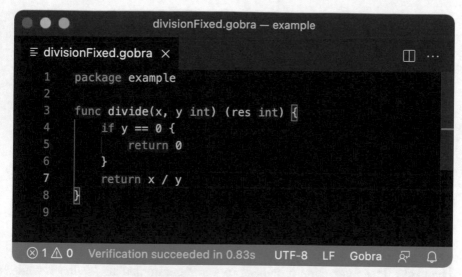

Figure 23.2: Successful verification of the function `divide` that performs a division only if y is not zero.

23.2.1 Proving a Program Correct

A program is correct if it performs the desired computations and in particular that it executes only safe operations (that is, does not exhibit crashes or undefined behavior). A verifier checks these properties at compile time and guarantees that they hold for all possible program executions. For this purpose, it produces a set of proof obligations that must be proved to hold for the program to successfully verify.

Figure 23.1 illustrates the interaction with a verifier. The function `divide` crashes if the parameter y is zero. The verifier employed here (Gobra, which will be introduced in the next section) verifies the function for *all possible* calls, irrespective of the calls that actually occur in the program. Therefore, it reports that the function might lead to a division by zero. Gobra's integration into Visual Studio Code (VSCode)[1] shows this verification error using a squiggly line below any problematic expression and provides a descriptive error message. The generated proof obligation for this example includes $\forall x, y \in int. \ y \neq 0$, which does not hold. Discharging proof obligations is in general an undecidable problem. Whenever in doubt, our verifier is conservative and reports an error. Therefore, our approach may lead to false positives, sometimes called spurious errors, but never to false negatives. For this reason, the error message in Figure 23.1 is phrased carefully to report that the program *might* contain a division by zero.

---

[1]https://code.visualstudio.com/

```
1  requires y != 0
2  func divide(x, y int) (res int) {
3      return x / y
4  }
```

Figure 23.3: Precondition restricting the possible arguments with which the function divide can be called.

Figure 23.2 shows a possible fix to the example discussed before by adding an if-statement that performs the division only if y is different from zero. As indicated in green in the status bar, this modified program verifies successfully.

As shown in Figures 23.1 and 23.2, the Gobra verifier is integrated into the VSCode IDE. Verification is performed just like syntax and type checking: The verifier is automatically invoked after the code changes and offers prompt feedback. This interaction provides programmers immediate feedback while writing code and supports them to quickly iterate over multiple versions of their program until verification succeeds.

23.2.2 Functions and Function Calls

Verifying that a function is correct for all possible calls requires that each function handles all possible parameter values. This may lead to defensive implementations like the arguably unnatural code in Figure 23.2 or extensive parameter validation, which clutters up the code. A better solution is to restrict the permitted parameter values and to check the restriction at each call site. Constraints on a function's parameters are expressed by equipping the function with a *precondition*. The verifier checks that the precondition is satisfied by each function call. In turn, during the verification of a function body, the verifier may assume that the precondition holds.

Figure 23.3 shows the program from Figure 23.1 with a precondition (indicated by the keyword requires). This precondition implies that no (valid) call to the function will lead to a division by zero. Hence, the function verifies even without the if statement. Figure 23.4 shows a client trying to call divide with an invalid value for parameter y. This client is correctly rejected by the verifier as the precondition is violated.

Preconditions decouple function verification from the verification of its callers. This provides modularity: We can verify a function without knowing its callers (which is, for instance, useful for library functions), and adding a caller to a program does not require re-verification of the function. To further increase modularity, we also decouple the verification of a caller from the implementation of the called function. Instead of using the implementation to reason about a call, we equip each function with a *postcondition*, which specifies the function's result and effects. The verifier checks that the function

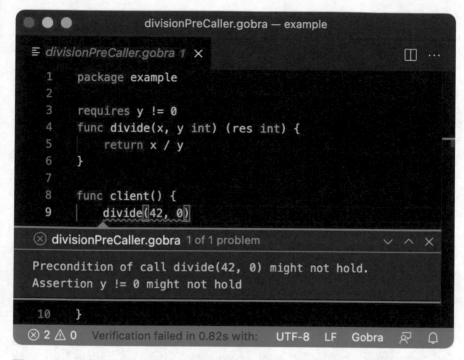

Figure 23.4: Verification failure because `caller` violates the precondition of `divide`.

```
1   requires y != 0
2   ensures   x % y == 0 ==> x == res * y
3   func divide(x, y int) (res int) {
4        return x / y
5   }
```

Figure 23.5: Postcondition specifying the result of the function `divide`.

implementation satisfies the postcondition when it terminates, and uses the postcondition to reason about calls.

To illustrate this concept, consider two nested calls to our `divide` function: `divide(42, divide(4, 2))`. With the implementation from Figure 23.3, this code snippet will fail to verify: Since the verifier does not consider the implementation of `divide`, it has no information about the result of the inner call `divide(4, 2)`. In particular, this call might return 0, violating the precondition of the outer call.

To fix this problem, we equip function `divide` with the postcondition shown in Figure 23.5 (indicated by the keyword `ensures`). This postcondition specifies the function's result `res` in terms of its arguments x and y and, in particular,

```
1   requires  n >= 0
2   ensures   res == (n + 1) * n / 2
3   func sum(n int) (res int) {
4       res = 0
5       invariant 0 <= i && i <= n + 1
6       invariant res == i * (i - 1) / 2
7       for i := 0; i <= n; i++ {
8           res = res + i
9       }
10  }
```

Figure 23.6: Iteratively summing the first n numbers. Verification needs an induction hypothesis stated with the keyword `invariant`.

implies that `res` is different from 0 when `x` is. This information is sufficient to verify the nested calls above.

A function's pre and postconditions together constitute its *specification* (sometimes also called its *contract*). The verifier checks that whenever a function is called in a state that satisfies the precondition, it will not lead to a memory error and, if it terminates, then it terminates in a state in which the postcondition holds. Whereas preconditions must be sufficiently strong to exclude memory errors such as program crashes, postconditions are often partial: They specify those aspects of a function's behavior that are needed to verify other parts of the program.

Function specifications make verification procedure-modular: Each function is verified in isolation, using only the specification of the functions it calls. Modular verification necessitates writing function specifications, but has several important benefits: Firstly, it can reason about calls even if the implementation of the called function is not available or not known statically, for instance, for dynamically-bound calls and calls to foreign functions. Secondly, it can verify functions without knowing their callers, which is crucial to give correctness guarantees for library functions. Thirdly, changes to parts of the implementation do not require re-verification of the entire program. This greatly reduces the effort during code maintenance. For instance, optimizing a function implementation in a way that does not affect its specification does not require re-verification of its callers. Moreover, the effort for modular verification grows roughly linearly in the number of functions and can thus scale to large code bases. Finally, modular verification also unlocks performance optimizations as functions can be verified in parallel.

23.2.3 Loops and Recursion

Verifying a loop requires additional specifications that are expressed as loop invariants. They must hold before entering the loop, and after each loop it-

```
1  requires  n >= 0
2  ensures   res == (n + 1) * n / 2
3  func sum(n int) (res int) {
4      if n == 0 {
5          return 0
6      } else {
7          return n + sum(n - 1)
8      }
9  }
```

Figure 23.7: Recursively summing the first n numbers. The function specification serves as induction hypothesis.

eration. Consider the program in Figure 23.6 that calculates the sum of the numbers from 1 to n. Two loop invariants are necessary here: The first one expresses that the index i stays within the given bounds. The second loop invariant on line 6 expresses that res stores the sum of the numbers from 1 to i - 1. When exiting the loop, the verifier is able to deduce from the first loop invariant and the negated loop condition that i has the value n + 1. Substituting i by its value in the loop invariant immediately leads to the stated postcondition. Technically, the conjunction of the loop invariants corresponds to an induction hypothesis which is then checked by the verifier to hold for the base case and for the induction step.

Reasoning about recursive function calls is similar to reasoning about loops. However, instead of using loop invariants, the induction hypothesis is stated as function specification. Figure 23.7 shows a recursive implementation of the program from Figure 23.6.

23.2.4 Reasoning about Heap Manipulations

So far, all variables in our examples were local to a function. We will now explain how to reason about programs that manipulate heap data structures. In case the heap memory is explicitly allocated and deallocated, as for example in C or C++, there are numerous safety properties that must be checked—e.g., preventing access to unallocated memory or freeing memory twice. Although Go automatically allocates memory on the heap and uses a garbage collector to free up memory, there are still two main challenges: Firstly, a procedure may, in principle, modify the entire heap. Callers require information about the side effects of a call in order to determine which heap properties are potentially affected by a call and which ones are definitely preserved (*framed*) across the call. Secondly, multiple threads can read and write to the same heap locations in parallel, resulting in data races. We discuss next how we address these challenges.

```
1   func setZero(x *int) {
2       *x = 0
3   }

5   func client(a *int) {
6       *a = 42
7       m := 1
8       b := &m
9       c := b
10      setZero(b)
11      assert *a == 42
12  }
```

Figure 23.8: Modular reasoning about heap manipulating programs is challenging because a procedure can have arbitrary side-effects.

Heap Separation. We illustrate the first challenge with the example in Figure 23.8. The function `client` gets pointer a as a parameter and sets the referenced value to 42. The variables b and c alias and point to the value m, which is stored on the heap. The function proceeds to call `setZero(b)`. The assert statement on the last line in function `client` expresses that the value pointed to by variable a has not been modified by a call to `setZero`.

Using modular verification, this assertion fails to verify because, without inspecting the implementation of `setZero`, the verifier does not know whether the function modifies the memory location pointed to a. For instance, that location could be reachable from a global data structure and thus be modified even though a is not directly passed to `setZero`. Note that a postcondition such as `*x == 0` would not solve the problem: This postcondition specifies what `setZero` modifies, but gives no guarantees which memory locations are definitely *not* modified.

To address this challenge, we conceptually split the heap into two disjoint parts: One part that is operated on, and thus potentially modified by, the call `setZero(b)`, and another part that is guaranteed to be left untouched. The heap location that variable a points to should be located in the second part, which allows us to verify the assertion at the end of function `client`.

The concept of heap separation can also be used to verify the absence of data races: If we can prove that two threads operate on disjoint parts of the heap then data races cannot occur. We will discuss next how to express which locations belong to each conceptual part of the heap.

Permissions. To address the challenges mentioned above, the verifier associates a *permission* with every heap location and uses a flavor of separation logic [436] to reason about them. Permissions are a concept that supports verification, but is not reflected in the compiled code or the actual program

```
1   requires  acc(x)
2   ensures   acc(x) && *x == 0
3   func setZero(x *int) {
4       *x = 0
5   }

7   requires acc(a)
8   ensures   acc(a) && *a == 42
9   func client(a *int) {
10      *a = 42
11      m@ := 1
12      b  := &m
13      c  := b
14      setZero(b)
15      assert *a == 42
16  }
```

Figure 23.9: Permissions represent the ownership of heap locations. They enable modular verification of heap-manipulating programs. The @
annotation in the declaration of variable m indicates that m is allocated on the heap (because its address is taken in the subsequent
line).

execution. Conceptually, permissions are created when a heap location is allocated; they cannot be duplicated or forged. Permissions are held by procedure
executions and may be transferred between them. We say that a procedure
execution *owns* a heap location if it holds the associated permission.

In their simplest form, permissions are binary: Having permission to a heap
location permits read and write access, whereas accessing a heap location without holding the associated permission leads to a verification failure.

A function precondition expresses which permissions it requires from its
caller and these permissions are transferred when the call takes place. A
verification error occurs if the caller does not hold the required permissions.
Conversely, the function postconditions expresses which permissions are transferred to the caller when the function terminates.

Figure 23.9 shows the program from Figure 23.8 with the necessary permissions such that verification succeeds. acc(x) expresses permission to access
the heap location referenced by x, where x is of a pointer type. Therefore, the
precondition of function setZero makes sure that the write operation in its
body will successfully verify as the necessary permissions are present. The
postcondition not only specifies that the heap value will be set to zero but also
that the permissions are returned to a caller. Similarly, the function client
requires a permission to the memory location pointed to by a from its caller,
transfers this permission to setZero (as per setZero's precondition), obtains

it back when the call terminates (as per setZero's postcondition), and finally returns it to its own caller.

Since permissions cannot be duplicated, there is at most one permission for each heap location in each program state. Consequently, when a state contains permissions to both a and b, these variables are known to no alias, that is, a and b point to different heap locations. This property is useful to prove that a heap modification does not affect certain heap properties. Function client in Figure 23.9 illustrates this: Variable m is declared as a heap-based value, which is indicated by the @ symbol. Hence, this declaration creates the corresponding permission, such that function client owns permissions to both acc(a) and acc(&m) after line 11. When performing the call to setZero, the latter is transferred to the called function, whereas the former is retained in the execution of client. This lets the verifier conclude that the call does not affect the heap location that a points to, which is sufficient to verify the assertion after the call.

This example illustrates how we achieve the heap separation discussed above: By transferring some permissions and retaining others, we conceptually split the heap in two parts, namely those locations owned by the caller and those owned by the called function. The different parts of the heap are dynamically split and re-combined whenever permissions are transferred. Our example shows the transfers for a function call, but similar transfers occur when forking and joining threads, acquiring and releasing locks, sending and receiving messages, etc. We will discuss these cases later.

Memory Safety. Permissions are useful to reason about side effects, but they also play a crucial role in proving memory safety. Our verification technique associates a permission with each allocated heap location. Therefore, holding a permission to a location pointed to by x implies that x is different from nil. Consequently, checking permissions for each heap access prevents dereferencing a nil pointer.

Similarly, by associating a permission with each element of an array or slice, we prevent out-of-bounds accesses by checking for each access whether the corresponding permission is held by the current function execution.

Fractional Permissions. The permissions discussed so far are binary: Functions have either read and write access, or no access at all to a heap location. There are, however, use cases that require us to differentiate between read and write accesses. One such use case is concurrency, where multiple threads may safely read the same heap location without causing a data race, but binary permissions do not support that.

Another common case is functions that only read from a heap data structure. In order to be able to read, such a function requires (binary) permission to the data structure. This has the negative effect that, due to modular verification, callers must now conservatively assume that the function might also mutate

```
1   type Cell struct {
2       Value int
3   }

5   requires acc(c, 1/2)
6   ensures  acc(c, 1/2)
7   ensures  res == c.Value
8   func getValue(c *Cell) (res int) {
9       return c.Value
10  }

12  func main() {
13      cell@ := Cell{Value: 42}
14      val := getValue(&cell)
15      assert cell.Value == 42
16  }
```

Figure 23.10: Fractional permissions can be used to express the immutability of heap locations.

the structure. In some cases, one can strengthen the postcondition with extra clauses that express that the data structure is indeed unchanged. However, writing such postconditions is cumbersome and not always possible for dynamic data structures. A more powerful solution is to differentiate between read and write permissions, such that callers can conclude from the permission specification, without additional postconditions, that a function will leave a data structure unchanged.

To differentiate between read and write accesses, we allow permissions to be split into fractions [85]. Each non-zero fraction permits reading, but the full permission is required for an update. Fractions can be specified by including a *permission amount*, that is, a fraction between 0 and 1, as a second argument as in acc(x, 1/2). Permissions are split and re-combined automatically when the verifier transfers permissions.

Figure 23.10 shows a function getValue that only reads from the simple heap data structure Cell. Since this function takes only a half instead of full permission in its precondition, callers such as function main may conclude that the cell's Value field remains unchanged. Technically, when function main calls getValue, its full permission to cell.Value is split into two halves. One half is passed to the callee while the other half, together with the information that cell.Value is 42, resides in the caller. Preserving heap information across calls is often called *framing permissions and properties around a function call*. Hence, the assertion on line 15 verifies.

Abstracting over Permissions.   The acc assertion for specifying permissions to a heap location has two shortcomings: Firstly, it breaks information

```
1   type node struct {
2        value int
3        next *node
4   }

6   pred list(ptr *node) {
7        acc(&ptr.value) && acc(&ptr.next) &&
8        (ptr.next != nil ==> list(ptr.next))
9   }

11  requires list(ptr)
12  ensures  list(ptr)
13  func (ptr *node) insert(value int) {
14       unfold list(ptr)
15       if (ptr.next == nil) {
16            ptr.next = &node{value: value}
17            fold list(ptr.next)
18       } else {
19            ptr.next.insert(value)
20       }
21       fold list(ptr)
22  }
```

Figure 23.11: Predicates abstract over a data structure's memory footprint.

hiding because clients see, based on the specification, which memory locations the implementation might access. Secondly, only a statically-known number of locations can be enumerated. It is, thus, not possible to use acc predicates to specify permissions to all locations of an unbounded heap structure, such as a linked list. Both shortcomings are addressed by supporting predicates that abstract over assertions and can be thought of as representing the memory footprint of an object or data structure. Predicate instances are resources like field permissions; that is, they are held by function executions and may be transferred between functions.

To illustrate the concept of predicates, let us consider a linked list and an insert operation as shown in Figure 23.11. The predicate list specifies permissions to the struct fields of the current node and permissions to the following list elements by the recursive predicate instance list(ptr.next). Therefore, a predicate instance list(head), where head is the first element of the list, recursively contains permissions for the list's entire footprint. The method insert requires the predicate instance to obtain permissions to the entire list.

Clients of this function need to know only that the predicate exists, but not its definition, which achieves information hiding. That is, for clients, the predicate may remain *abstract*. Clients typically obtain an instance of an abstract predicate by calling a constructor, and are then able to pass it (as a resource) to functions that manipulate the list (such as `insert`). Only those functions (which are part of the list implementation), but not the client, need to know how the predicate is defined. Returning the abstract predicate to the caller via the function's postcondition enables the client to make the next call to a function of the list implementation.

The implementation of `insert` shows an important technicality of predicates: The verifier does not automatically unfold a predicate to use its definition because automatic unfolding could cause the verifier to not terminate for recursive predicates. Therefore, programmers have to explicitly unfold and fold predicates via two statements. The `unfold` statement exchanges a predicate instance for its body, and `fold` performs the opposite operation. These additional statements are required for verification, but are not present in the executed programs. In Figure 23.11, the `unfold` statement on line 14 exchanges the predicate instance `list(ptr)` by its body. This operation makes available the necessary permissions to read and write the struct fields of `ptr`. The `fold` operation on line 21 exchanges these permissions for the predicate instance `list(ptr)`. After allocating and inserting a new list node on line 16, an additional `fold` statement is necessary such that the implication in the predicate's body holds when performing the `fold` operation on line 21.

23.3 Verification of Go Programs

SCION's open-source implementation is written in Go [23], a modern systems programming language with a focus on concurrent and distributed computing. It provides an uncommon combination of features which includes mutable heap-allocated data structures, subtyping based on method availability (structural subtyping), message-passing communication via channels, and concurrency through *goroutines*, which are a form of lightweight threads managed by the Go runtime.

The Go ecosystem includes advanced tooling to test concurrent Go programs [521] and to detect common errors [211]. However, even this is not enough to prove the absence of bugs: Go programs might still suffer from implementation errors such as `nil` pointer dereferences, out-of-bound array accesses, and functional errors. This problem motivated the development of Gobra, an automated program verifier for heap-manipulating, concurrent Go programs [546]. Gobra takes as input a Go program annotated with assertions such as pre and postconditions and loop invariants. Verification proceeds by encoding the annotated programs into the intermediate verification language Viper [378] and then applying an existing SMT-based verifier. When verification fails, Gobra provides helpful error messages and identifies the lines of code that cause verification to fail, as we illustrated with the screenshots in the previous section.

The current version of Gobra supports reasoning about programs that use advanced features of Go such as mutexes, goroutines, channels, and interfaces.

Gobra verifies various kinds of safety properties, including (a) memory safety, (b) crash-freedom, (c) absence of data races, (d) functional correctness, and (e) I/O behavior (see §23.4). In the future, we plan to improve Gobra by adding support to additional properties including information flow security (see §23.5), program termination, and deadlock freedom. Furthermore, we plan to extend Gobra to support other advanced features of Go, including function types and closures, as well as interoperability with C code.

In this section, we present the basic features available in Gobra to specify Go programs (§23.3.1). We build on top of these ideas in the following sections, where we show how to specify and verify concurrent Go programs (§23.3.2) and programs that use interfaces (§23.3.3).

23.3.1 Specifying Go Programs with Gobra

Gobra provides various mechanisms to specify Go programs. In this section, we discuss *function specifications*, *predicates*, and *ghost code*.

Function Specifications

Gobra allows users to specify functions in Go with pre and postconditions, as described in the previous section. In Figure 23.12, we show the annotated version of the UnmarshallText function from the SCION codebase[2]. This function parses a SCION AS identifier from a []byte and assigns it to *as. For readability, our examples use the full syntax of Gobra, which includes syntactic constructs not available in Go. Alternatively, Gobra supports Go programs with function specifications passed via comments. The latter approach allows users to directly compile the verified Go code without erasing all Gobra annotations.

In addition to the original Go code, the annotated version in Figure 23.12 contains specifications for the functions UnmarshalText and ASFromString in the form of pre and postconditions. The specification for UnmarshallText ensures that it has write permission to the memory location as (line 5). It also requires read permission to every entry in the slice text using *quantified permissions* [377], that is, by quantifying over every valid index of the slice and requiring access to the corresponding heap location (lines 6 and 7). These preconditions are sufficient to prove that UnmarshalText is memory safe. After the execution of UnmarshallText, all permissions are transferred back to the caller (lines 8–10). Besides memory safety, we also prove the functional property that if text does not contain a valid AS identifier then the value stored in location *as remains unchanged, i.e., at the end of executing UnmarshallText, as contains the same value as in the beginning of the

---

[2]All examples from the SCION codebase in this section are based on commit ae63a60 from github.com/scionproto/scion.

```
1   package addr

3   type AS uint64

5   requires acc(as)
6   requires forall i int ::
7     0 <= i && i < len(text) ==> acc(&text[i], 1/2)
8   ensures acc(as)
9   ensures forall i int ::
10    0 <= i && i < len(text) ==> acc(&text[i], 1/2)
11  ensures !validAS(string(text)) ==> *as == old(*as)
12  func (as *AS) UnmarshalText(text []byte) error {
13    newAS, err := ASFromString(string(text))
14    if err != nil {
15      return err
16    }
17    *as = newAS
18    return nil
19  }

21  ensures validAS(s) ==> err == nil
22  ensures !validAS(s) ==> err != nil
23  func ASFromString(s string) (AS, err error) {
24    // omitted implementation
25    ...
26  }
```

Figure 23.12: UnmarshalText parses an AS identifier from a slice of bytes and stores it in as. It calls ASFromString to parse the string obtained from the slice text. If the string does not contain a valid identifier, ASFromString returns a non-nil error (line 22) and UnmarshalText returns without changing the contents of as (line 11). This snippet was taken from the annotated version of github.com/scionproto/scion/go/lib/addr/isdas.go.

method (line 11). This property is expressed using the old keyword. In general, old(e) causes all heap-dependent subexpressions of e to be evaluated in the initial state of the corresponding function call. Variables stored in the stack are not affected by the old keyword. Function parameters occurring in postconditions are always evaluated to their initial values because these are the only values that are known to the callers.

The specification of ASFromString ensures that error is nil if and only if the argument string s contains a valid autonomous system identifier (lines 21 and 22). The postconditions use the pure function validAS which returns true if and only if its argument contains a valid AS identifier. *Pure functions* are checked not to produce side-effects such as modifying heap-allocated struc-

```
1   package slayers

3   // AddrLen indicates the length of a host address
4   // in the SCION header. The four possible lengths
5   // are 4, 8, 12, or 16 bytes.
6   type AddrLen uint8

8   // AddrLen constants
9   const (
10     AddrLen4  AddrLen  = 0
11     AddrLen8  AddrLen  = 1
12     AddrLen12 AddrLen  = 2
13     AddrLen16 AddrLen  = 3
14  )

16  pure func validAddrLen(addrLen AddrLen) bool {
17     return addrLen == AddrLen4  ||
18            addrLen == AddrLen8  ||
19            addrLen == AddrLen12 ||
20            addrLen == AddrLen16
21  }
```

Figure 23.13: Declaration of constants of type `AddrLen`. The pure function `validAddrLen` checks if its argument is one of the declared constants. This code was taken from the annotated version of `github.com/scionproto/scion/go/lib/slayers/scion.go`.

tures and must be deterministic. For that reason, calls to pure functions may be used in specifications, which makes them a powerful specification mechanism.

For example, Figure 23.13 shows the declaration of the type `AddrLen`. The constants `AddrLen4`, `AddrLen8`, `AddrLen12`, and `AddrLen16` enumerate all expected values for variables of type `AddrLen` and represent the four possible lengths of host addresses in a SCION header. Because the Go compiler allows values to be assigned to variables of type `AddrLen` other than the ones that were explicitly enumerated, it is useful to define a pure function `validAddrLen` that checks whether the provided value of type `AddrLen` is one of the expected values. `validAddrLen` can then be used in specifications, for example, to restrict values of parameters of type `AddrLen` to valid values.

Predicates

In Gobra, predicates are declared in the top-level scope of a program using the `pred` keyword. A predicate may have any number of parameters; its body is an assertion using only these parameters. Predicate bodies must be *self-framing*, that is, access only those heap locations for which the predicate contains some

```
1   package router

3   type DataPlane struct {
4     internal          BatchConn
5     internalIP        net.IP
6     localIA           addr.IA
7     neighborIAs       map[uint16]addr.IA
8     internalNextHops  map[uint16]net.Addr
9     mtx               sync.Mutex
10    running           bool
11  }

13  pred DataPlaneInv(d *DataPlane) {
14    acc(&d.internal) &&
15    acc(&d.internalIP) &&
16    acc(&d.neighborIAs) &&
17    acc(&d.localIA) &&
18    acc(&d.running) &&
19    d.neighborIAs != nil ==> acc(d.neighborIAs) &&
20    d.internalNextHops != nil ==> acc(d.
          internalNextHops)
21  }
```

Figure 23.14: Declaration of the type DataPlane and the predicate
DataPlaneInv in the annotated version of github.com/
scionproto/scion/go/pkg/router/dataplane.go. For
brevity, some fields of DataPlane are omitted.

permission. Figure 23.14 shows a simplified version of the DataPlane type,
which stores information required for the SCION border router's forwarding
logic. It defines the predicate DataPlaneInv, which abstracts over the permissions to the fields of a DataPlane d, as well as the permissions to modify
the underlying maps of fields d.neighborIAs and d.internalNextHops if
they are not nil.

Ghost Code

It is often useful to add code to a program that is used for verification, but is
not required for the program's execution. The unfold and fold statements
discussed in the previous section are an example; another example is maintaining additional variables and data structures that are useful for specification and
verification. Code that is used for verification only is often called *ghost code*.

Gobra supports ghost state through ghost variables and ghost parameters.
For example, Figure 23.15 defines a function contains, which receives a slice
s and a value x, and determines whether x occurs in s. One way to express
that an element is contained in a slice is to specify that there exists an index

```
1   requires forall k int :: 0 <= k && k < len(s) ==>
2     acc(&s[k], 1/2)
3   ensures   forall k int :: 0 <= k && k < len(s) ==>
4     acc(&s[k], 1/2)
5   ensures   isContained ==>
6     0 <= idx && idx < len(s) && s[idx] == x
7   func contains(s []int, x int) (isContained bool,
8     ghost idx int) {
9     invariant 0 <= i && i <= len(s)
10    invariant forall k int ::
11      0 <= k && k < len(s) ==> acc(&s[k], 1/2)
12    for i := 0; i < len(s); i += 1 {
13      if s[i] == x {
14        return true, i
15      }
16    }
17    return false, -1
18  }
```

Figure 23.15: The function `contains` checks if a value occurs in a slice. It returns the index of the first occurrence via an out-parameter.

within the slice at which the element is stored. In the example, we use the ghost out-parameter `idx` (line 8), annotated with the keyword ghost, to store a witness for this index, allowing us to write a useful postcondition (lines 5 and 6).

Gobra also supports ghost statements and ghost functions and methods, which are used to modify ghost state. Ghost conditionals and loops are often used to perform case splits and induction in proofs, respectively. Ghost statements are usually prefixed by the keyword `ghost`, except for `assert`, `assume`, `fold`, and `unfold` statements, and calls to ghost functions.

Since ghost code is not required for the program's execution, it is erased during compilation. To make sure this erasure does not affect the behavior of a (verified) program, ghost code is not allowed to interfere with non-ghost code, for instance, by assigning to a non-ghost variable. This rule is enforced by the Gobra tool.

23.3.2 Verifying Concurrent Programs with Gobra

Go supports concurrency through goroutines, which are lightweight threads started by prefixing a function call with the go keyword. When goroutines are launched, the permissions in the function's precondition are transferred to the launched function. Unlike for regular function calls however, the caller cannot immediately assume the postconditions of the functions after the call, since the goroutine may not have terminated yet. This fact ensures that Go-

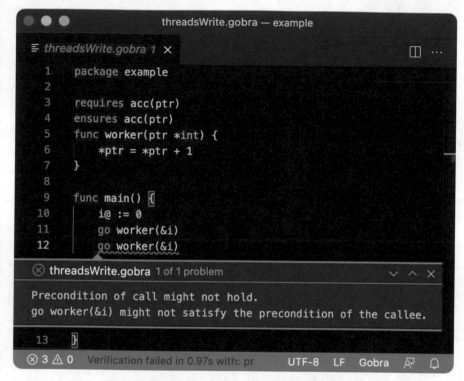

Figure 23.16: Verification fails for programs containing data races.

bra rejects programs that may suffer from race conditions. As an example, Figure 23.16 shows a program spawning two goroutines that each execute the function worker concurrently. These goroutines cause a potential data race because they update the same heap location without synchronization. This program legitimately fails to verify due to a permission error: Each worker requires the permission to the heap location referenced by parameter ptr in order to modify its content. The function main declares a local-variable i, obtaining the permission to write to this location. It then transfers this permission to the first worker on line 11. Afterwards, main has no permissions left for this heap location. Therefore, spawning the second goroutine fails with an insufficient permission error as shown in the error message.

Permission transfers from the launched function back to the caller (or any other function execution) need to be encoded via suitable synchronization operations. Go offers common synchronization primitives such as mutexes and waitgroups, but goroutines idiomatically synchronize via channels. In general, Gobra enables verification of concurrent programs by associating Go's synchronization primitives with *invariants*, which not only express properties of data but also express how permissions to shared memory get transferred between threads. For instance, mutex invariants may include properties as well as permissions to the data protected by the mutex, and channel invariants may include properties and permissions of the data sent over a channel. These invariants are specified via parameters of a *predicate type* (§23.3.2.1) to ghost operations when the synchronization primitive is initialized. The next sections demonstrate how these patterns are used to verify concurrent code that synchronizes via mutexes and via channels.

23.3.2.1 First-Class Predicates

Gobra allows passing predicates via parameters to initilization methods via so-called *first-class predicates*. Expressions denoting first-class predicates have predicate types of the form $pred(x_1\ T_1,\ \ldots,\ x_n\ T_n)$, where n is the arity of the predicate and the T_i are the corresponding parameter types. To create instances of first-class predicates, Gobra provides *predicate constructors*. A predicate constructor $P\{d_1,\ \ldots,\ d_n\}$ partially applies a declared predicate P with the arguments $d_1, \ldots, d_n$. Arguments to predicate constructors must be either an expression of the correct type or a wildcard (_), representing an unapplied argument. As an example, consider the predicate DataPlaneInv introduced in Figure 23.14. The predicate constructor DataPlaneInv{_} has type pred(*DataPlane), since the only argument of DataPlaneInv remains unaplied. Conversely, DataPlaneInv{d} has type pred() if d is an expression of type *DataPlane.

As with other predicates, first-class predicates may occur in assertions and in fold and unfold statements.

23.3.2.2 Synchronization via Mutexes

In Figure 23.16 on the preceding page, we presented a program that spawns two threads via goroutines, which concurrently modify a shared heap location via a pointer. This program contains a race condition and thus does not verify. In Figure 23.17, we provide a thread-safe version, which uses a mutex to enforce mutual exclusion and prove memory safety. The initialization of the mutex via the ghost method SetInv (line 12) associates the mutex with an invariant. This invariant, AccInv{&i}, contains a write permission to the heap location at &i to express that the mutex protects this location. This permission is transferred from the calling function to the mutex when it is initialized and then held by the mutex. Initializing a mutex mtx also produces the permission mtx.LockP(); a non-zero fraction of it is required to call methods Lock and Unlock. Therefore, these methods cannot be called on mutexes that might not have been initialized.

After initializing mtx, the main function launches two goroutines executing the function worker. This function receives a memory location ptr and an initialized mutex mtx, as specified on line 18. Moreover, line 19 requires that the mutex is associated with the invariant AccInv, which implies that the mutex holds the permission to ptr. The function worker acquires the mutex

```
1    import "sync"

3    pred AccInv(i *int) {
4      acc(i)
5    }

7    func main() {
8      i@ := 0
9      mtx := &sync.Mutex{}
10     fold AccInv{&i}()
11     // AccInv{&i}() holds here
12     ghost mtx.SetInv(AccInv{&i})
13     // AccInv{&i}() does not hold here
14     go worker(&i, mtx)
15     go worker(&i, mtx)
16   }

18   requires acc(mtx.LockP(), 1/2)
19   requires mtx.LockInv() == AccInv{ptr}
20   ensures acc(mtx.LockP(), 1/2)
21   func worker(ptr *int, mtx *sync.Mutex) {
22     mtx.Lock()
23     unfold AccInv{ptr}()
24     *ptr += 1
25     fold AccInv{ptr}()
26     mtx.Unlock()
27   }
```

Figure 23.17: Race-free version of the program from Figure 23.16. It uses
a mutex protecting the invariant iAccInv& to synchronize accesses to memory location &i.

(line 22), increments the value stored in ptr (line 24), and finally releases the
mutex (line 26). Acquiring the mutex transfers the permission from the mutex
to the acquiring function, and releasing the mutex transfers it back. Consequently, a function owns the location only while holding the mutex, which
prevents concurrent accesses and, thus, data races.

The same idea is applied in Figure 23.18 to prove the memory safety of the
method AddInternalInterface from the implementation of SCION's data
plane. This method sets the interface that the DataPlane uses to communicate
within the local AS.

23.3.2.3 Message Passing via Channels

Gobra associates channels with invariants to specify properties and permissions of the data sent over the channel. Like mutexes, the invariant is specified

```
1   requires acc(&d.mtx, 1/2) && d.mtx.LockP()
2   requires d.mtx.LockInv() == DataPlaneInv{d}
3   ensures acc(&d.mtx, 1/2) && d.mtx.LockP()
4   func (d *DataPlane) AddInternalInterface(
5     conn BatchConn, ip net.IP) error {
6     d.mtx.Lock()
7     // Gobra does not support defer statements yet
8     // defer d.mtx.Unlock()
9     unfold DataPlaneInv{d}()
10    if d.running {
11      fold DataPlaneInv{d}()
12      d.mtx.Unlock()
13      return serrors.New("modification␣not␣allowed")
14    }
15    if conn == nil {
16      fold DataPlaneInv{d}()
17      d.mtx.Unlock()
18      return serrors.New("empty␣value")
19    }
20    if d.internal != nil {
21      fold DataPlaneInv{d}()
22      d.mtx.Unlock()
23      return serrors.New("already␣set")
24    }
25    d.internal = conn
26    d.internalIP = ip
27    fold DataPlaneInv{d}()
28    d.mtx.Unlock()
29    return nil
30  }
```

Figure 23.18: AddInternalInterface sets the interface that the DataPlane
uses to communicate within the local AS. This code was taken
from the annotated version of github.com/scionproto/
scion/go/pkg/router/dataplane.go. In this snippet, the
defer statement in line 8 was replaced by repeating the call to
d.mtx.Unlock in line 12, line 17, line 22, and line 28 because
Gobra does not yet support deferred statements.

during initialization, as an argument of the Init method. For synchronous
channels, an additional predicate can specify permissions transferred in the opposite direction, from the receiver to the sender. Initializing a channel also
creates send and receive permissions for the channel, which are used to control
which threads may access it.

We demonstrate Gobra's support for channels in Figure 23.19. In this example, a thread allocates an unbuffered channel c and initializes it with the

```
1  pred sendInvariant(v *int) {
2    acc(v) && *v > 0
3  }

5  func main() {
6    var c@ = make(chan *int)

8    var x@ int = 42
9    var p *int = &x
10   c.Init(sendInvariant{_})
11   go worker(c)
12   assert *p == 42
13   fold sendInvariant{_}(p)
14   c <- p

16   res, ok := <- c
17   if (ok) {
18     unfold sendInvariant{_}(res)
19     assert *res > 0
20     // regained write access
21     *res = 1
22   }
23 }

25 requires acc(c.SendChannel(), 1/2)
26 requires acc(c.RecvChannel(), 1/2)
27 requires c.SendGivenPerm() == sendInvariant{_}
28 func worker(c chan *int) {
29   res, ok := <- c
30   if (ok) {
31     unfold sendInvariant{}(res)
32     // obtained write access to res after
33     // unfolding the invariant
34     *res = *res + 1
35     // fold the invariant and send pointer
36     // and permission back
37     fold sendInvariant{}(res)
38     c <- res
39   }
40 }
```

Figure 23.19: Communication of a pointer and corresponding write permission via a channel. For simplicity, some details were omitted.

sendInvariant predicate (line 10). This invariant contains a permission to the heap location pointed to by v, as well as the constraint that the value in this location must be positive. Therefore, each send operation (as in line 14) checks this constraint and transfers the permission for the sent pointer from the

```
1  type AbstractCell interface {
2    getCell() int
3    setCell(v int)
4  }
```

Figure 23.20: Declaration of type `AbstractCell` in Go, which represents a
memory cell that stores an integer. The method `getCell` returns
the value currently stored in the cell and `setCell` updates it.

```
1  func swap(c1 AbstractCell, c2 AbstractCell) {
2    tmp := c1.getCell()
3    c1.setCell(c2.getCell())
4    c2.setCell(tmp)
5  }
```

Figure 23.21: Definition of the polymorphic function `swap` in Go, which swaps
the contents of two `AbstractCell` regardless of their implementation.

sender to the sent message. Conversely, the receiver of the message obtains the
permission, together with the information that the stored value is positive. In
our example, this allows the receiver to access `*res` in line 34. Similarly, the
pointer `res` and its corresponding permission are sent back to the first thread
in line 38, which allows it to access `*res` and prove the assertion in line 19.

23.3.3 Interfaces in Gobra

Go interfaces allow programmers to write functions that operate on different
data types, a feature known as *polymorphism*. Consider the declaration for
the interface type `AbstractCell` in Figure 23.20, representing a memory cell
that stores an integer. The interface has two methods, `getCell` and `setCell`,
which return the value currently stored in the AbstractCell and update it, respectively.

Having defined the type `AbstractCell`, one can define the function `swap`
shown in Figure 23.21, which swaps the contents of two arbitrary values of
type `AbstractCell`, independently of their actual implementation.

In Gobra, interface methods can contain specifications just like other methods. A function specification on an interface method describes the behavior of
every allowed implementation of the respective interface type. In Figure 23.22,
we provide the annotated declaration of type `AbstractCell`. Given that the
concrete implementations of a data structure are not known in the interface declaration, the specifications of interface methods cannot refer to heap locations
directly. Instead, they typically introduce a predicate (such as `mem` in our ex-

```
1   type AbstractCell interface {
2      pred mem()

4      requires acc(mem(), 1/2)
5      pure getCell() int

7      requires mem()
8      ensures mem()
9      ensures getCell() == v
10     setCell(v int)
11  }
```

Figure 23.22: Specified version of the type `AbstractCell` shown in Figure 23.20. It contains specifications for the interface methods, which must be satisfied by every implementation of this type. Additionally, it contains a predicate `mem` which abstracts over the heap structure of any concrete implementation.

```
1   requires c1.mem() && c2.mem()
2   ensures c1.mem() && c2.mem()
3   ensures c1.getCell() == old(c2.getCell())
4   ensures c2.getCell() == old(c1.getCell())
5   func swap(c1 AbstractCell, c2 AbstractCell) {
6      tmp := c1.getCell()
7      c1.setCell(c2.getCell())
8      c2.setCell(tmp)
9   }
```

Figure 23.23: Definition of the polymorphic function `swap` in Gobra. Its specification only mentions methods and predicates declared in the interface `AbstractCell`. Thus, it is compatible with every implementation of `AbstractCell`.

ample, see line 2) to abstract over the concrete implementation. This predicate is left abstract in the interface, that is, it has no body. Its definition is provided by the concrete implementations of the interface.

Using the predicate `mem`, we can now specify that the method `getCell` needs read permission to the heap locations of the cell implementation (line 4). We also introduce a specification for the method `setCell`, which requires a write permission to these heap locations (line 7). Furthermore, we use the pure method `getCell` in the specification of `setCell` to express that, after the execution, the value stored in the cell is now `v` (line 9). The interface method specifications let us verify modularly—i.e., without knowing the concrete interface implementations—that the `swap` function shown in Figure 23.23 is correct.

```
1  type BackupCell struct {
2    cell int
3    backup int
4  }

6  pred (b *BackupCell) mem() {
7    acc(b)
8  }

10  requires acc(c.mem(), 1/2)
11  ensures unfolding c.mem() in res == c.cell
12  pure func (c *BackupCell) getCell() (res int){
13    return unfolding c.mem() in c.cell
14  }

16  requires acc(c.mem(), 1/2)
17  pure func (c *BackupCell) getBackup() (res int){
18    return unfolding c.mem() in c.backup
19  }

21  requires c.mem()
22  ensures c.mem()
23  ensures c.getCell() == v
24  func (c *BackupCell) setCell(v int) {
25    unfold c.mem()
26    c.backup = c.cell
27    c.cell = v
28    fold c.mem()
29  }
```

Figure 23.24: Implementation of type BackupCell in Gobra, which stores the current value of the cell and the value that it held before. It implements the type AbstractCell.

In Gobra, in order for a type to implement an interface I, it must provide an implementation of every method and predicate defined in I. Figure 23.24 shows the definition of BackupCell in Gobra, which stores two integers: the current value of the cell and the value held before the last update. This type implements all methods and predicates of AbstractCell. Notice that the definitions of the methods getCell and getBackup use an unfolding expression, which is used to temporarily unfold a predicate instance for the evaluation of an expression. For example, unfolding c.mem() in res == c.cell (line 11) unfolds the predicate c.mem() to evaluate res == c.cell. Without unfolding c.mem(), this postcondition would be ill-formed because there would be no permission to access the memory location c.cell. unfolding expressions are especially useful in contexts where statements are not allowed such as within specifications and pure functions' bodies.

As we explained earlier, each concrete implementation of an interface must satisfy the interface specification. For this purpose, Gobra requires an *implementation proof* whenever a Go program assigns a value to a variable of an interface type. This implementation proof shows that each method of the implementation type satisfies the specification of the corresponding method in the interface. This means that, for each method, the precondition of the interface method must entail the precondition of the implementation method and the postcondition of the implementation method must entail the postcondition of the interface method. Implementation proofs are inferred automatically by Gobra in simple cases. For complex cases that include ghost operations to manipulate predicates, the user must explicitly provide implementation proofs.

23.4 Verification of Protocol Implementations

Code verification proves that an implementation behaves as intended. Some of the intended behaviors, such as memory safety, race freedom, and deadlock freedom, apply uniformly to all implementations. Checks for those are built into our verification technique and applied to all programs. Others properties, such as functional correctness, are specific to each program and require explicit specifications to express what the intended behavior is. For SCION, the main correctness property of the code is to implement the SCION protocol correctly.

To implement the protocol of a distributed system correctly, each system component must perform the I/O operations prescribed by the protocol. For instance, the core functionality of a SCION router is to receive packets and forward them to the next router on the pre-determined path. In this section, we extend our specification and verification technique to reason about the I/O behavior of programs, and we explain how we extract the specification of the SCION implementation from the formal model of the SCION protocol presented in §22.4.

23.4.1 Specification and Verification of I/O Behavior

Our goal is to prove that each component of the SCION implementation performs only the I/O operations that are permitted according to the SCION protocol. To express which I/O operations a program is permitted to perform, we generalize the concept of permissions that we introduced in §23.2.4. Whereas the permissions discussed there guard memory accesses, we now introduce *I/O permissions* to guard the execution of I/O operations. An I/O permission expresses the right to perform a specific I/O operation (such as sending a message) with specific arguments (constraining the content of the message). Verification then checks that each component performs only those I/O operations it has permissions for, which prevents, for instance, sending messages that are not permitted by the protocol, manipulating the message headers in ways not foreseen by the protocol, or changing the payload. Note that we prove that all I/O operations performed by the implementation are actually permitted by the protocol, which is the essential correctness property of a SCION component. Proving the converse, that each I/O operation in the protocol is actually performed by the implementation, is left for future work.

We build on the I/O specification language by Penninckx et al. [408]. For each relevant I/O operation in Go's I/O library, we define a corresponding I/O permission, expressed as an abstract predicate. For instance, for a send operation, we define the abstract permission predicate send(...). The arguments of the predicate are the in- and out-parameters of the I/O operation. For send, the parameters could be the message payload, the target address, and the target port. For simplicity, we will often omit the address and port. Additionally to the in- and out-parameters, the first and last argument of the I/O permission is a so-called start and end *place*, respectively. A place is an abstract position in a sequence of I/O operations. We use the start and end places of I/O permissions to specify in which order I/O permissions can be processed. A token permission token(p) tracks the current place p. When an I/O operation is performed, the token changes from the start to the end place of the processed I/O permission.

The following I/O specification allows a method to receive an integer x and subsequently send x+1:

```
requires  token(p)
requires  recv(p, ?x, ?p1) && send(p1, x+1, ?p2)
```

The first precondition expresses that the method requires the token to be at place p. The method is then allowed to receive *some* value x, thereby advancing the token to *some* place p1, and then to send x+1 (advancing the token to p2). Variables starting with question marks are implicitly existentially quantified to abstract over the a-priori unknown received value and the end places.

To check that a program performs only I/O operations for which it has permission, we equip each method of the I/O library with specifications that require the corresponding permission. Hence, making a call in a state in which this permission is not available leads to a verification failure. The following specification illustrates a simplified contract for a send library method. Variables starting with question marks are implicitly existentially quantified.

```
requires  token(?p1) && send(p1, msg, ?p2)
ensures   ok  ==> token(p2)
ensures   !ok ==> token(p1) && send(p1, msg, p2)
func send(msg int) (ok bool)
```

With an analogous contract for receive, we can prove the specification of the following function:

```
requires token(p)
requires recv(p, ?x, ?p1) && send(p1, x+1, ?p2)
ensures   ok  ==> token(p2)
func demo() (ok bool) {
  x, ok := recv();
  if ok {
    ok = send(x+1)
  }
}
```

The I/O behavior of possibly non-terminating programs, such as a router, are specified by including the required I/O permissions in a corecursive predicate. For instance, the following predicate permits a program to receive a value; if that value is positive, the program may send the largest value received so far and repeat the process:

```
pred SendMax(p place, largestSoFar int) {
  recv(p, ?x, ?p1) &&
  (0 < x ==> send(p1, max(largestSoFar, x), ?p2) &&
             SendMax(p2, max(largestSoFar, x))
}
```

The above specification and verification technique for I/O behavior can easily be supported by any code verifier that supports abstract permissions and corecursive predicates. To fully integrate it into Gobra, we must still equip Go's I/O library with the required specifications and make minor adjustments to the verifier—e.g., to permit predicates that can be unrolled indefinitely, such as SendMax above.

23.4.2 From Protocol Models to I/O Specifications

Now that we have a technique to specify and verify the I/O behavior of a program, we can explain how we systematically derive an I/O specification from an abstract protocol model, to provide formal correctness guarantees all the way from the high-level protocol down to a concrete implementation. This systematic derivation consists of three main steps:

1. We decompose the monolithic protocol model into an environment model and a system model. The system model is then further decomposed into models for each component. Soundness of our approach requires that the environment model faithfully represents the real environment.

2. We map each component model to an I/O specification.

3. We verify the implementations of each component against its I/O specification.

To illustrate our approach, we continue the authentication protocol example from §22.3. So far, we have refined an abstract protocol model $\mathcal{E}_a$ down to a concrete protocol model $\mathcal{E}_c$. In the remainder of this subsection, we will show how we connect $\mathcal{E}_c$ to implementations of the initiator and the responder. Details of our approach, including a formalization and soundness argument, are presented by Sprenger et al. [486].

Interface Model. To prepare the concrete protocol model for the subsequent decomposition, we refine it to an *interface protocol model*. This refinement, like the initial development of the protocol model, is performed and proved correct within Isabelle/HOL. The interface model establishes the structural requirements of the decomposition and introduces I/O events that can be mapped to I/O library functions, in our example, UDP send and receive.

In the interface model, we split the overall state into a system state $state_S$ and an environment state $state_E$. We make the corresponding change to the events: Events of the protocol model may combine multiple operations, both modifying system state and interacting with the environment state. The events are split into *internal events*, which affect only the system state, and *I/O events*, which connect the system and the environment.

In our running example, we identify the network with the attacker's knowledge IK, reflecting that the attacker can observe every message that is sent on the network. Consequently, the environment state consists of only the attacker's knowledge. The system state extends the abstract state record with two functions that map run identifiers to the run's input and output buffer, storing the messages a component has received and sent, respectively. The input buffer additionally stores the sender's connection information used for replies.

$$state_{itf} = state_E \times state_S$$
$$state_E = (\!| IK \in \mathbb{P}(\mathbb{T}) |\!)$$
$$state_S = state_a + (\!| ibuf : \mathcal{R} \to \mathbb{P}(Connection \times \mathbb{T}), obuf : \mathcal{R} \to \mathbb{P}(\mathbb{T}) |\!)$$

Figure 23.25 shows a part of the interface model event system resulting from the refinement of the concrete protocol model (Figure 22.5). The *resp_recv* event of the concrete protocol model is split into the internal event *resp_get* and the I/O event *UDP_recv*. The argument a corresponds to the address and port of a received message and was added to align the I/O event with an I/O library function. Similarly, the previous *resp_send* event is split into *resp_put* and *UDP_send*. All other events are omitted. The internal events *resp_get* and *resp_put* are almost identical to *resp_recv* and *resp_send*, respectively, except that the attacker's knowledge is replaced with input and output buffers to make the events operate on system state only. The I/O events establish the connection between the buffers and the environment network, i.e., the attacker's knowledge. To reflect that communication via UDP may duplicate, reorder, or

$$UDP\_recv(r,m,a): \ m \in IK \ \triangleright \ ibuf := ibuf(r \cup= \{(a,m)\})$$

$$UDP\_send(r,m): \ m \in obuf(r) \ \triangleright \ IK := IK \cup \{m\}$$

$resp\_get(r_B,B,A,N_B,N_A):$
$\quad runs_{\mathsf{Resp}}(r_B) = \langle B \rangle \wedge N_B = \#[\mathsf{Resp},r_B,0] \wedge \langle A,B,N_A \rangle \in ibuf(r_B)$
$\quad \triangleright \ runs_{\mathsf{Resp}} := runs_{\mathsf{Resp}}(r_B := \langle B,A,N_A \rangle)$

$resp\_put(r_B,B,A,N_B,N_A):$
$\quad runs_{\mathsf{Init}}(r_B) = \langle B,A,N_A \rangle \wedge N_B = \#[\mathsf{Resp},r_B,0]$
$\quad \triangleright \ obuf := obuf(r_B \cup= \{[N_B,N_A,A]_{\mathsf{pri}(B)}\})$

Figure 23.25: Part of the event system of the interface protocol model. We use $\cup=$ to denote that a set is updated by adding the set on the right-hand side, i.e., $m(x \cup= S)$ is a shorthand for $m(x := m(x) \cup S)$.

$$UDP\_recv_{\mathsf{E}}(r,m,a): \ m \in IK \ \triangleright \ -$$
$$UDP\_send_{\mathsf{E}}(r,a,m): \ true \ \triangleright \ IK := IK \cup \{m\}$$
$$UDP\_recv_{\mathsf{S},r}(m,a): \ true \ \triangleright \ ibuf := ibuf(r \cup= \{(a,m)\})$$
$$UDP\_send_{\mathsf{S},r}(a,m): \ m \in obuf(r) \ \triangleright \ -$$

Figure 23.26: The I/O events of the decomposed models.

lose. messages, we model the network as sets of messages. Semantically, we must justify that the I/O library functions satisfy the environment model. In our current approach, this is part of the trust assumptions and not verified.

Decomposition. The decomposition splits the monolithic interface model into the parallel composition of an environment model and a system model, which is further decomposed in the models for each component. The events of the decomposed models only access their respective state. In particular, environment events do not involve system state and component events do not involve the attacker's knowledge. The decompositions are refinement steps in Isabelle/HOL that, like previous refinement steps, preserve all safety properties proved for the more abstract protocol model. For each component model, such as the initiator and responder in our example, there may be an unbounded number of instances.

Figure 23.26 shows the I/O events of the resulting event systems. The environment event system consists of only the events UDP_recv_E and UDP_send_E, as well as the attacker and trivial skip event (the latter two are not shown in the figure). The initiator and responder model have separate versions of the I/O events $UDP\_recv_{S,r}$ and $UDP\_send_{S,r}$, only accessing the component state, that synchronize with the corresponding I/O events of the environment. Furthermore, the component models consist of the internal events (not shown in the figure). The state of the initiator and responder model is the record $(\!| run: store, ibuf: \mathbb{P}(Connection \times \mathbb{T}), obuf: \mathbb{P}(\mathbb{T}) |\!)$. Generally, different component models can have different states.

Mapping Component Models to I/O Specifications. We establish a formal connection between the component model and the implementation by mapping the component model event system $\mathcal{E}(\overline{p})$ to the I/O specification token(?t) && P(t, $\overline{p}$, s_0) where $\overline{p}$ are the parameters and s_0 is the initial state of the event system. The predicate P contains for each event and all values satisfying the event's guard a permission to execute the I/O operation corresponding to the event and another instance of P with an updated state argument. The new state is the result of applying the event's update function to the old state argument of P.

Figure 23.27 shows the predicate P for the responder model of the authentication protocol. The parameters of the event system $\overline{p}$ are the run identifier r, the role B, and the private key pri(B). The state argument s is the responder's component state represented by the type State. The fields of the State structure use Gobra's mathematical sets to represent the types of the model state on the code level. The four events UDP_recv, UDP_send, $resp\_get$, and $resp\_put$ are each mapped to one conjunct. In the conjunct for the event UDP_recv (line 8), the state argument, as in the component model, is updated such that the input buffer contains the received connection info a and message m. In the conjunct of the event UDP_send (lines 9 and 10), the permission to execute the send operation is guarded by the fact that the sent message is contained in the output buffer. All other conjuncts are analogous.

The I/O permissions provided by an instance of P often depend on P's state argument s. For instance, sending a message m requires an instance P(t, ...,s), token(t), and the knowledge that m is contained in the output buffer s.obuf. To prove such conditions during code verification, we relate the value of P's state argument to the program state via suitable invariants. For instance, an implementation maintaining a queue Q of messages to be sent may relate the state of the queue to P's state argument with the invariant acc(Q) && token(?t) && P(t, ..., s) && s.obuf == toSet(Q), where toSet abstracts the content of a queue to a set. The invariants specify that the value of the state argument's output buffer is equal to an abstraction of the queue's content. Using such an invariant, if a message is taken from

```
1   ghost  type  State  struct{
2      run    Store
3      ibuf  set[(Connection,  Term)]
4      obuf  set[Term]
5   }

7   pred  P(t,p̄,s  State)  {
8      (UDP_recv(t,?m,?a,?t')  &&  P(t',p̄,s[ibuf+={(a,m)}]))
9   &&  (forall  m,a  ::  m  in  s.obuf
10            ==>  UDP_send(t,a,m,?t')  &&  P(t',p̄,s))
11  &&  (forall  A,N_A,N_B,a  ::  N_B  ==  nonce(r,  0)
12         &&  s.run  ==  ⟨B⟩  &&  (a,⟨A,B,N_A⟩)  in  s.ibuf
13            ==>  resp_get(t,B,A,N_B,N_A,?t')
14            &&  P(t',p̄,s[run:=⟨B,A,N_A⟩]))
15  &&  (forall  A,N_A,N_B  ::  N_B  ==  nonce(r,  0)
16         &&  s.run  ==  ⟨B,A,N_B⟩
17            ==>  resp_put(t,B,A,N_B,N_A,?t')
18            &&  P(t',p̄,s[obuf+={[N_B,N_A,A]_pri(B)}]))
19  &&  ...
20  }
```

Figure 23.27: I/O specification of the responder. Each conjunct of the predicate originates from an event of the component's event system. The I/O permissions of a conjunct are often guarded by facts about the state argument s. The type State represents the state of the component's event system. s[f := e] is a field update. Similarly, s[f += e] is a field update adding the set e to the value of f. nonce(r,0) corresponds to the fresh value #[Resp, r, 0]. We use syntactic sugar for message terms.

the queue, we can guarantee ownership of the permissions required to send the message.

Code Verification. Once we have obtained an I/O specification that reflects the component model's behavior, we can prove that the implementation satisfies this specification, as explained in §23.4.1. This final step of our verification process ensures that any I/O operation performed by the implementation is permitted by the protocol, which in turn guarantees that all safety properties proved for the design models also hold for the implementation. Proving liveness, that is, that the implementation always makes progress, is future work.

Figure 23.28 shows a simplified excerpt of the verification of a responder implementation. The method takes the role identifier, the private key, and the generated nonce as arguments. Furthermore, it takes the run identifier and the model state as ghost arguments. Recall that ghost arguments are present only for verification purposes, but not during the program execution.

```
1   requires token(?t) && P(t, r, B, key, s)
2   requires s.run == ⟨B⟩
3   requires N_B == nonce(r, 0)
4   ensures true
5   func code(B, key, N_B, ghost r, ghost s) (ok bool) {

7       request, address, ok := receive()
8       if !ok { return }

10      // checks that request is equal to ⟨A,B,N_A⟩
11      A, N_A, ok := parse(request, B)
12      if !ok { return }

14      // takes token(?t') && resp_get(t',B,A,N_B,N_A,?t")
15      // and returns the advanced token(t")
16      // store of state argument is updated to ⟨B,A,N_A⟩
17      ghost exec_resp_get(B, A, N_B, N_A)

19      reply := sign(pair(N_B, N_A, a), key)

21      // takes token(?t") && resp_put(t",B,A,N_B,N_A,?t"')
22      // and returns the advanced token(t"')
23      // reply is added to obuf
24      ghost exec_resp_put(B, A, N_B, N_A)

26      ok = send(address, reply)
27  }
```

Figure 23.28: Excerpt of the verification of a responder implementation. The specification ensures that any I/O operation performed by the implementation is permitted by the protocol. The expression $nonce(r,0)$ returns the fresh value $\#[Resp,r,0]$. For simplicity, we assume that send and receive operations work directly on cryptographic message terms. In the code, we use syntactic sugar for message terms.

We simplified the implementation for illustration purposes: We omit parameter types as well as folding and unfolding instances of predicate P (and, thus, the use of P's state argument). Moreover, we assume that send and receive operations work directly on cryptographic message terms. In a more realistic implementation, messages would be represented by byte arrays, which can be handled by our methodology [486].

In the following, we explain the verification of the example, step by step. This simple implementation does not maintain any concrete state and, thus, does not require invariants to relate the concrete and the model state. The invoked I/O operations directly manipulate the model state, in particular, the state argument of the predicate P.

According to its precondition (lines 1–3), the method code initially owns token(t) and P(t,r,B,key,s) where s.run == ⟨B⟩. An I/O operation changes the state argument in two steps, which we illustrate on the receive operation (line 7): First, we acquire the necessary I/O permissions. We unfold P(t,r,B,key,s) to get UDP_receive(t, request, address, t′) and P(t′,r,B,key,s[ibuf += (request,address)]). At this point, the place argument of the token and P is different, namely t and t′, respectively. Next, we process the I/O statement. A successful receive consumes token(t) as well as the receive permission and produces the advanced token(t′). Advancing the token makes the place argument of the token and P equal again. In conclusion, after the successful receive operation (line 10), the state argument was updated from s to s[ibuf += (request,address)]. The reasoning is analogous for the subsequent I/O operations.

The next executed I/O operation is the send at line 26. To get the send permission from P, the sent message must be contained in the state argument's output buffer. From the successful parsing (line 11) and the signature (line 19), we know that the sent message is $[N_B, N_A, A]_{key}$. Since this message is not contained in the output buffer of the state argument, we perform suitable updates of the model state via two auxiliary internal operations. Both operations are present in the component model (see Figure 23.25); on the code level, they are used for verification only, that is, ghost operations. The internal operation exec_resp_get (line 17) updates the store of the state argument to ⟨B,A,N_A⟩. The operation requires the resp_get permission, which is guarded in P by the facts (a,⟨A,B,N_A⟩) in s′.ibuf and s′.run == ⟨B⟩, where s′ is s[ibuf += (request,address)]. We get the former from the successful parsing, entailing request is equal to ⟨A,B,N_A⟩, and the latter from s.run == ⟨B⟩. Finally, the internal operation exec_resp_put (line 24) adds the message $[N_B, N_A, A]_{key}$ to the state argument's output buffer, after which we can get the send permission for the send operation at line 26. The operation exec_resp_put requires the resp_put permission, which we get directly from P with our knowledge of the state argument's store.

As we have seen, the guards in P enforce that the implementation conforms to the protocol to execute I/O operations. For instance, without checking that the received message is equal to ⟨A,B,N_A⟩, subsequent I/O operations cannot be verified. Similarly, the update of the state argument's output buffer prevents the implementation from sending messages other than $[N_B, N_A, A]_{key}$.

23.5 Secure Information Flow

Our combination of design-level and code-level verification ensures that the SCION protocol satisfies a wide range of security properties and that these properties carry over to the protocol implementation. However, it does not exclude security vulnerabilities that manifest themselves only on the code level. While many code-level vulnerabilities, such as buffer overruns, are prevented by the choice of a safe programming language or detected during the verification of memory safety, others are not detected by the verification approach described so far.

A particularly prominent code-level vulnerability is the leakage of secret information, such as cryptographic keys. An implementation may leak secrets directly (for instance, by sending them over the network or printing them on the screen), indirectly (for instance, by performing output that depends on secret information), or via side channels (for instance, where the execution time depends on secret information). In each case, an attacker can obtain secret information by observing the I/O behavior of the system and then use this information to mount an attack.

In this section, we explain how we can verify secure information flow, that is, that an implementation does not leak secrets. We will focus on the core ideas here and refer to our research papers for the treatment of advanced topics such as secure-information-flow verification in concurrent systems [176, 177]. The techniques described in the following have been developed in the context of the Python verifier Nagini [175]. Integrating them in Gobra and applying them to the SCION code base is ongoing work.

23.5.1 Non-interference

Information-flow security is often phrased as a non-interference property: Assume that we classify all variables of a program as either public (called *low*) or secret (*high*). Then, the program is *information-flow secure* if the values of the high variables do not affect the values of low variables. More formally, consider two executions of the program, where the initial states agree on all low variables, but may differ in the high variables. Then the final states after the executions also agree on all low variables.

The example in Figure 23.29 illustrates this concept. Assuming that low is low and high is high, the program snippet is not information-flow secure because the final value of low depends on the initial value of the secret high. So by observing the value of low, an attacker can obtain information about the (high) secret.

We assume that an attacker cannot directly observe the variables of a program, but can observe the program's I/O behavior (we will discuss other observable properties such as termination below). Once we have verified that a program satisfies the non-interference property, we can easily prevent it from leaking secret information via I/O operations via additional verification checks, for instance, to ensure that the arguments to a print function are low.

Non-interference is a so-called *hyperproperty*, that is, a property that relates multiple (here, two) executions of a program. Hyperproperties pose major challenges for quality assurance because they cannot be checked with standard

```
1  if high > 0 {
2     low = 0
3  } else {
4     low = 1
5  }
```

Figure 23.29: A conditional statement where the value of the secret variable high influences the final value of the public variable low.

```
1  // copy one
2  if high₁ > 0 {
3     low₁ = 0
4  } else {
5     low₁ = 1
6  }
7  // copy two
8  if high₂ > 0 {
9     low₂ = 0
10 } else {
11    low₂ = 1
12 }
```

Figure 23.30: Applying self-composition to the example from Figure 23.29. Since the two copies of high may have different initial values, the two copies of low may have different final values and, thus, non-interference fails to verify.

testing approaches. A simple program like the one in Figure 23.29 could be tested for non-interference by executing it multiple times with different values for high and comparing the results for low. However, this approach does not work for the common cases of non-deterministic and concurrent programs, where differences in the output might be caused, for instance, by different thread schedules, even for the same secret values.

Hyperproperties are also challenging for standard program verification techniques, which are able to prove properties of a single program execution. To be able to apply existing program verifiers such as Gobra to prove information-flow security, we construct a new program that simulates two executions of the original program. This constructed program will never be executed, but its correctness implies information-flow security of the original program. We explain the construction in the next subsection.

```
1  res := 1
2  for 0 < n {
3     res = res * n
4     n = n - 1
5  }
```

Figure 23.31: A factorial statement. The resulting value of res is low if the
initial value of n is low.

23.5.2 Self-Composition

A straightforward way to simulate two executions of a program is to duplicate all program variables and to sequentially execute the program twice, once
on each copy of the variables. For the example from Figure 23.29, this *self-
composition* [53] introduces the variables low_1 and low_2, as well as $high_1$
and $high_2$, and leads to the program in Figure 23.30.

One can easily see that the self-composed program indeed simulates two executions of the original program. Non-interference of the original program can
now be expressed easily on a single execution of the self-composed program:
if low_1 == low_2 in the initial state (the initial states of the two program executions agree on the low variables) then low_1 == low_2 in the final state (the final
states of the two program executions agree on the low variables) irrespective
of the initial values of $high_1$ and $high_2$. This condition is easy to check using
a standard program verifier such as Gobra; this check fails for our example, as
expected.

To express information-flow properties, we equip our specification language
with an assertion $low(e)$, which expresses that e evaluates to the same value
in both executions of the program. For instance, a postcondition low(low)
in the original program is desugared as low_1 == low_2 on the self-composed
program (and would fail to verify, as explained above). We use these specifications to express which variables (or expressions) are expected to contain
non-secret information. Variable names (such as high and low) are used for illustration purposes to indicate the intention of the programmer, but do not carry
any specific meaning. Hence, low(low) is *not* a tautology, but expresses that
the variable low, which is *intended* to contain non-secret information, indeed
contains no secret in the current state.

Self-composition works well for very simple programs, but has severe shortcomings for programs with loops and calls. The factorial example in Figure 23.31 demonstrates the problem. Assume we would like to prove the postcondition low(res) assuming the precondition low(n). The self-composed
version of the program contains two loops. However, even though low(res)
holds throughout the execution of the original program, it is not a valid loop
invariant for either of the two loops: Since the self-composed program first
computes res_1 and only later res_2, the two copies of the variable are out

of sync between the initial and the final program states. Consequently, the only way to verify the desired postcondition is to provide a specification that *uniquely* determines the final values of res_1 and res_2 in terms of the initial values of n_1 and n_2, respectively. For instance, if we verified that res_i contains the factorial of the initial value of n_i, we could use this information to derive $res_1 == res_2$ after both loops from $n_1 == n_2$ before the loops. However, verifying such a strong specification is unnecessary and often undesirable, for instance, when method specifications abstract over implementation details.

To address this problem, we replace self-composition by another program construction that also simulates two executions of the original program, but aligns the two executions closely. In particular, this construction does not duplicate loops and calls, which allows one to verify loop invariants and function specifications that express that variables are low. These specifications allow us to verify secure information flow modularly. Hence, we have dubbed the program construction *modular product programs*.

23.5.3 Modular Product Programs

The key idea of modular product programs is to track via two Boolean ghost variables p_1 and p_2 which of the two executions of the original program are active at any program point. These *activation variables* allow us to encode the original program's control flow without duplicating its control structures. Intuitively, if the original program contains a loop whose condition depends on a secret then one execution might continue iterating whereas the other execution might terminate. This behavior is reflected in the product program by *potentially different* values for p_1 and p_2 in the loop body. On the other hand, if the control flow does not depend on a secret then p_1 and p_2 definitely contain the same value.

During the construction of a modular product program, primitive statements such as assignments are duplicated, but each copy of the statement is executed only if the corresponding execution is active. For instance, the statement `res := 1` gets encoded as follows:

```
1  if p₁ { res₁ := 1 }
2  if p₂ { res₂ := 1 }
```

A conditional statement `if b { ST } else { SE }` is encoded by introducing new activation variables that reflect which execution is active in which branch (since b may evaluate to different values in the two executions). That is, we introduce four fresh activation variables

```
1  pt₁ := p₁ && b₁  // execution 1 takes then-branch
2  pt₂ := p₂ && b₂  // execution 2 takes then-branch
3  pf₁ := p₁ && !b₁  // execution 1 takes else-branch
4  pf₂ := p₂ && !b₂  // execution 2 takes else-branch
```

```
1  for p₁ && b₁ || p₂ && b₂ {
2    pt₁ := p₁ && b₁ // execution 1 enters the loop
3    pt₂ := p₂ && b₂ // execution 2 enters the loop
4    // encode S with activation vars pt₁ and pt₂
5    ...
6  }
```

Figure 23.32: Activation variables reflect which executions are active.

and then encode the then-branch ST using the activation variables pt_1 and pt_2 (instead of p_1 and p_2), and the else-branch SE with pf_1 and pf_2.

A loop for b { S } of the original program is encoded by a single loop in the product program, where activation variables reflect which executions are active:

Since activation variables reflect which of the two program executions are active at any point in the execution of the product program, we need to adjust the encoding of $low(e)$ in specifications. The two copies of expression e must evaluate to the same value only if both executions are currently active. That is, $low(e)$ is encoded as p_1 && p_2 ==> e_1 == e_2.

Figure 23.33 shows the full encoding of our factorial examples from Figure 23.31. Since the loops of the two executions are closely aligned in the product program, we can specify the simple loop invariant low(n) && low(res) to prove the desired postcondition low(res).

The encoding of calls is similar: Each call in the original program leads to only one call in the product program; the call in the product program takes the current activation variables as parameters and executes the body of the called function only for the currently active execution(s). Analogously to loops, this encoding avoids duplicating calls and, thus, allows function specifications to express which parameters and results are low.

23.5.4 Advanced Topics

The modular product programs presented in the previous subsection allow one to verify basic secure information flow for programs consisting of primitive statements, conditionals, loops, and calls. However, modular product programs are not limited to simple programs and properties. We summarize some advanced features here and refer the reader to the literature for details [176, 177].

Heap Data Structures. Our examples so far used variables of primitive types such as integers. Structs can be encoded in two different ways: either as two instances of the struct, each with the same fields, or as only one instance, which contains two copies of each field. Both options are possible for stack-allocated variables, but there is a crucial difference for heap-allocated

```
1   if p₁ { res₁ := 1 }
2   if p₂ { res₂ := 1 }
3   for p₁ && 0 < n₁ || p₂ && 0 < n₂ {
4     pt₁ := p₁ && 0 < n₁
5     pt₂ := p₂ && 0 < n₂

7     if pt₁ { res₁ = res₁ * n₁ }
8     if pt₂ { res₂ = res₂ * n₂ }

10    if pt₁ { n₁ = n₁ - 1 }
11    if pt₂ { n₂ = n₂ - 1 }
12  }
```

Figure 23.33: The product program for the factorial statement from Figure 23.31. Not duplicating the loop allows us to prove non-interference with a simple loop invariant that relates both executions of the original program.

variables. An encoding via two struct instances requires two allocation operations in the product programs; these yield different references. Consequently, the reference to a heap-allocated struct instance would always be considered as potentially secret information, such that any field read operation would in general also be considered to yield a potential secret, which is impractical. Therefore, we use the encoding with only one struct instance, but two copies of each field. The instance is created by a single allocation operation, which is specified to yield a non-secret result. The field values can still be secret if the two copies of a field contain different values.

Concurrency. Verifying standard secure information flow for concurrent programs is too strict. Control flow that depends on secrets may affect the run times of different threads, which may in turn affect the thread schedule and the results of the program. Therefore, it is common to instead verify a weaker notion: *Possibilistic non-interference* enforces that secrets do not influence the *possible* values of low outputs, i.e., if some combination of low output values is reachable from an initial state, then the same combination of low output values must still be reachable using *some* possible thread schedule after arbitrarily changing the high inputs. Alternatively, one can verify *probabilistic* non-interference, which requires that two executions from low-equivalent initial states will produce the same low outputs *with the same probabilities*.

Both notions of non-interference can be checked via modular product programs. The encoding leverages the fact that we verify programs to be data-race free. Consequently, two threads interact only at well-defined synchronization points, for instance, when acquiring or releasing a lock. It is, thus, possible to enforce that these operations do not depend on secret data. That is, a secret value must neither determine whether a thread acquires a lock nor which lock it acquires.

Declassification. Non-interference is too strict for practical applications. For instance, a program that checks whether an input matches a stored password leaks *some* information (here, one bit) because the subsequent computation depends on the result of the check. To handle such cases, it is necessary to declassify information, that is, to regard information as low even though it is technically not. A standard example of declassification is encryption: The encrypted message depends on the secret input, but is itself not considered secret.

There is much work on how to express declassification [446]. The actual declassification of an expression *e* can be easily encoded in product programs: It simply *assumes* (without a proof obligation) that *e* is low. Such an operation cannot lead to inconsistencies because product programs never express that an expression is *not low* (low assertions are not allowed to be negated). That is, the verifier knows an expression is low or it does not know that it is low; in the latter case it will consider the case that the expression might be high. Of course declassification operations must be reviewed carefully to avoid accidental leaking of secret information.

Side Channels. Secret information may be leaked directly via the output of a program, or indirectly via so-called side channels. For instance, the secret might influence whether the program terminates, how long it runs, how much memory it consumes, etc. Verifying the absence of side channels is difficult. Some of them can be prevented by instrumenting the program with counters, for instance, for the number of execution steps or the amount of allocated memory. However, the accuracy of such counters is questionable if they do not take into account how a program is compiled, optimized, and executed. Nevertheless, this approach can still be useful for SCION, where attackers will typically not have access to the machines that run the protocol, but only observe the transferred packets.

We verify the absence of termination side channels by proving for each loop and recursive call that its termination does not depend on a secret. To this end, the programmer provides an exact termination condition; we verify that the loop or call terminates if and only if this condition holds and that the condition is low [176].

24 Current Status and Plans

LINARD ARQUINT, DAVID BASIN, TOBIAS KLENZE, SI LIU, PETER MÜLLER, JOÃO PEREIRA, CHRISTOPH SPRENGER, FELIX WOLF

Chapter Contents

24.1 **Completed Work** **563**

    24.1.1 Methodology and Tools 563

    24.1.2 Design-Level Verification 565

    24.1.3 Code-Level Verification 565

24.2 **Ongoing Work** **566**

    24.2.1 Methodology and Tools 566

    24.2.2 Design-Level Verification 567

    24.2.3 Code-Level Verification 567

24.3 **Future Plans and Open Challenges** **567**

    24.3.1 Methodology and Tools 567

    24.3.2 Design-Level Verification 570

    24.3.3 Code-Level Verification 571

In the previous chapters we have motivated the need for formal methods—both at the design level and the code level—and discussed techniques for both levels. We have seen the results of verification efforts related to the SCION data plane and the N-Tube algorithm and discussed example code from the SCION border router.

But where are we in our quest for a fully verified Internet? Which parts of the architecture and code have been verified already and what are the plans for the future? We discuss these questions in this chapter.

24.1 Completed Work

24.1.1 Methodology and Tools

Refinement Infrastructure in Isabelle/HOL. We have embedded an infrastructure for developing distributed systems by stepwise refinement in Isabelle/HOL. This includes a formalization of event systems and their trace semantics as well as various proof rules for establishing invariants and other trace properties and for proving refinements between such models. We have also formally established the soundness of refinement and simulation for establishing trace inclusion and the preservation of trace properties across refinements.

Tamarin.   We have developed Tamarin [366], an advanced symbolic security protocol model checker. Protocols are modeled as multiset rewrite systems and security properties are formalized in a first-order logic expressing properties of protocol traces. The attacker's capabilities can be expressed using built-in or user-specified equational theories. Tamarin also enables the verification of observational equivalence properties, like those used to specify privacy. Tamarin has both automatic and interactive modes of interaction. It has been successfully applied to a wide spectrum of real-world protocols such as TLS 1.3, 5G, and the EMV payment protocol.

Gobra.   Our developed Go verifier supports a large subset of Go. Notably, Gobra is the first verifier to support all of Go's native heap data structures, Go's challenging structural subtyping via interfaces, and Go's channel-based concurrency. Regarding specifications, Gobra can verify memory safety, data-race freedom, crash safety, and user-provided specifications. Its support for quantifiers, first-class predicates, and pure functions enables users to specify sophisticated functional properties. In particular, Gobra can express the I/O specifications resulting from our Igloo methodology for protocol verification. While not yet implemented in Gobra, we have developed techniques, namely product programs, that can be implemented straightforwardly in Gobra to enable the verification of hyperproperties such as secure information flow.

Igloo: Linking Design-Level and Code-Level Verification.   We have developed a framework for soundly linking model development by stepwise refinement with code verification based on separation logics, called Igloo [486]. This framework combines methodological aspects with an infrastructure formalized in Isabelle/HOL. The methodology proposes the development of successively more detailed models by refinement, including abstract models, (distributed) protocol models, and interface models. An interface model is decomposed into an environment model and different component models (e.g., for the different protocol roles), which are then translated to I/O specifications in separation logic. Finally, each component is implemented and verified against its I/O specification. The Igloo formalization includes infrastructure to support all these steps: refinement (see also above), decomposition theorems, and the translation of component models to I/O specifications. Our formally established theoretical results imply the end-to-end soundness of our framework. Namely, the global properties established for the models are preserved down to the running system, assuming the code verifier's soundness and the faithfulness of the environment model.

24.1.2 Design-Level Verification

Path Authorization for Individual Segments in Isabelle/HOL. As presented in §22.4, we have successfully verified the path authorization security property of the data plane for individual up-segments. This result holds for arbitrary sets of authorized paths and network topologies.

Statistical Model Checking of N-Tube with PVeStA. As presented in §22.5, we have applied a rewriting-logic-based approach to formalize N-Tube and statistically analyze its core properties such as stability and fairness. Our statistical model checking results provide quantitative guarantees of N-Tube's correctness and assess its resistance to attacks in various malicious scenarios.

Verification of ARPKI Using Tamarin. The attack-resilient public-key infrastructure (ARPKI) [57] provides transparent and accountable certificate-related operations, such as certificate issuance, update, revocation, and validation. ARPKI offers very strong security guarantees: An impersonation attack requires compromising all n trusted signing and verifying entities involved. ARPKI was co-designed with a formal model and its core security property was verified using the Tamarin prover.

Verification of OPT and DRKey Using Coq. In [566], the authors prove source authentication and path validation properties of the OPT packet forwarding protocols as well as secrecy and authentication properties of the DRKey protocols [290]. They use LS^2, a logic for reasoning about secure systems, in combination with axioms from Protocol Composition Logic (PCL). They prove the protocols' properties based on a direct axiomatization of their logic in the Coq proof assistant (i.e., without proving the logic's soundness).

24.1.3 Code-Level Verification

So far, our code-level verification efforts have focused on proving memory safety of the SCION border router and its dependencies from the SCION repository. Furthermore, we specified (but did not verify) the third-party library gopacket[1], which is widely used in the border router implementation to serialize and deserialize network packets. We have verified a significant part of the SCION border router implementation and its dependencies, amounting to roughly 2500 lines of verified code.

In the process, we identified a memory safety issue[2] in the function HostFromRaw from the package github.com/scionproto/scion/go/lib/addr, for which we submitted a patch that has already been accepted. This function parses a host addresses from slices of bytes. Before our patch,

---

[1] https://pkg.go.dev/github.com/google/gopacket
[2] https://github.com/scionproto/scion/issues/4080

it was possible to trigger a runtime error when parsing SVC addresses due to a missing bounds check. We also note that, while trying to prove the memory safety of the border router, we engaged in conversations[3] with the SCION developers about implicit assumptions in the code, which we made explicit in our specifications.

24.2 Ongoing Work

24.2.1 Methodology and Tools

Code Verification in the Development Process.   Tool support, such as integrated development environments (IDEs) and bug-finders, is already well established in software engineering and improves the code development process. Similarly, tool support can improve the process of code verification in various ways. Counterexample generators help one understand why an annotated specification fails by generating a program state that is permitted by the program and violates the specification. Incremental verification reduces the verification time of subsequent verification runs by re-verifying only the parts of the code affected by changes to the code. Continuous integration goes one step further and makes incremental verification available to multiple contributors working on the verification in parallel. Furthermore, continuous verification guarantees that code remains verified in the presence of active code development where code is being modified. Finally, an IDE can make these features readily available to users. Developing such tool support that improves the productivity of Gobra users is ongoing work.

Linking Design-level and Code-level Verification.   Our Igloo methodology for connecting protocol and code verification is very expressive, but offers rather limited automation. We are currently addressing this issue in two ways. First, in the Isabelle/HOL framework, we are automating two steps of the Igloo methodology, namely, the decomposition of the interface models into different component models and the translation of the latter into I/O specifications. These steps require many boilerplate specifications and proofs. The automation of these steps is realized by extending Isabelle with new commands using the Isabelle/ML programming interface. Second, we are exploring an alternative, complementary approach, where we specify security protocols in the input language of automated security protocol model checkers such as Tamarin [366] and ProVerif [77] and then generate code-level I/O specifications from these protocol specifications.

---

[3]An example of such a discussion can be found in `https://github.com/scionproto/scion/issues/4094`.

24.2.2 Design-Level Verification

Data-Plane Verification.   Our verification of the data plane applies to single up-segments. Down-segments, core-segments, and segment combinations are not yet covered. We are currently extending our models to support full forwarding paths composed of multiple segments. The properties for full paths are expressed in terms of the authorization of all segments and of permitted segment combinations. We aim to show valley freedom and the absence of forwarding loops in the SCION data plane.

Proving N-Tube's Correctness.   In addition to quantitative guarantees, we also want to establish qualitative correctness and security properties about the N-Tube design. We are currently working on the labeled-transition-system-based formalization of N-Tube, a strong attacker model, and all its safety and security properties, as well as the inductive proofs establishing these properties.

24.2.3 Code-Level Verification

Despite what was already achieved, proving memory safety of the full implementation of the SCION border router is still ongoing work. Additionally, we are currently verifying that the border router processes packets in finite time, implying that it cannot enter an infinite loop due to malformed packets.

24.3 Future Plans and Open Challenges

Verifying the correctness and security of SCION is a huge undertaking. Even though we have already made substantial progress, there are still many open challenges that we plan to address in future work. We summarize some of the most prominent directions here.

24.3.1 Methodology and Tools

Formalization and Verification of Availability Properties.   Availability is the ability of a system to respond to requests and make progress. This includes the avoidance of livelocks, where a system is caught in a cycle without making progress. Availability properties are central to SCION's promise of a more secure Internet. So far we have not yet addressed the problem of how to formalize and reason about them. This problem has two aspects.

The first aspect is how to formalize availability properties. One solution is to treat them as liveness properties (i.e., "something good will eventually happen"). However, our design-level reasoning framework currently mainly focuses on safety properties (i.e., "nothing bad ever happens"). Proving that refinements preserve liveness properties requires quite intricate reasoning techniques [316, Chapter 8]. Moreover, it may not be sufficient to know that an event happens in finite time, but one may wish to indicate an upper bound. The advantage is that such bounded availability properties are safety properties, which are easier to reason about.

The second aspect is how to reason about availability properties at a global level. This is a challenging problem. For example, sending a packet from AS A to AS B potentially involves the interplay of a large number of components, including beacon servers, certificate servers, path servers, time synchronization, and routers. Ideally, we would like to derive global availability properties from a combination of a formalized analysis of the components' interdependencies (establishing in particular cycle freedom) with proofs of local availability properties for each component. The dependency analysis in itself can help to make the system more fault-tolerant by uncovering hidden dependencies (e.g., cyclic ones) and ensuring smooth startup and recovery procedures. For example, it should be able to answer questions like: Which subsystems are affected if there is a problem with a beaconing server or with time synchronization? How can we recover from such situations? An informal dependency analysis was presented in §6.1.

Verification as Protocols and Code Evolve.   A major challenge for the successful verification of real-world systems like SCION is to keep up with frequent changes in their design and implementation. Our Igloo approach to end-to-end verification has the advantage of decoupling the design and the code verification using I/O specifications as an interface between the two. Hence, implementation-only changes only require the code re-verification. In §24.2.1, we have already discussed code verification in the development process including the handling of changes. On the design-level, various techniques can help cope with changes. Modularity helps limit the scope of a necessary re-verification to changed components. Abstraction and parametrization can help verify an entire class of systems from which more concrete verified systems can be obtained by refinement or instantiation. In Isabelle/HOL, the structured (human-readable) Isar proof style is extremely helpful for proof maintenance, as one can quickly see which parts of a proof fail after a change and must be adapted. The future verification of SCION components will require a coherent and systematic application of appropriate instances of these techniques, adapted to our setting.

Scalability and Annotation Overhead.   Code-level verification of the SCION implementation is feasible, but requires effort that is substantially higher than writing the code in the first place. The main driver for this is the annotation burden imposed by state-of-the-art verification tools such as Gobra. An important goal for future work is to reduce this overhead substantially. One promising direction is to support verification by dedicated type systems or static analyzes that can check certain properties with low overhead on a large portion of the code base, and to focus the logic-based verification on those parts that are too intricate to be checked by type systems.

A promising candidate for this approach is secure information flow. The vast majority of code does not manipulate secrets. That is, all data is low and, thus, a simple type system suffices to check this code. The more expressive, but also more laborious verification approach described in §23.5 would then be applied only to those functions that manipulate secrets, leading to a reduced annotation overhead and speeding up verification.

Since much of the complexity of program verification is caused by pointer manipulations and side effects, another promising direction is to employ linear types to track which heap objects are owned exclusively by a function and which ones may be shared. Our work on verifying programs in the Rust language, which provides such a type system, has demonstrated that linear types have the potential to reduce the annotation overhead dramatically, especially since the permission specifications described in §23.2 can be generated automatically from linear type information [38]. Developing such a type system and corresponding verification technique for Go is promising future work.

Expressiveness of Gobra. The Gobra verifier covers a large subset of the Go programming language, but still has various limitations. In the SCION verification so far, we worked around these restrictions by slightly rewriting parts of the implementation to fall into the language subset supported by the tool. We are actively working on extending Gobra to support additional Go features and libraries.

We are also planning to extend Gobra so support the verification of additional properties, most prominently secure information flow, following the approach described in §23.5 (and possibly complemented by a type system as mentioned above). Another relevant property is *progress*, in particular, that each SCION component eventually performs all I/O operations prescribed by the protocol. Our approach for proving this liveness property is to associate I/O specifications with measures, which provide an upper bound on the number of function calls or loop iterations performed before the I/O operation is executed.

Usability of Code-level Verification. The verification of the SCION implementation, as well as most large code-level verification efforts, have been carried out by verification experts with considerable expertise and experience. However, relying on verification experts is an obstacle to the more widespread adoption of program verification such as the verification of the entire SCION code base. Our goal is to enable the developers of the SCION implementation to verify their code themselves. A stepping stone to this ambitious goal could be to enable them to maintain verified code, that is, to verify revisions of code that was initially annotated and verified by a verification expert.

Besides reducing the annotation overhead (as discussed above), engineers require better support to localize, understand, and fix verification errors. This task is tricky because developers need to understand whether a verification error reported by the tool is caused by an incorrect implementation, an incorrect or insufficient specification, or an unavoidable limitation of the verification tool. As future work, we are planning to equip Gobra with debugging features, such as counterexamples that illustrate the root cause of verification errors.

Linking Design-level and Code-level Verification. As we have explained in §23.4 and in §24.1.1 above, we have developed the Igloo methodology to formally integrate design-level and code-level verification and obtain end-to-end guarantees [486]. In §24.2.1, we have described on going work to automate Igloo's decomposition and translation steps. As future work, we plan to further improve the automation in two directions. First, we intend to automate the translation of the I/O specification from their Isabelle/HOL representation to the input languages of various code verifiers (including Gobra). Second, we plan to develop an incremental approach to reduce the re-verification effort when design-level models change. The goal is to require re-verification only of those parts of the implementation that are actually affected by the change.

Furthermore, our methodology for linking design-level and code-level verification does not yet support all concurrency patterns occurring in the SCION code. To address this issue, we plan to extend our approach to a more flexible I/O separation logic.

24.3.2 Design-Level Verification

Applying Igloo on the SCION Data Plane and Router. We have already verified a high-level model of the SCION data plane (§22.4) and some important code-level properties of the router implementation (§23.3). What is still missing is to connect the model to the code using an I/O specification for the router derived from the model using the Igloo methodology. This requires several steps. First, the model must be made feature-complete by including different segment types and their combinations as well as peering links (see ongoing work). Second, we must further refine the resulting complete model into an interface model, which is then decomposed into router components and an environment that includes both the intra-AS and inter-AS networks. From the resulting router model, we will derive an I/O specification against which the code can be verified. An open challenge for linking the I/O specifications to code is how to soundly connect the message terms on the modeling level with the bytestrings on the code level. The assumptions we have made about the message abstraction function in §23.4 is most likely too strong as it would require that every bytestring can be uniquely interpreted as a term. This is unrealistic, as in real systems collisions between, e.g.,a hash and a ciphertext are not excluded.

Other SCION Components. SCION is akin to a Swiss Army knife. At its heart are several big cutting blades, but there are many smaller extensions. The big blades are its core components for the control plane and data plane, featured in Part I of this book. The extensions were discussed in Part III and include hidden paths, time synchronization, high-speed traffic filtering, bandwidth reservation, and the like. Moreover, end users currently deploy the SCION–IP Gateway, as the computing devices they use do not natively support the SCION protocol and the gateway serves to translate between SCION and IP traffic. In the future, we envision that operating systems and smart devices will directly build SCION support into their protocol stacks, forgoing the need for such a translation.

As has been explained, our primary verification focus so far has been on the data plane. However, these other components and extensions are also critical and they are worthy targets for future verification efforts.

Compositionality. So far, we have focused on the verification of individual SCION protocols, such as the data plane's forwarding and the control plane's bandwidth reservation algorithm. In the future, we would like to connect our formal models of different components to obtain even stronger guarantees. Take for instance the *path authorization* property: It states that packets in the data plane can only traverse authorized paths. However, it assumes that the control plane correctly and securely created a set of authorized paths in accordance with the routing policies of all on-path ASes. A stronger property that connects control and data plane is: Packets can only traverse paths that are in accordance with the routing policies of all on-path ASes. This connection is challenging, as the models and proofs do not easily compose.

24.3.3 Code-Level Verification

Extending Code Coverage. Once we have completed the verification of the SCION border router, we will tackle the other components of SCION, such as the beacon server, path server, certificate server, and name server. A faulty implementation of any of these components can, in principle, invalidate core properties of the SCION protocol.

As mentioned in §24.1.3, third-party libraries are currently specified but not verified. We plan to extend our verification efforts to third-party libraries used in the SCION implementation. An alternative line of work consists of gradually integrating verified libraries in the SCION codebase and into our verification approach, such as EverCrypt [422] for verified cryptographic operations, and EverParse [427] to automatically generate correct packet serializers and deserializers.

Extending Property Coverage.   Our verification effort so far has focused on memory safety and termination. As future work, we plan to extend the covered properties to all relevant aspects of SCION, in particular, I/O behavior and secure information flow.

Part VII

Back Matter

25 Related Work

DAVID HAUSHEER AND ADRIAN PERRIG*

Chapter Contents

25.1 Future Internet Architectures 575
    25.1.1 Huawei's "New IP" 579
25.2 Deployment of New Internet Architectures 580
25.3 Inter-domain Multipath Routing Protocols 582

Given the extensive list of topics covered in this book, a wealth of related work exists. Balancing relevance, completeness and readability, we focus on related work specifically in the areas of future Internet architectures, deployment of new Internet architectures, and inter-domain multipath routing protocols. Current efforts to secure the current Internet inter-domain infrastructure are discussed in Section 1.1.3. For a more detailed treatment of related work pertaining to the various chapters of the book, we refer to the scientific papers of the respective topics, which all include a detailed discussion of related work.

25.1 Future Internet Architectures

Several efforts at redesigning the Internet have been made over the past three decades to fix problems appearing with the existing Internet infrastructure, and to satisfy new requirements of emerging Internet-based applications. New requirements include naming, routing, mobility, network efficiency, availability, manageability, and evolvability of the Internet.

The idea of partitioning the network into smaller parts has previously been considered for making network routing more scalable, for instance in hierarchical routing [277, 293], the Landmark hierarchy [512], hierarchies of nodes in Nimrod [99, 476], regions in NewArch [126], clusters of computers in FARA [125], isolated regions with independent routing protocols in HLP [496], realms and trust boundaries in the Postmodern Internet Architecture (POMO) [72, 95], and regions in NIRA [559].

The NewArch project [126] describes comprehensive requirements for a new Internet, such as separation of identity from location, late binding using

---

*This chapter reuses content from Barrera, Chuat, Perrig, Reischuk, and Szalachowski [49].

association, identity authenticity, and evolvability. However, it mostly emphasizes a new direction for endpoint entities while the packet delivery in the current IP network is left intact. NewArch uses the New Internet Routing Architecture (NIRA) [559] for inter-domain routing, which aims to introduce competition among ISPs in the core by providing route control to the end users, who can choose domain-level paths.

In Plutarch [145], the global network is divided into contexts, and interstitial functions are used to communicate between them. This allows interconnecting heterogeneous networks, instead of imposing the same L3 protocols everywhere, which might not always be feasible (e.g., for sensors) while the interstitial functions translate between the different contexts, still providing global connectivity. The authors argue that a model with explicit interaction at context boundaries is more accurate and extensible. SCION follows the same approach by clearly separating communication between ISDs (core-path segments) and within ISDs (down- and up-path segments). SCION also distinguishes between inter-AS communication (through AS and interface-level granularity path segments) and intra-AS communication (independent intra-AS name spaces).

HLP [496] is a proposed next-generation routing architecture which uses a hybrid approach between link-state and path-vector algorithms in order to improve the scalability, convergence and isolation properties compared to BGP. SCION follows their approach of hierarchically partitioning the network and opting for AS-based routing instead of prefix-based routing. HLP combines a link-state algorithm inside with a path-vector algorithm between hierarchies and obscures AS-path information present in BGP through a generic cost metric to reduce overhead. In SCION, the beaconing mechanism allows for stateless routers, and the beaconing scales fundamentally better than BGP as shown in §6.3 and §6.4. Even though a simplified version of the HLP protocol can be implemented in BGP-speaking routers by changing the current operational practice of BGP, allowing for incremental deployment, not all BGP policies are supported by HLP. The consequences of such a mixed deployment remain unclear.

Information-centric networking (ICN) or content-centric networking (CCN) architectures optimize content access through in-network content caches. Since content access across a user population frequently exhibits strong temporal and spatial locality, in-network content caches can serve the same requests made by nearby users. For instance, the Named Data Networking (NDN) [269, 383] architecture decouples location from identity and uses identity for locating the corresponding content. NDN relies on in-network caching of data and is useful for accessing popular static content. The CCNx project proposes a related implementation of content-centric networking, developing detailed specifications and prototype systems [398]. The Publish-Subscribe Internet Routing Paradigm (PSIRP) supports information-centric networking based on a publish-subscribe pattern [502]. It proposes an elegant approach to reduce the state on routers by having packets carry Bloom filters to encode the next hops of a multicast packet [275]. These architectures, however, have a high overhead for point-to-point communication, for ephemeral content (e.g., voice or video calls), or for per-user customized content. Our energy analysis presented in §16.1 suggests that content-centric approaches have higher energy utilization than fetching content directly from the origin server, due to the increased power consumption of routers with this architecture.

MobilityFirst [429] is an architecture with the main goal of providing connectivity to billions of mobile devices. At its core is the Auspice system, which provides a highly efficient global name resolution service that can quickly map billions of identities to their locations [468]. NEBULA [25] addresses security problems in the current Internet. NEBULA takes a so-called default-off approach to reach a specific service, where a sender can send packets only if an approved path to a service is available. The network architecture helps the service to verify whether the packet followed the approved path (i.e., supporting path verification). However, NEBULA achieves this property at a high cost. All routers on the path need to perform computationally expensive path verification for each packet and need to keep per-flow state. Serval [392] provides a service abstraction layer for service-ID-based resolution in NEBULA. Serval introduces a service-access layer that enables late binding of a service to its location, which provides flexibility in migrating and distributing services.

XIA [22, 227], which paved the way for the initial work on SCION, unifies different networking paradigms, such as content- and service-centric networking, using a generic principal-centric networking approach. XIA intends to be extensible and evolvable to support new types of communication and facilitate deployability. This is achieved by enabling applications or protocols to start using new principal types before the network develops inherent support for them. Instead, network entities unfamiliar with the new principal use a fallback mechanism, and still provide global connectivity.

The Framework for Internet Innovation (FII) [299] proposes a clean-slate redesign of the Internet to remove deployment barriers for innovations, such that the Internet can be composed of a heterogeneous conglomerate of architectures. It defines three primitives: an inter-domain routing architecture, a network API, and an interface for hosts to protect themselves against DoS attacks. The latter requires a trusted third party, with the ability to *shut up* a host which is attacking another host or network entity. While this party does not need to be globally valid, and there can be multiple parties, any misbehaving trusted third party would have a severe impact on the system. The authors propose to use pathlet [212] routing as the inter-domain routing architecture.

ChoiceNet [444, 547] introduces an "economy plane" that allows the establishment of dynamic business relationships to create a competitive marketplace for innovative solutions. The authors briefly touch on the scalability question of ChoiceNet, stating that its scalability would depend on the spectrum of choices provided, how the choices were made, the frequency with which choices change, and the threat model to be protected against. However, it is unclear if the system could scale to a network of the size of the Internet. ChoiceNet requires the network to generate alternatives for users to choose from and provides monetary rewards for alternatives that address users' needs. SCION could provide an instantiation in the form of paths with different properties presented to users.

Route Bazaar [100] introduces a contractual system based on a public ledger, where ASes and customers agree on QoS-aware routes in the form of BGP-overlay pathlets. Route Bazaar is orthogonal to SCION, as it proposes a new contractual system, which could be used to offer SCION paths (instead of pathlets) to customers, circumventing the disadvantages of overlay connections.

Several architecture proposals suggest the approach of better path control for senders and receivers, for example i3 [493], Platypus [424, 425], NIRA [559], SNAPP [401], Pathlets [212], and Segment Routing [187]. These proposals enable the source to embed a forwarding path into the packet header, a concept that we refer to as packet-carried forwarding state (PCFS). PCFS provides many beneficial properties, such as enabling multipath communication and protecting packets from unanticipated re-routing.

Forward [181] and SysSec [183] are proposing to build secure and trusted Information and Communication Technology (ICT) systems by engaging academia and industry. Forward is an initiative by the European Commission to promote the collaboration and partnership between industry and academia in their common goal of protecting ICT infrastructures. The Forward project categorizes security threats to various ICT systems including individual devices, social networks, critical infrastructures (such as smart electric grids), and the Internet infrastructure, and it aims at coordinating multiple research efforts to build secure and trusted ICT systems and infrastructures. SysSec aims to bring together the systems security research community in Europe, promoting cybersecurity education, engaging a think tank in discovering the threats and vulnerabilities of the current and future Internet, creating an active research road map in the area, and developing a joint working plan to conduct state-of-the-art collaborative research. Since Forward and SysSec currently focus on identifying and handling threats, we believe our proposed tasks to be a good addition to the projects in that they provide an architecture that would significantly reduce the attack surface.

RINA [530] is a recursive inter-network architecture that provides unified APIs across all protocol layers. In RINA, all layers have the same functions with different scope and range, where a layer is a distributed application that performs and manages inter-process communication.

Many researchers are currently studying software-defined networking (SDN), for example in the OpenFlow [359] project or with P4 [83]. These efforts mainly consider intra-domain communication, which SCION can leverage to communicate within a domain.

We have developed SCION with a focus on security and high availability for point-to-point communication, which is a unique perspective and can contribute to other future Internet efforts. For instance, content-centric networking also needs a routing mechanism to reach the data source. SCION can offer the routing protocol to support that functionality. Once a server is found in a service-based infrastructure or a nearby content cache is found in a content-centric architecture, point-to-point communication between the end host and the server will offer high communication efficiency, as pure forwarding is faster than server-based or content-based lookups. Similarly, SCION can provide the point-to-point communication fabric in a mobility-centric architecture. Consequently, SCION offers mechanisms that complement many previously proposed future Internet architectures.

25.1.1 Huawei's "New IP"

At an International Telecommunications Union (ITU) meeting, Huawei proposed radical changes to the way the Internet works under the name of "New IP". These changes intend to support future technologies such as holograms and self-driving cars, as well as provide improvements for data security and privacy. Huawei states that the way the current Internet works is "unstable" and "vastly insufficient," as they highlight the lack of control and security the current TCP/IP system already exhibits.

The China-based telecommunications company asserted that New IP is being developed purely to meet future technical requirements and that it does not have any type of built-in control mechanisms. To address these issues, Huawei proposes an array of requirements that new networks should meet:

1. A flexible and variable-length addressing scheme, which provides stronger privacy than today's location-based addressing.

2. A mechanism to provide different quality of service (QoS) for different traffic types mainly in relation to latency and bandwidth. This includes the use of multiple paths to meet high bandwidth requirements.

3. An improved key-exchange mechanism to mitigate man-in-the-middle attacks.

4. The possibility to audit and shut down connections in order to protect against denial-of-service (DoS) attacks.

Huawei suggests the following to meet these requirements:

1. In order to allow for flexible address lengths, and to improve privacy, New IP will decouple the identity and location of a user by assigning different IDs to both. An "Encrypted Identifier" would be assigned to each user binding the identity through an "Identity Manager". The inner

address would get obfuscated at the domain border router, providing additional privacy. However, these addresses would be traceable through the Accountability Manager which will have the power to shut down undesirable or unauthorized connections.

2. To enable QoS, they mention the possibility of encoding user preferences for traffic treatment in their New IP header.

3. To reinforce security, New IP includes an "Accountability Manager". This is labeled as a "Decentralized Public Key Database" and they explain that this module together with the Identity Manager can be used to audit and therefore trace a connection.

The final point of how New IP handles security has caused a wave of controversy around the world and from other members of the UN.

Internet users and companies around the world stand divided on the benefits and risks of New IP. The current vulnerabilities of the Internet leave a lot to be desired, and as Huawei points out, these vulnerabilities will cause massive problems for future technologies such as self-driving cars.

Whether one agrees with Huawei's approach to security or not, the problems they highlight about today's Internet are valid. Security and control are major hurdles for users of the Internet whose routing architecture has little changed over the past 30 years. As SCION not only satisfies Huawei's requirements but exceeds them, New IP can build on SCION to achieve additional properties.

25.2 Deployment of New Internet Architectures

Several future Internet efforts provide testbeds for running and testing a new architecture.

Future Internet Research and Experimentation (FIRE) [180] was a program funded by the European Union targeted at next-generation network research supporting numerous Future Internet research projects and facilities [510].

Following up on FIRE, Fed4FIRE [178] (and its successor Fed4FIRE+) is a federation of Next Generation Internet (NGI) testbeds combining the facilities from different FIRE research communities to perform innovative experiments.

Provided by the European-wide GÉANT [225] research network, the GÉANT Testbeds Service (GTS) [226] has been a network testbed spanning over the GÉANT core network allowing researchers to define, build, test virtual networks, facilitating next-generation Internet research across the GÉANT infrastructure

FIWARE [179] aims at a sustainable ecosystem driven by future Internet technologies. The foundation provides FIWARE Lab, a sandbox environment deployed over a geographically distributed network infrastructure to innovate and experiment with FIWARE Technologies.

The EU-funded SLICES [182] infrastructure is supporting large-scale, experimental networking research for various application domains, in particular cloud and edge-based computing.

GENI [65, 358] is a distributed virtual laboratory for the development, deployment, and validation of transformative, at-scale concepts in network science, services, and security, deployed at around 50 US sites. VINI [60] is a virtual network infrastructure that enables researchers to test new networking protocols in realistic but controlled settings. FABRIC [45] is a programmable networking infrastructure facilitating experiments with novel network designs and applications. The BRIDGES [382] facility consists of two geographically separate optical links across the Atlantic, capable of carrying up to 200 Gbps of traffic, enabling networking research over very high-speed links. GENI, VINI, FABRIC, and BRIDGES are testbeds that allow the evaluation of Internet architectures on a large-scale network, but have so far not been used to build the SCION production network. The vast effort on deployment concepts and test beds for next-generation routing infrastructures demonstrates the challenge of deploying a novel Internet architecture. The design choices of SCION have made it possible to overcome this challenge, without creating an overlay on today's Internet.

Trotsky [356] proposes a backward-compatible architectural framework to deploy new Internet architectures. The Internet is described as a collection of layered overlays, with the only exception of intra- and inter-domain communication. There is a single inter-domain communication protocol, the *narrow waist* of the Internet, hindering innovation at this layer so far. Trotsky describes how to design new inter-domain protocols, by constructing the inter-domain layer as an overlay on the intra-domain communication.

The abstractions used in Trotsky and SCION are similar, since both propose novel inter-AS control planes, while treating intra-AS connectivity as logical pipes. The main goals of Trotsky, incremental deployment and extensibility, are reflected in SCION's deployment model and flexible path dissemination approach, enabling different path selection algorithms per AS. Despite the conceptual similarities, the goals of these efforts are quite different, and SCION could benefit from additional deployment in Trotsky's infrastructure.

KREONet2 is the international network of KREONET (Korea Research Environment Open Network) [291], Korea's National Science & Research Network. KREONet2 is a global 10-100 Gbps high-speed research network that started with GLORIAD [210], the Global Ring Network for Advanced Applications Development, supported by the US National Science Foundation (NSF). The global span of the KREONet2 network, which is interconnected with more than 36 global research networks across the world including GÉANT, Internet2, and CERN, makes this network an excellent infrastructure to deploy new Internet architectures at global scale. We discuss how KREONet2 supports a SCION testbed in §14.2.6.

The China Next Generation Internet (CNGI) Project [554] supports the construction of China's next generation Internet backbones with an emphasis on IPv6 as the next-generation Internet technology. Following up on CNGI, the China Education Network (CERNET) has recently launched a new backbone network in China (called FITI) [511], which consists of 4096 new ASes.

25.3 Inter-domain Multipath Routing Protocols

Although numerous research teams worked on multipath routing, only few approaches were developed beyond a proof of concept, and even fewer were deployed. We compare the most relevant approaches to SCION and highlight the differences. We refer to Singh et al. [545] for an in-depth survey of approaches that were not further developed or deployed.

BGP-Based Approaches. BGP Add-Path is a deployed extension to BGP [435], which allows announcing additional paths for a certain prefix without implicitly revoking the existing path. The two main drawbacks of BGP Add-Path are increased border router memory requirements for storing additional paths, and lack of path control for endpoints. Other proposals using BGP include BGP-XM [97], which explores existing redundant routing paths provided by BGP, and Path Splicing [374] and STAMP [337], which provide multiple paths by running k ($k = 2$ for STAMP and configurable for Path Splicing) parallel BGP sessions to explore multiple routing paths. The main drawback of these approaches is the overhead of running multiple BGP sessions, which typically requires network operators to purchase additional hardware. A large number of BGP-based inter-domain multipath routing approaches have similar limitations: DIMR [561], AMIR [423], YAMR [194], BGP-XM [97], MIFO [569], D-BGP and B-BGP [527], and R-BGP [308].

Source-Based Routing. Source-based routing protocols allow a sender to (partially) control the packets' forwarding paths.

Resilient Overlay Network (RON) [24], is a proposal to improve both the resilience and the performance (with respect to application requirements) of the Internet. RON is an application-level (BGP-)overlay network composed of RON nodes, that monitor the performance of different paths and quickly reroute traffic in case of a link outage that would otherwise require reconvergence of BGP. However, due to the fact that RON is an overlay network, it cannot achieve the same guarantees as a natively deployed architecture.

BANANAS [286] encodes partial paths as PathIDs, and a packet specifying a PathID is sent along the specified path. BANANAS supports incremental deployment, by enriching the link-state tables of upgraded routers with the knowledge of which other routers are multipath-capable. This way, the set of available paths can be computed locally by enriched routers. In order to avoid the situation where a path computed by one router does not exist in another, they employ a distributed path validation algorithm. The additional information in upgraded routers increases the size of routing tables, since each PathID forwarding rule adds an additional entry, and the validation algorithm increases computational complexity.

Platypus [424, 425] enhances source routing with per-packet capabilities to enable policy compliance among operators. However, path exploration and path selection are not defined, and thus it is difficult to reason about its scalability. Wide-Area Relay Addressing Protocol (WRAP) [34] is based on loose source routing, i.e., WRAP packets specify a list of IP addresses of AS edge routers that packets should traverse. Since each AS edge router maintains at least two AS paths to each other AS, WRAP must maintain multiple routing paths per destination prefix, hampering scalability. New Inter-Domain Routing Architecture (NIRA) [559] constructs end-to-end paths from up- and down-segments connected at a core network similar to SCION, but only supports a single core network and no isolation domains which are essential for the scalability and sovereignty of SCION. Routing Deflection [558] allows endpoints to deflect their traffic at certain BGP routers to choose different paths. While this approach can be incrementally deployed with minimal changes to BGP, it only provides coarse-grained path control.

Multi-path Interdomain Routing (MIRO) [555], is a mix between source-based and tunneling-based routing. ASes can negotiate the advertisement of alternative paths pairwise, for the purpose of avoiding a specific AS (e.g., for security reasons). This keeps the increased state small and MIRO could in principle be incrementally deployed, as long as the most densely-interconnected ASes adopt it first. Pathlet Routing [212], allows (partial) paths (pathlets) to be constructed from a set of routers. These pathlets are then disseminated in a similar way as BGP disseminates routes to prefixes today. Incremental deployment is hindered by routers needing to understand the pathlet vocabulary. Additionally, policies in pathlet routing are no longer destination based, making it non-interoperable with BGP policies.

Bibliography

[1] Josh Aas, Daniel McCarney, and Roland Shoemaker. Multi-perspective validation improves domain validation security. https://perma.cc/ZHC2-BHRQ, February 2020. 420, 425

[2] Martín Abadi and Leslie Lamport. The existence of refinement mappings. *Theor. Comput. Sci.*, 82(2):253–284, 1991. doi: 10.1016/0304-3975(91)90224-P. URL https://doi.org/10.1016/0304-3975(91)90224-P. 479, 498

[3] Abhishta Abhishta, Roland van Rijswijk-Deij, and Lambert J. M. Nieuwenhuis. Measuring the impact of a successful ddos attack on the customer behaviour of managed dns service providers. *SIGCOMM Comput. Commun. Rev.*, 48(5):7076, 2019. 432

[4] Jean-Raymond Abrial. *Modeling in Event-B - System and Software Engineering*. Cambridge University Press, 2010. ISBN 978-0-521-89556-9. doi: 10.1017/CBO9781139195881. URL https://doi.org/10.1017/CBO9781139195881. 481, 482

[5] Jean-Raymond Abrial, Michael J. Butler, Stefan Hallerstede, Thai Son Hoang, Farhad Mehta, and Laurent Voisin. Rodin: an open toolset for modelling and reasoning in event-b. *STTT*, 12(6):447–466, 2010. 481

[6] Lada Adamic and Bernardo Huberman. Zipf's law and the internet. *Glottometrics*, 3:143–150, 11 2001. URL https://www.researchgate.net/publication/240426917_Zipf's_Law_and_the_Internet. 80

[7] Danny Adamitis, David Maynor, Warren Mercer, Matthew Olney, and Paul Rascagneres. Dns hijacking abuses trust in core internet service. https://blog.talosintelligence.com/, 2019. 8, 436

[8] Bernard Aebischer and Lorenz Hilty. *The Energy Demand of ICT: A Historical Perspective and Current Methodological Challenges*, volume 310, pages 71–103. 08 2015. ISBN 978-3-319-09227-0. doi: 10.1007/978-3-319-09228-7_4. 395, 399

[9] Yehuda Afek, Anat Bremler-Barr, and Lior Shafir. Nxnsattack: Recursive DNS inefficiencies and vulnerabilities. In *Proceedings of the USENIX Security Symposium*, 2020. 438, 441, 454

[10] Gul A. Agha, José Meseguer, and Koushik Sen. PMaude: Rewrite-based specification language for probabilistic object systems. *Electr. Notes Theor. Comput. Sci.*, 153(2), 2006. 511, 512

[11] William Aiello, John Ioannidis, and Patrick McDaniel. Origin authentication in interdomain routing. In *Proceedings of the ACM Conference on Computer and Communications Security (CCS)*, October 2003. 165

[12] Akamai. Dnssec targeted in dns reflection, amplification ddos attacks. https://community.akamai.com/, February 2016. 438

[13] Kahraman Akdemir, Martin Dixon, Wajdi Feghali, Patrick Fay, Vinodh Gopal, Jim Guilford, Erdinc Ozturk, Gil Wolrich, and Ronen Zohar. Breakthrough AES performance with Intel AES New Instructions. *White paper, June*, 2010. 12, 217

[14] S. Alexander and R. Droms. DHCP Options and BOOTP Vendor Extensions. RFC 2132, IETF, March 1997. URL http://tools.ietf.org/rfc/rfc2132.txt. 329

[15] Abdallah Alma'aitah and Zine-Eddine Abid. Transistor level optimization of sub-pipelined aes design in cmos 65nm. In *Proceedings of the International Conference on Microelectronics (ICM)*, pages 1–4. IEEE, 2011. 12

[16] Eihal Alowaisheq, Siyuan Tang, Zhihao Wang, Fatemah Alharbi, Xiaojing Liao, and XiaoFeng Wang. Zombie awakening: Stealthy hijacking of active domains through dns hosting referral. In *Proceedings of the ACM Conference on Computer and Communications Security (CCS)*, 2020. 8

[17] Bowen Alpern and Fred B. Schneider. Defining liveness. *Inf. Process. Lett.*, 21(4):181–185, 1985. doi: 10.1016/0020-0190(85)90056-0. URL https://doi.org/10.1016/0020-0190(85)90056-0. 485

[18] Musab AlTurki and José Meseguer. PVeStA: A parallel statistical model checking and quantitative analysis tool. In *CALCO'11*, volume 6859 of *LNCS*, pages 386–392. Springer, 2011. 511, 512, 514

[19] S. Amante, B. Carpenter, S. Jiang, and J. Rajahalme. IPv6 Flow Label Specification. RFC 6437, IETF, November 2011. URL http://tools.ietf.org/rfc/rfc6437.txt. 95

[20] American Registry for Internet Numbers (ARIN). Resource Public Key Infrastructure (RPKI). https://perma.cc/4GTS-QWHC. 36

[21] AMS-IX. Total traffic statistics. https://perma.cc/QQ84-CNUH, 2021. 214

[22] Ashok Anand, Fahad Dogar, Dongsu Han, Boyan Li, Hyeontaek Lim, Michel Machado, Wenfei Wu, Aditya Akella, David G. Andersen, John W. Byers, Srinivasan Seshan, and Peter Steenkiste. XIA: An architecture for an evolvable and trustworthy internet. In *Proceedings of the ACM Workshop on Hot Topics in Networks (HotNets)*, 2011. doi: 10.1145/2070562.2070564. URL https://doi.org/10.1145/2070562.2070564. 577

[23] Anapaya Systems and ETH Zurich. SCION open-source implementation. https://github.com/scionproto/scion, 2021. 533

[24] David G. Andersen, Hari Balakrishnan, M. Frans Kaashoek, and Robert Morris. Resilient overlay networks. In *Proceedings of ACM Symposium on Operating Systems Principles (SOSP)*, October 2001. 10, 28, 582

[25] Tom Anderson, Ken Birman, Robert Broberg, Matthew Caesar, Douglas Comer, Chase Cotton, Michael J. Freedman, Andreas Haeberlen, Zachary G. Ives, Arvind Krishnamurthy, William Lehr, BoonThau Loo, David Mazieres, Antonio Nicolosi, Jonathan M. Smith, Ion Stoica, Robbert Renesse, Michael Walfish, Hakim Weatherspoon, and Christopher S. Yoo. The NEBULA future Internet architecture. In *The Future Internet*, Lecture Notes in Computer Science. Springer-Verlag, 2013. 577

[26] Tore Anderson. Evaluating local dnssec validators. https://www.redpill-linpro.com/techblog/, 2019. 437

[27] L. Andersson and R. Asati. Multiprotocol Label Switching (MPLS) Label Stack Entry: "EXP" Field Renamed to "Traffic Class" Field. RFC 5462, IETF, February 2009. URL http://tools.ietf.org/rfc/rfc5462.txt. 281

[28] Anders Andrae. New perspectives on internet electricity use in 2030. *Engineering and Applied Science Letters*, 3:19–31, 06 2020. doi: 10.30538/psrp-easl2020.0038. 393, 395, 398, 399

[29] Anders Andrae and Tomas Edler. On Global Electricity Usage of Communication Technology: Trends to 2030. *Challenges*, 6(1):117–157, 2015. ISSN 2078-1547. doi: 10.3390/challe6010117. URL https://www.mdpi.com/2078-1547/6/1/117. 393, 395, 399

[30] Anonymous. The collateral damage of internet censorship by dns injection. *ACM SIGCOMM Computer Communication Review*, 42(3):2127, June 2012. 437

[31] Markku Antikainen, Tuomas Aura, and Mikko Särelä. Denial-of-service attacks in Bloom-filter-based forwarding. *IEEE/ACM Transactions on Networking*, 22(5):1463–1476, 2013. 206

[32] Maria Apostolaki, Aviv Zohar, and Laurent Vanbever. Hijacking Bitcoin: Routing attacks on cryptocurrencies. In *Proceedings of the IEEE Symposium on Security and Privacy (S&P)*, 2017. 381

[33] R. Arends, R. Austein, M. Larson, D. Massey, and S. Rose. DNS Security Introduction and Requirements. RFC 4033, IETF, March 2005. URL http://tools.ietf.org/rfc/rfc4033.txt. 6, 437

[34] Katerina Argyraki and David R. Cheriton. Loose source routing as a mechanism for traffic policies. In *Proceedings of the ACM SIGCOMM Workshop on Future Directions in Network Architecture*, FDNA 04, page 5764, New York, NY, USA, 2004. Association for Computing Machinery. ISBN 158113942X. doi: 10.1145/1016707.1016718. URL https://doi.org/10.1145/1016707.1016718. 583

[35] Katerina Argyraki and David R. Cheriton. Network capabilities: The good, the bad and the ugly. In *Proceedings of the ACM Workshop on Hot Topics in Networks (HotNets)*, 2005. 240, 264

[36] J. Arkko and S. Bradner. IANA Allocation Guidelines for the Protocol Field. RFC 5237, IETF, February 2008. URL http://tools.ietf.org/rfc/rfc5237.txt. 315

[37] Frank Arute, Kunal Arya, Ryan Babbush, Dave Bacon, Joseph C Bardin, Rami Barends, Rupak Biswas, Sergio Boixo, Fernando GSL Brandao, David A Buell, et al. Quantum supremacy using a programmable superconducting processor. *Nature*, 574(7779):505–510, 2019. 415

[38] V. Astrauskas, P. Müller, F. Poli, and A. J. Summers. Leveraging Rust types for modular specification and verification. In *Object-Oriented Programming Systems, Languages, and Applications (OOPSLA)*, volume 3, pages 147:1–147:30. ACM, 2019. doi: 10.1145/3360573. 569

[39] Jean-Philippe Aumasson and Daniel Bernstein. SipHash: a fast short-input PRF. In *International Conference on Cryptology in India*, pages 489–508. Springer, 2012. 206, 412

[40] AVFirewalls. Fortinet FortiGate 6300F. https://www.avfirewalls.com.au/FortiGate-6300F.asp, 2021. 207, 212

[41] AVFirewalls. Fortinet FortiGate 1100E. https://perma.cc/MRS3-X33S, 2021. 212

[42] Amazon AWS. AWS Transit Gateway. https://aws.amazon.com/transit-gateway/, 2021. 345

[43] Frédéric Badeau and Arnaud Amelot. Using B as a high level programming language in an industrial project: Roissy VAL. In Helen

Treharne, Steve King, Martin C. Henson, and Steve A. Schneider, editors, *ZB 2005: Formal Specification and Development in Z and B, 4th International Conference of B and Z Users, Guildford, UK, April 13-15, 2005, Proceedings*, volume 3455 of *Lecture Notes in Computer Science*, pages 334–354. Springer, 2005. doi: 10.1007/11415787_20. URL https://doi.org/10.1007/11415787_20. 478, 479

[44] Vaibhav Bajpai and Jürgen Schönwälder. A survey on Internet performance measurement platforms and related standardization efforts. *IEEE Communications Surveys & Tutorials*, 17(3):1313–1341, 2015. 368

[45] Ilya Baldin, Anita Nikolich, James Griffioen, Indermohan Inder Monga, Kuang-Ching Wang, Tom Lehman, and Paul Ruth. Fabric: A national-scale programmable experimental network infrastructure. *IEEE Internet Computing*, 23(6), 2019. URL https://doi.org/10.1109/MIC.2019.2958545. 581

[46] Jayant Baliga, Robert Ayre, Kerry Hinton, Wayne V. Sorin, and Rodney S. Tucker. Energy consumption in optical ip networks. *Journal of Lightwave Technology*, 27(13):2391–2403, 2009. doi: 10.1109/JLT.2008.2010142. 396, 397, 398

[47] Swiss National Bank. Payment transactions via swiss interbank clearing (sic), 2019. URL https://data.snb.ch/en/topics/finma#!/cube/zavesic. 376, 385

[48] R. Barnes, J. Hoffman-Andrews, D. McCarney, and J. Kasten. Automatic Certificate Management Environment (ACME). RFC 8555, IETF, March 2019. URL http://tools.ietf.org/rfc/rfc8555.txt. 444, 466

[49] David Barrera, Laurent Chuat, Adrian Perrig, Raphael M. Reischuk, and Pawel Szalachowski. *Introduction*, pages 3–15. In Perrig et al. [413], 2017. ISBN 978-3-319-67079-5. doi: 10.1007/978-3-319-67080-5_1. URL https://doi.org/10.1007/978-3-319-67080-5_1. 1, 575

[50] David Barrera, Laurent Chuat, Adrian Perrig, Raphael M. Reischuk, and Pawel Szalachowski. *The SCION Architecture*, pages 17–42. In Perrig et al. [413], 2017. ISBN 978-3-319-67079-5. doi: 10.1007/978-3-319-67080-5_2. URL https://doi.org/10.1007/978-3-319-67080-5_2. 17

[51] David Barrera, Laurent Chuat, Adrian Perrig, Raphael M. Reischuk, and Pawel Szalachowski. The SCION Internet architecture. *Communications of the ACM*, 60(6), June 2017. 374

[52] David Barrera, Tobias Klenze, Adrian Perrig, Raphael M. Reischuk, Benjamin Rothenberger, and Pawel Szalachowski. *Security Analysis*, pages 301–330. In Perrig et al. [413], 2017. ISBN 978-3-319-67079-5. doi: 10.1007/978-3-319-67080-5_13. URL https://doi.org/10. 1007/978-3-319-67080-5_13. 157

[53] Gilles Barthe, Pedro R. D'Argenio, and Tamara Rezk. Secure information flow by self-composition. *Mathematical Structures in Computer Science*, 21(6):1207–1252, 2011. doi: 10.1017/S0960129511000193. URL https://doi.org/10.1017/S0960129511000193. 558

[54] Cristina Basescu, Yue-Hsun Lin, Haoming Zhang, and Adrian Perrig. High-speed inter-domain fault localization. In *Proceedings of the IEEE Symposium on Security and Privacy (S&P)*, May 2016. 174

[55] Cristina Basescu, Raphael M. Reischuk, Pawel Szalachowski, Adrian Perrig, Yao Zhang, Hsu-Chun Hsiao, Ayumu Kubota, and Jumpei Urakawa. SIBRA: Scalable Internet bandwidth reservation architecture. In *Proceedings of the Symposium on Network and Distributed Systems Security (NDSS)*, February 2016. 237, 256

[56] David Basin. *The Cyber Security Body of Knowledge*, chapter Formal Methods for Security. University of Bristol, 2021. URL https://www. cybok.org/. Version 1.0. 479

[57] David Basin, Cas Cremers, Tiffany Hyun-Jin Kim, Adrian Perrig, Ralf Sasse, and Pawel Szalachowski. ARPKI: Attack resilient public-key infrastructure. In *Proceedings of the ACM Conference on Computer and Communications Security (CCS)*, November 2014. 565

[58] David A. Basin, Cas Cremers, and Catherine A. Meadows. Model checking security protocols. In Edmund M. Clarke, Thomas A. Henzinger, Helmut Veith, and Roderick Bloem, editors, *Handbook of Model Checking*, pages 727–762. Springer, 2018. doi: 10.1007/978-3-319-10575-8_22. URL https://doi.org/10.1007/978-3-319-10575-8_22. 475

[59] Tony Bates. CIDR report. https://perma.cc/H2BM-LGK9, 2020. 61, 243, 337, 395, 465

[60] Andy Bavier, Nick Feamster, Mark Huang, Larry Peterson, and Jennifer Rexford. In VINI veritas: realistic and controlled network experimentation. In *Proceedings of the 2006 conference on applications, technologies, architectures, and protocols for computer communications (SIGCOMM)*, 2006. 368, 581

[61] BBC News. Asia communications hit by quake. https://perma.cc/ 5GAS-AW39, December 2006. 11, 165

[62] Patrick Behm, Paul Benoit, Alain Faivre, and Jean-Marc Meynadier. Météor: A successful application of B in a large project. In Jeannette M. Wing, Jim Woodcock, and Jim Davies, editors, *FM'99 - Formal Methods, World Congress on Formal Methods in the Development of Computing Systems, Toulouse, France, September 20-24, 1999, Proceedings, Volume I*, volume 1708 of *Lecture Notes in Computer Science*, pages 369–387. Springer, 1999. doi: 10.1007/3-540-48119-2_22. URL https://doi.org/10.1007/3-540-48119-2_22. 478, 479

[63] Mihir Bellare, Joe Kilian, and Phillip Rogaway. The security of the cipher block chaining message authentication code. *Journal of Computer and System Sciences*, 61(3):362–399, 2000. 414

[64] Theophilus Benson, Aditya Akella, and David A. Maltz. Network traffic characteristics of data centers in the wild. In *IMC'10*, pages 267–280. ACM, 2010. 513, 514

[65] Mark Berman, Jeffrey S. Chase, Lawrence Landweber, Akihiro Nakao, Max Ott, Dipankar Raychaudhuri, Robert Ricci, and Ivan Seskar. GENI: A federated testbed for innovative network experiments. *Computer Networks*, 61:5–23, 2014. 368, 581

[66] Daniel Bernstein, Niels Duif, Tanja Lange, Peter Schwabe, and Bo-Yin Yang. High-speed high-security signatures. *Journal of Cryptographic Engineering*, 2(2):77–89, 2012. 413

[67] Daniel J. Bernstein and Tanja Lange. Post-quantum cryptography. *Nature*, 549(7671):188–194, 2017. 416

[68] David Besanko, Sachin Gupta, and Dipak Jain. Logit demand estimation under competitive pricing behavior: An equilibrium framework. *Management Science*, 44(11-part-1):1533–1547, 1998. 405

[69] Robert Beverly, Ryan Koga, and Kimberly C. Claffy. Initial longitudinal analysis of IP source spoofing capability on the Internet. *Internet Society*, 2013. 175

[70] Karthikeyan Bhargavan, Bruno Blanchet, and Nadim Kobeissi. Verified models and reference implementations for the TLS 1.3 standard candidate. In *2017 IEEE Symposium on Security and Privacy, SP 2017, San Jose, CA, USA, May 22-26, 2017*, pages 483–502. IEEE Computer Society, 2017. ISBN 978-1-5090-5533-3. doi: 10.1109/SP.2017.26. URL https://doi.org/10.1109/SP.2017.26. 478

[71] Karthikeyan Bhargavan, Barry Bond, Antoine Delignat-Lavaud, Cédric Fournet, Chris Hawblitzel, Catalin Hritcu, Samin Ishtiaq, Markulf Kohlweiss, Rustan Leino, Jay R. Lorch, Kenji Maillard, Jianyang Pan,

Bryan Parno, Jonathan Protzenko, Tahina Ramananandro, Ashay Rane, Aseem Rastogi, Nikhil Swamy, Laure Thompson, Peng Wang, Santiago Zanella Béguelin, and Jean Karim Zinzindohoue. Everest: Towards a verified, drop-in replacement of HTTPS. In Benjamin S. Lerner, Rastislav Bodík, and Shriram Krishnamurthi, editors, *2nd Summit on Advances in Programming Languages, SNAPL 2017, May 7-10, 2017, Asilomar, CA, USA*, volume 71 of *LIPIcs*, pages 1:1–1:12. Schloss Dagstuhl - Leibniz-Zentrum für Informatik, 2017. doi: 10.4230/LIPIcs. SNAPL.2017.1. URL https://doi.org/10.4230/LIPIcs.SNAPL. 2017.1. 474

[72] Bobby Bhattacharjee, Ken Calvert, Jim Griffioen, Neil Spring, and James Sterbenz. Postmodern internetwork architecture. Technical Report ITTC Technical Report ITTC-FY2006-TR-45030-01, University of Kansas, February 2006. 575

[73] Debopam Bhattacherjee and Ankit Singla. Network topology design at 27,000 km/hour. In *Proceedings of the International Conference on Emerging Networking Experiments and Technologies (CoNEXT)*, 2019. doi: 10.1145/3359989.3365407. URL https://dl.acm.org/doi/ 10.1145/3359989.3365407. 378

[74] Debopam Bhattacherjee, Waqar Aqeel, Sangeetha Abdu Jyothi, Ilker Nadi Bozkurt, William Sentosa, Muhammad Tirmazi, Anthony Aguirre, Balakrishnan Chandrasekaran, P. Brighten Godfrey, Gregory P. Laughlin, Bruce M. Maggs, and Ankit Singla. cISP: A speed-of-light Internet service provider. In *Proceedings of the USENIX Symposium on Networked Systems Design and Implementation (NSDI)*. USENIX Association, 2022. URL https://www.usenix.org/conference/ nsdi22/presentation/bhattacherjee. 379

[75] Henry Birge-Lee, Yixin Sun, Annie Edmundson, Jennifer Rexford, and Prateek Mittal. Bamboozling certificate authorities with BGP. In *Proceedings of the USENIX Security Symposium*, 2018. 8, 425

[76] Henry Birge-Lee, Liang Wang, Daniel McCarney, Roland Shoemaker, Jennifer Rexford, and Prateek Mittal. Experiences deploying multi-vantage-point domain validation at let's encrypt. In *Proceedings of the USENIX Security Symposium*, 2021. 8, 425

[77] Bruno Blanchet. An efficient cryptographic protocol verifier based on prolog rules. In *14th IEEE Computer Security Foundations Workshop (CSFW-14 2001), 11-13 June 2001, Cape Breton, Nova Scotia, Canada*, pages 82–96. IEEE Computer Society, 2001. doi: 10.1109/ CSFW.2001.930138. URL https://doi.org/10.1109/CSFW.2001. 930138. 494, 566

[78] Burton H. Bloom. Space/time trade-offs in hash coding with allowable errors. *Communications of the ACM*, 13(7):422–426, 1970. 205

[79] Rakesh Bobba, Jon Grov, Indranil Gupta, Si Liu, José Meseguer, Peter Csaba Ölveczky, and Stephen Skeirik. Survivability: Design, formal modeling, and validation of cloud storage systems using Maude. In *Assured Cloud Computing*, chapter 2, pages 10–48. Wiley-IEEE Computer Society Press, 2018. 511, 512

[80] Sergio Boixo, Sergei V Isakov, Vadim N Smelyanskiy, Ryan Babbush, Nan Ding, Zhang Jiang, Michael J Bremner, John M Martinis, and Hartmut Neven. Characterizing quantum supremacy in near-term devices. *Nature Physics*, 14(6):595–600, 2018. 415

[81] Raffaele Bolla, Roberto Bruschi, Franco Davoli, and Flavio Cucchietti. Energy efficiency in the future Internet: a survey of existing approaches and trends in energy-aware fixed network infrastructures. *IEEE Communications Surveys & Tutorials*, 13(2):223–244, 2011. 11

[82] Béla Bollobás and W Fernandez De La Vega. The diameter of random regular graphs. *Combinatorica*, 2(2):125–134, 1982. 344

[83] Pat Bosshart, Dan Daly, Glen Gibb, Martin Izzard, Nick McKeown, Jennifer Rexford, Cole Schlesinger, Dan Talayco, Amin Vahdat, George Varghese, and David Walker. P4: Programming protocol-independent packet processors. *SIGCOMM Comput. Commun. Rev.*, 44(3):8795, July 2014. ISSN 0146-4833. doi: 10.1145/2656877.2656890. URL https://doi.org/10.1145/2656877.2656890. 578

[84] Timm Böttger, Felix Cuadrado, Gianni Antichi, Eder Leão Fernandes, Gareth Tyson, Ignacio Castro, and Steve Uhlig. An empirical study of the cost of dns-over-https. In *Proceedings of the ACM Internet Measurement Conference (IMC)*, 2019. 442

[85] John Boyland. Checking interference with fractional permissions. In Radhia Cousot, editor, *Static Analysis, 10th International Symposium, SAS 2003, San Diego, CA, USA, June 11-13, 2003, Proceedings*, volume 2694 of *Lecture Notes in Computer Science*, pages 55–72. Springer, 2003. 531

[86] Markus Brandt, Tianxiang Dai, Amit Klein, Haya Shulman, and Michael Waidner. Domain validation++ for mitm-resilient pki. In *Proceedings of the ACM Conference on Computer and Communications Security (CCS)*, 2018. 445

[87] Peter E. Bulychev, Alexandre David, Kim Guldstrand Larsen, Marius Mikucionis, Danny Bøgsted Poulsen, Axel Legay, and Zheng Wang.

UPPAAL-SMC: statistical model checking for priced timed automata. In *QAPL'12*, volume 85 of *EPTCS*, pages 1–16, 2012. 510

[88] R. Bush. Origin Validation Operation Based on the Resource Public Key Infrastructure (RPKI). RFC 7115, IETF, January 2014. URL http://tools.ietf.org/rfc/rfc7115.txt. 344

[89] Kevin Butler, Toni R. Farley, Patrick McDaniel, and Jennifer Rexford. A Survey of BGP Security Issues and Solutions. *Proceedings of the IEEE*, 98(1), 2010. 165

[90] CA/Browser Forum. Baseline requirements for the issuance and management of publiclytrusted certificates. https://cabforum.org/baseline-requirements/. 444

[91] Matthew Caesar and Jennifer Rexford. BGP routing policies in ISP networks. *IEEE Network: The Magazine of Global Internetworking*, 2005. 3

[92] CAIDA. Topology research. https://www.caida.org/research/topology, 2021. 514

[93] CAIDA. Spoofer project. https://www.caida.org/projects/spoofer/, 2021. 175

[94] CAIDA. State of IP spoofing. https://spoofer.caida.org/summary.php, 2021. 175

[95] Kenneth L. Calvert, James Griffioen, and Leonid Poutievski. Separating routing and forwarding: A clean-slate network layer design. In *Proceedings of International Conference on Broadband Communications, Networks and Systems (BROADNETS)*, 2007. 575

[96] Kenneth L. Calvert, James Griffioen, Anna Nagurney, and Tilman Wolf. A vision for a spot market for interdomain connectivity. In *2019 IEEE 39th International Conference on Distributed Computing Systems (ICDCS)*, pages 1860–1867, 2019. doi: 10.1109/ICDCS.2019.00184. URL https://ieeexplore.ieee.org/abstract/document/8884969. 147

[97] Jose M. Camacho, Alberto García-Martínez, Marcelo Bagnulo, and Francisco Valera. BGP-XM: BGP eXtended Multipath for transit autonomous systems. *Computer Networks*, 57(4):954–975, 2013. ISSN 1389-1286. URL https://doi.org/10.1016/j.comnet.2012.11.011. 582

[98] Enrico Cambiaso, Gianluca Papaleo, Giovanni Chiola, and Maurizio Aiello. Slow dos attacks: definition and categorisation. *International*

Journal of Trust Management in Computing and Communications, 1(3-4):300–319, 2013. 299

[99] I. Castineyra, N. Chiappa, and M. Steenstrup. The Nimrod Routing Architecture. RFC 1992, IETF, August 1996. URL http://tools.ietf.org/rfc/rfc1992.txt. 575

[100] Ignacio Castro, Aurojit Panda, Barath Raghavan, Scott Shenker, and Sergey Gorinsky. Route Bazaar: Automatic interdomain contract negotiation. In *Proceedings of USENIX Conference on Hot Topics in Operating Systems (HotOS)*, 2015. 578

[101] Center for Applied Internet Data Analysis. AS relationships – with geographic annotations. https://www.caida.org/data/as-relationships-geo/, 2016. 151, 152, 156, 402

[102] Center for Applied Internet Data Analysis. AS-rank. https://asrank.caida.org/, 2021. 151, 153

[103] Center for Applied Internet Data Analysis (CAIDA). Macroscopic Internet Topology Data Kit (ITDK). https://www.caida.org/data/internet-topology-data-kit/ (last visited Apr 19, 2021), archived at https://perma.cc/CU8X-7GRU. 402

[104] Chen Chen and Adrian Perrig. PHI: Path-hidden lightweight anonymity protocol at network layer. In *Proceedings on Privacy Enhancing Technologies (PoPETs)*, July 2017. 457

[105] Chen Chen, Daniele Enrico Asoni, David Barrera, George Danezis, and Adrian Perrig. HORNET: High-speed onion routing at the network layer. In *Proceedings of the ACM Conference on Computer and Communications Security (CCS)*, October 2015. 180, 457

[106] Chen Chen, David Barrera, and Adrian Perrig. Modeling data-plane power consumption of future Internet architectures. In *Proceedings of the IEEE Conference on Collaboration and Internet Computing (CIC)*, November 2016. 394, 395

[107] Chen Chen, Daniele E. Asoni, Adrian Perrig, David Barrera, George Danezis, and Carmela Troncoso. TARANET: Traffic-analysis resistant anonymity at the network layer. In *Proceedings of the IEEE European Symposium on Security and Privacy (EuroS&P)*, April 2018. 180, 457

[108] Lily Chen, Lily Chen, Stephen Jordan, Yi-Kai Liu, Dustin Moody, Rene Peralta, Ray Perlner, and Daniel Smith-Tone. *Report on post-quantum cryptography*, volume 12. US Department of Commerce, National Institute of Standards and Technology, 2016. 416

[109] Xiaomin Chen, Mohit Chamania, Admela Jukan, André C. Drummond, and Nelson L. S. da Fonseca. On the benefits of multipath routing for distributed data-intensive applications with high bandwidth requirements and multidomain reach. In *2009 Seventh Annual Communication Networks and Services Research Conference*, pages 110–117, 2009. doi: 10.1109/CNSR.2009.26. 373

[110] S. Cheshire and M. Krochmal. Multicast DNS. RFC 6762, IETF, February 2013. URL http://tools.ietf.org/rfc/rfc6762.txt. 331

[111] S. Cheshire and M. Krochmal. DNS-Based Service Discovery. RFC 6763, IETF, February 2013. URL http://tools.ietf.org/rfc/rfc6763.txt. 330

[112] Stuart Cheshire and Daniel Steinberg. *Zero configuration networking: The definitive guide.* O'Reilly Media, Inc., 2006. 327

[113] Laurent Chuat, Adrian Perrig, Raphael M. Reischuk, and Pawel Szalachowski. *Authentication Infrastructure*, pages 61–92. In Perrig et al. [413], 2017. ISBN 978-3-319-67079-5. doi: 10.1007/978-3-319-67080-5_4. URL https://doi.org/10.1007/978-3-319-67080-5_4. 35

[114] Laurent Chuat, Adrian Perrig, Raphael M. Reischuk, and Pawel Szalachowski. *ISD Coordination*, pages 93–100. In Perrig et al. [413], 2017. ISBN 978-3-319-67079-5. doi: 10.1007/978-3-319-67080-5_5. URL https://doi.org/10.1007/978-3-319-67080-5_5. 35

[115] Laurent Chuat, Adrian Perrig, Raphael M. Reischuk, and Brian Trammell. *Isolation Domains (ISDs)*, pages 43–57. In Perrig et al. [413], 2017. ISBN 978-3-319-67079-5. doi: 10.1007/978-3-319-67080-5_3. URL https://doi.org/10.1007/978-3-319-67080-5_3. 17, 35

[116] Laurent Chuat, Cyrill Krähenbühl, Prateek Mittal, and Adrian Perrig. F-PKI: Enabling innovation and trust flexibility in the HTTPS public-key infrastructure. In *Proceedings of the Symposium on Network and Distributed Systems Security (NDSS)*, 2022. doi: 10.14722/ndss.2022.24241. URL https://dx.doi.org/10.14722/ndss.2022.24241. 419, 451

[117] Taejoong Chung, Roland van Rijswijk-Deij, Balakrishnan Chandrasekaran, David Choffnes, Dave Levin, Bruce M. Maggs, Alan Mislove, and Christo Wilson. A longitudinal, End-to-End view of the DNSSEC ecosystem. In *26th USENIX Security Symposium (USENIX Security 17)*, pages 1307–1322, 2017. URL https://www.usenix.org/conference/usenixsecurity17/technical-sessions/presentation/chung. 437, 439, 443

[118] Cisco. Field notice: Endless BGP convergence problem in Cisco IOS software releases, 2001. 142

[119] CISCO. Cisco CRS 16-Slot Single-Shelf System Data Sheet, 2016. URL https://perma.cc/V8YW-XD9X. 395

[120] Cisco. Chapter: Configuring weighted fair queueing, 2021. URL https://www.cisco.com/c/en/us/td/docs/ios-xml/ios/qos_conmgt/configuration/15-mt/qos-conmgt-15-mt-book/qos-conmgt-cfg-wfq.html. 283

[121] Luca Cittadini, Giuseppe Di Battista, Thomas Erlebach, Maurizio Patrignani, and Massimo Rimondini. Assigning AS relationships to satisfy the Gao-Rexford conditions. In *Proceedings of the IEEE Conference on Network Protocols (ICNP)*, 2010. 139

[122] Thomas Claburn. Akamai edge dns goes down, takes a chunk of the internet with it. https://www.theregister.com, 2021. 432, 443

[123] B. Claise. Cisco Systems NetFlow Services Export Version 9. RFC 3954, IETF, October 2004. URL http://tools.ietf.org/rfc/rfc3954.txt. 218

[124] D. Clark, L. Chapin, V. Cerf, R. Braden, and R. Hobby. Towards the Future Internet Architecture. RFC 1287, IETF, December 1991. URL http://tools.ietf.org/rfc/rfc1287.txt. 326

[125] David Clark, Robert Braden, Aaron Falk, and Venkata Pingali. FARA: Reorganizing the addressing architecture. In *Proceedings of the ACM SIGCOMM Workshop on Future Directions in Network Architecture*, 2003. 575

[126] David Clark, Karen Sollins, John Wroclawski, Dina Katabi, Joanna Kulik, Xiaowei Yang, Robert Braden, Ted Faber, Aaron Falk, Venkata Pingali, Mark Handley, and Noel Chiappa. NewArch: Future generation Internet architecture. Technical report, Air Force Research Labs, 2004. 575

[127] Manuel Clavel, Francisco Durán, Steven Eker, Patrick Lincoln, Narciso Martí-Oliet, José Meseguer, and Carolyn L. Talcott. *All About Maude*, volume 4350 of *LNCS*. Springer, 2007. 511

[128] David Clayton, Christopher Patton, and Thomas Shrimpton. Probabilistic data structures in adversarial environments. In *Proceedings of the ACM Conference on Computer and Communications Security (CCS)*, 2019. 206

[129] Cloudflare. Argo Smart Routing. `https://www.cloudflare.com/products/argo-smart-routing/`, 2021. 345

[130] Avichai Cohen, Yossi Gilad, Amir Herzberg, and Michael Schapira. Jumpstarting BGP security with path-end validation. In *Proceedings of the ACM SIGCOMM Conference*, 2016. doi: 10.1145/2934872.2934883. 350

[131] Angelo Coiro, Marco Polverini, Antonio Cianfrani, and Marco Listanti. Energy Saving Improvements in IP Networks Through Table Lookup Bypass in Router Line Cards. 01 2013. doi: 10.1109/ICCNC.2013.6504147. 394

[132] SNB Communications. SNB and SIX launch the communication network Secure Swiss Finance Network. `https://www.snb.ch/en/mmr/reference/pre_20210715/source/pre_20210715.en.pdf`, July 2021. 386

[133] Comodo. Report of incident: Comodo detected and thwarted an intrusion on 26-mar-2011. `https://perma.cc/AT8Q-TJJC`. 420

[134] The PlanetLab Consortium. PlanetLab, an open platform for developing, deploying, and accessing planetary-scale services. `https://www.planet-lab.org/`, 2016. 362, 368

[135] A. Conta, S. Deering, and M. Gupta. Internet Control Message Protocol (ICMPv6) for the Internet Protocol Version 6 (IPv6) Specification. RFC 4443, IETF, March 2006. URL `http://tools.ietf.org/rfc/rfc4443.txt`. 6

[136] D. Cooper, S. Santesson, S. Farrell, S. Boeyen, R. Housley, and W. Polk. Internet X.509 Public Key Infrastructure Certificate and Certificate Revocation List (CRL) Profile. RFC 5280, IETF, May 2008. URL `http://tools.ietf.org/rfc/rfc5280.txt`. 38, 365

[137] Danny Cooper, Ethan Heilman, Kyle Brogle, Leonid Reyzin, and Sharon Goldberg. On the risk of misbehaving RPKI authorities. In *Proceedings of the ACM Workshop on Hot Topics in Networks (HotNets)*, November 2013. 6, 163

[138] M. Cooper, Y. Dzambasow, P. Hesse, S. Joseph, and R. Nicholas. Internet X.509 Public Key Infrastructure: Certification Path Building. RFC 4158, IETF, September 2005. URL `http://tools.ietf.org/rfc/rfc4158.txt`. 421

[139] G. Cormode and S. Muthukrishnan. An improved data stream summary: The count-min sketch and its applications. *Journal of Algorithms*, 55(1):58–75, 2005. ISSN 01966774. doi: 10.1016/j.jalgor.2003.

12.001. URL http://linkinghub.elsevier.com/retrieve/pii/ S0196677403001913. 218, 221

[140] Vlad C. Coroama and Lorenz M. Hilty. Assessing Internet Energy Intensity: A Review of Methods and Results. *Environmental Impact Assessment Review*, 45:63–68, 2014. ISSN 0195-9255. doi: https://doi.org/10.1016/j.eiar.2013.12.004. URL https://www.sciencedirect.com/science/article/pii/S0195925513001121. 400

[141] Denis Cousineau, Damien Doligez, Leslie Lamport, Stephan Merz, Daniel Ricketts, and Hernán Vanzetto. TLA + proofs. In Dimitra Giannakopoulou and Dominique Méry, editors, *FM 2012: Formal Methods - 18th International Symposium, Paris, France, August 27-31, 2012. Proceedings*, volume 7436 of *Lecture Notes in Computer Science*, pages 147–154. Springer, 2012. doi: 10.1007/978-3-642-32759-9_14. URL https://doi.org/10.1007/978-3-642-32759-9_14. 481

[142] Jim Cowie. The new threat: Targeted Internet traffic misdirection. https://perma.cc/EVD6-ZM8S, November 2013. 7

[143] Cas Cremers, Marko Horvat, Jonathan Hoyland, Sam Scott, and Thyla van der Merwe. A comprehensive symbolic analysis of TLS 1.3. In Bhavani M. Thuraisingham, David Evans, Tal Malkin, and Dongyan Xu, editors, *Proceedings of the 2017 ACM SIGSAC Conference on Computer and Communications Security, CCS 2017, Dallas, TX, USA, October 30 - November 03, 2017*, pages 1773–1788. ACM, 2017. ISBN 978-1-4503-4946-8. doi: 10.1145/3133956.3134063. URL http://doi.acm.org/10.1145/3133956.3134063. 478

[144] Scott Crosby and Dan Wallach. Denial of service via algorithmic complexity attacks. In *Proceedings of the USENIX Security Symposium*, 2003. 207

[145] Jon Crowcroft, Steven Hand, Richard Mortier, Timothy Roscoe, and Andrew Warfield. Plutarch: An argument for network pluralism. *ACM SIGCOMM Computer Communication Review*, 33(4):258266, August 2003. doi: 10.1145/972426.944763. URL https://doi.org/10.1145/972426.944763. 576

[146] CZ.NIC Labs. The BIRD Internet Routing Daemon Project. https://bird.network.cz, 2021. 354

[147] David Dagon, Chris Lee, Wenke Lee, and Niels Provos. Corrupted dns resolution paths: The rise of a malicious resolution authority. 2008. 434

[148] Tianxiang Dai, Philipp Jeitner, Haya Shulman, and Michael Waidner. From ip to transport and beyond: Cross-layer attacks against applications. In *Proceedings of the ACM SIGCOMM Conference*, 2021. 9

[149] J. Damas, M. Graff, and P. Vixie. Extension Mechanisms for DNS (EDNS(0)). RFC 6891, IETF, April 2013. URL `http://tools.ietf.org/rfc/rfc6891.txt`. 430

[150] Eli Dart, Lauren Rotman, Brian Tierney, Mary Hester, and Jason Zurawski. The science dmz: A network design pattern for data-intensive science. *Scientific Programming*, 22(2):173–185, 2014. 389

[151] Dataportal. DIGITAL 2021: GLOBAL OVERVIEW REPORT. `https://perma.cc/7WJF-ELE9`, 2021. 396

[152] Andy Davidson. BGP traffic engineering. `https://archive.nanog.org/sites/default/files/tues.tutorial.davidson.multihoming.41.pdf`, 2013. The Africa Peering and Interconnection Forum. 137

[153] Spencer Dawkins. Path Aware Networking: Obstacles to Deployment (A Bestiary of Roads Not Taken). Internet-Draft draft-irtf-panrg-what-not-to-do-13, Internet Engineering Task Force, October 2020. URL `https://datatracker.ietf.org/doc/html/draft-irtf-panrg-what-not-to-do-13`. Work in Progress. 11

[154] Wladimir De la Cadena, Asya Mitseva, Jens Hiller, Jan Pennekamp, Sebastian Reuter, Julian Filter, Thomas Engel, Klaus Wehrle, and Andriy Panchenko. Trafficsliver: Fighting website fingerprinting attacks with traffic splitting. In *Proceedings of the ACM Conference on Computer and Communications Security (CCS)*, page 19711985, 2020. ISBN 9781450370899. doi: 10.1145/3372297.3423351. 179

[155] Joeri de Ruiter and Caspar Schutijser. Next-generation internet at terabit speed: Scion in p4. In *Proceedings of the International Conference on Emerging Networking Experiments and Technologies (CoNEXT)*, 2021. 94, 321, 367

[156] Wouter de Vries, José Jair Santanna, Anna Sperotto, and Aiko Pras. How asymmetric is the internet? In *IFIP International Conference on Autonomous Infrastructure, Management and Security*, 2015. 191

[157] S. Deering and R. Hinden. Internet Protocol, Version 6 (IPv6) Specification. RFC 2460, IETF, December 1998. URL `http://tools.ietf.org/rfc/rfc2460.txt`. 2

[158] S. Deering and R. Hinden. Internet Protocol, Version 6 (IPv6) Specification. RFC 8200, IETF, July 2017. URL `http://tools.ietf.org/rfc/rfc8200.txt`. 2, 95, 96, 124

[159] Martin Dehnel-Wild and Cas Cremers. Security vulnerability in 5G-AKA draft (3GPP TS 33.501 draft v0.7.0). Technical report, February 2018. 13

[160] Varun Deshpande, Hakim Badis, and Laurent George. BTCmap: Mapping bitcoin peer-to-peer network topology. In *IFIP/IEEE International Conference on Performance Evaluation and Modeling in Wired and Wireless Networks (PEMWN)*, pages 1–6, 2018. doi: 10.23919/PEMWN.2018.8548904. 344

[161] Omer Deutsch, Neta Rozen Schiff, Danny Dolev, and Michael Schapira. Preventing (network) time travel with chronos. In *NDSS*, 2018. 191

[162] Devon OBrien. Chromium's ct log policy update: Permissible rejection criteria 198 views. https://groups.google.com/a/chromium.org/g/ct-policy/, 2017. 447

[163] DFN. German national research and education network. https://www.dfn.de/. 367

[164] Amogh Dhamdhere, Kc Claffy, David D. Clark, Alexander Gamero-Garrido, Matthew Luckie, Ricky K. P. Mok, Gautam Akiwate, Kabir Gogia, Vaibhav Bajpai, and Alex C. Snoeren. Inferring persistent inter-domain congestion. In *Proceedings of the ACM SIGCOMM Conference*, 2018. doi: 10.1145/3230543.3230549. 265

[165] T. Dierks and E. Rescorla. The Transport Layer Security (TLS) Protocol Version 1.2. RFC 5246, IETF, August 2008. URL http://tools.ietf.org/rfc/rfc5246.txt. 413

[166] Wenxiu Ding, Zheng Yan, and Robert H. Deng. A survey on future Internet security architectures. *IEEE Access*, 4:4374–4393, July 2016. 181

[167] Danny Dolev and Andrew Chi-Chih Yao. On the security of public key protocols. *IEEE Trans. Information Theory*, 29(2):198–207, 1983. doi: 10.1109/TIT.1983.1056650. URL https://doi.org/10.1109/TIT.1983.1056650. 491

[168] Danny Dolev and Andrew Chi-Chih Yao. On the security of public key protocols. *IEEE Transactions on Information Theory*, 29(2):198–207, 1983. doi: 10.1109/TIT.1983.1056650. 162, 422

[169] Danny Dolev, Nancy A Lynch, Shlomit S Pinter, Eugene W Stark, and William E Weihl. Reaching approximate agreement in the presence of faults. *Journal of the ACM (JACM)*, 33(3):499–516, 1986. 195

[170] V. Dukhovni and W. Hardaker. The DNS-Based Authentication of Named Entities (DANE) Protocol: Updates and Operational Guidance. RFC 7671, IETF, October 2015. URL `http://tools.ietf.org/rfc/rfc7671.txt`. 429

[171] Zakir Durumeric, James Kasten, David Adrian, J. Alex Halderman, Michael Bailey, Frank Li, Nicolas Weaver, Johanna Amann, Jethro Beekman, Mathias Payer, and Vern Paxson. The matter of Heartbleed. In *Proceedings of the ACM Internet Measurement Conference (IMC)*, 2014. 13

[172] Ottmar Edenhofer. *Renewable Energy Sources and Climate Change Mitigation: Special Report of the Intergovernmental Panel on Climate Change.* Cambridge university Press, 2012. 402

[173] Toby Ehrenkranz and Jun Li. On the state of ip spoofing defense. *ACM Transactions on Internet Technology (TOIT)*, 9(2):1–29, 2009. 277

[174] Adam Eijdenberg, Ben Laurie, and Al Cutter. Verifiable data structures, 2015. URL `https://github.com/google/trillian/blob/master/docs/papers/VerifiableDataStructures.pdf`. 468

[175] Marco Eilers and Peter Müller. Nagini: A static verifier for python. In Hana Chockler and Georg Weissenbacher, editors, *Computer Aided Verification - 30th International Conference, CAV 2018, Held as Part of the Federated Logic Conference, FloC 2018, Oxford, UK, July 14-17, 2018, Proceedings, Part I*, volume 10981 of *Lecture Notes in Computer Science*, pages 596–603. Springer, 2018. doi: 10.1007/978-3-319-96145-3_33. URL `https://doi.org/10.1007/978-3-319-96145-3_33`. 556

[176] Marco Eilers, Peter Müller, and Samuel Hitz. Modular product programs. *ACM Trans. Program. Lang. Syst.*, 42(1):3:1–3:37, 2020. doi: 10.1145/3324783. URL `https://doi.org/10.1145/3324783`. 556, 560, 562

[177] Marco Eilers, Severin Meier, and Peter Müller. Product programs in the wild: Retrofitting program verifiers to check information flow security. In *Computer Aided Verification (CAV)*, pages 718–741. Springer International Publishing, 2021. ISBN 978-3-030-81685-8. doi: 10.1007/978-3-030-81685-8_34. URL `https://doi.org/10.1007/978-3-030-81685-8_34`. 556, 560

[178] European Commission. Fed4FIRE+: Federation for FIRE plus. `https://www.fed4fire.eu/`, . 367, 580

[179] European Commission. FIWARE: Core platform of the future Internet. `https://www.fiware.org`, . 580

[180] European Commission. FIRE: Future Internet research and experimentation. `https://www.ict-fire.eu`, . 580

[181] European Commission. FORWARD: Managing emerging threats in ICT infrastructures. `http://www.ict-forward.eu`, . 578

[182] European Commission. Slices: Scientific large-scale infrastructure for computing/communication experimental studies. `https://www.slices-ri.eu/`, . 581

[183] European Commission. SysSec: A European network of excellence in managing threats and vulnerabilities in the future Internet. `http://www.syssec-project.eu`, . 578

[184] Reinhard Exel. Mitigation of asymmetric link delays in ieee 1588 clock synchronization systems. *IEEE Communications Letters*, 18(3):507–510, 2014. 191

[185] D. Farinacci, V. Fuller, D. Meyer, and D. Lewis. The Locator/ID Separation Protocol (LISP). RFC 6830, IETF, January 2013. URL `http://tools.ietf.org/rfc/rfc6830.txt`. 28

[186] P. Ferguson and D. Senie. Network Ingress Filtering: Defeating Denial of Service Attacks which employ IP Source Address Spoofing. RFC 2827, IETF, May 2000. URL `http://tools.ietf.org/rfc/rfc2827.txt`. 174

[187] Clarence Filsfils, Stefano Previdi, Bruno Decraene, Stephane Litkowski, and Rob Shakir. Segment routing architecture. Internet-draft, February 2017. 578

[188] Simon Fischer and Berthold Vöcking. Adaptive routing with stale information. *Theoretical Computer Science*, 2009. 11

[189] Lester Randolph Ford and Delbert R Fulkerson. Maximal flow through a network. *Canadian journal of Mathematics*, 8:399–404, 1956. 156

[190] Romain Fouchereau. 2021 Global DNS Threat Report. Technical report, European Security, IDC, 2021. 432

[191] Linux Foundation. Af_xdp, 2021. URL `https://www.kernel.org/doc/html/latest/networking/af_xdp.html`. 383

[192] Mozilla Foundation. Public suffix list. `https://publicsuffix.org`. 427, 451

[193] Nick Galov. 39 jaw-dropping DDoS statistics to keep in mind for 2021. `https://hostingtribunal.com/blog/ddos-statistics/`, 2021. 265

[194] Igor Ganichev, Bin Dai, P. Brighten Godfrey, and Scott Shenker. YAMR: Yet another multipath routing protocol. *ACM SIGCOMM Computer Communication Review*, 40(5):1319, October 2010. ISSN 0146-4833. doi: 10.1145/1880153.1880156. URL https://doi.org/10.1145/1880153.1880156. 582

[195] Lixin Gao and Jennifer Rexford. Stable Internet routing without global coordination. *Networking, IEEE/ACM Trans on*, 9(6):681–692, 2001. 139, 140

[196] Kilian Gärtner, Jonghoon Kwon, and David Hausheer. SpeedCam: Towards efficient flow monitoring for multipath communication. In *Proceedings of the IFIP/IEEE International Symposium on Integrated Network Management (IM)*, 2021. URL https://netsec.ethz.ch/publications/papers/speedcam_annet21.pdf. 367

[197] Marten Gartner. Parts - path-aware transport over scion, 2021. URL https://github.com/netsys-lab/parts. 383

[198] Marten Gartner. Bittorrent over scion, 2021. URL https://github.com/martenwallewein/torrent-client. 384

[199] Marten Gartner. Scion dvb-t2 video streaming setup, 2021. URL https://github.com/netsys-lab/scion-video-setup. 384

[200] Marten Gartner, Thorben Krüger, and David Hausheer. Scion pathdiscovery, 2021. URL https://github.com/netsys-lab/scion-path-discovery. 383

[201] M. Gattulli, M. Tornatore, R. Fiandra, and A. Pattavina. Low-Carbon Routing Algorithms for Cloud Computing Services in IP-over-WDM Networks. In *IEEE International Conference on Communications (ICC)*, pages 2999–3003, 2012. doi: 10.1109/ICC.2012.6364347. 399

[202] GENI. The ExoGENI testbed. http://www.exogeni.net/. 367

[203] Thomas Gerbet, Amrit Kumar, and Cédric Lauradoux. The power of evil choices in Bloom filters. In *Proceedings of the IEEE/IFIP International Conference on Dependable Systems and Networks (DSN)*, 2015. 206

[204] Jonas Gessner. Leveraging application layer path-awareness with scion. Master's thesis, ETH Zurich, Zurich, 2021. 384

[205] Jonas Gessner. Scion-webrtc, 2021. URL https://github.com/JonasGessner/SCION-WebRTC. 384

[206] Phillipa Gill, Michael Schapira, and Sharon Goldberg. A survey of inter-domain routing policies. *ACM SIGCOMM Computer Communication Review*, 44(1):28–34, 2013. 140, 142

[207] Damien Giry. Cryptographic key length recommendation. `http://www.keylength.com`, 2016. 408

[208] Giacomo Giuliari, Tobias Klenze, Markus Legner, David Basin, Adrian Perrig, and Ankit Singla. Internet backbones in space. *ACM SIGCOMM Computer Communication Review*, 2020. doi: 10.1145/3390251.3390256. URL `https://netsec.ethz.ch/publications/papers/ccr-ibis-2020.pdf`. 145, 378, 379

[209] Giacomo Giuliari, Dominik Roos, Marc Wyss, Juan Angel García-Pardo, Markus Legner, and Adrian Perrig. Colibri: A cooperative lightweight inter-domain bandwidth-reservation infrastructure. In *International Conference on emerging Networking EXperiments and Technologies (CoNEXT 21)*, 2021. doi: 10.1145/3485983.3494871. URL `https://netsec.ethz.ch/publications/papers/2021_conext_colibri.pdf`. 227

[210] GLORIAD. Global ring network for advanced applications development. `https://www.gloriad.org/`. 581

[211] Go Authors. Vet. `https://perma.cc/N3UQ-CFPU`, 2021. 533

[212] P. Brighten Godfrey, Igor Ganichev, Scott Shenker, and Ion Stoica. Pathlet routing. In *Proceedings of the ACM SIGCOMM Conference on Data Communication*, SIGCOMM, 2009. ISBN 9781605585949. doi: 10.1145/1592568.1592583. URL `https://doi.org/10.1145/1592568.1592583`. 577, 578, 583

[213] Markus Goldstein, Christoph Lampert, Matthias Reif, Armin Stahl, and Thomas Breuel. Bayes optimal DDoS mitigation by adaptive history-based IP filtering. In *Proceedings of the International Conference on Networking (ICN)*, 2008. 209

[214] F. Gont. ICMP Attacks against TCP. RFC 5927, IETF, July 2010. URL `http://tools.ietf.org/rfc/rfc5927.txt`. 8

[215] F. Gont, S. Krishnan, T. Narten, and R. Draves. Temporary Address Extensions for Stateless Address Autoconfiguration in IPv6. RFC 8981, IETF, February 2021. URL `http://tools.ietf.org/rfc/rfc8981.txt`. 326

[216] Geoffrey Goodell, William Aiello, Timothy Griffin, John Ioannidis, Patrick D. McDaniel, and Aviel D. Rubin. Working around BGP: An incremental approach to improving security and accuracy in interdomain routing. In *Proceedings of the Symposium on Network and Distributed Systems Security (NDSS)*, February 2003. 165

[217] Jose Gracia, Patrick OConnor, Lawrence Markel, Rui Shan, Thomas Rizy, and Alfonso Tarditi. Hydropower plants as black start resources. *ORNL*, page 1077, 2018. 130

[218] Glenn Greenwald. *No place to hide: Edward Snowden, the NSA, and the US surveillance state*. Macmillan, 2014. 179

[219] Grid5000. Large-scale and flexible testbed for experiment-driven research. https://www.grid5000.fr/. 367

[220] T. Griffin and G. Huston. BGP Wedgies. RFC 4264, IETF, November 2005. URL http://tools.ietf.org/rfc/rfc4264.txt. 4, 140

[221] Timothy Griffin and Gordon Wilfong. An analysis of BGP convergence properties. *ACM SIGCOMM Computer Communication Review*, 29(4): 277–288, 1999. 4, 140

[222] Timothy G. Griffin, F. Bruce Shepherd, and Gordon Wilfong. The stable paths problem and interdomain routing. *IEEE/ACM Transactions on Networking*, 10(2):232–243, 2002. 142

[223] S. Gueron, A. Langley, and Y. Lindell. AES-GCM-SIV: Nonce Misuse-Resistant Authenticated Encryption. RFC 8452, IETF, April 2019. URL http://tools.ietf.org/rfc/rfc8452.txt. 414

[224] A. Gulbrandsen, P. Vixie, and L. Esibov. A DNS RR for specifying the location of services (DNS SRV). RFC 2782, IETF, February 2000. URL http://tools.ietf.org/rfc/rfc2782.txt. 330

[225] GÉANT. Pan-european research and education network. https://www.geant.org/. 580

[226] GÉANT Association. GÉANT testbeds service (GTS). https://www.geant.org/Services/Connectivity_and_network/GTS. 367, 580

[227] Dongsu Han, Ashok Anand, Fahad Dogar, Boyan Li, Hyeontaek Lim, Michel Machado, Arvind Mukundan, Wenfei Wu, Aditya Akella, David G. Andersen, John W. Byers, Srinivasan Seshan, and Peter Steenkiste. XIA: Efficient support for evolvable internetworking. In *Proceedings of the USENIX Symposium on Networked Systems Design and Implementation (NSDI)*, 2012. 577

[228] Yihua He, Michalis Faloutsos, Srikanth Krishnamurthy, and Bradley Huffaker. On routing asymmetry in the internet. In *GLOBECOM'05. IEEE Global Telecommunications Conference, 2005.*, volume 2, pages 6–pp. IEEE, 2005. 191

[229] Ward Heddeghem, Filip Idzikowski, Willem Vereecken, Didier Colle, Mario Pickavet, and Piet Demeester. Power Consumption Modeling in Optical Multilayer Networks. *Photonic Network Communications*, 24, Oct 2012. doi: 10.1007/s11107-011-0370-7. 402

[230] Amir Herzberg and Haya Shulman. Dnssec: Security and availability challenges. In *Proceedings of the IEEE Conference on Communications and Network Security (CNS)*, 2013. 440

[231] Stephen Herzog. Revisiting the Estonian cyber attacks: Digital threats and multinational responses. *Journal of Strategic Security*, 4(2):49–60, 2011. 8

[232] Kerry Hinton, Jayant Baliga, Michael Feng, Robert Ayre, and Rodney S. Tucker. Power consumption and energy efficiency in the Internet. *IEEE Network*, 25(2):6–12, 2011. 394

[233] Kerry Hinton, Fatemeh Jalali, and Ashrar Matin. Energy consumption modelling of optical networks. *Photonic Network Communications*, 03 2015. doi: 10.1007/s11107-015-0491-5. 401

[234] Muks Hirani, Sarah Jones, and Ben Read. Global dns hijacking campaign: Dns record manipulation at scale. `https://www.fireeye.com/blog/threat-research/`, 2019. 439, 446

[235] Samuel Hitz. Demonstrating the reliability and resilience of Secure Swiss Finance Network. `https://perma.cc/4H3Q-WZNG`, February 2021. Anapaya Blog. 28

[236] Samuel Hitz. Anapaya connect. `https://perma.cc/3L42-D5BH`, October 2021. Anapaya Website. 322

[237] Samuel Hitz, Adrian Perrig, Stephen Shirley, and Pawel Szalachowski. *Control Plane*, pages 119–160. In Perrig et al. [413], 2017. ISBN 978-3-319-67079-5. doi: 10.1007/978-3-319-67080-5_7. URL `https://doi.org/10.1007/978-3-319-67080-5_7`. 65

[238] Paul E. Hoffman and Peter van Dijk. Recursive to Authoritative DNS with Unauthenticated Encryption. Internet-draft, Internet Engineering Task Force, 2021. 440

[239] Toke Høiland-Jørgensen, Jesper Dangaard Brouer, Daniel Borkmann, John Fastabend, Tom Herbert, David Ahern, and David Miller. The express data path: Fast programmable packet processing in the operating system kernel. In *Proceedings of the 14th international conference on emerging networking experiments and technologies*, pages 54–66, 2018. 313

[240] M. Holdrege and P. Srisuresh. Protocol Complications with the IP Network Address Translator. RFC 3027, IETF, January 2001. URL http://tools.ietf.org/rfc/rfc3027.txt. 326

[241] Thomas Holterbach, Edgar Costa Molero, Maria Apostolaki, Alberto Dainotti, Stefano Vissicchio, and Laurent Vanbever. Blink: Fast connectivity recovery entirely in the data plane. In *16th {USENIX} Symposium on Networked Systems Design and Implementation ({NSDI} 19)*, pages 161–176, 2019. 191

[242] Hans Hoogstraaten, Ronald Prins, Daniël Niggebrugge, Danny Heppener, Frank Groenewegen, Janna Wettink, Kevin Strooy, Pascal Arends, Paul Pols, Robbert Kouprie, Steffen Moorrees, Xander van Pelt, and Yun Zheng Hu. Black Tulip: Report of the investigation into the diginotar certificate authority breach. Technical report, August 2012. 420, 446

[243] Austin Hounsel, Paul Schmitt, Kevin Borgolte, and Nick Feamster. Encryption without centralization: Distributing dns queries across recursive resolvers. In *Proceedings of the ACM Applied Networking Research Workshop (ANRW)*, 2021. 437

[244] Rebekah Houser, Shuai Hao, Zhou Li, Daiping Liu, Chase Cotton, and Haining Wang. A Comprehensive Measurement-based Investigation of DNS Hijacking. In *Proceedings of the International Symposium on Reliable Distributed Systems (SRDS)*, 2021. 8, 446

[245] R. Housley. Cryptographic Message Syntax (CMS). RFC 5652, IETF, September 2009. URL http://tools.ietf.org/rfc/rfc5652.txt. 41

[246] R. Housley. Guidelines for Cryptographic Algorithm Agility and Selecting Mandatory-to-Implement Algorithms. RFC 7696, IETF, November 2015. URL http://tools.ietf.org/rfc/rfc7696.txt. 408, 412

[247] R. Housley. TLS 1.3 Extension for Certificate-Based Authentication with an External Pre-Shared Key. RFC 8773, IETF, March 2020. URL http://tools.ietf.org/rfc/rfc8773.txt. 467

[248] Yih-Chun Hu, Adrian Perrig, and Marvin Sirbu. SPV: Secure path vector routing for securing BGP. In *Proceedings of the ACM SIGCOMM Conference*, September 2004. 165

[249] Yih-Chun Hu, Adrian Perrig, and David Johnson. Wormhole attacks in wireless networks. *IEEE Journal on Selected Areas in Communications (JSAC)*, 24(2), February 2006. 170, 507

[250] Yih-Chun Hu, Tobias Klausmann, Adrian Perrig, Raphael M. Reischuk, Stephen Shirley, Pawel Szalachowski, and Ercan Ucan. *Deployment and Operation*, pages 191–239. In Perrig et al. [413], 2017. ISBN 978-3-319-67079-5. doi: 10.1007/978-3-319-67080-5_10. URL https://doi.org/10.1007/978-3-319-67080-5_10. 129, 317

[251] Bradley Huffaker, Marina Fomenkov, Daniel J. Plummer, David Moore, and Kimberly Claffy. Distance metrics in the Internet. In *Proceedings of the IEEE International Telecommunications Symposium (ITS)*, 2002. 236

[252] G. Huston. Autonomous System (AS) Number Reservation for Documentation Use. RFC 5398, IETF, December 2008. URL http://tools.ietf.org/rfc/rfc5398.txt. 32

[253] G. Huston, G. Michaelson, and R. Loomans. A Profile for X.509 PKIX Resource Certificates. RFC 6487, IETF, February 2012. URL http://tools.ietf.org/rfc/rfc6487.txt. 465

[254] Geoff Huston. The QoS emperor's wardrobe. https://labs.ripe.net/Members/gih/the-qos-emperors-wardrobe, 2012. 265

[255] Geoff Huston. Measuring dnssec performance. https://labs.apnic.net, 2013. 438

[256] Geoff Huston. BGP in 2014. https://perma.cc/GKK2-6PNV, January 2015. 11, 165

[257] IANIX. Major dnssec outages and validation failures. https://ianix.com/pub/dnssec-outages.html, 2021. 437, 438, 440, 443

[258] IEA. Global Energy & CO2 Status Report 2019, IEA. https://perma.cc/PLV6-XRAF. 393

[259] IEEE. Ieee standard for a precision clock synchronization protocol for networked measurement and control systems. 190

[260] IEEE. IEEE standard for local and metropolitan area networks–link aggregation. *IEEE Std 802.1AX-2020 (Revision of IEEE Std 802.1AX-2014)*, pages 1–333, 2020. doi: 10.1109/IEEESTD.2020.9105034. 322

[261] IETF. Charter for Working Group. Technical Report charter-ietf-taps-02, Internet Engineering Task Force, March 2021. URL `https://datatracker.ietf.org/doc/charter-ietf-taps/`. Work in Progress. 310

[262] imec and Ghent University. The Virtual Wall testbed. `https://doc.ilabt.imec.be/ilabt/virtualwall/`. 367

[263] International Energy Agency (IEA). Data & Statistics. `https://www.iea.org/data-and-statistics?country=WORLD&fuel=Electricity%20and%20heat&indicator=ElecGenByFuel` (last visited Apr 19, 2021), archived at `https://perma.cc/72QW-26D9`. 402

[264] Internet Assigned Numbers Authority (IANA). Autonomous system (AS) numbers. `https://perma.cc/TEG2-8D3Z`, . 32

[265] Internet Assigned Numbers Authority (IANA). Protocol numbers. `https://perma.cc/FBE8-S2W5`, . 315

[266] ITU. Fixed-broadband subscriptions. `https://perma.cc/789Q-S3TL`, 2021. 396

[267] J. Iyengar and M. Thomson. QUIC: A UDP-Based Multiplexed and Secure Transport. RFC 9000, IETF, May 2021. URL `http://tools.ietf.org/rfc/rfc9000.txt`. 243

[268] Dennis Jackson, Cas Cremers, Katriel Cohn-Gordon, and Ralf Sasse. Seems legit: Automated analysis of subtle attacks on protocols that use signatures. In *Proceedings of the ACM Conference on Computer and Communications Security (CCS)*, 2019. ISBN 978-1-4503-6747-9. 53

[269] Van Jacobson, Diana K. Smetters, James D. Thornton, Michael F. Plass, Nicholas H. Briggs, and Rebecca L. Braynard. Networking named content. In *Proceedings of the International Conference on Emerging Networking Experiments and Technologies (CoNEXT)*, 2009. 576

[270] Sushant Jain, Alok Kumar, Subhasree Mandal, Joon Ong, Leon Poutievski, Arjun Singh, Subbaiah Venkata, Jim Wanderer, Junlan Zhou, Min Zhu, Jonathan Zolla, Urs Hölzle, Stephen Stuart, and Amin Vahdat. B4: Experience with a globally-deployed software defined WAN. In *Proceedings of the ACM SIGCOMM Conference*, August 2013. 382

[271] Jian Jiang, Jinjin Liang, Kang Li, Jun Li, Haixin Duan, Jianping Wu, et al. Ghost domain names: Revoked yet still resolvable. In *The 19th Annual Network & Distributed System Security Symposium (NDSS 2012)*, 2012. 441

[272] Tony John and David Hausheer. S3mp: A scion based secure smart metering platform. In *2021 IFIP/IEEE International Symposium on Integrated Network Management (IM)*, pages 944–949, 2021. 320, 380

[273] Tony John, Piet De Vaere, Caspar Schutijser, Adrian Perrig, and David Hausheer. Linc: Low-cost inter-domain connectivity for industrial systems. In *Proceedings of the SIGCOMM '21 Poster and Demo Sessions*, SIGCOMM '21, page 6870, New York, NY, USA, 2021. Association for Computing Machinery. ISBN 9781450386296. doi: 10.1145/3472716.3472850. URL https://doi.org/10.1145/3472716.3472850. 320, 380

[274] Wolfgang John and Sven Tafvelin. Analysis of Internet backbone traffic and header anomalies observed. In *Proceedings of the ACM Internet Measurement Conference (IMC)*, 2007. 216

[275] Petri Jokela, András Zahemszky, Christian Esteve Rothenberg, Somaya Arianfar, and Pekka Nikander. LIPSIN: Line speed publish/subscribe inter-networking. In *Proceedings of the ACM SIGCOMM Conference*, 2009. 577

[276] Sangeetha Abdu Jyothi. Solar superstorms: Planning for an internet apocalypse. In *Proceedings of Sigcomm*, 2021. 191, 193

[277] Farouk Kamoun and Leonard Kleinrock. Stochastic performance evaluation of hierarchical routing for large networks. *Computer Networks*, 3: 337–353, November 1979. 575

[278] Min Suk Kang and Virgil D. Gligor. Routing bottlenecks in the Internet: Causes, exploits, and countermeasures. In *Proceedings of the ACM Conference on Computer and Communications Security (CCS)*, 2014. 265

[279] Min Suk Kang, Soo Bum Lee, and Virgil D. Gligor. The Crossfire attack. In *Proceedings of the IEEE Symposium on Security and Privacy (S&P)*, May 2013. 9, 269, 276

[280] Pravein Govindan Kannan, Raj Joshi, and Mun Choon Chan. Precise time-synchronization in the data-plane using programmable switching asics. In *Proceedings of the 2019 ACM Symposium on SDN Research*, pages 8–20, 2019. 191

[281] M. Kapor. Building the open road: The NREN as test-bed for the national public network. RFC 1259, IETF, September 1991. URL http://tools.ietf.org/rfc/rfc1259.txt. 389

[282] Aqsa Kashaf, Vyas Sekar, and Yuvraj Agarwal. Analyzing third party service dependencies in modern web services: Have we learned

from the Mirai-Dyn incident? In *Proceedings of the ACM Internet Measurement Conference (IMC)*, 2020. ISBN 9781450381383. doi: 10.1145/3419394.3423664. 131, 433

[283] D. Katz and D. Ward. Bidirectional Forwarding Detection (BFD). RFC 5880, IETF, June 2010. URL http://tools.ietf.org/rfc/rfc5880.txt. 123

[284] Jonathan Katz, Alfred Menezes, Paul Van Oorschot, and Scott Vanstone. *Handbook of applied cryptography*. CRC press, 1996. 56

[285] Ethan Katz-Bassett, Colin Scott, David Choffnes, Italo Cunha, Vytautas Valancius, Nick Feamster, Harsha Madhyastha, Thomas Anderson, and Arvind Krishnamurthy. LIFEGUARD: Practical repair of persistent route failures. In *Proceedings of the ACM SIGCOMM Conference*, August 2012. 28

[286] H. Tahilramani Kaur, S. Kalyanaraman, A. Weiss, S. Kanwar, and A. Gandhi. BANANAS: An evolutionary framework for explicit and multipath routing in the internet. In *Proceedings of the ACM SIGCOMM Workshop on Future Directions in Network Architecture*, FDNA '03, page 277288, New York, NY, USA, 2003. Association for Computing Machinery. ISBN 1581137486. doi: 10.1145/944759.944766. URL https://doi.org/10.1145/944759.944766. 582

[287] S. Kent. IP Authentication Header. RFC 4302, IETF, December 2005. URL http://tools.ietf.org/rfc/rfc4302.txt. 61

[288] S. Kent and K. Seo. Security Architecture for the Internet Protocol. RFC 4301, IETF, December 2005. URL http://tools.ietf.org/rfc/rfc4301.txt. 163

[289] Kalevi Kilkki and Benjamin Finley. In search of lost QoS. https://arxiv.org/abs/1901.06867, 2019. 265

[290] Tiffany Hyun-Jin Kim, Cristina Basescu, Limin Jia, Soo Bum Lee, Yih-Chun Hu, and Adrian Perrig. Lightweight source authentication and path validation. In *Proceedings of the ACM SIGCOMM Conference*, August 2014. 53, 565

[291] KISTI. Korea research environment open network. https://www.kreonet.net/. 368, 581

[292] Gerwin Klein, Kevin Elphinstone, Gernot Heiser, June Andronick, David Cock, Philip Derrin, Dhammika Elkaduwe, Kai Engelhardt, Rafal Kolanski, Michael Norrish, Thomas Sewell, Harvey Tuch, and Simon Winwood. sel4: formal verification of an OS kernel. In

Jeanna Neefe Matthews and Thomas E. Anderson, editors, *Proceedings of the 22nd ACM Symposium on Operating Systems Principles 2009, SOSP 2009, Big Sky, Montana, USA, October 11-14, 2009,* pages 207–220. ACM, 2009. doi: 10.1145/1629575.1629596. URL https://doi.org/10.1145/1629575.1629596. 474, 479

[293] Leonard Kleinrock and Farouk Kamoun. Hierarchical routing for large networks: Performance evaluation and optimization. *Computer Networks,* 1:155–174, 1977. 575

[294] J. Klensin, N. Freed, and K. Moore. SMTP Service Extension for Message Size Declaration. RFC 1653, IETF, July 1994. URL http://tools.ietf.org/rfc/rfc1653.txt. 2

[295] Tobias Klenze. *Formal Development of Secure Data Plane Protocols.* PhD thesis, ETH Zürich, 2021. URL https://doi.org/10.3929/ethz-b-000506662. 170

[296] Tobias Klenze, Giacomo Giuliari, Christos Pappas, Adrian Perrig, and David Basin. Networking in heaven as on earth. In *Proceedings of the ACM Workshop on Hot Topics in Networks (HotNets),* November 2018. URL https://netsec.ethz.ch/publications/papers/ibis_hotnets.pdf. 378

[297] Tobias Klenze, David Basin, and Christoph Sprenger. Formal verification of secure forwarding protocols. In *34th IEEE Computer Security Foundations Symposium (CSF 2021),* 2021. 30, 170, 496

[298] Zeno Koller. Measuring loss and reordering with few bits. Master's thesis, ETH Zurich, 2018. 366

[299] Teemu Koponen, Scott Shenker, Hari Balakrishnan, Nick Feamster, Igor Ganichev, Ali Ghodsi, P. Brighten Godfrey, Nick McKeown, Guru Parulkar, Barath Raghavan, Jennifer Rexford, Somaya Arianfar, and Dmitriy Kuptsov. Architecting for innovation. *ACM SIGCOMM Computer Communication Review,* July 2011. 577

[300] Martin Koppehel. Bittorrent over scion, 2021. URL https://github.com/martin31821/torrent. 384

[301] Brian Krebs. DDoS on Dyn impacts Twitter, Spotify, Reddit. https://perma.cc/XJ7Q-CY5X, October 2016. 8

[302] Brian Krebs. Its way too easy to get a .gov domain name. https://perma.cc/NAM2-LSX9, November 2019. 425

[303] Thorben Krüger and David Hausheer. Towards an api for the path-aware internet. In *Proceedings of the ACM SIGCOMM 2021 Work-shop on Network-Application Integration*, NAI'21, page 6872, New York, NY, USA, 2021. Association for Computing Machinery. ISBN 9781450386333. doi: 10.1145/3472727.3472808. URL https://doi.org/10.1145/3472727.3472808. 310

[304] Cyrill Krähenbühl, Markus Legner, Silvan Bitterli, and Adrian Per-rig. Pervasive Internet-wide low-latency authentication. In *Proceed-ings of the International Conference on Computer Communications and Networks (ICCCN)*, 2021. URL https://netsec.ethz.ch/publications/papers/kraehenbuehl2021pila.pdf. 461, 462, 467

[305] Cyrill Krähenbühl, Seyedali Tabaeiaghdaei, Christelle Gloor, Jonghoon Kwon, Adrian Perrig, David Hausheer, and Dominik Roos. Deployment and scalability of an inter-domain multi-path routing infrastructure. In *The 17th International Conference on emerging Networking EXperi-ments and Technologies (CoNEXT '21)*, December 2021. doi: 10.1145/3485983.3494862. 65, 129, 151, 317

[306] Sanjeev Kumar. Smurf-based distributed denial of service (DDoS) at-tack amplification in Internet. In *Second International Conference on Internet Monitoring and Protection (ICIMP)*, July 2007. 8

[307] Nate Kushman, Srikanth Kandula, and Dina Katabi. Can you hear me now?! it must be BGP. *ACM SIGCOMM Computer Communication Review*, April 2007. 4, 28

[308] Nate Kushman, Srikanth Kandula, Dina Katabi, and Bruce M. Maggs. R-BGP: Staying connected in a connected world. In *Proceedings of the 4th USENIX Conference on Networked Systems Design and Implemen-tation*, NSDI '07, page 25, USA, 2007. USENIX Association. URL https://dl.acm.org/doi/10.5555/1973430.1973455. 582

[309] M. Kwiatkowska, G. Norman, and D. Parker. PRISM 4.0: Verification of probabilistic real-time systems. In *Proceedings of the International Conference on Computer Aided Verification (CAV)*, 2011. 511

[310] Jonghoon Kwon, Juan A. García-Pardo, Markus Legner, François Wirz, Matthias Frei, David Hausheer, and Adrian Perrig. SCIONLab: A next-generation Internet testbed. In *Proceedings of the IEEE Conference on Network Protocols (ICNP)*, 2020. URL https://netsec.ethz.ch/publications/papers/icnp2020_scionlab.pdf. 361

[311] Jonghoon Kwon, Taeho Lee, Claude Hähni, and Adrian Perrig. SVLAN: Secure & scalable network virtualization. In *Proceedings of the Sympo-*

sium on Network and Distributed Systems Security (NDSS), 2020. doi: 10.14722/ndss.2020.24162. 367

[312] Jonghoon Kwon, Claude Hähni, Patrick Bamert, and Adrian Perrig. MONDRIAN: Comprehensive inter-domain network zoning architecture. In *Proceedings of the Symposium on Network and Distributed System Security (NDSS)*, 2021. doi: 10.14722/ndss.2021. 24378. URL https://netsec.ethz.ch/publications/papers/ndss21_mondrian.pdf. 382

[313] Craig Labovitz, Abha Ahuja, Abhijit Bose, and Farnam Jahanian. Delayed Internet routing convergence. In *Proceedings of the ACM SIGCOMM Conference*, 2000. 4, 191

[314] APNIC Labs. Use of dnssec validation for world. https://stats.labs.apnic.net/dnssec/, 2021. 440

[315] Mohit Lad, Ricardo Oliveira, Beichuan Zhang, and Lixia Zhang. Understanding resiliency of Internet topology against prefix hijack attacks. In *International Conference on Dependable Systems and Networks (DSN)*, pages 368–377. IEEE, 2007. 273

[316] Leslie Lamport. *Specifying Systems, The TLA+ Language and Tools for Hardware and Software Engineers*. Addison-Wesley, 2002. ISBN 0-3211-4306-X. 481, 568

[317] Bob Lantz, Brandon Heller, and Nick McKeown. A network in a laptop: rapid prototyping for software-defined networks. In *Proceedings of the 9th ACM SIGCOMM Workshop on Hot Topics in Networks*, 2010. 368

[318] B. Laurie, A. Langley, and E. Kasper. Certificate Transparency. RFC 6962, IETF, June 2013. URL http://tools.ietf.org/rfc/rfc6962.txt. 420, 464

[319] Ben Laurie, Pierre Phaneuf, and Adam Eijdenberg. Certificate transparency over dns. https://perma.cc/TW3W-Y4ZN, March 2016. 429

[320] Ben Laurie, Adam Langley, Emilia Kasper, Eran Messeri, and Rob Stradling. Certificate Transparency Version 2.0. Internet-Draft draft-ietf-trans-rfc6962-bis-42, Internet Engineering Task Force, 2021. URL https://datatracker.ietf.org/doc/html/draft-ietf-trans-rfc6962-bis-42. Work in Progress. 447

[321] Hyunwoo Lee, Zach Smith, Junghwan Lim, Gyeongjae Choi, Selin Chun, Taejoong Chung, and Ted Taekyoung Kwon. maTLS: How to make TLS middlebox-aware? In *Proceedings of the Symposium on Network and Distributed Systems Security (NDSS)*, 2019. 430

[322] Jason Lee, Adrian Perrig, and Pawel Szalachowski. *Host Structure*, pages 179–190. In Perrig et al. [413], 2017. ISBN 978-3-319-67079-5. doi: 10.1007/978-3-319-67080-5_9. URL https://doi.org/10.1007/978-3-319-67080-5_9. 303

[323] Ki Suh Lee, Han Wang, Vishal Shrivastav, and Hakim Weatherspoon. Globally synchronized time via datacenter networks. In *Proceedings of the 2016 ACM SIGCOMM Conference*, pages 454–467, 2016. 190

[324] Taeho Lee, Christos Pappas, David Barrera, Pawel Szalachowski, and Adrian Perrig. Source accountability with domain-brokered privacy. In *Proceedings of the International Conference on Emerging Networking Experiments and Technologies (CoNEXT)*, December 2016. 326

[325] Taeho Lee, Christos Pappas, Pawel Szalachowski, and Adrian Perrig. Communication based on per-packet one-time addresses. In *Proceedings of the IEEE Conference on Network Protocols (ICNP)*, November 2016. 180, 326

[326] Taeho Lee, Christos Pappas, Adrian Perrig, Virgil Gligor, and Yih-Chun Hu. The case for in-network replay suppression. In *Proceedings of the ACM Asia Conference on Computer and Communications Security (ASIACCS)*, April 2017. doi: 10.1145/3052973.3052988. URL https://doi.org/10.1145/3052973.3052988. 204, 205

[327] Markus Legner, Tobias Klenze, Marc Wyss, Christoph Sprenger, and Adrian Perrig. EPIC: Every packet is checked in the data plane of a path-aware Internet. In *Proceedings of the USENIX Security Symposium*, August 2020. URL https://netsec.ethz.ch/publications/papers/Legner_Usenix2020_EPIC.pdf. 30, 227, 228, 237

[328] M. Lepinski and S. Kent. An Infrastructure to Support Secure Internet Routing. RFC 6480, IETF, February 2012. URL http://tools.ietf.org/rfc/rfc6480.txt. 5, 163, 336, 337

[329] M. Lepinski and K. Sriram. BGPsec Protocol Specification. RFC 8205, IETF, September 2017. URL http://tools.ietf.org/rfc/rfc8205.txt. 5, 152, 163

[330] Matt Lepinski. BGPsec protocol specification. Internet-Draft, June 2016. 137, 165

[331] Matt Lepinski and Sean Turner. An overview of BGPsec. Internet-Draft, June 2016. 137, 165

[332] Xavier Leroy. Formal certification of a compiler back-end or: programming a compiler with a proof assistant. In J. Gregory Morrisett and Simon L. Peyton Jones, editors, *Proceedings of the 33rd ACM SIGPLAN-SIGACT Symposium on Principles of Programming Languages, POPL*

2006, Charleston, South Carolina, USA, January 11-13, 2006, pages 42–54. ACM, 2006. doi: 10.1145/1111037.1111042. URL https://doi.org/10.1145/1111037.1111042. 474, 479

[333] Martin Lévesque and David Tipper. Improving the ptp synchronization accuracy under asymmetric delay conditions. In *2015 IEEE International Symposium on Precision Clock Synchronization for Measurement, Control, and Communication (ISPCS)*, pages 88–93. IEEE, 2015. 191

[334] Bingyu Li, Jingqiang Lin, Fengjun Li, Qiongxiao Wang, Qi Li, Jiwu Jing, and Congli Wang. Certificate Transparency in the Wild: Exploring the Reliability of Monitors. In *Proceedings of the ACM Conference on Computer and Communications Security (CCS)*, 2019. 426

[335] Ming Li, Andrey Lukyanenko, Zhonghong Ou, Antti Ylä-Jääski, Sasu Tarkoma, Matthieu Coudron, and Stefano Secci. Multipath transmission for the internet: A survey. *IEEE Communications Surveys Tutorials*, 18 (4):2887–2925, 2016. doi: 10.1109/COMST.2016.2586112. 373

[336] Qi Li, Yih-Chun Hu, and Xinwen Zhang. Even rockets cannot make pigs fly sustainably: Can BGP be secured with BGPsec? In *Workshop on Security of Emerging Networking Technologies (SENT)*, 2014. 6, 13

[337] Yong Liao, Lixin Gao, Roch Guerin, and Zhi-Li Zhang. Reliable inter-domain routing through multiple complementary routing processes. In *Proceedings of the 4th International Conference on emerging Networking EXperiments and Technologies*, CoNEXT '08, New York, NY, USA, 2008. Association for Computing Machinery. ISBN 9781605582108. doi: 10.1145/1544012.1544080. URL https://doi.org/10.1145/1544012.1544080. 582

[338] libcurl. The multiprotocol file transfer library. https://curl.se/libcurl. 311

[339] J. Littlefield. Vendor-Identifying Vendor Options for Dynamic Host Configuration Protocol version 4 (DHCPv4). RFC 3925, IETF, October 2004. URL http://tools.ietf.org/rfc/rfc3925.txt. 329

[340] Si Liu, Peter Csaba Ölveczky, and José Meseguer. Modeling and analyzing mobile ad hoc networks in Real-Time Maude. *J. Log. Algebraic Methods Program.*, 85(1):34–66, 2016. doi: 10.1016/j.jlamp.2015.05.002. URL https://doi.org/10.1016/j.jlamp.2015.05.002. 511

[341] Si Liu, Peter Csaba Ölveczky, Qi Wang, Indranil Gupta, and José Meseguer. Read atomic transactions with prevention of lost updates: ROLA and its formal analysis. *Formal Asp. Comput.*, 31(5):503–540, 2019. 511

[342] Si Liu, Atul Sandur, José Meseguer, Peter Csaba Ölveczky, and Qi Wang. Generating correct-by-construction distributed implementations from formal maude designs. In Ritchie Lee, Susmit Jha, and Anastasia Mavridou, editors, *NASA Formal Methods - 12th International Symposium, NFM 2020, Moffett Field, CA, USA, May 11-15, 2020, Proceedings*, volume 12229 of *Lecture Notes in Computer Science*, pages 22–40. Springer, 2020. doi: 10.1007/978-3-030-55754-6_2. URL https://doi.org/10.1007/978-3-030-55754-6_2. 512

[343] Wenjing Lou and Yuguang Fang. A multipath routing approach for secure data delivery. In *2001 MILCOM Proceedings Communications for Network-Centric Operations: Creating the Information Force (Cat. No.01CH37277)*, volume 2, pages 1467–1473, 2001. doi: 10.1109/ MILCOM.2001.986098. 375

[344] Gavin Lowe. An attack on the Needham–Schroeder public-key authentication protocol. *Information Processing Letters*, 56(3):131–133, 1995. ISSN 0020-0190. doi: https://doi.org/10.1016/0020-0190(95)00144-2. URL https://www.sciencedirect.com/science/article/pii/ 0020019095001442. 475

[345] Gavin Lowe. A hierarchy of authentication specification. In *10th Computer Security Foundations Workshop (CSFW '97), June 10-12, 1997, Rockport, Massachusetts, USA*, pages 31–44. IEEE Computer Society, 1997. doi: 10.1109/CSFW.1997.596782. URL https://doi.org/10. 1109/CSFW.1997.596782. 491

[346] Matthew Luckie, Robert Beverly, Ryan Koga, Ken Keys, Joshua A. Kroll, and k claffy. Network hygiene, incentives, and regulation: Deployment of source address validation in the Internet. In *Proceedings of the ACM Conference on Computer and Communications Security (CCS)*, 2019. ISBN 9781450367479. doi: 10.1145/3319535.3354232. 175

[347] Robert Lychev, Sharon Goldberg, and Michael Schapira. BGP security in partial deployment: Is the juice worth the squeeze? In *Proceedings of the ACM SIGCOMM Conference*, pages 171–182, 2013. 5

[348] Nancy A. Lynch and Frits W. Vaandrager. Forward and backward simulations: I. untimed systems. *Inf. Comput.*, 121(2):214–233, 1995. doi: 10.1006/inco.1995.1134. URL https://doi.org/10.1006/ inco.1995.1134. 479, 485

[349] C. Lynn, S. Kent, and K. Seo. X.509 Extensions for IP Addresses and AS Identifiers. RFC 3779, IETF, June 2004. URL http://tools. ietf.org/rfc/rfc3779.txt. 465

[350] Damien Magoni and Jean-Jacques Pansiot. Internet topology modeler based on map sampling. In *Proceedings of the International Symposium on Computers and Communications (ISCC)*, 2002. doi: 10.1109/iscc.2002.1021797. 236

[351] Keyu Man, Zhiyun Qian, Zhongjie Wang, Xiaofeng Zheng, Youjun Huang, and Haixin Duan. Dns cache poisoning attack reloaded: Revolutions with side channels. In *Proceedings of the ACM Conference on Computer and Communications Security (CCS)*, 2020. 8, 434

[352] Z. Morley Mao, Randy Bush, Timothy G Griffin, and Matthew Roughan. BGP beacons. In *Proceedings of the 3rd ACM SIGCOMM conference on Internet measurement*, 2003. 368

[353] et al. Marten Seemann, Lucas Clemente. quic-go, 2021. URL https://github.com/lucas-clemente/quic-go. 306, 309

[354] Celso Martinho and Tom Strickx. Understanding how facebook disappeared from the internet. https://blog.cloudflare.com/, 2021. 432, 443

[355] Vasileios Mavroeidis, Kamer Vishi, Mateusz D. Zych, and Audun Jøsang. The impact of quantum computing on present cryptography. *International Journal of Advanced Computer Science and Applications*, 9(3), 2018. doi: 10.14569/IJACSA.2018.090354. URL http://dx.doi.org/10.14569/IJACSA.2018.090354. 415

[356] James McCauley, Yotam Harchol, Aurojit Panda, Barath Raghavan, and Scott Shenker. Enabling a Permanent Revolution in Internet Architecture. In *Proceedings of the ACM SIGCOMM Conference on Data Communication*, 2019. doi: 10.1145/3341302.3342075. URL https://doi.org/10.1145/3341302.3342075. 581

[357] Steve McConnell. *Code complete - a practical handbook of software construction, 2nd Edition*. Microsoft Press, 2004. ISBN 9780735619678. URL https://www.worldcat.org/oclc/249645389. 520

[358] Rick McGeer, Mark Berman, Chip Elliott, and Robert Ricci, editors. *The GENI Book*. Springer International Publishing, 2016. doi: 10.1007/978-3-319-33769-2. URL https://doi.org/10.1007/978-3-319-33769-2. 581

[359] Nick McKeown, Tom Anderson, Hari Balakrishnan, Guru Parulkar, Larry Peterson, Jennifer Rexford, Scott Shenker, and Jonathan Turner. Openflow: enabling innovation in campus networks. *ACM SIGCOMM Computer Communication Review*, 38(2):69–74, March 2008. 578

[360] D. McPherson, V. Gill, D. Walton, and A. Retana. Border Gateway Protocol (BGP) Persistent Route Oscillation Condition. RFC 3345, IETF, August 2002. URL http://tools.ietf.org/rfc/rfc3345.txt. 142

[361] M. Mealling and R. Daniel. The Naming Authority Pointer (NAPTR) DNS Resource Record. RFC 2915, IETF, September 2000. URL http://tools.ietf.org/rfc/rfc2915.txt. 330

[362] Deepankar Medhi and Karthikeyan Ramasamy. *Network Routing: Algorithms, Protocols, and Architectures*. Morgan Kaufmann Publishers Inc., San Francisco, CA, USA, 2007. ISBN 0120885883. 241, 255, 279

[363] Angelique Medina. One small config change for cloudflare, one giant outage for the internet. https://www.thousandeyes.com/, 2020. 443

[364] Angelique Medina. Facebook outage analysis. https://perma.cc/HLN5-RXQL, 2021. 131

[365] Jiten Mehta and Eric Kinnear. Boost performance and security with modern networking. https://developer.apple.com/videos/play/wwdc2020/10111/. URL https://developer.apple.com/videos/play/wwdc2020/10111/. 310

[366] Simon Meier, Benedikt Schmidt, Cas Cremers, and David A. Basin. The TAMARIN prover for the symbolic analysis of security protocols. In Natasha Sharygina and Helmut Veith, editors, *Computer Aided Verification - 25th International Conference, CAV 2013, Saint Petersburg, Russia, July 13-19, 2013. Proceedings*, volume 8044 of *Lecture Notes in Computer Science*, pages 696–701. Springer, 2013. doi: 10.1007/978-3-642-39799-8_48. URL https://doi.org/10.1007/978-3-642-39799-8_48. 494, 564, 566

[367] Microsoft. Azure Virtual WAN. https://docs.microsoft.com/en-us/azure/virtual-wan/virtual-wan-about, 2021. 345

[368] D. Mills, J. Martin, J. Burbank, and W. Kasch. Network Time Protocol Version 4: Protocol and Algorithms Specification. RFC 5905, IETF, June 2010. URL http://tools.ietf.org/rfc/rfc5905.txt. 190

[369] Stephen A. Misel. Wow, AS7007! https://perma.cc/MW92-2L3X, April 1997. NANOG mailing list. 4

[370] J. Mitchell. Autonomous System (AS) Reservation for Private Use. RFC 6996, IETF, July 2013. URL http://tools.ietf.org/rfc/rfc6996.txt. 32

[371] Mitre Corporation. CVE-2008-1447: "DNS insufficient socket entropy vulnerability" or "the Kaminsky bug". https://perma.cc/BB4P-MWYL, 2008. 80

[372] K. Moriarty, B. Kaliski, and A. Rusch. PKCS 5: Password-Based Cryptography Specification Version 2.1. RFC 8018, IETF, January 2017. URL http://tools.ietf.org/rfc/rfc8018.txt. 411

[373] Reynaldo Morillo, Justin Furuness, Cameron Morris, James Breslin, Amir Herzberg, and Bing Wang. ROV++: Improved deployable defense against BGP hijacking. In *Proceedings of Network and Distributed Systems Security (NDSS)*, February 2021. 7

[374] Murtaza Motiwala, Megan Elmore, Nick Feamster, and Santosh Vempala. Path splicing. In *Proceedings of the ACM SIGCOMM Conference on Data Communication*, SIGCOMM '08, page 2738, New York, NY, USA, 2008. Association for Computing Machinery. ISBN 9781605581750. doi: 10.1145/1402958.1402963. URL https://doi.org/10.1145/1402958.1402963. 582

[375] Giovane C. M. Moura, Sebastian Castro, John Heidemann, and Wes Hardaker. Tsuname: Exploiting misconfiguration and vulnerability to ddos dns. In *Proceedings of the ACM Internet Measurement Conference (IMC)*, 2021. 454

[376] Moritz Müller, Jins de Jong, Maran van Heesch, Benno Overeinder, and Roland van Rijswijk-Deij. Retrofitting post-quantum cryptography in internet protocols: A case study of dnssec. *ACM SIGCOMM Computer Communication Review*, 50(4):4957, 2020. 438

[377] P. Müller, M. Schwerhoff, and A. J. Summers. Automatic verification of iterated separating conjunctions using symbolic execution. In S. Chaudhuri and A. Farzan, editors, *Computer Aided Verification (CAV)*, volume 9779 of *LNCS*, pages 405–425. Springer-Verlag, 2016. 534

[378] Peter Müller, Malte Schwerhoff, and Alexander J. Summers. Viper: A verification infrastructure for permission-based reasoning. In Barbara Jobstmann and K. Rustan M. Leino, editors, *Verification, Model Checking, and Abstract Interpretation*, pages 41–62, Berlin, Heidelberg, 2016. Springer Berlin Heidelberg. ISBN 978-3-662-49122-5. 533

[379] Jürg Mägerle and Robert Oleschak. The swiss interbank clearing (sic) payment system, 2019. URL http://www.snb.ch/en/mmr/reference/sic_system/source. 376, 385

[380] Alireza Nafarieh, Yashar Fazili, and William Robertson. Dynamic Inter-domain Negotiation for Green Algorithms in Optical Networks. *Procedia Computer Science*, 21:25–32, 2013.

ISSN 1877-0509. doi: https://doi.org/10.1016/j.procs.2013.09. 006. URL http://www.sciencedirect.com/science/article/ pii/S1877050913007990. The 4th International Conference on Emerging Ubiquitous Systems and Pervasive Networks (EUSPN-2013) and the 3rd International Conference on Current and Future Trends of Information and Communication Technologies in Healthcare (ICTH). 399

[381] Jad Naous, Michael Walfish, Antonio Nicolosi, David Mazieres, Michael Miller, and Arun Seehra. Verifying and enforcing network paths with ICING. In *Proceedings of the International Conference on Emerging Networking Experiments and Technologies (CoNEXT)*, 2011. 97, 228

[382] National Science Foundation. Bridges: Binding research infrastructures for the deployment of global experimental science. https:// cpopoviciu.github.io/BRIDGES/. 581

[383] NDN. Named Data Networking (NDN) - A Future Internet Architecture. https://named-data.net, June 2015. 576

[384] Roger M. Needham and Michael D. Schroeder. Using encryption for authentication in large networks of computers. *Communications of the ACM*, 21(12):993–999, December 1978. ISSN 0001-0782. doi: 10.1145/359657.359659. URL https://doi.org/10.1145/359657. 359659. 475

[385] Netcope. Netcope NFB-200G2QL FPGA platform equipped with Virtex Ultrascale+ FPGA chip, Accessed September 2020. URL https://www.netcope.com/getattachment/bb2b8efa-9925- 438d-b895-897d7c1e4745/NFB-200G2QL-product-brief.aspx. 224

[386] Cédric Neukom. High-performance file transfer in scion. Master's thesis, ETH Zurich, 2020. 383

[387] C. Neuman, T. Yu, S. Hartman, and K. Raeburn. The Kerberos Network Authentication Service (V5). RFC 4120, IETF, July 2005. URL http: //tools.ietf.org/rfc/rfc4120.txt. 57

[388] Chris Newcombe, Tim Rath, Fan Zhang, Bogdan Munteanu, Marc Brooker, and Michael Deardeuff. How amazon web services uses formal methods. *Commun. ACM*, 58(4):66–73, 2015. doi: 10.1145/ 2699417. URL https://doi.org/10.1145/2699417. 478

[389] K. Nichols, S. Blake, F. Baker, and D. Black. Definition of the Differentiated Services Field (DS Field) in the IPv4 and IPv6 Headers. RFC

2474, IETF, December 1998. URL `http://tools.ietf.org/rfc/rfc2474.txt`. 95, 238, 281

[390] Tobias Nipkow, Lawrence C. Paulson, and Markus Wenzel. *Isabelle/HOL - A Proof Assistant for Higher-Order Logic*, volume 2283 of *Lecture Notes in Computer Science*. Springer, 2002. ISBN 3-540-43376-7. doi: 10.1007/3-540-45949-9. URL `https://doi.org/10.1007/3-540-45949-9`. 481, 495

[391] NIST. FIPS PUB 180-2, Secure Hash Standard (SHS), 2008. 413

[392] Erik Nordstrom, David Shue, Prem Gopalan, Rob Kiefer, Matvey Arye, Steven Ko, Jennifer Rexford, and Michael J. Freedman. Serval: An end-host stack for service-centric networking. In *Proceedings of the USENIX Symposium on Networked Systems Design and Implementation (NSDI)*, 2012. 577

[393] nsnam. ns-3 Network Simulator. `https://www.nsnam.org/`, archived at `https://perma.cc/4S4N-8VGV`, 2021. 151, 152, 402

[394] University of Oregon. Route Views project. `http://www.routeviews.org/routeviews`. 151, 368

[395] Ookla. Speedtest Global Index. `https://perma.cc/MK9P-CCYC`, 2021. 396

[396] H. Orman and P. Hoffman. Determining Strengths For Public Keys Used For Exchanging Symmetric Keys. RFC 3766, IETF, April 2004. URL `http://tools.ietf.org/rfc/rfc3766.txt`. 408

[397] OVGU Magdeburg. Deployment and evaluation of the SCION secure internet architecture on Fed4FIRE+ testbeds. `https://www.fed4fire.eu/demo-stories/oc5/scion-on-fed4fire/`. 367

[398] Palo Alto Research Center (PARC). Project CCNx. `http://blogs.parc.com/ccnx`. 576

[399] Giorgos Papastergiou, Gorry Fairhurst, David Ros, Anna Brunstrom, Karl-Johan Grinnemo, Per Hurtig, Naeem Khademi, Michael Tüxen, Michael Welzl, Dragana Damjanovic, et al. De-ossifying the internet transport layer: A survey and future perspectives. *IEEE Communications Surveys & Tutorials*, 19(1):619–639, 2016. doi: 10.1109/COMST.2016.2626780. 2

[400] Christos Pappas, Adrian Perrig, Raphael M. Reischuk, Stephen Shirley, and Pawel Szalachowski. *Data Plane*, pages 161–177. In Perrig et al. [413], 2017. ISBN 978-3-319-67079-5. doi: 10.1007/978-3-319-67080-5_8. URL `https://doi.org/10.1007/978-3-319-67080-5_8`. 93

[401] Bryan Parno, Adrian Perrig, and David Andersen. SNAPP: Stateless network-authenticated path pinning. In *Proceedings of the ACM Symposium on Information, Computer, and Communications Security (ASIACCS)*, March 2008. 578

[402] Gaurav Parthasarathy, Peter Müller, and Alexander J. Summers. Formally validating a practical verification condition generator. In *CAV (1)*, volume 12760 of *Lecture Notes in Computer Science*, pages 704–727. Springer, 2021. 522

[403] Abhinav Pathak, Himabindu Pucha, Ying Zhang, Y Charlie Hu, and Z Morley Mao. A measurement study of internet delay asymmetry. In *International Conference on Passive and Active Network Measurement*, 2008. 191

[404] Vern Paxson. End-to-end routing behavior in the internet. *IEEE/ACM transactions on Networking*, 5(5):601–615, 1997. 143

[405] Paul Pearce, Ben Jones, Frank Li, Roya Ensafi, Nick Feamster, Nick Weaver, and Vern Paxson. Global measurement of DNS manipulation. In *26th USENIX Security Symposium (USENIX Security 17)*, pages 307–323, 2017. URL https://www.usenix.org/conference/usenixsecurity17/technical-sessions/presentation/pearce. 434

[406] Edwin Pednault, John Gunnels, Dmitri Maslov, and Jay Gambetta. On quantum supremacy. https://perma.cc/MZ4C-MPJ7, October 2019. 415

[407] Tao Peng, Christopher Leckie, and Kotagiri Ramamohanarao. Protection from distributed denial of service attacks using history-based IP filtering. In *IEEE International Conference on Communications (ICC)*, 2003. 209, 212

[408] Willem Penninckx, Bart Jacobs, and Frank Piessens. Sound, modular and compositional verification of the input/output behavior of programs. In *ESOP*, volume 9032 of *Lecture Notes in Computer Science*, pages 158–182. Springer, 2015. 548

[409] Nicole Perlroth. Hackers used new weapons to disrupt major websites across U.S. https://perma.cc/L7US-MK8P, October 2016. The New York Times. 8

[410] Adrian Perrig and Pawel Szalachowski. *Cryptographic Algorithms*, pages 381–386. In Perrig et al. [413], 2017. ISBN 978-3-319-67079-5. doi: 10.1007/978-3-319-67080-5_17. URL https://doi.org/10.1007/978-3-319-67080-5_17. 407

[411] Adrian Perrig, Raphael M. Reischuk, Dominik Roos, and Pawel Szalachowski. *OPT and DRKey*, pages 279–297. In Perrig et al. [413], 2017. ISBN 978-3-319-67079-5. doi: 10.1007/978-3-319-67080-5_12. URL https://doi.org/10.1007/978-3-319-67080-5_12. 228

[412] Adrian Perrig, Pawel Szalachowski, Raphael M Reischuk, and Laurent Chuat. *SCION: a secure internet architecture*. Springer, 2017. ISBN 978-3-319-67079-9. 517

[413] Adrian Perrig, Pawel Szalachowski, Raphael M. Reischuk, and Laurent Chuat. *SCION: A Secure Internet Architecture*. Springer International Publishing, 2017. ISBN 978-3-319-67079-5. doi: 10.1007/978-3-319-67080-5. URL https://doi.org/10.1007/978-3-319-67080-5. 36, 38, 53, 81, 86, 122, 124, 589, 590, 596, 607, 609, 616, 623, 624, 625

[414] Yarin Perry, Neta Rozen-Schiff, and Michael Schapira. A devil of a time: How vulnerable is ntp to malicious timeservers? In *NDSS*, 2021. 191

[415] Hanna Pihkola, Mikko Hongisto, Olli Apilo, and Mika Lasanen. Evaluating the Energy Consumption of Mobile Data TransferFrom Technology Development to Consumer Behaviour and Life Cycle Thinking. *Sustainability*, 10:2494, 07 2018. doi: 10.3390/su10072494. 398

[416] Michele Polese, Federico Chiariotti, Elia Bonetto, Filippo Rigotto, Andrea Zanella, and Michele Zorzi. A survey on recent advances in transport layer protocols. *IEEE Communications Surveys & Tutorials*, 21 (4):35843608, October 2019. ISSN 1553-877X. doi: 10.1109/COMST. 2019.2932905. 2

[417] J. Postel. Internet Protocol. RFC 791, IETF, September 1981. URL http://tools.ietf.org/rfc/rfc0791.txt. 2, 6

[418] John Preskill. Quantum computing and the entanglement frontier, 2012. 415

[419] Penny Pritzker and Patrick D. Gallagher. SHA-3 standard: Permutation-based hash and extendable-output functions. *Information Tech Laboratory National Institute of Standards and Technology*, pages 1–35, 2014. 413

[420] 2STIC Programme. 2STiC: Security, Stability and Transparency in internetwork Communication. https://www.2stic.nl/, 2021. 367

[421] DPDK Project. Data Plane Development Kit. https://dpdk.org, 2020. 217, 223, 235

[422] Jonathan Protzenko, Bryan Parno, Aymeric Fromherz, Chris Hawblitzel, Marina Polubelova, Karthikeyan Bhargavan, Benjamin Beurdouche, Joonwon Choi, Antoine Delignat-Lavaud, Cédric Fournet, Tahina Ramananandro, Aseem Rastogi, Nikhil Swamy, Christoph M. Wintersteiger, and Santiago Zanella Béguelin. Evercrypt: A fast, verified, cross-platform cryptographic provider. *IACR Cryptol. ePrint Arch.*, 2019:757, 2019. URL https://eprint.iacr.org/2019/757. 571

[423] Donghong Qin, Jiahai Yang, Zhuolin Liu, Jessie Wang, Bin Zhang, and Wei Zhang. AMIR: Another multipath interdomain routing. In *Proceedings of the IEEE 26th International Conference on Advanced Information Networking and Applications*, AINA '12, pages 581–588, 03 2012. doi: 10.1109/AINA.2012.83. URL https://doi.org/10.1109/AINA.2012.83. 582

[424] Barath Raghavan and Alex C. Snoeren. A system for authenticated policy-compliant routing. In *Proceedings of the ACM SIGCOMM Conference*, 2004. 578, 583

[425] Barath Raghavan, Patric Verkaik, and Alex C. Snoeren. Secure and policy-compliant source routing. *IEEE/ACM Transactions on Networking*, 17(3), 2009. 578, 583

[426] K. Ramakrishnan, S. Floyd, and D. Black. The Addition of Explicit Congestion Notification (ECN) to IP. RFC 3168, IETF, September 2001. URL http://tools.ietf.org/rfc/rfc3168.txt. 95

[427] Tahina Ramananandro, Antoine Delignat-Lavaud, Cédric Fournet, Nikhil Swamy, Tej Chajed, Nadim Kobeissi, and Jonathan Protzenko. Everparse: Verified secure zero-copy parsers for authenticated message formats. In Nadia Heninger and Patrick Traynor, editors, *28th USENIX Security Symposium, USENIX Security 2019, Santa Clara, CA, USA, August 14-16, 2019*, pages 1465–1482. USENIX Association, 2019. URL https://www.usenix.org/conference/usenixsecurity19/presentation/delignat-lavaud. 571

[428] Audrey Randall, Enze Liu, Gautam Akiwate, Geoffrey M Voelker, Stefan Savage, and Aaron Schulman. Home is where the hijacking is: Understanding dns interception by residential routers. 2021. 437

[429] Dipankar Raychaudhuri, Kiran Nagaraja, and Arun Venkataramani. MobilityFirst: A robust and trustworthy mobility-centric architecture for the future Internet. *ACM SIGMOBILE Mobile Computing and Communications Review*, July 2012. 577

[430] Red Hat, Inc. Ansible. https://www.ansible.com. 363

[431] Y. Rekhter, B. Moskowitz, D. Karrenberg, G. J. de Groot, and E. Lear. Address Allocation for Private Internets. RFC 1918, IETF, February 1996. URL http://tools.ietf.org/rfc/rfc1918.txt. 138, 326, 333

[432] Y. Rekhter, T. Li, and S. Hares. A Border Gateway Protocol 4 (BGP-4). RFC 4271, IETF, January 2006. URL http://tools.ietf.org/rfc/rfc4271.txt. 2, 152

[433] E. Rescorla. The Transport Layer Security (TLS) Protocol Version 1.3. RFC 8446, IETF, August 2018. URL http://tools.ietf.org/rfc/rfc8446.txt. 6, 467

[434] Réseaux IP Européens Network Coordination Centre (RIPE NCC). YouTube hijacking: A RIPE NCC RIS case study. https://perma.cc/4DK6-FKR3, March 2008. 7

[435] Alvaro Retana. Advertisement of multiple paths in BGP: Implementation report. Internet-Draft draft-ietf-idr-add-paths-implementation-00, IETF Secretariat, February 2015. URL http://www.ietf.org/internet-drafts/draft-ietf-idr-add-paths-implementation-00.txt. 582

[436] John C. Reynolds. Separation logic: A logic for shared mutable data structures. In *LICS*, pages 55–74. IEEE Computer Society, 2002. 528

[437] S. Ricciardi, D. Careglio, F. Palmieri, U. Fiore, G. Santos-Boada, and J. Sole-Pareta. Energy-Aware RWA for WDM Networks with Dual Power Sources. In *IEEE International Conference on Communications (ICC)*, pages 1–6, 2011. doi: 10.1109/icc.2011.5962432. 399

[438] RIPE NCC. RIPE Atlas. https://atlas.ripe.net. 362, 368

[439] Richard Roberts, Yaelle Goldschlag, Rachel Walter, Taejoong Chung, Alan Mislove, and Dave Levin. You Are Who You Appear to Be: A Longitudinal Study of Domain Impersonation in TLS Certificates. In *Proceedings of the ACM Conference on Computer and Communications Security (CCS)*, 2019. doi: 10.1145/3319535.3363188. 427

[440] Dominik Roos. COLIBRI A Cooperative Lightweight Inter-domain Bandwidth Reservation Infrastructure. Master's thesis, ETH Zurich, Zurich, 2018. 517

[441] Benjamin Rothenberger, Daniele E. Asoni, David Barrera, and Adrian Perrig. Internet kill switches demystified. In *Proceedings of EuroSec*, April 2017. URL http://www.netsec.ethz.ch/publications/papers/killswitch2017.pdf. 6, 36, 163, 176, 439, 450

[442] Benjamin Rothenberger, Dominik Roos, Markus Legner, and Adrian Perrig. PISKES: Pragmatic Internet-scale key-establishment system. In *Proceedings of the ACM Asia Conference on Computer and Communications Security (ASIACCS)*, 2020. doi: 10.1145/3320269.3384743. 35, 53

[443] Benjamin Rothenberger, Markus Legner, Marc Frei, Jonas Gude, Florian Jacky, Pascal Sprenger, and Adrian Perrig. Lightning Filter: New Cryptography-Based Approaches for High-Speed Traffic Filtering, To appear. 203

[444] George Rouskas, Ilia Baldine, Ken Calvert, Rudra Dutta, Jim Griffioen, Anna Nagurney, and Tilman Wolf. ChoiceNet: Network innovation through choice. In *Proceedings of International Conference on Optical Networking Design and Modeling (ONDM)*, 2013. URL https://ieeexplore.ieee.org/document/6524925. 577

[445] Leopold Ryll. Development and evaluation of a scion-enhanced distributed storage solution. Master's thesis, OVGU Magdeburg, 2018. URL http://www.netsys.ovgu.de/netsys_media/ Downloads/ThesisRyll2018.pdf. 384

[446] Andrei Sabelfeld and David Sands. Dimensions and principles of declassification. In *18th IEEE Computer Security Foundations Workshop, (CSFW-18 2005), 20-22 June 2005, Aix-en-Provence, France*, pages 255–269, 2005. 562

[447] Amit Sahoo, Krishna Kant, and Prasant Mohapatra. BGP convergence delay under large-scale failures: Characterization and solutions. *Computer Communications*, 32(7), May 2009. 4, 11, 165

[448] Sally Adee. The global internet is disintegrating. what comes next? https://www.bbc.com/future/article, May 2019. 451, 452

[449] Jerome H. Saltzer, David P. Reed, and David D. Clark. End-to-end arguments in system design. *ACM Transactions on Computer Systems*, 2(4), November 1984. 12

[450] Sandvine. The Global Internet Phenomena Report. https://www. sandvine.com/global-internet-phenomena-report-2019, 2019. 266

[451] Arish Sateesan, Jo Vliegen, Simon Scherrer, Hsu-Chun Hsiao, Adrian Perrig, and Nele Mentens. Speed records in network flow measurement on FPGA. In *Proceedings of the International Conference on Field-Programmable Logic (FPL)*, 2021. URL https://netsec.ethz.ch/ publications/papers/sateesan2021speed.pdf. 225

[452] Simon Scherrer, Markus Legner, Adrian Perrig, and Stefan Schmid. Incentivizing stable path selection in future internet architectures. In *Proceedings of the International Symposium on Computer Performance, Modeling, Measurements and Evaluation (PERFORMANCE)*, August 2020. URL `https://netsec.ethz.ch/publications/papers/scherrer_incentivizing_stability_2020.pdf`. 11

[453] Simon Scherrer, Markus Legner, Adrian Perrig, and Stefan Schmid. Enabling novel interconnection agreements with path-aware networking architectures. In *Proceedings of the IEEE/IFIP International Conference on Dependable Systems and Networks (DSN)*, 6 2021. doi: 10.1109/DSN48987.2021.00027. URL `https://doi.org/10.1109/DSN48987.2021.00027`. 129, 142

[454] Simon Scherrer, Markus Legner, Tobias Schmidt, and Adrian Perrig. Footprints on the path: how routing data could reduce the Internet's carbon toll, 2021. URL `https://perma.cc/PRW7-675A/`. 320

[455] Simon Scherrer, Che-Yu Wu, Yu-Hsi Chiang, Benjamin Rothenberger, Daniele Asoni, Arish Sateesan, Jo Vliegen, Nele Mentens, Hsu-Chun Hsiao, and Adrian Perrig. Low-rate overuse flow tracer (LOFT): An efficient and scalable algorithm for detecting overuse flows. *Proceedings of the Symposium on Reliable Distributed Systems (SRDS)*, 2021. URL `https://netsec.ethz.ch/publications/papers/loft2021.pdf`. 203, 218, 225

[456] Simon Scherrer, Che-Yu Wu, Yu-Hsi Chiang, Benjamin Rothenberger, Daniele E. Asoni, Arish Sateesan, Jo Vliegen, Nele Mentens, Hsu-Chun Hsiao, and Adrian Perrig. Low-rate overuse flow tracer (loft): An efficient and scalable algorithm for detecting overuse flows, 2021. 222

[457] D. Schinazi and T. Pauly. Happy Eyeballs Version 2: Better Connectivity Using Concurrency. RFC 8305, IETF, December 2017. URL `http://tools.ietf.org/rfc/rfc8305.txt`. 310

[458] Brandon Schlinker, Kyriakos Zarifis, Italo Cunha, Nick Feamster, and Ethan Katz-Bassett. PEERING: An AS for us. In *Proceedings of the 13th ACM Workshop on Hot Topics in Networks (HotNets)*, 2014. 369

[459] Paul Schmitt, Anne Edmundson, Allison Mankin, and Nick Feamster. Oblivious dns: Practical privacy for dns queries: Published in popets 2019. In *Proceedings of Privacy Enhancing Technologies (PETS)*, 2019. 435, 457

[460] Max Schuchard, Eugene Y. Vasserman, Abdelaziz Mohaisen, Denis Foo Kune, Nicholas Hopper, and Yongdae Kim. Losing control of the Internet: Using the data plane to attack the control plane. In *Proceedings of*

the Symposium on Network and Distributed Systems Security (NDSS), February 2011. 3, 12, 53, 165

[461] L.-C. Schulz and D. Hausheer. Towards SCION-enabled IXPs: The SCION peering coordinator. In *Proceedings of the Conference on Networked Systems*, NetSys '21, September 2021. URL https://doi.org/10.14279/tuj.eceasst.80.1159. 325

[462] Lorenz Schwittmann, Matthäus Wander, and Torben Weis. Domain impersonation is feasible: A study of ca domain validation vulnerabilities. In *Proceedings of the IEEE European Symposium on Security and Privacy (EuroS&P)*, 2019. 445

[463] Thilo Schöndienst and Vinod Vokkarane. Renewable Energy-Aware Grooming in Optical Networks. *Photonic Network Communications*, 28:71–81, Aug 2014. doi: 10.1007/s11107-014-0436-4. 399

[464] Chuck Semeria. Supporting differentiated service classes: Queue scheduling disciplines. Technical report, 2001. URL https://www.cse.iitb.ac.in/~varsha/allpapers/packet-scheduling/wfqJuniper.pdf. 283

[465] Koushik Sen, Mahesh Viswanathan, and Gul Agha. On statistical model checking of stochastic systems. In Kousha Etessami and Sriram K. Rajamani, editors, *Computer Aided Verification, 17th International Conference, CAV 2005, Edinburgh, Scotland, UK, July 6-10, 2005, Proceedings*, volume 3576 of *Lecture Notes in Computer Science*, pages 266–280. Springer, 2005. doi: 10.1007/11513988_26. URL https://doi.org/10.1007/11513988_26. 511, 512

[466] Anapaya Systems Sergiu Costea. dispatcher: Replace with scmp daemon, 2021. URL https://github.com/scionproto/scion/issues/3961. 313

[467] Anees Shaikh, Jennifer Rexford, and Kang Shin. Evaluating the impact of stale link state on quality-of-service routing. *IEEE/ACM Transactions on Networking*, 2001. 11

[468] Abhigyan Sharma, Xiaozheng Tie, Hardeep Uppal, Arun Venkataramani, David Westbrook, and Aditya Yadav. A global name service for a highly mobile internetwork. In *Proceedings of the ACM SIGCOMM Conference*, August 2014. 577

[469] Y. Sheffer, R. Holz, and P. Saint-Andre. Recommendations for Secure Use of Transport Layer Security (TLS) and Datagram Transport Layer Security (DTLS). RFC 7525, IETF, May 2015. URL http://tools.ietf.org/rfc/rfc7525.txt. 408

[470] Micah Sherr, Michael Greenwald, Carl Gunter, Sanjeev Khanna, and Santosh Venkatesh. Mitigating DoS attack through selective bin verification. In *Proceedings of the IEEE ICNP Workshop on Secure Network Protocols*, 2005. 209

[471] R.B. Shoemaker. Automated Certificate Management Environment (ACME) IP Identifier Validation Extension. RFC 8738, IETF, February 2020. URL http://tools.ietf.org/rfc/rfc8738.txt. 466

[472] Haya Shulman and Michael Waidner. One key to sign them all considered vulnerable: Evaluation of DNSSEC in the internet. In *Proceedings of the USENIX Symposium on Networked Systems Design and Implementation (NSDI)*, 2017. 439

[473] SIDN. Stichting internet domeinregistratie nederland. https://www.sidn.nl/. 367

[474] Sudheesh Singanamalla, Suphanat Chunhapanya, Marek Vavruša, Tanya Verma, Peter Wu, Marwan Fayed, Kurtis Heimerl, Nick Sullivan, and Christopher Wood. Oblivious dns over https (odoh): A practical privacy enhancement to dns. In *Proceedings of the Symposium on Network and Distributed Systems Security (NDSS)*, 2020. 457

[475] Ankit Singla, Balakrishnan Chandrasekaran, P. Brighten Godfrey, and Bruce Maggs. The Internet at the speed of light. In *Proceedings of the ACM Workshop on Hot Topics in Networks (HotNets)*, 2014. ISBN 9781450332569. doi: 10.1145/2670518.2673876. 204

[476] Karen E. Sirois and Stephen T. Kent. Securing the Nimrod routing architecture. In *Proceedings of the Symposium on Network and Distributed Systems Security (NDSS)*, February 1997. 575

[477] SIX. SIX SCION PKI. https://www.six-group.com/scionpki, 2022. 36

[478] Nigel P. Smart, Vincent Rijmen, Bogdan Warinschi, Gaven Watson, and Rodica Tirtea. Algorithms, key size and parameters report. Technical report, European Union Agency for Network and Information Security Agency (ENISA), November 2014. 408

[479] Jean-Pierre Smith, Prateek Mittal, and Adrian Perrig. Website fingerprinting in the age of QUIC. In *Proceedings on Privacy Enhancing Technologies (PoPETs)*, July 2021. URL https://netsec.ethz.ch/publications/papers/smith2021website.pdf. 374

[480] João Luís Sobrinho and Miguel Alves Ferreira. Routing on multiple optimality criteria. In *Proceedings of the ACM SIGCOMM Conference*, 2020. ISBN 9781450379557. doi: 10.1145/3387514.3405864. 75

[481] João Luís Sobrinho, Franck Le, and Laurent Vanbever. DRAGON simulator environment. `https://github.com/network-aggregation/dragon_simulator`, 2014. 151, 152, 402

[482] Abolfazl Soltani and Saeed Sharifian. An ultra-high throughput and fully pipelined implementation of AES algorithm on FPGA. *Microprocessors and Microsystems*, 39(7):480–493, 2015. ISSN 0141-9331. doi: https://doi.org/10.1016/j.micpro.2015.07. 005. URL `http://www.sciencedirect.com/science/article/pii/S0141933115001040`. 395

[483] JH. Song, R. Poovendran, J. Lee, and T. Iwata. The AES-CMAC Algorithm. RFC 4493, IETF, June 2006. URL `http://tools.ietf.org/rfc/rfc4493.txt`. 411, 412

[484] Mansi Sood and Osman Yagan. Tight bounds for connectivity of random k-out graphs, 2020. 344

[485] Christoph Sprenger and David A. Basin. Refining security protocols. *Journal of Computer Security*, 26(1):71–120, 2018. doi: 10.3233/JCS-16814. URL `https://doi.org/10.3233/JCS-16814`. 489

[486] Christoph Sprenger, Tobias Klenze, Marco Eilers, Felix Wolf, Peter Müller, Martin Clochard, and David Basin. Igloo: soundly linking compositional refinement and separation logic for distributed system verification. *Proc. ACM Program. Lang.*, 4(OOPSLA):152:1–152:31, 2020. 481, 550, 554, 564, 570

[487] Supraja Sridhara, François Wirz, Joeri de Ruiter, Caspar Schutijser, Markus Legner, and Adrian Perrig. Global distributed secure mapping of network addresses. In *Proceedings of the ACM SIGCOMM Workshop on Technologies, Applications, and Uses of a Responsible Internet (TAURIN)*, 2021. doi: 10.1145/3472951.3473503. 317, 340

[488] P. Srisuresh and K. Egevang. Traditional IP Network Address Translator (Traditional NAT). RFC 3022, IETF, January 2001. URL `http://tools.ietf.org/rfc/rfc3022.txt`. 326

[489] P. Srisuresh and M. Holdrege. IP Network Address Translator (NAT) Terminology and Considerations. RFC 2663, IETF, August 1999. URL `http://tools.ietf.org/rfc/rfc2663.txt`. 326

[490] Emily Stark, Ryan Sleevi, Rijad Muminovi, Devon O'Brien, Eran Messeri, Adrienne Porter Felt, Brendan McMillion, and Parisa Tabriz. Does Certificate Transparency break the web? Measuring adoption and error rate. In *Proceedings of the IEEE Symposium on Security and Privacy (S&P)*, May 2019. 420

[491] Statista. Global mobile data traffic from 2017 to 2022. `https://perma.cc/4B9Q-QL9X`, 2020. 398

[492] Daniel Sternberg. Users of curl. `https://everything.curl.dev/project/users`, 2021. 310

[493] Ion Stoica, Daniel Adkins, Shelley Zhuang, Scott Shenker, and Sonesh Surana. Internet indirection infrastructure. *IEEE/ACM Transactions on Networking*, April 2004. 578

[494] Fabio Streun, Joel Wanner, and Adrian Perrig. Evaluating Susceptibility of VPN Implementations to DoS Attacks Using Adversarial Testing. In *Proceedings of the Symposium on Network and Distributed Systems Security (NDSS)*, 2022. doi: 10.14722/ndss.2022.24043. URL `https://doi.org/10.14722/ndss.2022.24043`. 207, 277, 375

[495] Ahren Studer and Adrian Perrig. The Coremelt attack. In *Proceedings of the European Symposium on Research in Computer Security (ESORICS)*, September 2009. 9, 269, 276

[496] Lakshminarayanan Subramanian, Matthew Caesar, Cheng Tien Ee, Mark Handley, Morley Mao, Scott Shenker, and Ion Stoica. HLP: A next generation inter-domain routing protocol. In *Proceedings of the ACM SIGCOMM Conference*, 2005. 575, 576

[497] SWITCH. Swiss national research and education network. `https://www.switch.ch/`. 367

[498] Pawel Szalachowski and Adrian Perrig. Short paper: on deployment of DNS-based security enhancements. In *Proceedings of the International Conference on Financial Cryptography and Data Security*, 2017. 429

[499] Pawel Szalachowski, Laurent Chuat, Taeho Lee, and Adrian Perrig. RITM: Revocation in the middle. In *Proceedings of IEEE International Conference on Distributed Computing Systems (ICDCS)*, June 2016. 430

[500] Seyedali Tabaeiaghdaei. Green inter-domain routing in the SCION internet architecture. Master's thesis, ETH Zurich, 2021. 393

[501] Seyedali Tabaeiaghdaei and Christelle Gloor. ns-3_beaconing_simulator. `https://gitlab.inf.ethz.ch/OU-PERRIG/seyedali/ns-3_beaconing_simulator/-/tree/green_routing`, 2021. 402

[502] Sasu Tarkoma, Mark Ain, and Kari Visala. The Publish/Subscribe Internet Routing Paradigm (PSIRP): Designing the future Internet architecture. In *Future Internet Assembly*, 2009. 576

[503] Gavin Thomas. A proactive approach to more secure code. `https://perma.cc/9PB3-WKP8`, July 2019. 520

[504] Olaf Titz. Why TCP over TCP is a bad idea. `https://perma.cc/QK43-7LGZ`, 2001. 332

[505] Andree Toonk. Massive route leak causes Internet slowdown. `https://perma.cc/32TV-N6Y2`, June 2015. 377

[506] Emin Topalovic, Brennan Saeta, Lin-Shung Huang, Collin Jackson, and Dan Boneh. Towards short-lived certificates. In *Proceedings of the IEEE Web 2.0 Security and Privacy Workshop*, 2012. 446

[507] Tor Blog. Breaking through censorship barriers, even when Tor is blocked. `https://perma.cc/7TBT-6G7M`. 180

[508] Brian Trammell. RAINS (another Internet naming service) protocol specification. Internet-Draft, January 2019. 432

[509] Brian Trammell, Michael Welzl, Theresa Enghardt, Gorry Fairhurst, Mirja Kühlewind, Colin Perkins, Philipp S. Tiesel, Christopher A. Wood, Tommy Pauly, and Kyle Rose. An Abstract Application Layer Interface to Transport Services. Internet-Draft draft-ietf-taps-interface-12, Internet Engineering Task Force, April 2021. URL `https://datatracker.ietf.org/doc/html/draft-ietf-taps-interface-12`. Work in Progress. 310

[510] Georgios Tselentis, John Domingue, Alex Galis, Anastasius Gavras, David Hausheer, Srdjan Krco, Volkmar Lotz, and Theodore Zahariadis. Towards the future internet - a european research perspective. 05 2009. 580

[511] Tsinghua University. Launch ceremony of future internet technology infrastructure. `https://www.tsinghua.edu.cn/en/info/1399/10187.htm`. 582

[512] Paul F. Tsuchiya. The landmark hierarchy: a new hierarchy for routing in very large networks. In *Proceedings of the ACM SIGCOMM Conference*, 1988. 575

[513] Jonathan Turner. New directions in communications (or which way to the information age?). *IEEE Communications Magazine*, 1986. 220

[514] Abraão Aires Urquiza, Musab A. AlTurki, Max I. Kanovich, Tajana Ban Kirigin, Vivek Nigam, Andre Scedrov, and Carolyn L. Talcott. Resource-bounded intruders in denial of service attacks. In *CSF*, pages 382–396. IEEE, 2019. doi: 10.1109/CSF.2019.00033. URL `https://doi.org/10.1109/CSF.2019.00033`. 511

[515] Vytautas Valancius, Cristian Lumezanu, Nick Feamster, Ramesh Johari, and Vijay Vazirani. How Many Tiers? Pricing in the Internet Transit Market. In *Proceedings of the ACM SIGCOMM 2011 Conference, SIGCOMM'11*, volume 41, pages 194–205, Aug 2011. doi: 10.1145/2018436.2018459. 405

[516] Kannan Varadhan, Ramesh Govindan, and Deborah Estrin. Persistent route oscillations in inter-domain routing. *Computer networks*, 32(1): 1–16, 2000. 142

[517] G. Van de Velde, T. Hain, R. Droms, B. Carpenter, and E. Klein. Local Network Protection for IPv6. RFC 4864, IETF, May 2007. URL http://tools.ietf.org/rfc/rfc4864.txt. 326

[518] Marcos AM Vieira, Matheus S Castanho, Racyus DG Pacífico, Elerson RS Santos, Eduardo PM Câmara Júnior, and Luiz FM Vieira. Fast packet processing with ebpf and xdp: Concepts, code, challenges, and applications. *ACM Computing Surveys (CSUR)*, 53(1):1–36, 2020. 313

[519] Arun Vishwanath, Kerry Hinton, Robert Ayre, and Rodney Tucker. Modeling Energy Consumption in High-Capacity Routers and Switches. *IEEE Journal on Selected Areas in Communications*, 32:1524–1532, 08 2014. doi: 10.1109/JSAC.2014.2335312. 397, 401

[520] Thomas Vissers, Timothy Barron, Tom Van Goethem, Wouter Joosen, and Nick Nikiforakis. The wolf of name street: Hijacking domains through their nameservers. In *Proceedings of the ACM Conference on Computer and Communications Security (CCS)*, 2017. 434

[521] Dmitry Vyukov and Andrew Gerrand. Introducing the Go race detector. https://perma.cc/ZXC7-W67R, 2013. 533

[522] Tao Wan, Evangelos Kranakis, and Paul C. van Oorschot. Pretty secure BGP, psBGP. In *Proceedings of the Symposium on Network and Distributed Systems Security (NDSS)*, 2005. 165

[523] A. Wang, X. Huang, C. Kou, Z. Li, and P. Mi. Scenarios and Simulation Results of PCE in a Native IP Network. RFC 8735, IETF, February 2020. URL http://tools.ietf.org/rfc/rfc8735.txt. 446

[524] Anduo Wang, Alexander J. T. Gurney, Xianglong Han, Jinyan Cao, Boon Thau Loo, Carolyn L. Talcott, and Andre Scedrov. A reduction-based approach towards scaling up formal analysis of internet configurations. In *2014 IEEE Conference on Computer Communications, INFOCOM 2014, Toronto, Canada, April 27 - May 2, 2014*, pages 637–645. IEEE, 2014. doi: 10.1109/INFOCOM.2014.6847989. URL https://doi.org/10.1109/INFOCOM.2014.6847989. 511

[525] Bing Wang, Wei Wei, Zheng Guo, and Don Towsley. Multipath live streaming via tcp: Scheme, performance and benefits. *ACM Trans. Multimedia Comput. Commun. Appl.*, 5(3), aug 2009. ISSN 1551-6857. doi: 10.1145/1556134.1556142. URL https://doi.org/10.1145/1556134.1556142. 373

[526] Cun Wang, Zhengmin Li, Xiaohong Huang, and Pei Zhang. Inferring the average AS path length of the Internet. In *Proceedings of the IEEE International Conference on Network Infrastructure and Digital Content (IC-NIDC)*, September 2016. doi: 10.1109/icnidc.2016.7974603. 236

[527] Feng Wang and Lixin Gao. Path diversity aware interdomain routing. In *Proceedings of the 28th IEEE Conference on Computer Communications*, INFOCOM '09, pages 307–315, 05 2009. doi: 10.1109/INFCOM.2009.5061934. URL https://doi.org/10.1109/INFCOM.2009.5061934. 582

[528] Peng Wang, Dengguo Feng, Changlu Lin, and Wenling Wu. Security of truncated macs. In *Proceedings of the International Conference on Information Security and Cryptology*, pages 96–114. Springer, 2008. 414

[529] Qi Wang, Pubali Datta, Wei Yang, Si Liu, Adam Bates, and Carl A. Gunter. Charting the attack surface of trigger-action IoT platforms. In Lorenzo Cavallaro, Johannes Kinder, XiaoFeng Wang, and Jonathan Katz, editors, *CCS*, pages 1439–1453. ACM, 2019. doi: 10.1145/3319535.3345662. URL https://doi.org/10.1145/3319535.3345662. 511

[530] Yuefeng Wang, Flavio Esposito, Ibrahim Matta, and John Day. RINA: An architecture for policy-based dynamic service management. Technical Report BUCS-TR-2013-014, November 2013. 578

[531] Z. Wang and J. Crowcroft. A Two-Tier Address Structure for the Internet: A Solution to the Problem of Address Space Exhaustion. RFC 1335, IETF, May 1992. URL http://tools.ietf.org/rfc/rfc1335.txt. 326

[532] Zheng Wang. The availability and security implications of glue in the domain name system. *arXiv:1605.01394*, 2016. 436

[533] Nicholas Weaver, Christian Kreibich, and Vern Paxson. Redirecting DNS for ads and profit. In *Proceedings of the USENIX Workshop on Free and Open Communications on the Internet (FOCI 11)*, 2011. 434

[534] Thilo Weghorn. *Qualitative and Quantitative Guarantees for Access Control*. PhD thesis, ETH Zürich, 2019. URL https://www.

research-collection.ethz.ch/handle/20.500.11850/397549.
227

[535] J. Weil, V. Kuarsingh, C. Donley, C. Liljenstolpe, and M. Azinger.
IANA-Reserved IPv4 Prefix for Shared Address Space. RFC 6598,
IETF, April 2012. URL http://tools.ietf.org/rfc/rfc6598.
txt. 326

[536] Jennifer Lundelius Welch and Nancy Lynch. A new fault-tolerant algorithm for clock synchronization. *Information and computation*, 77(1):
1–36, 1988. 195

[537] Michael Welzl, Safiqul Islam, Michael Gundersen, and Andreas Fischer.
Transport services: A modern api for an adaptive internet transport layer.
IEEE Communications Magazine, 59(4):16–22, 2021. 310

[538] Dan Wendlandt, David G. Andersen, and Adrian Perrig. Perspectives:
Improving SSH-style Host Authentication with Multi-Path Probing. In
Proceedings of USENIX Annual Technical Conference, June 2008. 7

[539] Brian White, Jay Lepreau, Leigh Stoller, Robert Ricci, Shashi Gu-
ruprasad, Mac Newbold, Mike Hibler, Chad Barb, and Abhijeet
Joglekar. An integrated experimental environment for distributed systems and networks. In *Proceeding of the 5th Symposium on Operating System Design and Implementation (OSDI)*. USENIX Association,
2002. 368, 514

[540] N. Williams. JavaScript Object Notation (JSON) Text Sequences. RFC
7464, IETF, February 2015. URL http://tools.ietf.org/rfc/
rfc7464.txt. 439

[541] D. Wing and A. Yourtchenko. Happy Eyeballs: Success with Dual-
Stack Hosts. RFC 6555, IETF, April 2012. URL http://tools.ietf.
org/rfc/rfc6555.txt. 310, 458

[542] Wired. GitHub survived the biggest DDoS attack ever recorded. https:
//perma.cc/5TG2-UCZY, 2018. 211

[543] François Wirz. Network performance evaluation and prediction on
SCION. Master's thesis, ETH Zurich, 2018. 366

[544] Nicolas With and Thorben Krüger. Implementation of a Transport-
Agnostic, High-Level Networking API. https://code.ovgu.de/
hausheer/taps-api, 2021. 310

[545] Robert Wójcik, Jerzy Domundefinedał, Zbigniew Duliński, Grzegorz
Rzym, Andrzej Kamisiński, Piotr Gawłowicz, Piotr Jurkiewicz, Jacek
Rzundefinedsa, Rafał Stankiewicz, and Krzysztof Wajda. A survey

on methods to provide interdomain multipath transmissions. *Computer Networks*, 108(C):233259, October 2016. ISSN 1389-1286. doi: 10.1016/j.comnet.2016.08.028. URL https://doi.org/10.1016/j.comnet.2016.08.028. 582

[546] Felix Wolf, Linard Arquint, Martin Clochard, Wytse Oortwijn, João Pereira, and Peter Müller. Gobra: Modular specification and verification of go programs. In *CAV (1)*, volume 12759 of *Lecture Notes in Computer Science*, pages 367–379. Springer, 2021. 533

[547] Tilman Wolf, James Griffioen, Kenneth L. Calvert, Rudra Dutta, George N. Rouskas, Ilia Baldine, and Anna Nagurney. ChoiceNet: toward an economy plane for the Internet. *ACM SIGCOMM Computer Communication Review*, 44(3):58–65, July 2014. 577

[548] Paul Wouters and Wes Hardaker. The DELEGATION_ONLY DNSKEY flag. Internet-Draft draft-ietf-dnsop-delegation-only-02, Internet Engineering Task Force, 2021. URL https://datatracker.ietf.org/doc/html/draft-ietf-dnsop-delegation-only-02. Work in Progress. 439, 449, 450

[549] Charles Wright, Lucas Ballard, Fabian Monrose, and Gerald Masson. Language identification of encrypted voip traffic: Alejandra y roberto or alice and bob? In *Proceedings of the USENIX Security Symposium*, volume 3, pages 43–54, 2007. 179

[550] Charles V. Wright, Lucas Ballard, Scott E. Coull, Fabian Monrose, and Gerald M. Masson. Spot me if you can: Uncovering spoken phrases in encrypted VoIP conversations. In *Proceedings of the IEEE Symposium on Security and Privacy (S&P)*, pages 35–49, 2008. doi: 10.1109/SP.2008.21. 179

[551] J. Wroclawski. The Use of RSVP with IETF Integrated Services. RFC 2210, IETF, September 1997. URL http://tools.ietf.org/rfc/rfc2210.txt. 238

[552] Bo Wu, Ke Xu, Qi Li, Zhuotao Liu, Yih-Chun Hu, Martin J. Reed, Meng Shen, and Fan Yang. Enabling efficient source and path verification via probabilistic packet marking. In *Proceedings of the IEEE/ACM International Symposium on Quality of Service (IWQoS)*, June 2018. URL http://iwqos2018.ieee-iwqos.org/files/2018/05/Enabling_Efficient_Source_and_Path_Varification.pdf. 236

[553] Hao Wu, Hsu-Chun Hsiao, and Yih-Chun Hu. Efficient large flow detection over arbitrary windows: An algorithm exact outside an ambiguity region. In *Proceedings of the ACM Internet Measurement Conference (IMC)*, pages 209–222. ACM, 2014. 218

[554] Jianping Wu, Gang Ren, and Xing Li. Source address validation: Architecture and protocol design. pages 276–283, 11 2007. ISBN 978-1-4244-1588-5. doi: 10.1109/ICNP.2007.4375858. 582

[555] Wen Xu and Jennifer Rexford. MIRO: Multi-path interdomain routing. *ACM SIGCOMM Computer Communication Review*, 36(4):171182, August 2006. ISSN 0146-4833. doi: 10.1145/1151659.1159934. URL https://doi.org/10.1145/1151659.1159934. 583

[556] Tong Yang, Jie Jiang, Peng Liu, Qun Huang, Junzhi Gong, Yang Zhou, Rui Miao, Xiaoming Li, and Steve Uhlig. Elastic sketch: Adaptive and fast network-wide measurements. In *Proceedings of the ACM SIGCOMM Conference*, pages 561–575. ACM, 2018. 224

[557] Tong Yang, Haowei Zhang, Jinyang Li, Junzhi Gong, Steve Uhlig, Shigang Chen, and Xiaoming Li. HeavyKeeper: An accurate algorithm for finding top-*k* elephant flows. *IEEE/ACM Transactions on Networking*, 27(5):1845–1858, 2019. 224

[558] Xiaowei Yang and David Wetherall. Source selectable path diversity via routing deflections. *ACM SIGCOMM Computer Communication Review*, 36(4):159170, August 2006. ISSN 0146-4833. doi: 10.1145/1151659.1159933. URL https://doi.org/10.1145/1151659.1159933. 583

[559] Xiaowei Yang, David Clark, and Arthur W. Berger. NIRA: A new interdomain routing architecture. *IEEE/ACM Transactions on Networking*, 2007. 575, 576, 578, 583

[560] Ping Yi, Zhoulin Dai, Yiping Zhong, and Shiyong Zhang. Resisting flooding attacks in ad hoc networks. In *International Conference on Information Technology: Coding and Computing (ITCC'05)*, volume 2, pages 657–662, 2005. doi: 10.1109/ITCC.2005.248. 299

[561] Xia Yin, Dan Wu, Zhiliang Wang, Xingang Shi, and Jianping Wu. DIMR. *Computer Networks*, 91(C):356375, November 2015. ISSN 1389-1286. doi: 10.1016/j.comnet.2015.08.028. URL https://doi.org/10.1016/j.comnet.2015.08.028. 582

[562] Omer Yoachimik and Vivek Ganti. Network-layer DDoS attack trends for Q3 2020. https://blog.cloudflare.com/network-layer-ddos-attack-trends-for-q3-2020/, 2020. 265

[563] Håkan L. S. Younes and Reid G. Simmons. Statistical probabilistic model checking with a focus on time-bounded properties. *Inf. Comput.*, 204(9):1368–1409, 2006. 511

[564] Ryan Zarick, Mikkel Hagen, and Radim Bartoš. Transparent clocks vs. enterprise ethernet switches. In *2011 IEEE International Symposium on Precision Clock Synchronization for Measurement, Control and Communication*, pages 62–68. IEEE, 2011. 190

[565] Fan Zhang, Wenbo He, Xue Liu, and Patrick G. Bridges. Inferring users' online activities through traffic analysis. In *Proceedings of the ACM Conference on Wireless Network Security*, page 5970, 2011. ISBN 9781450306928. doi: 10.1145/1998412.1998425. 179

[566] Fuyuan Zhang, Limin Jia, Cristina Basescu, Tiffany Hyun-Jin Kim, Yih-Chun Hu, and Adrian Perrig. Mechanized network origin and path authenticity proofs. In Gail-Joon Ahn, Moti Yung, and Ninghui Li, editors, *Proceedings of the 2014 ACM SIGSAC Conference on Computer and Communications Security, Scottsdale, AZ, USA, November 3-7, 2014*, pages 346–357. ACM, 2014. ISBN 978-1-4503-2957-6. doi: 10.1145/2660267.2660349. URL http://doi.acm.org/10.1145/2660267.2660349. 565

[567] M. Zhang, C. Yi, B. Liu, and B. Zhang. GreenTE: Power-Aware Traffic Engineering. In *The 18th IEEE International Conference on Network Protocols*, pages 21–30, 2010. doi: 10.1109/ICNP.2010.5762751. 399

[568] Xiaofeng Zheng, Chaoyi Lu, Jian Peng, Qiushi Yang, Dongjie Zhou, Baojun Liu, Keyu Man, Shuang Hao, Haixin Duan, and Zhiyun Qian. Poison over troubled forwarders: A cache poisoning attack targeting DNS forwarding devices. In *Proceedings of the USENIX Security Symposium*, 2020. 8

[569] Ming Zhu, Dan Li, Ying Liu, Dan Pei, Kadangode Ramakrishnan, Lili Liu, and Jianping Wu. MIFO: Multi-path interdomain forwarding. In *Proceedings of the 44th International Conference on Parallel Processing*, ICPP '15, pages 180–189, 2015. doi: 10.1109/ICPP.2015.27. URL https://doi.org/10.1109/ICPP.2015.27. 582

[570] Netsec ETH Zurich. Hercules - bulk data transfer over scion, 2021. URL https://www.scion-architecture.net/pages/scion_day/slides/Hercules%20-%20Bulk%20Data%20Transfer%20over%20SCION.pdf. 383

[571] Netsec ETH Zurich. Nesquic, 2021. URL https://github.com/netsec-ethz/scion-apps/tree/nesquic. 383

[572] Netsec ETH Zurich. Scion apps repository, 2021. URL https://github.com/netsec-ethz/scion-apps. 383

Glossary

Autonomous System (AS).   An autonomous system is a network under a common administrative control. For example, the network of an Internet service provider, company, or university can constitute an AS. If an organization operates multiple networks that are not directly connected together, then the different networks are considered different ASes. In SCION, each **isolation domain (ISD)**⋆ is composed of several ASes, and administered by **core ASes**⋆.

Base TRC.   A base TRC is a **trust root configuration (TRC)**⋆ that **relying parties**⋆ trust axiomatically. In other words, trust for a base TRC is assumed, not derived from another cryptographic object. Each ISD must create and sign a base TRC when the ISD is established. A base TRC is either the first TRC of the ISD or the result of a **trust reset**⋆.

Beacon Service (BS).   The beacon service is part of each SCION **AS**⋆ control plane. It is responsible for generating, receiving, and propagating **path-segment construction beacons (PCBs)**⋆. SCION relies on two types of beaconing: intra-ISD beaconing (to construct path segments from core ASes to non-core ASes within an ISD) and core beaconing (to construct path segments among core ASes within an ISD and across ISDs).

Certificate Service (CS).   The certificate service keeps cached copies of certificates in SCION's control plane. It also manages the keys and certificates used for securing inter-AS communication. The certificate service may be queried by the **beacon service**⋆ when validating the authenticity of control-plane messages (e.g., when the beacon service needs a certificate to verify the signature on a PCB).

Control Plane.   The SCION control plane is responsible for the propagation and discovery of network paths, i.e., for the exchange of routing information between network nodes. The control plane thus determines where traffic can be sent and deals with questions such as how routes are established, which paths exist, what quality individual links offer, etc. Within a SCION **AS**⋆, such functionalities are carried out by the **control service**⋆. Packet forwarding is instead a task pertaining to the **data plane**⋆.

Control Service. The main services that SCION ASes must provide (i.e., **beacon service (BS)**⋆, **certificate service (CS)**⋆, and **path service (PS)**⋆) can be combined into one or more control services. These services can be hosted on multiple servers for failover and load-balancing.

Control-Plane PKI (CP-PKI). The control-plane PKI is the public-key infrastructure upon which SCION's **control plane**⋆ relies for the authentication of messages such as PCBs. It is a set of policies, roles, and procedures that are used to manage **trust root configurations (TRCs)**⋆ and certificates.

Core AS. Each **isolation domain (ISD)**⋆ is administered by a set of distinguished **autonomous systems (ASes)**⋆ called core ASes, which are responsible for initiating the beaconing process.

Data Plane. The data plane (sometimes also referred to as the forwarding plane) is responsible for forwarding data packets that end hosts have injected into the network. After routing information has been disseminated by the **control plane**⋆, packets are forwarded according to such information by the data plane.

Forwarding Path. A forwarding path is a complete end-to-end path between two SCION hosts, which is used to transmit packets in the data plane and can be created with a combination of up to three **path segments**⋆ (an up-segment, a core-segment, and a down-segment).

Grace Period. The grace period is an interval during which the previous version of a **trust root configuration (TRC)**⋆ is still considered active after a new version has been published.

Hop Field (HF). As they traverse the network, **path-segment construction beacons (PCBs)**⋆ accumulate cryptographically protected AS-level path information in the form of hop fields. In the data plane, hop fields are used for packet forwarding: they contain the incoming and outgoing interfaces of the ASes on the forwarding path.

Info Field (INF). Each **path-segment construction beacon (PCB)**⋆ contains a single info field, which provides basic information about the PCB. Together with **hop fields (HFs)**⋆, info fields are used to create **forwarding paths**⋆.

IP Address Prefix. In an IP address, the prefix (or network number) is the group of most significant bits that identifies the network or subnetwork. The remaining bits in the IP address are used to identify hosts. For instance, in CIDR notation, the IPv4 prefix 10.0.0.0/8 denotes that only the first 8 bits are used to identify the network, and the remaining 24 bits are used to address hosts within the network. IPv6 prefixes are analogous. SCION does not rely on IP-prefix matching for inter-domain forwarding, but on **Packet-Carried Forwarding State (PCFS)**⋆, making routers more efficient.

Isolation Domain (ISD). In SCION, **autonomous systems (ASes)**⋆ are organized into logical groups called isolation domains or ISDs. Each ISD consists of ASes that span an area with a uniform trust environment (i.e., a common jurisdiction). A possible model is for ISDs to be formed along national boundaries or federations of nations.

Packet-Carried Forwarding State (PCFS). Rather than relying on costly inter-domain forwarding tables, SCION data packets contain all the necessary path information, and that information is cryptographically protected. We refer to this property as packet-carried forwarding state or PCFS.

Path Authorization. A requirement for the **data plane**⋆ is that end hosts can only use paths that were constructed and authorized by ASes in the **control plane**⋆. End hosts should not be able to craft **hop fields (HFs)**⋆ themselves, modify HFs in authorized **path segments**⋆, or combine HFs of different path segments. We refer to this property as path authorization.

Path Control. Path control is a property of a network architecture (which is stronger than **path transparency**⋆) that gives end hosts the ability to select how their packets travel through the network.

Path Segment. Path segments are derived from **path-segment construction beacons (PCBs)**⋆ and registered at **path services**⋆. A path segment can be (1) an up-segment (i.e., a path between a non-core AS and a core AS in the same ISD), (2) a down-segment (i.e., the same as an up-segment, but in the opposite direction), or (3) a core-segment (i.e., a path between core ASes). Up to three path segments can be used to create a **forwarding path**⋆.

Path Service (PS). The path service is responsible for storing mappings between AS identifiers and a set of announced path segments. The path service is organized as a hierarchical caching system similar to that of DNS. Through the **beacon service**⋆, ASes select the set of path segments through which they want to be reached.

Path Transparency. Path transparency is a property of a network architecture (which is weaker than **path control***) that gives end hosts full visibility over the network paths their packets are taking.

Path-Segment Construction Beacon (PCB). Core ASes generate PCBs to explore paths within their **isolation domain (ISD)*** and among the different ISDs. ASes further propagate selected PCBs to their neighboring ASes. As a PCB traverses the network, it carries path segments, which can subsequently be used for traffic forwarding.

Relying Party. Any entity that uses a public key contained in a certificate or **trust root configuration (TRC)*** (e.g., to verify a signature) is considered a relying party. In SCION, a relying party must hold at least one **base TRC*** and must be able to verify TRC updates and certificates.

Trust Reset. A trust reset is the action of announcing a new **base TRC*** for an existing ISD. A trust reset should only be triggered after a catastrophic event involving the loss or compromise of several important private keys.

Trust Root Configuration (TRC). A trust root configuration or TRC is a signed collection of certificates pertaining to an **isolation domain (ISD)***. TRCs also contain ISD-specific policies.

Voting Quorum. The voting quorum is a **trust root configuration (TRC)*** field that indicates the number of votes needed on a successor TRC for it to be verifiable. A voting quorum greater than one will thus prevent a single entity from creating a malicious TRC update.

Abbreviations

ACME	Automatic Certificate Management Environment
AEAD	Authenticated encryption with associated data
AES	Advanced Encryption Standard
AFRINIC	African Network Information Centre
AH	Authentication header
APNIC	Asia Pacific Network Information Centre
ARIN	American Registry for Internet Numbers
AS	Autonomous system
ASE	AS entry
ASM	Assumptions
ASN	AS number
BFD	Bidirectional forwarding detection
BGP	Border Gateway Protocol
BGPsec	BGP security
BR	Border router
BS	Beacon service
CA	Certification authority
CASA	Cryptographic Agility for SCION ASes
CCN	Content-centric networking
CDF	Cumulative distribution function
CDN	Content delivery network
CEPB	CO_2 emission per bit
CIDR	Classless inter-domain routing
CME	Coronal mass ejection
CMS	Cryptographic message syntax
COLIBRI	Collaborative lightweight inter-domain bandwidth-reservation infrastructure
COND	Conditions
CP	Control plane
CP-PKI	Control-plane PKI
CPE	Customer premise equipment
CS	Certificate service
CSR	Certificate signing request
CT	Certificate Transparency
DAG	Directed acyclic graph

DDoS	Distributed denial of service
DHCP	Dynamic Host Configuration Protocol
DNS	Domain Name System
DNSSEC	DNS Security Extensions
DoC	Denial of capability
DoH	DNS over HTTPS
DoQ	DNS over QUIC
DoS	Denial of service
DoT	DNS over TLS
DP	Data plane
DPDK	Data Plane Development Kit
DRKey	Dynamically recreatable key
E2E	End-to-end
eBPF	Extended Berkeley Packet Filter
EER	End-to-end reservation
EPIC	Every packet is checked
ETHZ	Federal Institute of Technology in Zürich
FIA	Future Internet architecture
FII	Framework for Internet innovation
FPGA	Field-programmable gate array
GNSS	Global navigation satellite system
HBH	Hop-by-hop
HE	Hop entry
HF	Hop field
HPS	Hidden-path service
HTTP	Hypertext Transfer Protocol
HVF	Hop validation field
IA	ISD and AS number
IANA	Internet Assigned Numbers Authority
ICANN	Internet Corporation for Assigned Names and Numbers
ICMP	Internet Control Message Protocol
ICN	Information-centric networking
ICS	Industrial control system
ICT	Information and communication technology
IDE	Integrated development environment
IETF	Internet Engineering Task Force
INF	Info field
IoT	Internet of things
IP	Internet Protocol
IS-IS	Intermediate System to Intermediate System
ISD	Isolation domain

ISP	Internet service provider
ITU	International Telecommunication Union
IXP	Internet exchange point
LACNIC	Latin America and Caribbean Network Information Centre
LEO	Low Earth orbit
LOFT	Low-rate overuse flow tracer
Lo-RaWAN	Long Range Wide Area Network
MAC	Message-authentication code
MHT	Merkle hash tree
MITM	Man in the middle
MMD	Maximum merge delay
MPLS	Multiprotocol Label Switching
MS	Mapping service
MTU	Maximum transmission unit
NAT	Network address translation
NIRA	New Internet routing architecture
NTP	Network Time Protocol
OCSP	Online Certificate Status Protocol
ODNS	Oblivious DNS
OFD	Overuse flow detector
OSPF	Open shortest path first
OWD	One-way delay
PBKDF2	Password-based key-derivation function 2
PCB	Path-segment construction beacon
PCFS	Packet-carried forwarding state
PE	Peer entry
PGN	Publishing gossip node
PILA	Pervasive Internet-wide low-latency authentication
PKI	Public-key infrastructure
PLN	Publishing list node
POMO	Postmodern Internet architecture
PoP	Point of presence
PRF	Pseudorandom function
PRNG	Pseudorandom number generator
PS	Path service
PSIRP	Publish-subscribe Internet routing paradigm
QoS	Quality of service
RCert	RHINE certificate
RFC	Request for Comments

RHINE	Robust and high-performance Internet naming for end-to-end security
RIPE	Réseaux IP Européens
RIR	Regional Internet registry
ROA	Route Origination Authorization
ROV	Route Origination Validation
RPC	Remote procedure call
RPKI	Resource Public Key Infrastructure
RR	Resource record
RS	Replay suppression
RSVP	Resource Reservation Protocol
RTGS	Real-time gross settlement
RTT	Round-trip time
SAPV	Source authentication and path validation
SAV	Source address validation
SBAS	Secure backbone AS
SCI-ED	SCION for the ETH domain
SCION	Scalability, Control, and Isolation On Next-Generation Networks
SCMP	SCION Control Message Protocol
SD-WAN	Software-defined wide area network
SDN	Software-defined networking
SegR	Segment reservation
SGRP	SCION gateway routing protocol
SIAM	SCION–IP address-mapping system
SIC	Swiss Interbank Clearing
SIG	SCION-IP gateway
SLA	Service-level agreement
SLD	Second-level domain
SMC	Statistical model checking
SNB	Swiss National Bank
SNC	Setup-less neighbor-based communication
SPAO	SCION packet authenticator option
SSFN	Secure Swiss Finance Network
TCAM	Ternary content-addressable memory
TCP	Transmission Control Protocol
TI	Traffic isolation
TLD	Top-level domain
TLS	Transport Layer Security
TOFU	Trust on first use
TOMA	Trust on multiple announcements

TRC	Trust root configuration
TRS	Telescoped reservation setup
TS	Time service
TTL	Time to live
UDP	User Datagram Protocol
UTC	Coordinated Universal Time
VM	Virtual machine
VoIP	Voice over IP
VPN	Virtual private network
VSCode	Visual Studio Code
WAN	Wide area network
WDM	Wavelength-division multiplexing
XDP	eXpress Data Path
XOR	Exclusive or

Index

CO_2-aware inter-domain routing
· calculation of CO_2 emissions,
    400
· dissemination, 401
5G networks, 381

Abbreviations, 645
Address spoofing, 174
· defense, 176
Adversary model, *see* Security
    analysis
AES, 12, 411
Algorithm agility, 171, 407–415
Android, 307
AS, *see* Autonomous system
AS interface, 68, 94, 355
Attack, 7, 10
· beacon theft, 167
· creation of spurious ASes, 167
· ICMP, 8
· kill switch, *see* Kill switch
· network attack, 10
· path hijacking, 166
· path manipulation, 165
· path preference, 168
· path selection, 168
· path splicing, 171
· prefix hijacking, 7
· source-address spoofing, 174
Authentication, 35–63
· control-plane PKI, 36–52
· crude authentication, 463
· DRKey, 52–61
· PILA, 461–469
· SPAO, 61–63
Autonomous system (AS), 641
· certification, 294
Availability, 9, 377

· defense systems, 271–278
· guarantees, 267–300
· properties, 269

Bandwidth allocation
· N-Tube, 256–262
Bandwidth reservations
· COLIBRI, 237–266
Base TRC, 641
Beacon server, 24, 27
Beacon service (BS), 20, 74, 289,
    641
· discovery, 87
Beacon store, 73
Beaconing, 22, 24, 27, 51, 69–71
· initiation, 70
BGP policy, 136–144
· comparison with SCION,
    136–138
· consistent export routing policy,
    142
· next-hop routing policy, 142
BGPsec, 5–6, 36
Block cipher, 12
Blockchain, 381
Bloom filters, 205
Border Gateway Protocol (BGP),
    3, 165
· prefix hijacking, 7
· routing policies, 7
· stability issues, 11
Bottleneck routing, 4
BS, *see* Beacon service

Censorship, 180
Certificate service (CS), 20, 290,
    641
· discovery, 87

Certificates, 38
· AS, 21, 40
· CA, 40
· root, 39
· voting, 38
Certification, 294
Circular dependencies, 131
Cloud connectivity, 376
CMAC, 411
CO_2-aware inter-domain routing,
    399–402
COLIBRI, 237–266
· control plane, 243–245
· data plane, 245–246
· end-to-end reservations (EER),
    241, 245
· packet authentication, 249
· packet format, 248
· security analysis, 263–264
· Segment reservations (SegR),
    241, 243
COLIBRI gateway, 243
COLIBRI service, 242, 292
Content-centric networking, 576
Control plane, 10, 23–28, **66–91**,
    641
· messages, 48
· security, 165–170
· service discovery, 87–88
Control service, 642
Control service (CS), 20
Control-plane extensions, 185–201
· hidden paths, 185–189
· time synchronization, 190–197
Control-plane PKI (CP-PKI),
    21–23, 35–52, 462, 642
Core AS, 17, 642
CP-PKI, *see* Control-plane PKI
Cryptographic algorithms
· agility, 407
· asymmetric, 410
· post-quantum, 415
· symmetric, 409
Cryptographic hash function, 410

CS, *see* Certificate service

Data plane, 10, 28–31, **93–125**,
    642
· efficient path construction, 112
· path construction, 104–115
· security, 172–174
Data-plane extensions, 124,
    227–266
· end-to-end (E2E), 124
· EPIC, 228–237
· hop-by-hop (HBH), 124
Defense systems, 271–278
· bootstrapping, 277
Deployment, 12
· customer site, 323
· end host, 327–332
· global, 319–327
· incentives, 12
· ISP core network, 321
· IXP scenarios, 325
Discovery service, 291
DNS, 333, 432
DNSSEC, 6, 432
· fragility, 438–439
DRKey, 52–61, 283, 461
· bootstrapping, 283

Ed25519, 413
Efficiency, 11
Encapsulation, 332, 334
End users
· use cases, 373–374
End-host deployment, 327–332
End-to-end principle, 12
End-to-end reservations (EER),
    241, 245
EPIC, 228–237
· performance, 235
EPIC-HP
· hidden paths, 231–234
EPIC-SAPV, 234–235

F-PKI, 419–430
Forwarding, 31

· inter-domain, 94
· intra-domain, 94
Forwarding path, 9, 28, 642
· creation, 28
Forwarding plane, *see* Data plane
Forwarding policy, 10
Forwarding table, 3, 31
Future Internet architecture, 580
Future Internet architecture (FIA), 575

Gao-Rexford model, 139–142
Gateway, *see* SCION–IP Gateway (SIG)
Geofencing, 10, 374
GeoVITe
· SCI-ED, 390
Glossary, 641
Grace period, 23, 42, 642
Green networking, 393–406
· core routers, 394
· edge routers, 397
· metro routers, 397
· WDM devices, 396

Happy Eyeballs, 310
Hercules, 383
· SCI-ED, 389
HF, *see* Hop field
Hidden paths, 185–189
· EPIC-HP, 231–234
High-Speed Multipath File Transfer, 383
Home office, 373
Hop entry, 67
Hop field (HF), 19, 30, 68, 642
Host structure, **303–315**
Hot-potato routing, 143
Huawei New IP, 579–580
Hybrid addressing, 28

ICMP, 6, 8, 27
Incentives, 12
INF, *see* Info field
Info field (INF), 30, 171, 642

Information-centric networking, 576
Interface, 68, 94, 355
Internet exchange point (IXP), 109
Internet of Things (IoT), 379–381
Internet Protocol (IP), 2
Internet service provider (ISP), 3
IP address prefix, 643
IP prefix
· hijacking, 7
· longest-prefix match, 31
ISD, *see* Isolation domain
Isolation domain (ISD), 17, 643
· related work, 575
IXP
· deployment of SCION, 325

Kill switch, 36, 176–178

Leased lines
· replacement, 374
LEO Satellite Networks, 378
LightningFilter, 207–217
Link failure, 27
LOFT, 217–225
Low-latency communication, 378

Maximum transmission unit (MTU), 66–68, 332
Message auth. code (MAC), 410
· algorithm agility, 171
· brute-force attack, 171
Microwave networks, 378
MONDRIAN, 382
MPLS, 31
Multipath communication, 4, 10, 27

N-Tube, 256–262
Name resolution, **432**
· DNS, 333
· DNSSEC, 6
· interoperability, 333
· RHINE, 333
NAT, 326–327, 467

Nesquic, 309

OpenFlow, 578

Packet
· format, 95–96
· forwarding, 115
· initialization, 115
Packet-carried forwarding state
    (PCFS), **29**, 643
Packet-carried forwarding
    state (PCFS), 19
Path attack
· path hijacking, 166
· path manipulation, 165
· path preference attack, 168
· path selection, 168
· path splicing, 171
· source-address spoofing, 174
Path authorization, 30, 170–172,
    228, 643
Path construction, 104–115
· segment combination, 28
Path control, 9, 168, 382, 643
Path discovery, 69–71
Path exploration, 24, 69–71, 135,
    136, 142, 144, 147
Path lookup, 80–87
Path quality
· beaconing overhead, 150
· evaluation, 154–156
Path registration, 24, 412
· core, **72**
· intra-ISD, **71**
Path resolution, 104–112, 306
· path combination, 28, 104–112
· path lookup, **26**, 306, 412
Path reversal, 116
Path revocation, 27, 120–124
Path segment, 27, 643
· combination, 28, 104–115
· core-segment, 24, 29
· down-segment, 24
· registration, 71
· selection, 27, 73

· up-segment, 24
Path selection, 27, 168
· diversity-based, 76
· malicious, 147
Path server, 24, 27
Path service (PS), 20, 289, 643
· discovery, 87
Path transparency, 9, 644
Path validation, 230
Path-aware networking, 239
Path-segment construction beacon
    (PCB), 19, 66–69, 644
· components, 66
· extensions, 197–201
· filtering, 76
· path metadata, 197–201
· propagation, 70
· selection properties, 74
PCB, *see* Path-segment
    construction beacon
PCFS, *see* Packet-carried
    forwarding state
Peering link, 25, 29
Peering path, 109
PILA, 461–469
PKI
· control-plane PKI, 21–23, 36–52
· end entity PKI, 419
· F-PKI, 419–430
· Web PKI, 419
Post-quantum cryptography, 415
Private lines
· replacement, 374
Property
· availability, 377
Pseudorandom function (PRF),
    410, 411
Pseudorandom number generator,
    409, 411

Recovery, 23
Relying party, 644
Replay-suppression, 204–207
Reservations
· COLIBRI, 237–266

Revocation
· alternatives, 51
· paths, 120–124
· trust roots, 23
RHINE, 333, 432–459
Router on a stick, 321
RPKI, 5–6, 36

SBAS, 345–354
Scalability, 10, 11
· analysis, 148–150
· COLIBRI, 237–266
SCI-ED, 389–392
· GeoVITe, 390
· Hercules, 389
SCION daemon, 304, **305**
SCION dispatcher, **304**, 304
SCION gateway, 355
SCION Gateway Routing Protocol
    (SGRP), 336–337
SCION native applications, 383
SCION packet, 354
· format, 95
· header, 30, 95, 354
SCION path
· construction, 28
· forwarding path, 642
· peering path, 109
· shortcut path, 110
SCION path policy, 135–148
· beaconing control, 138
· explicit path policy, 138
· hop-field encryption, 139
· implementation, 138
· secrecy, 145
SCION suite, 272
· DoS defense systems, 272
· supporting systems, 272
SCION–IP Gateway (SIG), 324,
    332–336
· coordination systems, 336–345
· encapsulation, 332, 334
· routing protocol (SGRP), 336
ScionLab, 361–369
· architecture, 362

· Fed4FIRE+, 367
· KREONET, 368
· research projects, 366–368
SCMP, 27, 54, 89–91
· authentication, 90
· format, 89
Secure Backbone AS (SBAS),
    345–354
Secure Swiss Finance Network
    (SSFN), 385–389
Security
· control plane, 165–170
· data plane, 172–174
Security analysis
· censorship, 180
· COLIBRI, 263–264
· confidentiality, 179
· DRKey bootstrapping, 287
· security goals, **158**
· surveillance, 179
· threat model, **161**
Security properties
· achieved by SCION, 164–165
· anonymity, 180
· AS, 159
· censorship, 180
· confidentiality, 179
· end hosts, 160
· overall, 158
Segment reservations (SegR), 241,
    243
Segment routing, 578
Setup-Less Neighbor-Based
    Communication (SNC),
    279–280
Shortcut path, 110
SIAM, 337–345
SIC, *see* Swiss Interbank Clearing
SIG, *see* SCION-IP gateway
SNB, *see* Swiss National Bank
snet, 306
Software-defined netw. (SDN), 31,
    382, 578
Source Authentication, 229

Source authentication, 174
Source routing, 3
Source-address spoofing, 174
SPAO, 61–63
SSFN, 385–389
Surveillance, 179
Swiss Interbank Clearing (SIC),
    385
Swiss National Bank (SNB), 385

TAPS API, 310
TCAM, 3, 11
Telescoped reservation setup
    (TRS), 285–287
Time synchronization, 190–197
TLS, 6
TOMA, *see* Trust on multiple
    announcements
Traffic
· classes, 278
· filtering, 207–217
· marking, 281
· monitoring, 217–225
· priority processing, 282
· priorization, 278–283

· SNC, 279–280
Transparency, 9
· for paths, 9
TRC, *see* Trust root configuration
TRS, *see* Telescoped reservation
    setup
Trust on first use (TOFU), 22, 461
Trust on multiple announcements
    (TOMA), 49
Trust reset, 23, 37, 42, 52, 644
Trust root configuration (TRC), **21**,
    37, 38, 41, 644
· attestation, 50
· bootstrapping, 49
· discovery, 51
· dissemination, 22, 51
· format, 41
· revocation, 23, 51
· signing ceremony, 44
· update, 22, 36, 46, 52

Video
· SCION-WebRTC, 384
· streaming, 384
Voting quorum, 23, 43, 644

Printed in the United States
by Baker & Taylor Publisher Services